NOTICE: Return or renew all Library Materials! The *Minimum* Fee for each Lost Book is $50.00.

The person charging this material is responsible for its return to the library from which it was withdrawn on or before the **Latest Date** stamped below.

Theft, mutilation, and underlining of books are reasons for disciplinary action and may result in dismissal from the University.
To renew call Telephone Center, 333-8400
UNIVERSITY OF ILLINOIS LIBRARY AT URBANA-CHAMPAIGN

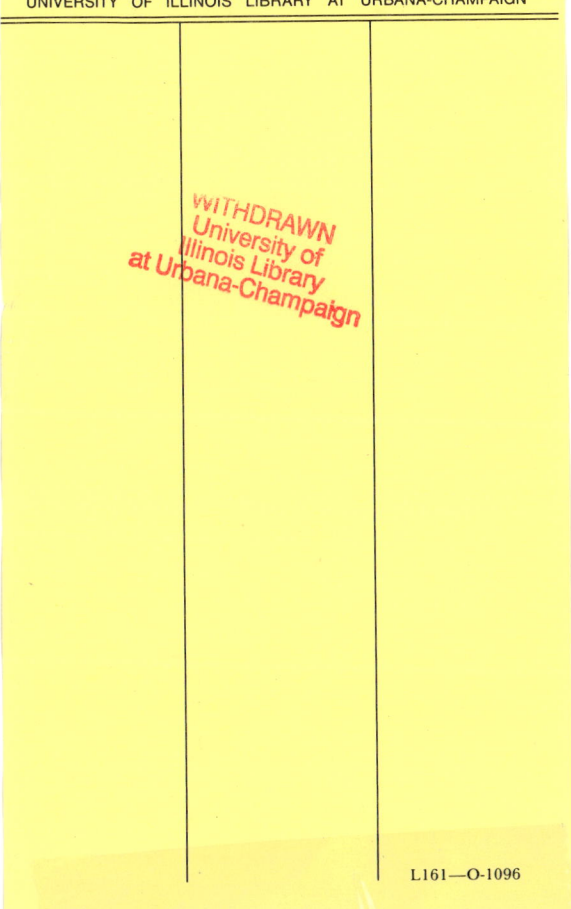

L161—O-1096

Food Microbiology Protocols

METHODS IN BIOTECHNOLOGY™

John M. Walker, Series Editor

15. **Enzymes in Nonaqueous Solvents:** *Methods and Protocols*, edited by *Evgeny N. Vulfson, Peter J. Halling, and Herbert L. Holland, 2001*
14. **Food Microbiology Protocols,** edited by *John F. T. Spencer and Alicia Leonor Ragout de Spencer, 2001*
13. **Supercritical Fluid Methods and Protocols,** edited by *John R. Williams and Anthony A. Clifford, 2000*
12. **Environmental Monitoring of Bacteria,** edited by *Clive Edwards, 1999*
11. **Aqueous Two-Phase Systems,** edited by *Rajni Hatti-Kaul, 2000*
10. **Carbohydrate Biotechnology Protocols,** edited by *Christopher Bucke, 1999*
9. **Downstream Processing Methods,** edited by *Mohamed A. Desai, 2000*
8. **Animal Cell Biotechnology,** edited by *Nigel Jenkins, 1999*
7. **Affinity Biosensors:** *Techniques and Protocols,* edited by *Kim R. Rogers and Ashok Mulchandani, 1998*
6. **Enzyme and Microbial Biosensors:** *Techniques and Protocols,* edited by *Ashok Mulchandani and Kim R. Rogers, 1998*
5. **Biopesticides:** *Use and Delivery,* edited by *Franklin R. Hall and Julius J. Menn, 1999*
4. **Natural Products Isolation,** edited by *Richard J. P. Cannell, 1998*
3. **Recombinant Proteins from Plants:** *Production and Isolation of Clinically Useful Compounds,* edited by *Charles Cunningham and Andrew J. R. Porter, 1998*
2. **Bioremediation Protocols,** edited by *David Sheehan, 1997*
1. **Immobilization of Enzymes and Cells,** edited by *Gordon F. Bickerstaff, 1997*

METHODS IN BIOTECHNOLOGY™

Food Microbiology Protocols

Edited by

John F. T. Spencer
and
Alicia L. Ragout de Spencer

*Planta Piloto de Procesos Industriales Microbiologicos,
San Miguel Tucumán, Argentina*

Humana Press ✺ Totowa, New Jersey

© 2001 Humana Press Inc.
999 Riverview Drive, Suite 208
Totowa, New Jersey 07512

All rights reserved. No part of this book may be reproduced, stored in a retrieval system, or transmitted in any form or by any means, electronic, mechanical, photocopying, microfilming, recording, or otherwise without written permission from the Publisher. Methods in Biotechnology™ is a trademark of The Humana Press Inc.

All authored papers, comments, opinions, conclusions, or recommendations are those of the author(s), and do not necessarily reflect the views of the publisher.

This publication is printed on acid-free paper. ∞
ANSI Z39.48-1984 (American Standards Institute) Permanence of Paper for Printed Library Materials.

Cover design by Patricia F. Cleary.

For additional copies, pricing for bulk purchases, and/or information about other Humana titles, contact Humana at the above address or at any of the following numbers: Tel: 973-256-1699; Fax: 973-256-8341; E-mail: humana@humanapr.com, or visit our Website at www.humanapress.com

Photocopy Authorization Policy:
Authorization to photocopy items for internal or personal use, or the internal or personal use of specific clients, is granted by Humana Press Inc., provided that the base fee of US $10.00 per copy, plus US $00.25 per page, is paid directly to the Copyright Clearance Center at 222 Rosewood Drive, Danvers, MA 01923. For those organizations that have been granted a photocopy license from the CCC, a separate system of payment has been arranged and is acceptable to Humana Press Inc. The fee code for users of the Transactional Reporting Service is: [0-89603-867-X/01 $10.00 + $00.25].

Printed in the United States of America. 10 9 8 7 6 5 4 3 2 1

Library of Congress Cataloging in Publication Data

Main entry under title: Food microbiology protocols.

Methods in biotechnology™.

Food microbiology protocols./edited by John F. T. Spencer and Alicia L. Ragout de Spencer.
    p. cm.—(Methods in biotechnology; 14)
  Includes bibliographical references and index.
  ISBN 0-89603-867-X (alk. paper)
  1. Food—Microbiology. I. Spencer, John F. T.   II. Ragout de Spencer, Alicia L.   III. Series.
  QR115.F658 2001
  664'.001'579—dc21   00-021718

Preface

Two of the recent books in the *Methods in Molecular Biology* series, *Yeast Protocols* and *Pichia Protocols*, have been narrowly focused on yeasts and, in the latter case, particular species of yeasts. *Food Microbiology Protocols*, of necessity, covers a very wide range of microorganisms. Our book treats four categories of microorganisms affecting foods: (1) Spoilage organisms; (2) pathogens; (3) microorganisms in fermented foods; and (4) microorganisms producing metabolites that affect the flavor or nutritive value of foods. Detailed information is given on each of these categories.

There are several chapters devoted to the microorganisms associated with fermented foods: these are of increasing importance in food microbiology, and include one bacteriophage that kills the lactic acid bacteria involved in the manufacture of different foods—cottage cheese, yogurt, sauerkraut, and many others. The other nine chapters give procedures for the maintenance of lactic acid bacteria, the isolation of plasmid and genomic DNA from species of *Lactobacillus*, determination of the proteolytic activity of lactic acid bacteria, determination of bacteriocins, and other important topics.

A substantial number of the chapters deal with yeasts, microorganisms which, after all, have also been associated with human foods and beverages for many thousands of years. The emphasis in *Food Microbiology Protocols* is on techniques for the improvement of methods for yeast hybridization and isolation, and for improvement of strains of industrially important yeasts, to be used in food and beverage production. For instance, the chapters by Katsuragi describe techniques for isolation of hybrids obtained by protoplast fusion and conventional mating, by the use of fluorescent staining, and by separation using flow cytometry. Other chapters discuss the identification of strains by analysis of mitochondrial DNA and other techniques. There are chapters on the isolation of strains of starches used in the production of human foods, and an important chapter on obtaining and isolating thermotolerant strains for the high temperature production of beverage and industrial alcohol. Finally, there are methods for the production of polyhydroxy alcohols for low-calorie sweeteners. The material on yeasts overlaps only slightly with that in the excellent book, *Yeast Protocols*, edited by Ivor H. Evans, so investigators interested in industrial yeasts should avail themselves of both volumes.

The chapters on spoilage organisms and pathogens include valuable information on the isolation and identification of most important species in these areas. Several of these are concerned with bacteria, yeasts, and molds, causing spoilage of poultry products, as well as causing disease in humans. Methods for identification by molecular biology techniques and by conventional plate counts are given. There are two reviews on topics of immediate interest.

Finally, the editors and the publishers would like to thank all those authors who gave so freely of their time and energy in preparing these chapters.

The editors wish especially to thank Dr. Faustino Siñeriz, Director of PROIMI, for allowing us to use the facilities at PROIMI in the preparation of this book, and for his kind encouragement in the work at all times. We also thank Dr. María E. Lucca for her able assistance in correcting the final version.

John F. T. Spencer
Alicia L. Ragout de Spencer

Contents

Preface v
Contributors xi

PART I SPOILAGE ORGANISMS

1 Psychrotrophic Microorganisms: *Agar Plate Methods, Homogenization, and Dilutions*
   Anavella Gaitán Herrera 3
2 Biochemical Identification of Most Frequently Encountered Bacteria That Cause Food Spoilage
   Maria Luisa Genta and Humberto Heluane 11
3 Mesophilic Aerobic Microorganisms
   Anavella Gaitán Herrera 25
4 Yeasts and Molds
   Anavella Gaitán Herrera 27
5 Coliforms
   Anavella Gaitán Herrera 29
6 Genetic Analysis of Food Spoilage Yeasts
   Stephen A. James, Matthew D. Collins, and Ian N. Roberts 37

PART II PATHOGENS

7 Conductimetric Method for Evaluating Inhibition of *Listeria monocytogenes*
   Graciela Font de Valdez, Graciela Lorca, and María Pía de Taranto 55
8 Molecular Detection of Enterohemorrhagic *Escherichia coli* O157:H7 and Its Toxins in Beef
   Kasthuri J. Venkateswaran 61
9 Detection of *Listeria monocytogenes* by the Nucleic Acid Sequence-Based AmplificationTechnique
   Burton W. Blais and Geoff Turner 67
10 Detection of *Escherichia coli* O157:H7 by Immunomagnetic Separation and Multiplex Polymerase Chain Reaction
   Ian G. Wilson 85

11 Detection of *Campylobacter jejuni* and Thermophilic
    Campylobacter spp. from Foods by Polymerase Chain Reaction
    Haiyan Wang, Lai-King Ng, and Jeff M. Farber 95
12 Magnetic Capture Hybridization Polymerase Chain Reaction
    Jinru Chen and Mansel W. Griffiths 107
13 Enterococci
    Anavella Gaitán Herrera 111
14 Salmonella
    Anavella Gaitán Herrera 113
15 *Campylobacter*
    Anavella Gaitán Herrera 119
16 *Listeria monocytogenes*
    Anavella Gaitán Herrera 125

PART III FERMENTED FOODS

17 Methods for Plasmid and Genomic DNA Isolation from Lactobacilli
    M. Andrea Azcárate-Peril and Raúl R. Raya 135
18 Methods for the Detection and Concentration of Bacteriocins
    Produced by Lactic Acid Bacteria
    **Sergio A. Cuozzo, Fernando J. M. Sesma,
    Aída A. Pesce de R. Holgado, and Raúl R. Raya** 141
19 Meat Protein Degradation by Tissue and Lactic Acid Bacteria
    Enzymes
    Silvina Fadda, Graciela Vignolo, and Guillermo Oliver 147
20 Maintenance of Lactic Acid Bacteria
    Graciela Font de Valdez 163
21 Probiotic Properties of Lactobacilli: *Cholesterol Reduction and
    Bile Salt Hydrolase Activity*
    Graciela Font de Valdez and María Pía de Taranto 173
22 Identification of Exopolysaccharide-Producing Lactic Acid Bacteria:
    A Method for the Isolation of Polysaccharides in Milk Cultures
    **Fernanda Mozzi, María Inés Torino, and
    Graciela Font de Valdez** 183
23 Differentiation of Lactobacilli Strains by Electrophoretic
    Protein Profiles
    **Graciela Savoy de Giori, Elvira María Hébert, and
    Raúl R. Raya** 191

24 Methods to Determine Proteolytic Activity of Lactic Acid Bacteria
    Graciela Savoy de Giori and Elvira María Hébert *197*
25 Methods for Isolation and Titration of Bacteriophages
    from *Lactobacillus*
    Lucía Auad and Raúl R. Raya *203*
26 Identification of Yeasts Present in Sour Fermented Foods
    and Fodders
    Wouter J. Middelhoven *209*

PART IV ORGANISMS IN THE MANUFACTURE OF OTHER FOODS AND BEVERAGES

27 Protein Hydrolysis: *Isolation and Characterization of
    Microbial Proteases*
    Marcela A. Ferrero *227*
28 Production of Polyols by Osmotolerant Yeasts
    Lucía I. C. de Figueroa and María E. Lucca *233*
29 Identification of Yeasts from the Grape/Must/Wine System
    Peter Raspor, Sonja Smole Mozina, and Neza Cadez *243*
30 Carotenogenic Microorganisms: *A Product-Based Biochemical
    Characterization*
    José Domingos Fontana *259*
31 Genetic and Chromosomal Stability of Wine Yeasts
    **Matthias Sipiczki, Ida Miklos, Leonora Leveleki, and
    Zsuzsa Antunovics** *273*
32 Prediction of Prefermentation Nutritional Status of Grape Juice:
    The Formol Method
    **Barry H. Gump, Bruce W. Zoecklein, and
    Kenneth C. Fugelsang** *283*
33 Enological Characteristics of Yeasts
    **Fabio Vasquez, Lucía I. C. de Figueroa, and
    Maria Eugenia Toro** *297*
34 Utilization of Native Cassava Starch by Yeasts
    **Lucía I. C. de Figueroa, Laura Rubenstein, and
    Claudio González** *307*

PART V SPECIAL METHODS AND EQUIPMENT

35 Reactor Configuration for Continuous Fermentation in Immobilized
    Systems: *Application to Lactate Production*
    **José Manuel Bruno-Bárcena, Alicia L. Ragout de Spencer,
    Pedro R. Córdoba, and Faustino Siñeriz** *321*

36 Molecular Characterization of Yeast Strains by Mitochondrial DNA Restriction Analysis
   Maria Teresa Fernández-Espinar, Amparo Querol, and Daniel Ramón 329
37 Selection of Yeast Hybrids Obtained by Protoplast Fusion and Mating, by Differential Staining, and by Flow Cytometry
   Tohoru Katsuragi 335
38 Selection of Hybrids by Differential Staining and Micromanipulation
   Tohoru Katsuragi 341
39 Flotation Assay in Small Volumes of Yeast Cultures
   Sandro Rogério de Sousa, Maristela Freitas Sanches Peres, and Cecilia Laluce 349
40 Obtaining Strains of *Saccharomyces* Tolerant to High Temperatures and Ethanol
   Maristela Freitas Sanches Peres, Sandro Rogério de Sousa, and Cecilia Laluce 355
41 Multilocus Enzyme Electrophoresis
   Timothy Stanley and Ian G. Wilson 369
42 Bacteriocin Production Process by a Mixed Culture System
   Suteaki Shioya and Hiroshi Shimizu 395

PART VI REVIEWS

43 Nutritional Status of Grape Juice
   Bruce W. Zoecklein, Barry H. Gump, and Kenneth C. Fugelsang 415
44 Problems with the Polymerase Chain Reaction: *Inhibition, Facilitation, and Potential Errors in Molecular Methods*
   Ian G. Wilson 427
45 Problems with Genetically Modified Foods
   José Manuel Bruno-Bárcena, M. Andrea Azcarate-Peril, and Faustino Siñeriz 481
Index 485

Contributors

ZSUZSA ANTUNOVICS • *Department of Genetics, University of Debrecen, Debrecen, Hungary*
LUCÍA AUAD • *Centro de Referencia para Lactobacilos (CERELA), San Miguel Tucumán, Argentina*
M. ANDREA AZCÁRATE-PERIL • *Centro de Referencia para Lactobacilos (CERELA), San Miguel Tucumán, Argentina*
BURTON W. BLAIS • *Laboratory Services Division, Canadian Food Inspection Agency, Ottawa, Ontario, Canada*
JOSÉ MANUEL BRUNO-BÁRCENA • *Departamento de Biotecnologia, Instituto Agrochimica y Tecnologia de Alimentos, Valencia, Spain*
NEZA CADEZ • *Food Science and Technology Department, University of Ljubljana, Ljubljana, Slovenia*
JINRU CHEN • *Department of Food Science and Technology, University of Guelph, Guelph, Ontario, Canada*
MATTHEW D. COLLINS • *Institute of Food Research, Norwich, UK*
PEDRO R. CÓRDOBA • *Planta Piloto de Procesos Industriales Microbiologicos (PROIMI), San Miguel Tucumán, Argentina*
SERGIO A. CUOZZO • *Centro de Referencia para Lactobacilos (CERELA), San Miguel Tucumán, Argentina*
SILVINA FADDA • *Centro de Referencia para Lactobacilos (CERELA), San Miguel Tucumán, Argentina*
JEFF M. FARBER • *Microbiology Research Division, Bureau of Microbial Hazards, Food Directorate, Health Canada, Ottawa, Ontario, Canada*
MARIA TERESA FERNÁNDEZ-ESPINAR • *Departamento de Biotecnologia, Instituto Agroquimia y Tecnologia de Alimentos, Valencia, Spain*
MARCELA A. FERRERO • *Microbiology Laboratory, Universität de Lles Illes Balears, Illes Balears, Spain*
LUCÍA I. C. DE FIGUEROA • *Planta Piloto de Procesos Industriales Microbiologicos (PROIMI), San Miguel Tucumán, Argentina*
JOSÉ DOMINGOS FONTANA • *Biomass Chemo/Biotechnology Laboratory, Curitiba, Pr., Brazil*
KENNTH C. FUGELSANG • *Department of Viticulture and Enology, California State University-Fresno, Fresno, CA*
MARIA LUISA GENTA • *PROIMI-MIRCEN, San Miguel Tucumán, Argentina*

GRACIELA SAVOY DE GIORI • *Centro de Referencia para Lactobacilos (CERELA), San Miguel Tucumán, Argentina*
CLAUDIO GONZÁLEZ • *PROIMI-MIRCEN, San Miguel de Tucumán, Argentina*
BARRY F. GUMP • *Department of Viticulture and Enology, California State University-Fresno, Fresno, CA*
ELVIRA M. HÉBERT • *Centro de Referencia para Lactobacilos (CERELA), San Miguel Tucumán, Argentina*
HUMBERTO HELUANE • *PROIMI-MIRCEN, San Miguel Tucumán, Argentina*
ANAVELLA GAITÁN HERRERA • *Pontificia Universidad Javeriana, Bogotá, Colombia*
AIDA A. DE R. PESCE HOLGADO • *Centro de Referencia para Lactobacilos (CERELA), San Miguel Tucumán, Argentina*
STEPHEN A. JAMES • *Institute of Food Research, Norwich, UK*
TOHORU KATSURAGI • *Nara Institute of Science and Technology, Graduate School of Biological Sciences, Ikoma, Nara, Japan*
CECILIA LALUCE • *Instituto Quimica de Universidade Estadual Paulista "Julio de Mesquita Filho" (UNESP), Araquara, SP, Brazil*
LEONORA LEVELEKI • *Department of Genetics, University of Debrecen, Debrecen, Hungary*
GRACIELA LORCA • *Centro de Referencia para Lactobacilos (CERELA), San Miguel Tucumán, Argentina*
MARÍA E. LUCCA • *Planta Piloto de Procesos Industriales Microbiologicos (PROIMI), San Miguel Tucumán, Argentina*
WOUTER J. MIDDELHOVEN • *Laboratorium voor Microbiologie, Department Biomoleculaire Wetenschappen, Universiteit Wageningen, Wageningen, The Netherlands*
IDA MIKLOS • *Department of Genetics, University of Debrecen, Debrecen, Hungary*
SONJA SMOLE MOZINA • *Food Science and Technology Department, University of Ljubljana, Ljubljana, Slovenia*
FERNANDA MOZZI • *Centro de Referencia para Lactobacilos (CERELA), San Miguel Tucumán, Argentina*
LAI-KING NG • *Bureau of Microbiology, Laboratory Centre for Disease Control, Health Canada, Winnipeg, Manitoba, Canada*
GUILLERMO OLIVER • *Centro de Referencia para Lactobacilos (CERELA), San Miguel Tucumán, Argentina*
MARISTELA FREITAS SANCHES PERES • *Programa de Pós Graduação da Faculdade de Ciências e Letras de Ribeirão Preto USP, SP, Brazil*
AMPARO QUEROL • *Departamento de Biotecnologia, Instituto Agrochimica y Tecnologia de Alimentos, Valencia, Spain*

Contributors

DANIEL RAMÓN • *Departamento de Biotecnologia, Instituto Agrochimica y Tecnologia de Alimentos, Valencia, Spain*
PETER RASPOR • *Food Science and Technology Department, University of Ljubljana, Ljubljana, Slovenia*
RAÚL R. RAYA • *Centro de Referencia para Lactobacilos (CERELA), San Miguel Tucumán, Argentina*
IAN N. ROBERTS • *Institute of Food Research, Norwich, UK*
LAURA RUBENSTEIN • *Planta Piloto de Procesos Industriales Microbiologicos (PROIMI), San Miguel Tucumán, Argentina*
FERNANDO J. M. SESMA • *Centro de Referencia para Lactobacilos (CERELA), San Miguel Tucumán, Argentina*
HIROSHI SHIMIZU • *Department of Biotechnology, Graduate School of Engineering, Osaka University, Osaka, Japan*
SUTEAKI SHIOYA • *Department of Biotechnology, Graduate School of Engineering, Osaka University, Osaka, Japan*
FAUSTINO SIÑERIZ • *Planta Piloto de Procesos Industriales Microbiologicos (PROIMI), San Miguel Tucumán, Argentina*
MATTHIAS SIPICZKI • *Department of Genetics, University of Debrecen, Debrecen, Hungary*
SANDRO ROGÉRIO DE SOUSA • *Instituto Quimica de UNES, Araquara, SP, Brazil*
ALICIA L. RAGOUT DE SPENCER • *Planta Piloto de Procesos Industriales Microbiologicos (PROIMI), San Miguel Tucumán, Argentina*
JOHN F. T. SPENCER • *Planta Piloto de Procesos Industriales Microbiologicos (PROIMI), San Miguel Tucumán, Argentina*
TIMOTHY STANLEY • *Bacteriology Department, Belfast City Hospital, Belfast, Northern Ireland*
MARÍA PÍA DE TARANTO • *Centro de Referencia para Lactobacilos (CERELA), San Miguel Tucumán, Argentina*
MARÍA INÉS TORINO • *Centro de Referencia para Lactobacilos (CERELA), San Miguel Tucumán, Argentina*
MARIA EUGENIA TORO • *Planta Piloto de Procesos Industriales Microbiologicos (PROIMI),San Miguel Tucumán, Argentina*
GEOFF TURNER • *Laboratory Services Division, Canadian Food Inspection Agency, Ottawa, Ontario, Canada*
GRACIELA FONT DE VALDEZ • *Centro de Referencia para Lactobacilos (CERELA), San Miguel Tucumán , Argentina*
FABIO VAZQUEZ • *Planta Piloto de Procesos Industriales Microbiologicos (PROIMI), San Miguel Tucumán, Argentina*

KASTHURI J. VENKATESWARAN • *Planetary Protection Sciences, Jet Propulsion Laboratory, NASA, Pasadena, CA*

GRACIELA VIGNOLO • *Centro de Referencia para Lactobacilos (CERELA), San Miguel Tucumán, Argentina*

HAIYAN WANG • *Microbiology Research Division, Bureau of Microbial Hazards, Food Directorate, Health Canada, Ottawa, Ontario, Canada*

IAN G. WILSON • *Bacteriology Department, Belfast City Hospital, Belfast, Northern Ireland*

BRUCE W. ZOECKLEIN • *Department of Food Science and Technology, Virginia Tech, Blacksburg, VA*

I

SPOILAGE ORGANISMS

1

Psychrotrophic Microorganisms

Agar Plate Methods, Homogenization, and Dilutions

Anavella Gaitán Herrera

1. Introduction

Isolation and enumeration of the microorganisms present in foods usually demands preliminary treatment of samples to release into a liquid medium those microorganisms present that may be included within the food. In a mixing procedure known as "stomaching," the food sample (*see* **Note 1**) and diluent are put into a sterile plastic bag that is vigorously struck on its outer surfaces by paddles inside a stomacher, the compression and shearing forces break up solid pieces of food. After samples are removed for analysis, the bag and its remaining contents can be discarded and the equipment is ready for use. If a stomacher is not available, an electric blender with cutting blades revolving at high speed can be used (*see* **Note 2**).

The nature of the diluent used is another highly important factor. Diluents such as tap or distilled water, saline solutions, phosphate buffers, and Ringer's solution are toxic to microorganisms, especially if the time of contact is unduly prolonged. For that reason, we suggest that using 0.1% peptone solution, physiological saline solution with 0.1% peptone added, is the most reliable source. Peptone 0.1% in saline solution in 0.85% NaCl is recommended by the International Standards Organization (*see* **Notes 3** and **4**).

The number of microorganisms by plate count has been one of the more commonly used microbiological methods for determination food quality. This method indicates the adequacy of sanitation and formation of an opinion on incipient spoilage (*see* **Note 5**). The plate count has a special application to imported foods for control of the standard of sanitation practiced in the manufacturing establishments *(1)*.

The three methods in common use for enumerating microorganisms are the standard-plate count, also called the aerobic plate count or pour plate, the "surface plate," or "spread-drop" method and the drop-plate method (*see* **Note 6**). None of these procedures can be depended on to list all types of organisms present within the test specimens. Many cells may not grow because of specifically unfavorable conditions of nutrition, aeration, temperature, or duration of incubation *(1)*.

The temperature chosen depends on the purpose of the examination. The incubation temperature of microorganisms have different growth temperature ranges such as 0–7°C for psychrotrophs, 30–35°C for mesophiles, and 55°C for thermophiles. No single incubation temperature will absolutely exclude all organisms from another group (*see* **Notes 7** and **8**). Incubators are required with minimal fluctuation or variation of temperature throughout the incubation chamber. All incubators should be checked and calibrated frequently *(2)*.

The psychrotrophic or psychrophilic microorganisms are able to grow for 7–10 d (*see* **Note 9**) at commercial refrigeration temperatures (0–7°C). Species of *Achromobacter, Acinetobacter, Alcaligenes, Bacillus, Flavobacterium, Streptococcus,* and *Pseudomonas* are included among the psychrotrophic bacteria. Furthermore, some yeasts and molds, including *Penicillium, Aspergillus, Geotrichum,* and *Botrytis,* are able to grow well in refrigerated foods (*see* **Note 10**) in large numbers. These microorganisms can cause off flavors amd physical defects in foods. Their growth rates are highly temperature dependent (*see* **Note 11**). Their presence indicates a high potential for spoilage during extended storage. Most of these microorganisms are destroyed by mild heat treatment, but some heat-resistant types such as some species of *Bacillus* and *Clostridium* may survive. The presence of psychrotrophic microorganisms in heat-processed foods implies post-processing contamination. They are sources of heat-resistant proteolytic and lipolytic enzymes *(3)* that affect adversely the quality of the food during storage after heat treatment (*see* **Note 12**). The presence of psychrotropic microorganisms is also important in such foods as frozen turkey or chicken when they are thawed.

2. Homogenization and Dilution: Method 1
2.1. Materials

1. Mechanical blender, two-speed model or single speed with rheostat control.
2. Glass or metal blending jars of 1 L capacity, with covers, resistant to autoclave temperatures. One sterile jar (autoclaved at 121°C) for 15 min) is required for each sample to be analyzed.
3. Balance with weights. Capacity at least 2500 g, sensitivity 0.1 g.
4. Instruments for preparing samples: knives, forks, forceps, scissors, spoons, spatulas, and tongue depressors, all sterilized for use by autoclaving or by hot air.
5. Pipets: 1, 5, and 10 mL.

Psychrotrophic Microorganisms

6. Refrigerator cooled to 2–5°C.
7. Peptone dilution fluid or peptone salt dilution fluid, sterilized in the autoclave for each sample 450 mL in flask or bottle, 90 or 99 mL blanks in dilution bottles or similar containers.
8. Mechanical mincer, mechanical blender, operating at not less than 8000 rpm and not more than 45,000 rpm.
9. For Method 1, items 2–4, and 6:
    a. Pipets, bacteriological
    b. Sterile culture tubes for dilution fluid, 15–20 mL capacity
    c. Peptone salt dilution fluid

3. Microogranisms: Agar Plate Method

1. Petri dishes, glass (100 × 15 mm) or plastic (90 × 15 mm) (*see* **Note 7**).
2. Pipets, bacteriological, 1, 5, and 10 mL sterile.
3. Water bath or incubator for tempering agar, 44–46°C.
4. Incubator, 29–31°C.
5. Colony counter.
6. Tally register.
7. Plate count agar (standard-methods agar) (*see* **Notes 5** and **6**).
8. Drying cabinet or incubator for drying the surface of agar plates, preferably at 50°C.
9. Glass spreaders (Drigalski spatulas; hockey-stick–shaped glass rods).
10. Pipets, bacteriological, sterile, with divisions of 0.1 mL or less.

3.1. Materials

1. Low-temperature incubator capable of maintaining a temperature of 1–7°C.
2. Nonselective agar media such as standard or trypticase soy broth.
3. Selective agar, crystal violet, or tetrazolium agar.
4. Pipets, bacteriological, 0.1 mL sterile
5. Materials for preparation and dilution of the food homogenate, as listed in **Subheading 2.1.**
6. Petri dishes, as described previously.
7. Colony counter.

4. Methods
4.1. Homogenization and Dilutions: Method 1

1. Begin the examination as soon as possible after the sample is taken. Refrigerate the sample at 0–5°C if the examination cannot be started immediately after it reaches the laboratory. If the sample is frozen, thaw it in its original container (or in the container in which it was received at the laboratory) for a maximum of 18 h in a refrigerator at 2–5°C. If the sample can be easily comminuted (as in ice cream), proceed without thawing.
2. Tare the empty sterile blender jar, then weigh into it, 50 g representative of the food sample. If the contents of the package are obviously not homogenous

(frozen dinner), take a 50-g sample from a macerate of the whole dinner, or analyze each different food portion separately, depending on the purpose of the test.

3. Add to the blender jar, 450 mL of the peptone dilution fluid or peptone salt dilution fluid. This provides a dilution of 10 (+/−) 1.
4. Blend the food and dilute promptly. Start at low speed and then switch to high speed within a few seconds. Time the blending carefully to permit 2 min at high speed. Wait 2 or 3 min for foam to disperse.
5. Measure 1 mL of the 10^{-1} dilution of the blended material, avoiding foam, into a 99-mL dilution blank. Shake this and all subsequent dilutions vigorously 25 times in a 30-cm arc. Repeat this process using the progressively increasing dilutions 10^{-2}, 10^{-3}, 10^{-4}, and 10^{-5}, or dilutions that experience indicates are desirable for the food being tested.

4.2. Homogenization and Dilutions: Method 2

1. Begin the examination as soon as possible after the sample is taken. Preferably, start analysis of unfrozen samples within 1 h after receipt. If the sample is frozen, thaw in the original container (or in the container in which it was received in the laboratory) in the refrigerator at 2.5°C and examine as soon as possible after thawing is complete or at least sufficient to permit suitable subsamples to be taken (maximum thawing time 18 h).
2. Proceed to **step 3** if the sample is difficult to blend, grind, and mix twice in the mechanical mincer.
3. Weigh into a tared blended jar at least 10 g of sample, representative of the food. Add nine times as much dilution fluid as sample. This provides a dilution of 10:1.
4. Time of grinding must not exceed 2.5 min.
5. Mix the contents of the jar by shaking and pipet duplicate portions of 1 mL each into separate tubes containing 9 mL of dilution fluid. Carry out **steps 7** and **8** on each of the diluted portions.
6. Mix the liquids carefully by aspirating 10 times with a sterile pipet.
7. Transfer with the same pipet, 1.0 mL to another dilution tube containing 9 mL of dilution fluid and mix with a fresh pipet.
8. Repeat **steps 7** and **8** until the required number of dilutions are made. Each successive dilution will decrease the concentration 10-fold.

4.2.1. Example 1

$$\text{Dilution } 10^{-1}, 350, \text{ and } 330: (350 + 330)/2 \times 10 = 3400$$

$$\text{Dilution } 10^{-2}, 26, \text{ and } 28: (26 + 28) 2 \times 100 = 2700$$

$$X = (3400 + 2700)2 = 3050$$

$$30 \times 10^2 \, \text{CFU/g}$$

Psychrotrophic Microorganisms

4.2.2. Example 2

$$90 \text{ colonies on dilution } 10^{-1} = 2900$$
$$40 \text{ colonies on dilution } 10^{-2} = 4000$$
$$4000/2900 = <2. \text{ Report the average of the two counts: } 35 \times 10^2$$
$$170 \text{ colonies on dilution } 10^{-1} = 1700$$
$$35 \text{ colonies on dilution } 10^{-2} = 3500$$
$$3500/1700 = >2 \text{ Report the lower count } 17 \times 10^{-2}$$

4.2.3. Computing the Estimated Standard Plate Count (ESPC)

1. If counts on individual plates do not fall within the range 30–300 colonies, report the calculated count as ESPC. Calculate the count as directed in **steps 2–4**.
2. If plates of all dilutions show more than 300 colonies, divide each of the duplicate plates for the highest dilution into convenient radial sections (e.g., 2, 4, 8) and count all of the colonies in one or more sections. Multiply the total in each case by the appropriate factor to obtain an estimate of the total number of colonies for the entire plate. Average the estimates for the two plates, multiply the dilution, and report the resulting count as the ESPC.
3. If the three are more than 200 colonies per one-eighth section of the plates made from the most dilute suspension, multiply 1600 (i.e., 200 X 8) by that dilution, and express ESPC as more than (>) the resulting number. In all such cases, it is advisable to report the dilution used in parentheses.

The Standard Plate Count

5.1. Methods

1. Prepare the sample.
2. To duplicate sets of Petri dishes, pipet 1 mL aliquots from 10^{-1}, 10^{-2}, 10^{-3}, 10^{-4}, and 10^{-5} dilutions, and a 0.1-mL aliquot from the 10^{-5} dilution to give 10^{-1} to 10^{-6} g of food per Petri dish.
3. Promptly pour into Petri dishes 10–15 mL of melted and tempered agar. Immediately mix the aliquots with the agar medium by tilting and rotating the Petri dishes. The sequence of steps is:
   a. Tilt dish to and fro five times, in one direction.
   b. Rotate it clockwise five times.
   c. Tilt it to and fro again five times in a direction at right angles to that used the first time.
   d. Rotate it counterclockwise five times.
4. When the agar is solidified, invert the Petri dishes and incubate at 29–31°C for 48 +/- 3 h.

5.2. The Surface-Spread Plate

1. Add 15 mL of melted, cooled (45–60°C Plate count agar to each Petri dish used and allow to solidify.

2. Transfer 0.1 mL of each of the dilutions to the agar surface. Using the same pipet for each dilution. Test at least three dilutions, even if the approximate range of numbers of microorganisms in the food is known. Start with the highest dilutions and proceed to the lowest, filling and emptying the pipet three times before transferring the 0.1-mL portion to the plate.
3. Promptly spread the 0.1-mL portions on the surface of the agar plates using glass spreaders (Drigalsky spatulas). Use a separate spreader for each plate. Allow the surfaces of the plates to dry for 15 min.
4. Incubate the plates in an inverted position, for 3 d.

5.2.1. Computing the Standard Plate Count and the Estimated Plate Count

5.2.1.1. THE STANDARD PLATE COUNT (SPC)

1. Select two plates corresponding to one dilution and showing between 30 and 300 colonies per plate. Count all colonies on each plate, using the colony counter and tally register. Take the average of the two counts and multiply by the dilution factor. Report the resulting number as the SPC.
2. Two plates should be counted even if one of them should give a count of fewer than 30 or more than 300 colonies. Again, take the average of the two counts and multiply by the dilution factor and report the resulting number as the SPC.
3. If the plates from two consecutive decimal dilutions fall into the countable range of 30–300 colonies compute the SPC for each of the dilutions as directed above, and report the average of the two values obtained, unless the higher computed count is more than twice the lower one, in which case report the lower computed count as the SPC.
4. Report only two significant digits as the colony-forming unit per gram or milliliter.

5.2.1.2. EXAMPLE 1

$$1600 \times 10^3 = >1600\,000 > 16 \times 10^3$$

Dilution 10^{-1}: No colonies: in cases when there are no colonies in the plates made from the most concentrated suspension, report the ESPC as less than (<) 1 times the dilution.

For instance, if dilution 10^4 = no colonies, report ESPC < 10/g if sampling 0.1 < 100/g.

1. Reporting and interpreting:
    a. Report the counts as SPC or ESPC per gram or milliliter of food.
    b. Only the SPC should be considered in determining the acceptance or rejection of a batch of food, never the ESPC. This is used only as a first approximation in the assessment of the bacteriological quality of a food.
2. Repeating: A second count on a given plate should be within 5% of the first when done by the same person., or within 10% of the first, when done by another per-

Psychrotrophic Microorganisms

son. If counts differ by more than these limits, the cause is sometimes poor eyesight, failure to recognize minute colonies as such, or failure to differentiate them from food particles.

3. Computing: Only two significant digits should be used in reporting the SPC or the ESPC. These are the first and second digits (starting from the left) of the average of the counts. The other digits should be replaced by zeros.

5.2.1.3. Example 2

1. If the actual count were 143,000, report the count as 140,000 (14×10^4). If the third digit from the left is 5 or greater, add one unit to the second digit (rounding off). If the actual count were 53,000, report 54,000 (54×10^3)
2. Spreading colonies:
   a. If spreading colonies are present on a plate, count the colonies lying outside the area of spreading, providing that the total of the repressed area does no exceed one-half the area of the plate. Correct the count to allow for the area not counted.
   b. Report the presence of the spreaders (spr) whenever the area affected exceeds one-quarter of the area of the plate. When more than 1 plate out of every 20 shows spreading colonies, take measures to reduce the occurrence, such as reducing the amount of moisture in the incubator and ensuring a thorough mixing of the dilution fluid and tempered agar.
3. Inhibitors: When counts for the lower dilutions are markedly lower than they normally should be, inhibitors may be present. Use tests to detect possible inhibitors.
4. Methods: Plate count methods using selective media.
   a. The agar plate method using nonselective media such as standard-methods agar or trypticase soy agar *(1)*.
   b. Prepare dilutions, pour plates ,or spread plates as described previously.
   c. Incubate plates at 7 ±1°C for 10 d.
   d. Count the colonies and compute the counts.
   e. Report the counts as psychrotrophic place counts per milliliter, gram, or square centimeter, as applicable.

5.2.1.4. Plate-Count Method Using Selective Media

1. Prepare dilutions and plates as described for the agar plate count except that crystal violet tetrazolium (CVT) agar is used.
2. Incubate plates as 22°C for 5 d or at 30°C for 48 h.
3. Count red colonies and compute as for plates with nonselective media *(4)*.

6. Notes

1. When the Colworth stomacher is used, a supply of thin-walled (about 200 gauge) polyethylene bags, about 18×30 cm, are required.
2. Weigh into a tared bag at least 10 g of sample and operate for 60 s.
3. If pathogenic microorganisms are known or suspected to be present in the food, place the bag inside another as a precaution against breakage. Work in laboratories having the required grade of containment.

4. For high-fat foods (over 20%), such as smoked meats, add 1% Tween-80 or some other nontoxic surfactant.
5. It is not essential that the medium be translucent, as all colonies grow and develop on the surface of the agar.
6. Melt plate-count agar in flowing steam or boiling water, but do not expose it to great heat for a prolonged period. Temper the agar to 44–46°C and control its temperature carefully to avoid killing bacteria in the diluted medium.
7. Promptly pour into the Petri dishes 10–15 mL of melted and tempered agar. Fewer than 20 min should elapse between making the dilution and pouring the agar.
8. One of several plates with inoculated agar medium and with inoculated dilution fluid should be made as a control.
9. In some refrigerated foods, a majority of the psychrotrophic bacteria responsible for quality loss and subsequent spoilage are Gram-negative rods.
10. A keeping-quality test, in which the agar plate count is determined prior to, and following, preliminary incubation of the food (5–7 d at commercial refrigeration temperature), can provide important information about the potential for the development of flora in a refrigerated food.
11. Samples should not be frozen because of possible death or injury to microorganisms during freezing. This may extend the lag phase of growth and the time of incubation is insufficient to detect them.
12. The incubation temperature used for counting may be different from that at which the food is actually stored.

References

1. Speck, M. L. (1984) *Compendium of Methods for the Microbiological Examination of Foods.* Compiled by the APHA Technical Committee on Microbiological Methods for Foods, 2nd ed. (Speck, M. L., ed.), American Public Health Association, Washington, DC.
2. Elliot, R .P., Clark, D. S., and Michener, H. D. (1978) *Microorganisms in Food 1. Significance and Methods of Enumeration*, 2nd ed., a publication of the International Commission on Microbiological Specifications for Foods (ICMSF) of the International Association of Microbiological Societies, University of Toronto Press, Toronto, Ontario, Canada.
3. Griffiths, M. W., Phillips, J. D., and Muir, D. D. (1981) Thermostability of proteases and lipases from a number of species of psychrophilic bacteria of dairy origin. *J. Appl. Bacteriol.* **50,** 289–303.
4. American Public Health Association (1972) *Standard Methods for the Examination of Dairy Products*, 13th ed., American Public Health Association, New York.

2

Biochemical Identification of Most Frequently Encountered Bacteria That Cause Food Spoilage

Maria Luisa Genta and Humberto Heluane

1. Introduction

When the microbial flora invades food, two major problems arise. First is the pathogenicity of several microbes, and second are the changes on the food characteristics, such as contents of nutrients (hydrocarbons, vitamins, aminoacids, metals, etc.), bad smell, color and flavor, texture modification, etc.

This chapter presents the identification techniques to the most usually encountered bacteria that could be present on food.

Salmonella may be present on raw or non-heat-treated food. Eggs, milk, mayonnaise, chicken, hamburger, and creams are the most frequent vehicles for this bacteria. *Salmonella* is sensitive to high temperatures and can survive at very low temperatures. It is a Gram negative, mobile, nonsporulated, and facultative anerobic bacteria. Only total absence of this bacteria on food is normally accepted by legal regulations.

As a consequence of the nature of the contamination, salmonellas are usually present together with large numbers of other enterobacteria. Therefore the use of enrichment and selective media is required.

Bacillus cereus is a gram positive, aerobic, and catalase-positive bacteria. The optimal growth for this microorganism is obtained at temperatures oscillating between 25 and 75°C and pHs between 6 and 7, even though this microorganism survives temperatures as low as $-5°C$. The spores are very resistant to heat. At pHs lower than 4 the growth of *B. cereus* is inhibited. This species is highly lipolytic, saccharolytic and proteolytic and it is pathogenic for humans. It may be present on flour, milk, dairy products, rice, chicken, spices, and herbs. Only total absence of this species is normally accepted by legal regulations.

Coliform is the name given to those microorganisms that have the following characteristics: rod shaped, gram negative, mobile or nonmobile, aerobic or facultative anerobic nonsporulating. They ferment lactose, producing acid and gas in presence of bile salts, at 30–38°C. Coliforms are usually found in the intestine of humans and animals.

The *Escherichia coli* genus belongs to the enterobacteracea family and it is usually a sign of fecal contamination. *E. coli* is a bacillus with the following characteristics: Gram negative, mobile or nonmobile, oxidase negative, catalase positive, and glucose and lactose fermentation positive. *E. coli* is found in the human intestine, synthesizes vitamin K, and is involved in the production of vitamin B. Humans eliminate *E. coli* through fecal residues. Some strains are enterohemorrhagic pathogens.

Staphylococci are Gram positive, nonmobile, nonsporulated, noncapsulated, catalase positive, oxidase negative and most of the strains grow in media containing 10% NaCl. They are anaerobically facultative, but they grow better in aerobiosis. Staphylococci are capable of growing at temperatures oscillating between 8 and 45°C, although the optimum growth temperature is 37°C and pHs oscillating between 4 and 9.3, but the optimum growth pHs are 7.0–7.5.

Staphylococci produce heat-resistant toxins that act at the digestive level. There are seven kinds of enterotoxins, A, B, C_1, C_2, C_3, D, and E. Toxins A and D are more frequently present in food intoxication. *Staphylococcus aureus* may cause skin infection (acne). Staphylococci could be present in, e.g., dairy, meat, sausages, fish, and eggs *(1–6)*.

2. Materials
2.1. Method 1: Salmonellas

1. 500 mL and 250 mL sterile Erlenmeyer flasks.
2. Sterile Petri dishes.
3. 10 mL sterile pipets.
4. Loop.
5. Sterile blender ("Minipimer" type).
6. Buffered peptone water: Ingredients per liter: peptone 10 g, sodium chloride 5 g, sodium phosphate dibasic 9 g, potassium phosphate monobasic 1.5 g, pH 7.0.
7. Tetrathionate broth: Ingredients per liter: proteose peptone 5 g, bacto bile salts 1 g, sodium thiosulfate 30 g, calcium carbonate 10 g.

2.1.1. Method of Preparation

  a. Prepare 100 mL of tetrathionate broth with distilled or deionized water and heat to boiling. Cool below 60°C.

b. Add 2 mL iodine solution (prepared by dissolving 6 g iodine crystals and 5 g potassium iodide in 20 mL distilled or deionized water) to medium. Do not heat after adding iodine.

c. Dispense 10–12 mL quantities into sterile test tubes. Use medium the same day it is prepared.

8. Selenite broth: Ingredients per liter: bacto tryptone 5 g, bacto lactose 4 g, sodium selenite 4 g, sodium phosphate 10 g, pH 7.0.

   Dissolve the ingredients 1 L distilled or deionized water and heat to boiling to pasteurize. Avoid excessive heating. Do not sterilize in the autoclave.

9. Wilson–Blair medium: Medium A: Ingredients per liter: bacto beef extract 5 g, proteose peptone 10 g, bacto dextrose 10 g, sodium chloride 5 g, bacto agar 30 g, pH 7.3.

   Solution 1: 40 g sodium sulfite anhydrous in 100 mL distilled water.

   Solution 2: 21 g sodium phosphate dibasic anhydrous in 100 mL distilled water.

   Solution 3: 12.5 g bismuth ammonium citrate granular in 100 mL distilled water.

   Solution 4: 0.96 g ferrous sulfate dried in 20 mL distilled water with two drops of hydrochloric acid.

2.1.1.1. Selective Reagent

The selective reagent consists of a combination of solutions 1–4. Each solution is made up separately, dissolved, then combined. Heat combined solution to boiling until a slate-gray color develops. Allow to cool and store at room temperature in a closed rubber-stoppered container. It is stable for up to 1 mo.

2.1.1.2. Basal Medium

a. Prepare 1 L of medium A with distilled or deionized water and heat to boiling to dissolve completely.

b. Sterilize in autoclave for 15 minutes at 121°C.

c. To prepare the complete medium, aseptically add 70 mL of selective reagent and 4 mL of a 1% titrate solution of Brillant Green. Mix thoroughly.

d. Dispense as desired.

10. *Salmonella–Shigella* agar (SS agar): Ingredients per liter: bacto beef extract 5 g, proteose peptone 5 g, bacto lactose 10 g, bacto bile salts 8.5 g, sodium citrate 8.5 g, sodium thiosulfate 8.5 g, ferric citrate 1 g, bacto agar 13.5 g, bacto Brillant Green 0.33 mg, bacto Neutral Green 0.025 g, pH 7.0.

    a. Prepare 1 L of SS agar in distilled or deionized water and heat to boiling.

    b. Boil for 2–3 min with frequent and careful swirling to dissolve completely. Avoid overheating. Do not autoclave. Cool to 55–60°C.

    c. Dispense into sterile Petri dishes. Allow the surface of the medium to become quite dry by partially removing the covers while the medium solidifies (about 2 h) *(3,7,8)*.

2.2. Method 2: Salmonellas

1. 500 mL and 250 mL sterile Erlenmeyer flasks.
2. Sterile Petri dishes.
3. 10 mL sterile pipets.
4. Loop.
5. Sterile blender ("Minipimer" type).
6. Test tubes.
7. Buffered peptone water: Ingredients per liter: peptone 10 g, sodium chloride 5 g, sodium phosphate dibasic 9 g, potassium phosphate monobasic 1.5 g, pH 7.
8. Chromogenic *Salmonella* esterease agar (CSE)
    a. Basal medium: Ingredients per liter: peptone 4 g, Lab-Lemco powder (Oxoid Ltd.) 3 g, tryptone 4 g, lactose 14.65 g, L-cisteine 0.128 g, trisodium citrate dihydrate 0.5 g, Tris base 0.06 g, Tween-20.3 g, Roko agar (Industrias Roko S.A., La Coruna, Spain) 12 g, pH basal medium 7.
    b. Chromogenic substrate: SLPA-octanoate [bromide form], 0.3223 g per liter in the final medium formulation (*see* **Note 1**).
    c. UV-absorbing compound: ethyl 4-dimethylamilobenzoate (0.035%, wt/vol) dissolved in 8 mL of methanol per liter of medium.
    d. Novobiocin: 70 mg per liter (Sigma-Aldrich Co., Ltd.) (*see* **Note 2**).

2.2.1. Preparation

a. Prepare required volume of basal medium.
b. Sterilize in autoclave for 15 minutes at 121°C.
c. Cool down to about 55°C.
d. Add other compounds (e.g., chromogenic substrate, UV-absorbing compound, and Novobiocin).
e. Pour complete agar medium into Petri dishes.
f. After setting, let plates to be surface dried. Use immediately or store in dark at room temperature for up to 2 wk.

9. Saline solution: Sterile solution: 0.85% NaCl in distilled water *(8,9)*.

2.3. Bacillus cereus

1. 1 mL sterile pipets.
2. Drigalski spatula.
3. Loop.
4. 500 mL and 250 mL Erlenmeyer flasks.
5. Sterile Petri dishes.
6. Mossel agar: Ingredients per liter: bacto beef extract 1 g, peptone 10 g, D-mannitol 10 g, sodium chloride 10 g, phenol red 0.025 g, bacto agar 12 g, pH 7.0
    a. Make up 90 mL of the medium and dispense into a flask. Autoclave 15 min at 121°C. Cool to 50–55°C.
    b. Add 10 µL of a suspension 1:1 of egg yolk in physiological solution. This suspension must be maintained at 50°C.

Biochemical Identification

    c. Add 100 µg of polymixin B sulfate per milliliter of medium.
    d. Use immediately.
7. Gelatinase reaction medium: Ingredients per liter: bacto beef extract 3 g, peptone 5 g, gelatin 120 g, pH 7.0. Make up the desired volume of medium and autoclave 15 min at 121°C.
8. Clark–Lubs medium (peptone broth): Ingredients per liter: peptone 7 g, glucose 5 g, dibasic potassium 5 g, pH 7.5.
    a. Dissolve ingredients in distilled or deionized water.
    b. Adjust pH to 7.5.
    c. Sterilize in the autoclave for 15 min at 121° Gay Lussac.
9. Methyl red: Make up a 0.5% solution in 60°C ethanol.
10. Diluted acetic acid.
11. Malachite green.
12. Sudan black: Make up a 0.3% solution in 70% ethanol.
13. Xylol.
14. Safranin *(3,7,8,10)*.

2.4. Coliforms

1. 1 mL Sterile pipets.
2. Loop.
3. Fermentation vials (Durham tubes).
4. Brilliant Green lactose bile broth media: Ingredients per liter: bacto peptone 10 g, bacto lactose 10 g, bile salts 20 g, Brillant Green 0.0133 g, pH 6.9.

2.4.1. Preparation

    a. Suspend ingredients in distilled or deionized water and warm slightly to dissolve completely.
    b. Dispense required amount in test tubes.
    c. Place an inverted fermentation vial (Durham tube) in each tube.
    d. Place caps on tubes and sterilize in the autoclave for 15 min at 121°C.
    e. Before opening the autoclave, allow the temperature to drop below 75°C to avoid entrapment of air bubbles in the inverted vials.
5. Tryptone water: Ingredients per liter: tryptone 10 g, sodium chloride 5 g.
6. Levine agar: Ingredients per liter: bacto peptone 10 g, bacto lactose 10 g, dipotassium phosphate 2 g, bacto agar 15 g, bacto eosin Y 0.4 g, bacto methylene blue 0.065 g, pH 6.8–7.0.
    a. Suspend the ingredients in distilled or deionized water and heat to boiling to dissolve completely.
    b. Sterilize in the autoclave for 15 min at 121°C.
7. Simmons citrate agar: Ingredients per liter: magnesium sulfate 0.2 g, ammonium dihydrogen phosphate 1 g, dipotassium phosphate 1 g, sodium citrate 2 g, sodium chloride 5 g, bacto agar 15 g, bacto brom thymol blue 0.08 g.

a. Suspend the ingredients in distilled or deionized water and heat to boiling to dissolve completely.
b. Sterilize in the autoclave for 15 min at 121°C.
8. Clark–Lubs medium (peptone broth): Ingredients per liter: peptone 7 g, glucose 5 g, phosphate bipotasic 5 g, pH 7.5.
   a. Dissolve ingredients in distilled or deionized water.
   b. Adjust pH to 7.5.
   c. Sterilize in the autoclave for 15 min at 121°C.
9. Kovacs reagent: Ingredients per liter: paradimethyl-aminobenzaldehyde 5 g, amyl alcohol 75 mL, hydrohydrochloric acid 25 mL.

    Dissolve the aldehyde in the amyl alcohol heating in a water bath at 50°C. Let it cool down. Add the hydrochloric acid very slowly (in drops). The yellow-golden solution obtained must be stored in a dark bottle with a ground glass stopper.
10. Methyl red solution: Prepare a 0.5% solution of methyl red in 60° Gay Lussac ethanol.
11. Creatinine.
12. Potassium hydroxide: 40% Solution in water *(3,7,8,13)*.

2.5. Staphylococcus aureus

1. 1 mL sterile pipets.
2. Drigalsky spatula.
3. Sterile Petri dishes.
4. Test tubes.
5. Hemolysis tubes.
6. Loop.
7. Hydrochloric acid 1 N.
8. Rabbit citratade plasma.
9. Baird–Parker agar: Ingredients per liter: bacto tryptone 10 g, bacto beef extract 5 g, bacto yeast extract 1 g, glycine 12 g, sodium pyruvate 10 g, lithium chloride 5 g, bacto agar 20 g, pH 6.8–7.0.

2.5.1. Preparation

a. Suspend the ingredients in distilled or deionized water.
b. Heat to melt agar and adjust pH.
c. Autoclave for 15 min at 121°C.
d. Cool down to 45–50°C.
e. Add 50 mL of a suspension 1:1 of egg yolk in physiological solution. This suspension must be maintained at 45–50°C.
f. Add 3 mL of a 3.5% potassium tellurite solution sterilized by filtration. The solution must be warmed up to 45–50°C.

10. DNase test agar: Ingredients per liter: bacto tryptose 20 g, deoxyribonucleic acid (DNA) 2 g, sodium chloride 5 g, bacto agar 15 g, pH 7.3.
    a. Suspend compounds in distilled or deionized water. Heat to boiling to dissolve completely.
    b. Sterilize in the autoclave for 15 min at 121°C.

Biochemical Identification

11. **Brain–heart infusion:** Ingredients per liter: Infusion from calf brains 200 g, infusion from beef heart 250 g, proteose peptone 10 g, bacto dextrose 2 g, sodium chloride 5 g, disodium phosphate 2.5 g, pH 7.4.
    a. Suspend the ingredients in distilled or deionized water.
    b. Dispense as desired.
    c. Sterilize in the autoclave for 15 min at 121°C *(3,7,8,10)*.

3. Methods
3.1. Salmonellas: Method 1

1. The sample of the solid or liquid food must be representative and weight not less than 25 g.
2. Mix 225 mL of the buffered peptone water with 25 g of the food sample. Blend the mixture with a Minipimer-type blender during 2 min. Afterward, the pH of the mixture must be adjusted to 6–7 with a buffer solution.
3. Transfer aseptically the mixture obtained in **step 2** to a 500-mL sterile Erlenmeyer flask. Incubate at 37°C for 16–20 h.
4. After incubation, transfer 10 mL of the culture to a flask containing 100 mL of tetrathionate broth (culture A). Incubate at 42–43 °C for 48 h.
5. Transfer 10 mL of the culture obtained in **step 3** to a flask containing 100 mL of selenite broth (culture B). Incubate at 37°C for 48 h.
6. After 24 h incubation in both tetrathionate broth (culture A) and selenite broth (culture B), samples of each culture must be plated on to Wilson–Blair medium and SS agar to obtain isolated colonies. Petri dishes must be incubated at 37°C for 48 h (*see* **Note 3**).
7. The same plating procedure indicated in **step 6** must be followed after 48 h of cultivation in both cultures (tetrathionate broth and selenite broth), but this time Petri dishes are incubated for 24 h.
8. *Salmonella* colonies on Wilson–Blair agar will appear as brown or black with shiny colonies surrounded by a dark halo. Some strains develop green colonies and do not darken the medium.

Salmonella colonies on SS agar appear as colorless or pinkish colonies at 18 h of incubation. Afterward, they become bigger and opaque and they can develop a gray or black central spot *(1,3,5,10,11)* (*see* **Note 4**).

3.2. Salmonellas: Method 2

1. Follow **steps 1–3** of **Subheading 3.1.**
2. Transfer a sample of the culture into test tubes containing buffered peptone water.
3. Incubate for 4–6 h at 37°C.
4. In order to obtain plates showing well-isolated colonies, the cultures must be serially diluted in saline solution.
5. Spread 100 µL of appropriate dilutions onto chromogenic medium.
6. Incubate plates at 37 or 42°C.
7. Observe for colony coloration for up to 48 h.

8. Salmonella spp. could be differentiated from nonsalmonellae by the production of burgundy-colored colonies. Nonsalmonellae appeared as white or colorless colonies *(3,9)*.

3.3. Bacillus cereus

1. Take 25 g of the sample and add 225 mL of peptone broth. Mix thoroughly for 1–2 min. The solution thus obtained is a 10^{-1} dilution of the original sample. The amount of peptone broth to add to the sample depends on the product to test. In case of flour it is convenient to predetermine the volume of medium to use.
2. Add 1 mL of the 10^{-1} dilution to 9 mL of sterile peptone broth to obtain 10^{-2} dilution. Make up a 10^{-3} and 10^{-4} dilutions following the same procedure.
3. Plate 0.1 mL of the dilutions on Mossel agar. Spread the inocula using a Drigalski spatula. The plates must be completely dry.
4. Incubate for 24–48 h at 30°C. Examine plates to find colonies that have grown (*see* **Note 5**).

3.3.1. Biochemical Confirmation

The colonies must be transferred to plates containing gelatinase reaction medium. Incubate for 2–3 d at 30°C. Cover the plates with diluted acetic acid. Colonies that are gelatinase positive will show a halo.

The doubtful colonies must also be transferred to test tubes containing 5 mL of Clark–Lubs medium (peptone broth). Incubate for 3 d at 30°C and then add to the cultures 4–5 drops of methyl red indicator, mix to homogenize, and check color. If the colonies are methyl red positives the culture color will turn to red, otherwise the culture color will remain yellow. *B. cereus* is methyl red (+), gelatinase (+), and lecitinase (++) (*see* **Note 6**).

The number of cells per gram of product is obtained multiplying the number of colonies developed on Mossel agar by the dilution factor used for the suspension plated on the Petri dishes *(3,4,10–12)*.

3.4. Coliforms

1. Add 225 mL peptone water to 25 g of the sample. Shake during 1–2 min. The final suspension is a 1:10 dilution of the sample.
2. Prepare 9 sterile test tubes containing 9 mL each one of BGBL medium and the inverted fermentation tubes (*see* BGBL culture medium, **Subheading 2.4.**, **item 4**).
3. Add with sterile pipet 1 mL of the 1:10 dilution of the sample to each of three of the test tubes mentioned earlier. The suspension obtained will be a 1:100 dilution of the original sample. Follow the same procedure to obtain a 1:1000 and a 1:10000 dilutions of the original sample. Note that three test tubes containing each dilution will be obtained.
4. Incubate the test tubes for 48 h at 30°C.

Biochemical Identification

5. After the incubation period, observe the test tubes. Tubes where 10% of the total volume of the inverted vial is occupied by gas are considered as coliform positive. The results of these test tubes are useful to determine the more probable number of microorganisms per gram or milliliter in the original sample. For each dilution (1:100, 1:1000, and, 1:10000) the tubes with positive gas production are counted. The number of positive tubes are used to obtain the more probable number from **Table 1** *(1,2,4,10,12,13)*. The positive tubes will also allow the identification of *E. coli*.

3.5. E. Coli

3.5.1. General Procedure

E. coli must be investigated in all test tubes where gas is present when the coliform identification technique was followed (*see* **Subheading 3.4.**). *E. coli* has the following biochemical characteristics: gas production when incubated in BGBL broth for 48 h at 44°C and indole production when incubated in tryptone water during 48 h at 44°C.

1. Mix each test tube where gas is observed when the coliform identification technique was followed. Transfer a loopful of the suspension from each tube to a new sterile tube containing 10 mL of BGBL medium and an inverted vial.
2. Repeat the foregoing procedure with test tubes containing 10 mL of tryptone water.
3. Incubate for 48 h at 44°C.
4. After the incubation period check the gas production. Write down the number of positive tubes.
5. Investigate indole presence in the tubes containing tryptone water with the following technique:
   a. Add 1 mL of Kovacs reagent.
   b. An indole positive reaction will develop a red ring on the surface of the medium after 5 min.
6. Consider *E. coli* positive subcultures that, having produced gas during the 30°C incubation, give an indole-positive reaction in tryptone water and gas production in BGBL at 44°C.

3.5.2. Specific Media for Confirmation of E. coli

In order to reinforce the identification of *E. coli,* use the following procedures:

1. Take samples with a loop from the *E. coli* positive test tubes containing BGBL medium. Plate the samples onto Levine agar.
2. Incubate plates for 48 h at 37°C.
3. Observe colonies. Different characteristics between *E. coli* and *Enterobacter aerogenes* are given in **Table 2**. Colonies of *Salmonella* and *Shigella* are transparent, amber colored.
4. Confirm the presence of *E. coli* by the IMVIC (test using the following techniques: indole, methyl red, Voges-Proskauer, and sodium citrate).
   a. Indole: Proceed as indicated in **step 5** of protocol for *E. coli* identification.

Table 1
More Probable Number of Microorganisms per Gram or Milliliter (*see* Note 7)

Positive tubes			Number per gram milliliter
10^{-2}	10^{-3}	10^{-4}	
0	0	1	3
0	1	0	2
1	0	0	4
1	0	1	7
1	1	0	7
1	1	1	11
1	2	0	11
2	0	0	9
2	0	1	14
2	1	0	15
2	1	1	20
2	2	0	21
2	2	1	28
3	0	0	23
3	0	1	39
3	0	2	64
3	1	0	43
3	1	1	75
3	1	2	120
3	2	0	93
3	2	1	150
3	2	2	210
3	3	0	240
3	3	1	460
3	3	2	1,100

  b. Methyl red:
    (1) Prepare a pure culture from the doubtful colonies using peptone water as the culture medium. This culture must be incubated for 6–8 h. Use a loopful of the latter culture as the inoculum to a test tube containing 10 mL of Clark–Lubs medium.
    (2) Incubate for 72 h at 30°C.
    (3) Divide culture in two identical volumes, one of which will be used in the methyl red test and the other one in the Voges–Proskauer test.
    (4) Add to one of them five drops of methyl red solution.
    (5) Observe results: Red color: methyl red +; Yellow color: methyl red –.

Biochemical Identification

Table 2
Different Characteristics between Colonies of *E. coli* and *A. aerogenes*

	E. coli	*A. aerogenes*
Size	Well-isolated colonies are 2–3 mm in diameter	Well-isolated colonies are usually larger than *E. coli*
Elevation	Colonies are slightly raised; surface flat or slightly	Colonies considerably raised and markedly convex concave
Appearance by reflected light	Colonies dark, buttonlike, often concentrically ringed with a greenish metallic sheen	Much lighter than *E. coli*, centers are deep brown Metallic sheen is not observed

    c. Voges–Proskauer test: This test is also known as of acethyl-methyl-carbinol production.
    (1) Add to the other portion of the culture in the Clark–Lubs medium a small amount of creatinine and 5 mL of the solution 40% of KOH.
    (2) Mix during 2 min.
    (3) Observe results: Pink Color: Voges–Proskauer +; no color change: Voges–Proskauer –.
    d. Sodium citrate test: Some microorganisms are able to grow with citrate as the sole carbon source. This characteristic is used to differentiate the Enterobacteriaceae.
    (1) Inoculate the doubtful culture with a loop in the middle of the plate.
    (2) Incubate for 48 h at 37°C.
    (3) Observe results. Blue color: citrate +. No color change: citrate – *(3,4,6,8,13–15)*.

The results of the IMVIC tests for different bacteria genus are given in **Table 3**.

3.6. Staphylococcus

1. Add 225 mL of sterile water to 25 g of the sample. Shake during 1–2 min. The final suspension is a 1:10 dilution of the sample.
2. Prepare 1:100, 1:1000, and 1:10000 dilutions of the sample.
3. Spread 0.1 mL of each dilution onto Petri dishes containing Baird–Parker agar. Use a Drigalski spatula to spread properly. Make duplicates of the plates.
4. Incubate for 24–48 h at 37°C.
5. Coagulase-positive colonies of staphylococci are black with a sheen surrounded by a clear zone due to the action of the enzymes on the egg yolk.
6. Investigate the presence of enzymes in the doubtful colonies using the following techniques.
    a. DNase test: *S. aureus* is DNase positive.
    (1) Plates containing DNase test agar are inoculated by streaking or spotting with the material or culture being tested. Make only three streaks or spots per plate.

Table 3
IMVIC Test Results for Different Bacteria

	Indole	Methyl red	Voges–Proskauer	Citrate
Shigella	±	+	–	–
Escherichia	±	+	–	–
Salmonella	–	+	–	+
Arizona	–	+	–	+
Citrobacter	–	+	–	+
Klebsiella	±	–	+	+
Enterobacter	–	–	+	+
Hafnia	–	–	+	+
Proteus	+	+	–	+
Providencia	+	+	–	+

(2) Incubate for 18–24 h at 37°C.
(3) Flood plates with 1 N hydrochloric acid and observe for clearing around the streak or spot indicating DNase activity.
  b. Coagulase test: Most of the *S. aureus* strains are coagulase positive, although in very few cases it could be coagulase negative.
(1) Test tubes containing 5 mL of brain–heart infusion are inoculated with the material being tested.
(2) Incubate for 24 h at 37°C.
(3) Add 2–3 drops of the culture to a hemolysis tube containing 0.5 mL rabbit plasma.
(4) Incubate at 37°C, periodically observing coagulation. Coagulation usually takes place before 4 h of inoculation.
  c. Thermonuclease test: *S. aureus* DNase is thermoresistant.
(1) Plates containing Baird–Parker agar are inoculated by streaking or spotting with the material or culture being tested.
(2) Incubate at 37°C for 24 h.
(3) The dishes that show colony development are then incubated at 65°C for 150 min. After this treatment the cells become inviable.
(4) Add to the plates DNase test agar containing toluidine blue. Incubate at 37°C for 4 h.
(5) DNase hydrolysis is shown by the presence of pinkish brilliant halos. The clearing is due to the action of *S. aureus* DNase that is thermoresistant.

Although *S. aureus* cells could have been destroyed in the sample by any reason, some toxins could remain active. Therefore, even though *S. aureus* cells are not detectable, it is important to investigate the presence of toxins. Specific kits are the more usual way to detect these toxins.

To isolate a toxin from the food sample, proceed as follows.

(1) Take 10 g of the food and add to it 10 mL of sterile 0.85% NaCl solution. Cool to 4°C.
(2) Centrifuge at 2000g for 30 min at 4°C.
(3) The supernatant is then filtered through a 0.2-μm filter.
(4) The filtrate is used to detect the presence of the toxin using adequate kits *(2,4,5,11,13–15)*.

4. Notes

1. Chromogenic *Salmonella* esterase agar is based on the detection of C_8–esterase activity in salmonellae.
2. Alternatively, the chromogenic substrate and UV-absorbing compound could be added directly to the basal medium if this is heated only to the boiling point. UV-absorbing compound is added to protect the substrate against photochemical degradation. The use of novobiocin reduce the growth of nonsalmonella strains in the chromogenic medium.
3. *Salmonella*: following the procedure four plates will be obtained:
    a. Culture A on Wilson–Blair medium.
    b. Culture A on SS agar.
    c. Culture B on Wilson–Blair medium.
    d. Culture B on SS agar.
4. *Salmonella*: It is recommended to confirm the presence of *Salmonella* that either the colonies are typical or they are doubtful. The appearance of the colonies varies not only for the different species but also with different batches of culture media. An API-20E kit could be used as a method to confirm *Salmonella*.
5. *B. cereus:* the polymixin added to the Mossel agar allows the growth of *B. cereus* but inhibits the growth of the secondary flora. The lecithin of the egg yolk will precipitate by the action of the lecithinase of *B. cereus*. *B. cereus* does not produce an acid from mannitol, therefore its presence does not turn the phenol red to yellow. The colonies of *B. cereus* are pink surrounded by an opaque zone due to the action of the lecithinase on the egg yolk. The colonies are invasive, irregular, and rough.
6. *B. cereus:* the other test to confirm the presence of *B. cereus* consists of transferring the doubtful colonies onto any nutrient agar for bacteria and incubating them for 24 h. The cells are then fixed by heat on a slide. The spore staining is done by adding the malachite green solution to the slide and heating with direct flame until vapor emission. Keep heating for 5–10 min. Wash the sample with water, dry it, and add the Sudan Black solution for lipid staining. Keep the Sudan Black solution in contact with the preparation for 15 min, wash it with xylitol for 5 s, and let dry. Add safranin and keep in contact for 1 min. Afterward, wash with water, dry, and observe. Spores show green, lipid compounds black, and the cytoplasm will show red.
7. Coliforms: use **Table 1**, e.g.,
    a. Dilution 1:100 positive tubes 1.
    b. Dilution 1:1000 positive tubes 2.
    c. Dilution 1:10000 positive tubes 0.
    Then the more probable number of microorganisms is 11 per gram or milliliters of the original sample.

References

1. Banwart, G. J. (1981) *Microbiología Básica de los Alimentos,* Ediciones Bellatterra S.A. Anthropos, Barcelona, Spain.
2. Cliver, D. O. (1990) *Foodborne Diseases,* Academic Press, San Diego, CA.
3. International Commission on Microbiological Specifications for Foods (ICMSF). *Microorganismos de los Alimentos. Vol. 1. Técnicas de Análisis Microbiológicos,* Editorial Acribia S.A., Zaragoza, Spain.
4. Sharf, J. M. (1965) Recomended methods for the microbiological examination of foods. American Public Health Association Inc. Washington, DC.
5. Vila Aguilar, R. (1994) Instituto de Agroquímica y Tecnología de Alimentos (IATA), Valencia, Spain.
6. Martialay Valle, F. (1989) *Prontuario de Técnicas en Microbiología de Alimentos,* Ministerio de Defensa, Sec. Gral. Técnica, Madrid, Spain.
7. Marchal, N., Bourdon, J. L., and Richard, C. L. (1982) Les milieux de culture pour l'isolement et l'identification biochimique des bactéries. Doin éditeurs. París.
8. Difco Manual, Difco Laboratories, USA (1995).
9. Cooke, V. M., Miles, R. J., Price, R. G., and Richardson, A.C. (1999) A novel chromogenic ester agar medium for detection of Salmonellae. *Appl. Environ. Microbiol.* **65,** 2.
10. Pascual Anderson, M. R. (1982) *Técnicas para el Análisis Microbiológico de Alimentos y Bebidas,* Ministerio de Sanidad y Consumo, Madrid, Spain.
11. Larpent, J. P. and Larpent Gourgaud, M. (1975) *Memento. Techniques de microbiologie,* Technique et Documentation, París.
12. Fung, D. Y. C. and Matthews, R. F. (1991) *Instrumental Methods for Quality Assurance in Foods,* Marcel Dekker, ASQC Quality Press, New York.
13. Collins, C. H. and Lyne, P. M. (1989) *Métodos Microbiológicos,* Editorial Acribia S.A., Zaragoza, Spain.
14. Speck, M. L. (1976) *Compendium of Methods for the Microbial Examination of Foods,* American Public Health Association, Inc., Washington, DC.
15. Thatcher, F. S. and Clark, D. S. (1973) *Análisis Microbiológico de los Alimentos,* Editorial Acribia S.A., Zaragoza, Spain.

3

Mesophilic Aerobic Microorganisms

Annavella Gaitán Herrera

1. Introduction

The mesophilic microorganism is ones of the more general and extensively microbiological indicators of food quality, *(1)* indicating the adequacy of temperature and sanitation control during processing, transport, and storage, and revealing sources of contamination during manufacture *(2)*.

2. Materials

1. Mechanical blender, two-speed model or single speed with rheostat control.
2. Glass or metal blending jars of 1 L capacity, with covers, that are resistant to autoclave temperatures. One sterile jar (autoclaved at 121°C for 15 min) for each sample to be analyzed.
3. Balance with weights. Capacity should be at least 2500 g, sensitivity 0.1 g.
4. Instruments for preparing samples: knives, forks, forceps, scissors, spoons, spatulas, or tongue depressors, sterilized previous to use by autoclaving or by hot air.
5. A supply of 1, 10, and 11 mL pipet(s).
6. Refrigerator, at 2–5°C.
7. Peptone dilution fluid or peptone salt dilution fluid sterilized in autoclave for each sample, 450 mL in flask or bottle, 90 or 99 mL blanks in dilution bottles or similar containers.

3. Methods

1. Begin the examination as soon as possible after the sample is taken. Refrigerate the sample at 0–5°C whenever the examination cannot be started immediately after it reaches the laboratory. If the sample is frozen, thaw it in its original

container (or on the container in which it was received at the laboratory) for a maximum of 18 h in a refrigerator at 2–5°C. If the frozen sample can be easily comminuted (ice cream), proceed without thawing.
2. Tare the empty sterile blender jar, the weigh into it 50 ± 0.1 g representative of the food sample. If the contents of the package are obviously not homogeneous (frozen dinner), take a 50-g sample from a macerate of the whole dinner, or analyze each different food portion separately, depending on the purpose of the test.
3. Add to the blender jar 450 mL of peptone dilution fluid or peptone salt dilution fluid. This provides a dilution of 10^{-1}.
4. Blend the food and dilute promptly. Start at low speed, then switch to high speed within a few seconds; Time the blending carefully to permit 2 min at high speed. Wait 2 or 3 min for foam to disperse.
5. Measure 1 mL of the 10^{-1} dilution of the blended material, avoiding foam, into a 99-mL dilution blank, or 10 mL into a 90-mL blank. Shake this and all subsequent dilutions vigorously 25 times in a 30-cm arc. Repeat this process using the progressively increasing dilutions to prepare dilutions of 10^{-2}, 10^{-3}, 10^{-4}, and 10^{-5}, or dilutions that are desirable for the food under test *(3)*.

4. References

1. Elliot, R. P., Clark, D. S., et al. (1978) *Microorganisms in Foods 1. Significance and Methods of Enumeration,* 2nd ed. A publication of the International Commission on Microbiological Specifications for Foods (ICMSF) of the International Association of Microbiological Societies. University of Toronto Press, Canada.
2. Speck M. L. (1984) *Compendium of Methods for the Microbiological Examination of Foods*, 2nd ed. Compiled by the APHA Technical Commitee on Microbiological Methods for Foods, American Public Health Association, Washington, DC.
3. Marth, E. H., ed (1978) *Standard Methods for the Examination of Dairy Products,* 14th ed., American Public Health Association, Washington, DC.

4

Yeasts and Molds

Anavella Gaitán Herrera

1. Introduction

Yeast and molds (nonfilament and filament molds) may be found as part of the normal flora of a food product on inadequately sanitized equipment or as airborne contaminants. They can produce toxic metabolites, resistance to freezing environments, and cause off odors and off flavors of foods. Media for the enumeration of molds use antibiotics, e.g., penicillin + streptomycin, chloramphenicol, chlortetracycline, oxytetracycline, and gentamicin; oxytetracycline has been useful. The incubation temperature is 22°C over 5–8 d *(1)*.

2. Materials

1. Petri dishes.
2. Bacteriological pipets 1, 5, and 10 mL.
3. Incubator 20–24°C. Water bath at 44–46°C for tempering agar.
4. Oxytetracycline gentamicin yeast extract glucose (OGY) agar.
5. Diluent, buffered peptone water.
6. Potato dextrose agor (PDA), Difco.

3. Methods

3.1. Antibiotic Method

Use plate count agar, temper the medium to 45°C and aseptically add 2 mL of antibiotic solution (add 500 mg each of chlortetracycline HCl and chloramphenicol [U.S. Biochemical Corp., Cleveland, OH] to 100 mL sterile phosphate-buffered distilled water and mix) per 100 mL medium.

1. Pipet 1 mL aliquots from 10^{-1} to 10^{-5} dilutions. Promptly put into Petri dishes 10–15 mL of OGY agar, melted and tempered to 45°C.
2. Incubate the Petri dishes at 22°C for 5–8 d.
3. Count all colonies on plates containing 30–300 colonies, compute the number of yeast and molds per gram or milliliter of food. Report as colony-forming units (CFU) per g or mL of sample *(2)*.

3.2. Acidified Method

Acidify PDA or malt agar with sterile 10% tartaric acid to pH 3.5 ± 0.1. Acidify the sterile and temperated medium with a quantity of acid solution immediately before pouring the agar onto plates. Do not reheat medium once acid has been added.

References

1. Speck, M. L. (1984) *Compendium of Methods for the Microbiological Examination of Foods.* 2nd ed. Compiled by the APHA Technical Commitee on Microbiological Methods for Foods. American Public Health Association, Washington, DC.
2. Elliot, R. P., Clark, D. S. et al (1978) *Microorganisms in Foods 1. Significance and Methods of Enumeration.* 2nd ed. A publication of the International Commission on Microbiological Specifications for Foods (ICMSF) of the International Association of Microbiological Societies. University of Toronto Press, Canada.

5

Coliforms

Anavella Gaitán Herrera

1. Introduction

The coliform group of indicator organisms includes some members of this family that are capable of fermenting lactose with the production of acid and gas within 48 h at 35°C. The family Enterobacteriaceae are Gram-negative oxigenic and facultatively anoxigenic rods, nonspore forming, that produce acid from glucose and other carbohydrates.

The fecal coliform group is restricted to organisms that grow in the gastrointestinal tract of humans and warm-blooded animals and includes members of at least three genera *Escherichia, Klebsiella,* and *Enterobacter.* Elevated temperatures tests for the differentation of organisms of the coliform group into those fecal origin and those of nonfecal origin have been used (*see* **Table 1**). The presence of *Escherichia coli* may be attributed to contamination from environmental sources and subsequent growth in the product *(1,2).*

Standard methods for the examination of water and wastewater describes two approaches to fecal coliform determinations:

1. Subculture from positive presumptive lauryl sulfate tryptose (LST) broth to *E. coli* broth with incubation at 45.5 ± 0.2°C for 24 ± 2 h.
2. A membrane filter procedure using medium-fecal coliform broth with incubation at 44.5 ± 0.2°C for 24 ± 2 h. Foods authorities specify incuation specify incubation *E. coli* broth 44.5 ± 0.2°C for 24 h for fecal coliform counts of fish, fish products, and shellfish, but 44.5 ± 0.2°C for other foods.

1.1. Enterobacteriacae Group

A brilliant green-Oxgall broth with glucose substituted for the lactose is used for enrichment. Cultures are recovered from violet red bile agar (VRBA) fortified with 1% glucose.

1.2. Escherichia coli

The detection and enumeration of *E. coli* of sanitary significant isolates must conform to the coliform and fecal coliform group determinations. The isolates were identified by indol, motility, Voges–Proskauer, citrate (IMViV):

1. IMViC pattern: ++ – – Type I.
2. IMViC pattern: –+ – – Type II.

It is important considering that this profile is inadequate for the idenification of the species. The relatively high incidence of type II in some specimens is explained by the fact that cultures require 48 h to produce a detectable amount of indole and need additional tests for speciation.

1.3. Enteropathogenic E. coli (EEC)

Enteropathogenic *E. coli* is an etiological agent of gastroenteritis in humans and domestic animals. Identifications of these cultures requires serological, pathological, and biochemical tests. Some cultures are unable to ferment lactose within 48 h or do not produce gas. Tests for enteropathogenicity include: production of one or more enterotoxins, colonization of epithelial cells, or invasiveness, and possession of somatic, capsular, and flagellar antigens.

2. Materials

1. Water baths capable of maintaining temperatures of 44 ± 0.2, 44.5 ± 0.2, and $49 \pm 1°C$.
2. Thermometer approx 45–55 cm long, with a range of 1.0–55°C.
3. Circulating air incubator, maintaining a temperature of $44 \pm 0.3°C$.
4. Ultraviolet lamp filtered with a Woods filter, emitting rays of 365 nm wavelength.
5. Glass spreader.
6. 5% CO_2 incubator, maintained at $36 \pm 1°C$, moisture saturated.
7. Serological racks.
8. Glass Petri dishes and pipets (0.1 mL, 0.025 mL, and Pasteur).
9. Vertical laminar flow hood, biological contaminant hood equipped with HEPA™ filter.
11. McFarland nephelometer.

2.1. Media

1. Lauryl sulfate tryptose (LST) broth.
2. Brilliant green bile (BGB) broth.

3. *E. coli* broth.
4. Levine eosin methylene blue (EMB) agar.
5. Tryptone (trypticase) broth.
6. Buffered glucose broth (MR-VP) medium.
7. Koser citrate medium.
8. Violet red-bile agar (VRBA).
9. Phosphate-buffered dilution water.
10. A-1 broth.
11. Tryptone bile agar.
12. Peptone water.
13. Minerals-modifed glutamate agar.
14. Brain–heart infusion (BHI).
15. Tryptone phosphate (TP) broth.
16. MacConkey agar.
17. Veal infusion agar.
18. Malonate broth.
19. Bromocresol purple carbohydrate broth supplemented with the following carbohydrates: 0.5% glucose, 0.5% sorbitol, 0.5% cellobiose, 0.5% mannitol, 0.5% lactose, and 0.5% adonitol.
20. KCN broth.
21. Blood agar base.
22. Urea agar.
23. Motility agar.
24. Plate count agar (PCA).

2.2. Reagents

1. Kovacs indole reagent.
2. Methyl red indicator.
3. Voges–Proskauer reagents.
4. Gram-stain reagents.
5. NaCl 0.5%, sterile.
6. pH test paper, range 5.0–8.0.
7. *E. coli* antisera.
8. *Shigella* antisera.
9. Pathogenic biotypes of *E. coli*: enteroinvasive, enterotoxigenic (producing heat-labile toxin), enterotoxigenic (producing heat-stable toxin), and classical (infantile enteropathogenic).
10. Cytochrome oxidase reagent.

3. Methods
3.1. Preparation of Sample

1. Hold in the refrigerator at 2–5°C for 18 h before analysis.
2. Weigh 25 g of regular or thawed food sample aseptically into a sterile blender jar.

3. Add 225 mL of diluent (buffered peptone water) and blend for 2 min.
4. Prepare decimal dilutions in the range 1:10, 1:100 by adding 10 mL of the previous dilution to 90 mL of the sterile diluent, shake all dilutions for 7 s.

3.2. Presumptive Test for Coliform Group Most Probable Number (MPN)

1. Inoculate three replicate tubes of LST broth per dilution with 1 mL of the previously prepared 1:10, 1:100, and 1:1000 dilutions.
2. Incubated tubes for 24 and 48 ± 2 h at 35 ± 0. 5°C.
3. Observe all tubes for gas production either in the inverted vial (Durham) or by effervescence produced. Read tubes for gas production after 24 h. Reincubate negative tubes for an additional 24 h.
4. Record LST tubes showing gas within 48 h, refer to **Table 2** (three-tube dilutions) and report as the presumptive MPN of coliform bacteria per gram or milliter of food.
5. To obtain the MPN, determine from each of the three selected dilutions the number of tubes that provided a confirmed coliform result. Refer to **Table 2** and note the MPN based on the levels of sample dilution and the number of confirmed postive tubes of each dilution selected.
6. Refer results to **Table 2** calculated from the data of the Man (1975).

3.3. Confirmed Test for Coliform Group

1. Subculture positive LST tubes (gas production) into BGB broth, incubate at 35 ± 0.5°C for 48 ± 2 h.
2. Record LST tubes showing gas within 48 h, refer to **Table 1** (three- or five tube dilutions) and report as confirmed MPN of coliform bacteria per g or mL of food.

3.3.1. Coliform Group (VRBA)

1. Homogenize 25 g or mL sample for 2 min in 225 mL phosphate buffer.
2. Prepare serial tenfold dilutions.
3. Transfer two 1 mL aliquots of each solution to Petri dishes. Pour 10 mL of VRBA mix; overlay with 5 mL VRBA. Incubate 24 h at 35°C.
4. Count purple-red colonies surrounded by zone of precipitated bile acids.
5. Confirmation: select colonies and transfer each to a tube of BGB broth, incubate at 35°C for 24 or 48 h. Colonies producing gas are confirmed as coliform organisms. Multiply the number of coliform organisms per gram of sample.

3.3.2. Fecal Coliform: The Mackenzie Test

This procedure should differentiate between coliforms of fecal origin (intestines) and coliforms from other sources.

Table 1
Differentiation of *E. coli* from Related Enterobacteriaceae

Culture	IMViC	Motility 37°C	Cytochrome oxidase	Glucose gas	KCN	Lactose
Typical *E. coli*	++ − −	±	−	+	−	+
Inactive *E. coli*	++ − −	−	−	−	−	−
Plesiomonas	++ − −	+	+	−	−	−
Klebsiella	− ± ± −	−	+	−	−	±
Hafnia	− ± ± −	+	−	+	+	−
Aeromonas	+++ ±	+	+	+	±	−

Culture	Mannitol	Sorbitol	Cellobiose	Malonate	Adonitol	Urease
Typical *E. coli*	+	±	−	−	−	−
Inactive *E. coli*	+	−	−	−	−	−
Plesiomonas	±	−	−	−	−	−
Klebsiella	±	−	+	±	±	±
Hafnia	+	−	±	±	−	−
Aeromonas	+	±	±	−	−	−

Table 2
Most Probable Number of Bacteria (MPN)

Number of positive tubes at each dilution level				Confidence limits			
10^{-1} (g)	10^{-2}	10^{-3}	MPN (g)	99%	MPN (g)	95%	MPN
0	1	0	3	<1	23	<1	17
1	0	0	4	<1	28	1	21
1	0	1	7	1	35	2	27
1	1	0	7	1	36	2	28
1	2	0	11	2	44	4	35
2	0	0	9	1	50	2	38
2	0	1	14	3	62	5	48
2	1	0	15	3	65	5	50
2	1	1	20	5	77	8	61
2	2	0	21	5	80	8	63
3	0	0	23	4	177	7	129
3	0	1	40	10	230	10	180
3	1	0	40	10	290	20	210
3	1	1	70	20	370	20	280
3	2	0	90	20	520	30	390
3	2	1	150	30	660	50	510
3	2	2	210	50	820	80	640
3	3	0	200	<100	1900	100	1400
3	3	1	500	100	3200	200	2400
3	3	2	1100	200	6400	300	4800

Inoculate 1 mL into each of three tubes of media. Multiply the MPN by the appropriate factor of 10, 100, or 1000.

Example: tubes selected come from 10^{-2}, 10^{-3}, and 10^{-4} multiply by 10, 10^{-3}, 10^{-4} and 10^{-5} multiply by 100.

1. Subculture positive LST tubes showing gas within 48 h to *E. coli* broth.
2. Incubate 24 ± 2 h at 4.5 ± 0.2°C.
3. Examine tubes for gas.
4. Calculate MPN values according to **Table 2**. Report results as MPN of fecal coliforms per gram or milliliter.

3.4. Enumeration of E. coli

1. Subculture positive LST tubes showing gas within 48 ± 2 h into *E. coli* broth by means of the 30-mm loop.
2. Incubate all *E. coli* tubes in a water bath for 48 ± 2 h at 45 ± 0.2°C.
3. Subculture all *E. coli* tubes showing gas on EMB plates and incubate 24 ± 2 h at 35°C.

4. If typical colonies are present (nucleated, dark-centered, with or without sheen) pick two colonies from EMB plate and transfer each to a PCA slant. Incubate slants at 35°C from 18–24 h.

3.4.1. Test for Indentification

1. Tryptone broth incubate 24 ± 2 h at 35°C and test for indole. Add 0.3 mL Kovacs reagent. The test is positive if the upper layer becomes red.
2. Incubate MR-VP medium for 48 ± 2 h at 35°C, transfer 1 mL of culture to tube to test for acetylmethylcarbinol. Add 0.6 mL alcoholic alpha-naphthol and 0.2 mL 40% KOH. Mix and add a few crystals of. Incubate at room temperature for 2 h. Test is positive if eosin pink develops. Incubate the remainder of MR-VP culture for 48 h and test for methyl red reaction by adding five drops of methyl red solution to the culture. Test is positive if the culture turns red, negative if yellow or orange.
3. Citrate broth. Incubate 48 h at 35°C and record growth as + or –.
4. LST broth: Incubate 48 ± 2 h at 35°C. Examine tubes for gas and formation from lactose.
5. IMViC reaction: + + – – or – + – –.
6. Compute MPN of *E. coli* per gram or milliliter.

3.5. Rapid Method for Enumeration of E. coli Biotype I

This method is modified by Anderson and Parker *(3)*.

1. Prepare serial tenfold dilutions in peptone water of food.
2. Aseptically transfer sterile cellulose acetate membranes to the surface or dried glutamate agar.
3. Transfer duplicate 1.0 mL aliquots of each dilution to cellulose acetate membranes, using a sterile glass Drigalski spreader, distribute the fluid over the entire membrane (except the periphery).
4. Incubate plates 4 h at 35°C (facilitate resuscitation). Transfer each membrane to tryptone bile agar plate and incubate for 18 h at 44°C.
5. Remove lids of plates. Add 3 mL Kovacs reagent into each lid. Remove membrane from agar and immerse in reagent for 5 min. Remove membrane and drain excess reagent. Dry the membranes under a UV lamp.
6. Count pink-stained colonies within 30 min.
7. Calculate the number of type I *E. coli* by selecting the dilution giving an average of 20–50 pink colonies and both membranes and multiply total by the dilution factor.

References

1. Speck, M. L. (1984) *Compendium of Methods for the Microbiological Examination of Foods,* 2nd ed. Compiled by the APHA Technical Committee on Microbiological Methods for Foods. American Public Health Association, Washington, DC.

2. Elliot, R. P., Clark, D. S., et al. (1978) *Microorganisms in Foods. Significance and Methods of Enumeration,* 2nd. ed. A publication of International Commission on Microbiological Specifications for Foods (ICMSF) of the International Association of Microbiological Societies, University of Toronto Press, Canada.
3. Anderson, J. M. and Baird Parker, A. C. (1975) A rapid and direct plate method for enumerating *Escherichia coli* biotype I in food. *J. Appl. Bacteriol.* **39,** 111–117.

6

Genetic Analysis of Food Spoilage Yeasts

Stephen A. James, Matthew D. Collins, and Ian N. Roberts

1. Introduction

Yeasts have been associated with foods since earliest times, both as beneficial agents and as major causes of spoilage and economic loss. Current losses to the food industry caused by yeast spoilage are estimated at several million pounds annually in the UK alone. As new food ingredients and new food manufacturing technologies are introduced, novel food spoilage yeasts are emerging to present additional problems. Consumer demand for milder food preservation regimes and tougher regulatory constraints on hygiene in the production environment all serve to increase the economic severity of problems caused by yeasts.

In order to better understand the processes by which yeast species (including novel preservative-resistant strains) emerge to cause spoilage, more information is needed concerning their biological diversity, natural habitats, and genetic interrelatedness. This information can be used to predict the potential for yeast spoilage in any particular food environment and to provide methods for rapid identification of the species involved. To date, over 700 biological species of yeasts have been described and thousands of different varieties have been shown to exist in all kinds of natural and artificial habitats. Rapid identification is therefore essential in order to establish at an early stage which species is involved and thus to predict the likely severity of the problem and sensible course of action.

The methods presented here are based on DNA sequence analysis of the ribosomal RNA (rRNA) genes. The small subunit rRNA gene (also referred to as 18S rDNA) has proved to be an excellent evolutionary chronometer for establishing species interrelationships *(1)*. The neighboring internal transcribed spacer (ITS) regions have been shown to be far more variable and have proved valuable for designing polymerase chain reaction (PCR) primers for rapid species identification *(2)*. These methods are extremely rapid and can take less than 24 h in contrast with conventional methods that may take 3–4 wk. Detailed protocols associated with both methodologies are provided as follows.

2. Materials

Caution: the procedures described here involve the use of potentially hazardous materials. The relevant safety regulations (e.g., Control of Substances Hazardous to Health [COSHH] regulations *[3]*) should be consulted prior to the use of these procedures.

1. Difco yeast malt (YM) broth: Difco dehydrated YM broth (ref. no. 0711-01, Difco Inc., Detroit, MI), 21 g/L. Alternatively, use YM medium: 3 g yeast extract, 3 g malt extract, 5 g peptone, and 1 g glucose, made up to 1 L. After mixing, the pH should be between 5 and 6. Sterilize by autoclaving for 15 min at 15-lb pressure (121°C).
2. YM agar: Add 2% (w/v) agar to Difco YM broth or YM media before sterilization. After mixing, the pH should be between 5 and 6. Sterilize by autoclaving for 15 min at 15-lb pressure (121°C). Dispense in 20 mL aliquots into sterile Petri dishes and leave to cool.
3. *AmpliTaq* DNA polymerase and 10X buffer (Perkin-Elmer, PE Biosystems, Warrington, UK).
4. Deoxynucleotide triphospate (dNTP) stock mix (40 μM each dNTP: ref. no. U1240, Promega UK Ltd., Southampton, UK). For a working stock solution, mix 10 µL of each dNTP (dATP, dCTP, dGTP, dTTP) with 60 µL sterile distilled water (SDW), to give a final concentration of 10 mM.
5. Thermocycler: Omnigene thermocycler (Hybaid Ltd., Ashford, Middlesex, UK).
6. Oligonucleotide primers: the sequences of the 18S rDNA and ITS amplification and sequencing primers are listed in **Tables 1** and **2**. All oligonucleotide primers were synthesized using a model 394A DNA synthesizer (PE Biosystems) and phosphoramidite chemistry. For DNA amplification, dilute primers to a final working stock concentration of 20 pmol/µL. For sequencing, dilute primers to a final working stock concentration of 4 pmol/µL.
7. Mineral oil (ref. no. M5904, Sigma Chemical Co. Ltd., Dorset, UK).
8. *Taq* DyeDeoxy terminator cycle sequencing kit (PE Biosystems).
9. PE Biosystems model 373A DNA sequencer (PE Biosystems).
10. QIAquick PCR purification kit (ref. no. 28104, QIAGEN Ltd., Crawley, West Sussex, UK).
11. Agarose (ref. no. A-0169, Sigma Chemical Co. Ltd., Poole, Dorset, UK).
12. 10X TBE: 121 g Tris Base, 55 g boric acid, and 7.4 g EDTA, made up to 1 L with SDW.
13. Benchtop microcentrifuge (e.g., Eppendorf centrifuge model 5415C).
14. Minigel apparatus (e.g., GNA-100 "mini-submarine" gel apparatus: ref. no. 18-2400-02, Pharmacia Biotech, St. Albans, Herts, UK).

Food Spoilage Yeasts

Table 1
Oligonucleotide Primers Used for DNA Amplification and Sequencing of Yeast 18S rDNA

Primer	Primer sequence 5'– 3'	Approx position (*S. cerevisiae* numbering)
P108	ACCTGGTTGATCCTGCCAGT	2–21
P130	GTCTCAAAGATTAAGCCATG	34–53
WIL1	ATTTCTGCCCTATCAACT	301–318
P1190	CAATTGGAGGGCAAGTCTGG	543–562
P2130	GGTGAAATTCTTGGATTTATTG	900–921
P2540	GGAGTATGGTCGCAAGGCTG	1108–1127
P3490	CCGCACGCGCGCTACACTGA	1454–1473
WIL2	AGTTGATAGGGCAGAAAT	318–301
M1190	CCAGACTTGCCCTCCAATTG	562–543
M2130	CAATAAATCCAAGAATTTCACC	921–900
M2540	CAGCCTTGCGACCATACTCC	1121–1108
M3490	TCAGTGTAGCGCGCGTGCGG	1473–1454
M3989	CTACGGAAACCTTGTTACGACT	1775–1754

Table 2
Oligonucleotide Primers used for DNA Amplification and Sequencing of the Yeast Internal Transcribed Spacer (ITS) Region

Primer	Primer sequence 5'–3'	Approx position (*S.cerevisiae* numbering)
pITS1	TCCGTAGGTGAACCTGCGG	1769–1787 [a]
p5.8Sr	ATGACRCTCAAACAGGCAT	156–138 [b]
pITS3	GCATCGATGAAGAACGCAG	31–49 [b]
pITS4	TCCTCCGCTTATTGATATG	68–50 [c]

[a] 18S rRNA gene, [b] 5.8S rRNA gene, [c] 26S rRNA gene.
R = A/G.

15. Gilson micropipetters (Anachem Ltd., Luton, Beds, UK).
16. Power pack (e.g., EPS 200: ref. no. 19-0200-00, Pharmacia Biotech).
17. Transilluminator (Anachem Ltd.).
18. Multiple-sequence alignment program (e.g., PILEUP *[4]*, contained in the Genetics Computer Group software package *[5]*).
19. Phylogeny inference software (e.g., PHYLIP *[6]*).

3. Procedures

Caution: follow good laboratory practice throughout.

3.1. Strain Cultivation

Streak out yeast isolates on YM agar plates, and grow at 24°C for 2–3 d to produce well-separated colonies. Authenticated yeast strains can be obtained from a number of culture collections, and their addresses are detailed in the Appendix.

3.2. DNA Amplification of Yeast 18S rDNA

1. Yeast 18S rDNA is amplified as two overlapping fragments by using the PCR *(7)* and the oligonucleotide primer combinations P108:M3490 and P1190:M3989 (for primer sequences refer to **Table 1**).
2. Carry out DNA amplification directly on yeast cell suspensions. Resuspend cells from a well-isolated single yeast colony in 50 µL SDW. Dilute 5 µL of this cell suspension in a further 45 µL SDW. Boil the diluted cell suspension for 5 min and place on ice. To the cooled "heat-treated" cell suspension add 1 µL of each amplification primer (20 pmol/µL), 2 µL of dNTP working solution, 10 µL *Taq* polymerase buffer (10X), and 1.5 U of *AmpliTaq* DNA polymerase. Make up the reaction mix to a final volume of 100 µL with SDW, and overlay the reaction mix with two to three drops (approx 50 µL) of sterile mineral oil.
3. Carry out DNA amplification in an appropriate thermocycler. When using the Hybaid Omnigene thermocycler the following cycling parameters should be used:
   a. 1 cycle of 94°C for 2 min (initial denaturation)
   b. 2 cycles of 94°C for 2 min (denaturation), 54°C for 1 min (primer annealing), and 72°C for 2 min (primer extension)
   c. 33 cycles of 92°C for 2 min, 54°C for 1 min, and 72°C for ? min
   d. 1 cycle of 7?°C for 5 min
4. Following DNA amplification, analyze 5 µL aliquots of each PCR sample by 1.0% agarose-TBE gel electrophoresis (125 V for 30–40 min), using a minigel apparatus (e.g., GNA-100 "minisubmarine" gel apparatus). Stain the agarose gel in ethidium bromide (0.5 µg/mL) for approx 20 min and visualize using a UV transilluminator.
5. Caution: ethidium bromide is a powerful mutagen and gloves should be worn when handling gels in the staining solution, and eyes should be protected (with suitable mask or goggles) at all times when analysing stained gels under UV light.

3.3. DNA Amplification of the ITS Region from Yeasts

The general protocol used to amplify the ITS region from yeast cells is identical to that detailed in **Subheading 3.2.** (DNA amplification of yeast 18S rDNA), with the exception that the ITS region is amplified as a single PCR fragment using the primer combination pITS1:pITS4 (for primer sequences refer to **Table 2** and ref. *[8]*).

3.4. Purification of PCR-Amplified DNA Fragments

Purify PCR-amplified DNA fragments using a QIAquick PCR purification kit following the manufacturer's protocol and resuspend DNA in 50 µL SDW. Analyze 1 to 2 µL of each purified PCR-amplified fragment by agarose gel electrophoresis (refer to **Subheading 3.2.**) and quantify against a known amount of DNA standard (e.g. uncut λ DNA).

3.5. Direct Sequencing of PCR-Amplified Fragments

1. The purified PCR-amplified fragments are sequenced directly using a *Taq* Dye-Deoxy terminator cycle sequencing kit. For each sequencing reaction use 1–2 µL (approx 50 to 100 ng) of purified PCR-amplified fragment and add 8 µL DyeDeoxy terminator ready-reaction mix, 1 µL sequencing primer (4 pmol/µL), and make up to a final volume of 20 µL with SDW. Overlay each sequencing reaction mix with one drop of sterile mineral oil.
2. Carry out cycle sequencing in an appropriate thermocycler such as the Hybaid Omnigene thermocycler, using the following thermal cycling parameters: 1 cycle of 96°C for 2 min, followed by 25 cycles of 96°C for 30 s, 50°C for 15 s, and 60°C for 4 min.
3. To purify, transfer each completed sequencing reaction to a 0.5 mL Eppendorf microcentrifuge tube containing 50 µL 95% ethanol and 2 µL 3 *M* sodium acetate (pH 4.6). Vortex briefly and store samples on ice for 10 min. Centrifuge samples in a benchtop microcentrifuge (e.g., Eppendorf centrifuge model 5415 C) for 20 min. Carefully remove ethanol solution using a Gilson micropipetter. Rinse pellet by adding 200 µL ice-cold 70% ethanol. Carefully remove ethanol solution with a micropipetter and dry pellet under vacuum (e.g., using a Speedivac centrifuge).
4. Following the manufacturer's instructions, electrophorese purified extension products using an PE Biosystems model 373A DNA sequencer.

3.6. Analysis of 18S rDNA Sequences

1. Using the complete set of primers listed in **Table 1**, approx 95% of the 18S rDNA from an individual yeast isolate can be double-strand sequenced.
2. Once determined, the 18S rDNA sequence of the yeast isolate can either be aligned with known yeast 18S rDNA sequences or used in a similarity search (e.g., performing a FASTA search *[9]*) to match with 18S rDNA sequences from a databank such as EMBL or GENBANK. In the latter case, the best matching sequences can then be extracted from the databank and included in the phylogenetic analysis.
3. To align the resulting set of 18S rDNA sequences, use a multiple sequence alignment program such as GCG PILEUP *(4)* to generate a multiple sequence format (MSF) file. Depending on how closely related the set of 18S rDNA sequences are, manual editing of the resulting sequence alignment may be required to accommodate for sequence insertions and deletions.

4. In order for a multiple sequence alignment produced using the PILEUP program to be compatible for use with the PHYLIP software package, two alterations need to be made to it.
5. First, the MSF file needs to be reformatted into PHYLIP interleaved format. To do this, the standalone program READSEQ (included in the PHYLIP software package) is used.
6. To reformat, the following command line is used:

    readseq -a yeast.msf -format=phylip -output=yeast.phy

    where, "yeast.msf" refers to the name given to the multiple sequence alignment file, and "yeast.phy" to the name assigned to the PHYLIP reformatted file.
7. Second, once reformatted all "periods" within the alignment need to be recoded as either "unknown" bases (i.e., using the letter N), or as "deletions" (i.e., using the symbol -). This manual editing of the reformatted sequence alignment is essential, as the PHYLIP software package is not compatible with all of the sequence symbols used by the GCG software package.

3.7. Phylogenetic Tree Construction Using the Neighbor-Joining Distance-Based Method

For generating phylogenetic trees from closely related organisms, the Neighbor-joining (NJ) method *(10)* is used, as recommended by Murray and co-workers *(11)*. This is a distance-based method that generates a phylogenetic tree using a distance matrix calculated from the sequence data.

1. To generate the distance matrix, use the DNADIST program contained in the PHYLIP software package, with the reformatted sequence alignment as the input file (e.g., "yeast.phy"). The DNADIST program offers a choice of four models for calculating pairwise distances between DNA sequences. The two models most commonly used with 18S rDNA sequences are the Jukes–Cantor and Kimura 2-parameter models. For short sequences, the Jukes-Cantor model is recommended (which assumes that all nucleotide positions along a DNA sequence display the same rate of base substitution).
2. The appropriate model for calculating pairwise distance is selected by typing in "d" until the model of choice is displayed in the program settings. Once the selection is made and an output file name is assigned for the resulting distance matrix (e.g., "yeast.dist"), the program can be run by typing "y" (i.e., "yes") to the question "are these settings correct?".
3. From the resulting DNA distance matrix file (e.g., "yeast.dist"), a phylogenetic tree can be constructed by using the Neighbor-joining method *(10)*.
4. The Neighbor-joining method is run by selecting the "Neighbor joining" option (N) of the NEIGHBOR program. To root the tree, select the outgroup option (O) and specify the species (by number) to be used as the outgroup. (*N.B.* Species are numbered depending on where their sequences are located in the original sequence alignment, with the uppermost sequence being assigned as species 1.) In

Food Spoilage Yeasts   43

the phylogenetic tree shown in **Fig. 1**, *Kluyveromyces lactis* was used as the outgroup. As with the DNADIST program, the NEIGHBOR program is run by typing "y" to the question "are these settings correct?" and hitting the return key.

5. Once the program has run (and this takes a matter of seconds), two files are created. These are the output and tree files (e.g., "yeast.out" and "yeast.tree," respectively). To visualize the results, type in the output file name, and the tree topology and branch lengths will be displayed. This data combined can then be used to draw the phylogenetic tree generated from the 18S rDNA sequence alignment (e.g., such as is shown in **Fig. 1**).

3.8. Statistical Analysis of the Phylogenetic Tree

Once the phylogenetic tree has been generated, the next step is to statistically test its topology. One method that is commonly used to test phylogenetic tree topologies is the statistical method of bootstrapping *(12)*. This method calculates the variability of the tree topology by resampling (with replacement) the original multiple sequence alignment a set number of times (i.e., a number between 100 and 2000) to generate a new set of artificial multiple alignments. This new data set is in turn used to generate a set of trees, from which a "consensus" tree is produced. In the resulting consensus tree, at each branch node on the tree, the fraction of bootstrap trials (referred to as the bootstrap value) that confirmed that node is shown. For example in **Fig. 1**, 96 of 100 bootstrap trials found that the asexual yeast species (referred to as the anamorph) *Candida holmii* and its sexual form (referred to as the teleomorph) *Saccharomyces exiguus* were distinct from all other species examined in the study. Consequently, the higher the bootstrap value, the greater the statistical support for an individual branch within a phylogenetic tree. The PHYLIP programs used to calculate the bootstrap values for a phylogenetic tree are SEQBOOT, DNADIST, NEIGHBOR, and CONSENSE (used in the order shown).

1. To generate the multiple dataset of artificial sequence alignments, select the SEQBOOT program and use the original multiple sequence alignment ("yeast.phy") as the input file (*N.B.* choose an appropriate output file name such as "yeast.seqs"). In the program settings, use option (J) to designate the number of multiple datasets (referred to in SEQBOOT as replicates) to be generated. In the bootstrapped tree shown in **Fig. 1**, 100 datasets were used.
2. Once the number of datasets has been selected, run the program by typing "y" to the question "are these settings correct?" and hitting the return key.
3. The output file "yeast.seqs" can now be used to generate a set of trees by using the programs DNADIST and NEIGHBOR as described in **Subheading 3.7**. The only difference that needs to be made is to alter option (M) in the settings of both programs so that they will analyze multiple datasets (*N.B.:* it is also necessary to indicate how many datasets are to be analyzed).

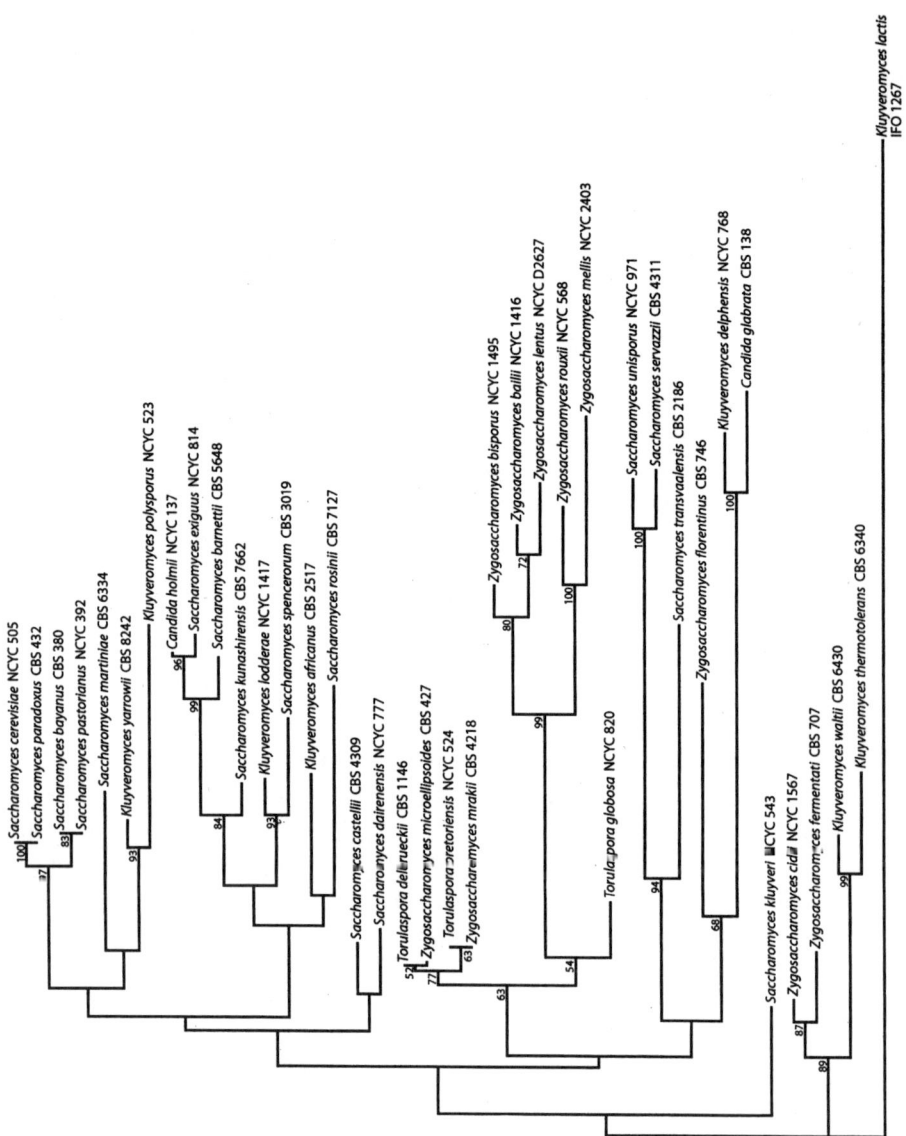

Fig. 1. Dendrogram showing the phylogenetic relationship between species of the yeast genera *Kluyveromyces*, *Saccharomyces*, *Torulaspora*, and *Zygosaccharomyces*, based on 18S rDNA sequences. The tree was constructed by using the neighbor-joining method *(10)*. Bootstrap values, expressed as percentages of 100 replications, are given at branch points (only values >50% are shown).

Food Spoilage Yeasts

4. To generate the consensus tree, run the CONSENSE program, using the tree file generated by the NEIGHBOR program as the input file. As with the NEIGHBOR program, the resulting consensus tree can be rooted by using option (O) to select the species to be used as the outgroup, and option (R) to root the tree based on this selection. Once these settings are changed, run the program.
5. CONSENSE generates two files, an output file and a tree file. The output file is the file to view, as this will show the calculated bootstrap values for all branch nodes of the consensus tree. The bootstrap data can then be combined with the originally generated phylogenetic tree to produce a bootstrapped tree, such as is shown in **Fig. 1**. For further detailed reading on all aspects of phylogenetic analysis, we recommend the reader refers to the chapter entitled "phylogenetic inference" of ref. *(13)*.

4. Notes

4.1. Species Identification and Differentiation Based on Partial rDNA Sequencing

As well as being invaluable for investigating the phylogenetic relationships between yeast species and genera, 18S rDNA sequence analysis is also proving to be extremely useful for both species identification and differentiation. For these purposes it is not necessary to analyse the entire 18S rDNA sequence, as it has been found to be comprised of conserved, semiconserved, and variable regions of sequence *(1,14)*. The best regions of the gene to analyze for identification purposes are the variable regions. The region most suitable for such purposes is the V4 region (where "V" stands for variable *[14]*), which has been found to display the greatest level of sequence variation between different yeast species *(15)*. This region is located between nucleotide positions 635 and 860 in the gene (*S. cerevisiae* numbering *[16]*).

To use this region for species differentiation and identification the 18S rDNA can be amplified from different yeast isolates as described in **Subheading 3.2.**, using the primer combination P108:M3490. Following the protocols as detailed in **Subheading 3.4. and 3.5.**, the V4 region of the 18S rDNA can be completely double-strand sequenced using the primers P1190 and M2130 (*see* **Table 1** for primer sequences). In our experience the majority of yeast species can be differentiated based on their V4 region sequences *(15,17–19)*. In most cases, strains of the same species possess identical V4 sequences, whereas strains of even closely related species (e.g., *Z. bailii* and *Z. bisporus*; *see* **Fig. 1**) tend to differ by at least one nucleotide in this region. Notable exceptions to this are the species pairs of *S. bayanus*/*S. pastorianus* and *S. cerevisiae*/*S. paradoxus*. In the case of these yeasts, which cannot be reliably distinguished apart using conventional physiological testing *(20)*, the V4 region cannot be used as each pair of species has identical V4 sequences. In fact, sequence analysis of the entire 18S rDNA molecule has revealed that each pair of species have identical sequences along the entire length *(17)*.

A second rDNA region routinely used for species differentiation is the 5' D2 domain *(21)* of the 25S rDNA *(22,23)*. In comparison to the V4 region of the 18S rDNA, the D2 domain appears to display a greater level of interspecies sequence variation *(23)*, permitting better resolution of closely related species, including some species that cannot be differentiated based on V4 18S rDNA sequences (e.g., *S. cerevisiae* and *S. paradoxus [22,23]*). In practice, strains established as being conspecific (i.e., belonging to the same species) by both physiological testing and nuclear (n) DNA/nDNA hybridization studies (*see* ref. *[24]*) have been found to exhibit less than 1% sequence variation within this region *(23)*. Whereas strains belonging to distantly related species have been found to differ by as much as 47% in this region (e.g., *Pichia bimundalis* and *Schizosaccharomyces japonicus* var *versatilis [23]*). For a detailed discussion of using this region of rDNA for species differentiation, we recommend the reader refers to refs. *(22,23,25)*.

To illustrate the value of these two rDNA regions for species differentiation, **Table 3** shows the nucleotide differences observed between the 18S rDNA and 25S rDNA sequences of the recently described species *Z. lentus (26)*, and its two closest relatives, *Z. bailii* and *Z. bisporus* (*see* **Fig. 1**). All three species are commonly associated with food spoilage *(20,26)*, but are frequently difficult to distinguish from one another using conventional physiological testing [15, 26]. Indeed until recently, strains of *Z. lentus* were routinely misidentified as *Z. bailii (26)*. However despite possessing very similar physiological profiles, as can be seen from **Table 3**, all three species can be readily differentiated from one another based on their partial sequences from the V4 region of the 18S rDNA and the D2 domain of the 25S rDNA.

4.2. Use of the Internal Transcribed Spacer (ITS) Region for Rapid Species Identification

An alternative for species identification to the two rDNA regions discussed in **Subheading 4** is the Internal Transcribed Spacer (ITS) region. In yeast, as with other eukaryotes, this spacer region separates the 5.8S rDNA from the 18S and 25S rDNA, with the ITS1 region located between the 18S rDNA and the 5.8S rDNA, and the ITS2 region between the 5.8S rDNA and the 25S rDNA (*see* **Fig. 2**). In a number of recent studies conducted on a variety of different yeasts, the ITS region was found to exhibit far greater levels of sequence between species than either the 18S rDNA or the 25S rDNA *(18,27,28)*. Indeed the level of sequence variation is such that in a study of the genus *Williopsis*, ITS1 and ITS2 sequences were found to provide resolution to the subspecies level (differentiating between the five varieties of *W. saturnus*), which could not be fully achieved using either 18S rDNA (same study) or 25S rDNA sequences *(29)*.

In contrast to the 18S and 25S rDNA, which in yeast are approx 1800 base pairs (bp) and 3000 bp, respectively, in size, the entire ITS region (including

Table 3
A Comparison of Nucleotide differences in the V4 region of the 18S rDNA (upper right) and D2 Domain of the 25S rDNA (lower left) Between the Food Spoilage Yeasts Z. bailii, Z. bisporus, and Z. lentus

Strain[a]	1	2	3
1. Z. bailii NCYC 1416[b]	—	2	3
2. Z. bisporus NCYC 1495[b]	18	—	5
3. Z. lentus NCYC D2627[b]	23	22	—

[a] National Collection of Yeast Cultures, Norwich, UK.
[b] Type strain.

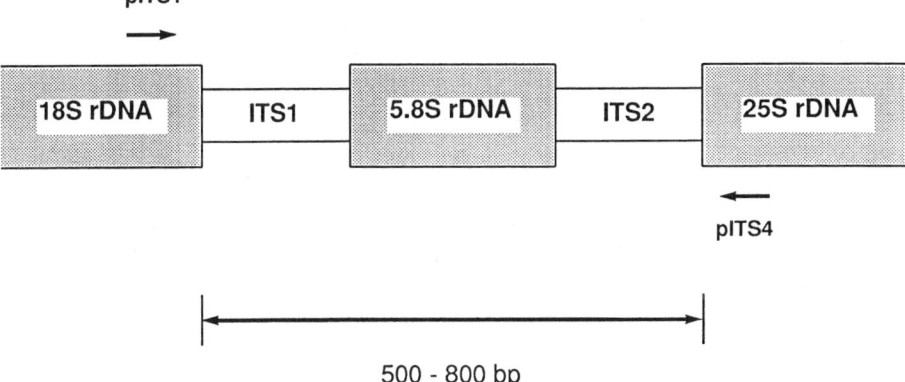

Fig.2. Diagram showing the location of the ITS region. Arrows indicate the approximate positions of the DNA amplification primers pITS1 and pITS4 (8).

the 5.8S rDNA) is far smaller, typically ranging from 500–800 bp. Consequently, this DNA spacer region is easy to both PCR amplify, using the primer combination pITS1:pITS4 (8) and the protocols as detailed in **Subheadings 3.2.** and **3.3.** and double-strand sequence (using the primers listed in **Table 2**). Hence, as with the 18S and 25S rDNA, ITS sequences (ITS1 and/or ITS2) represent a valuable means both for rapid species identification, and for possible differentiation at the subspecies level (18,26,27).

Due to the high levels of sequence variation observed between ITS sequences of closely related species (26–28), not only can species identification be achieved by direct sequence analysis of this region, but specific PCR primers can also be designed for rapidly screening for yeast species of particular interest (e.g., those most commonly associated with food spoilage). Such an approach was developed

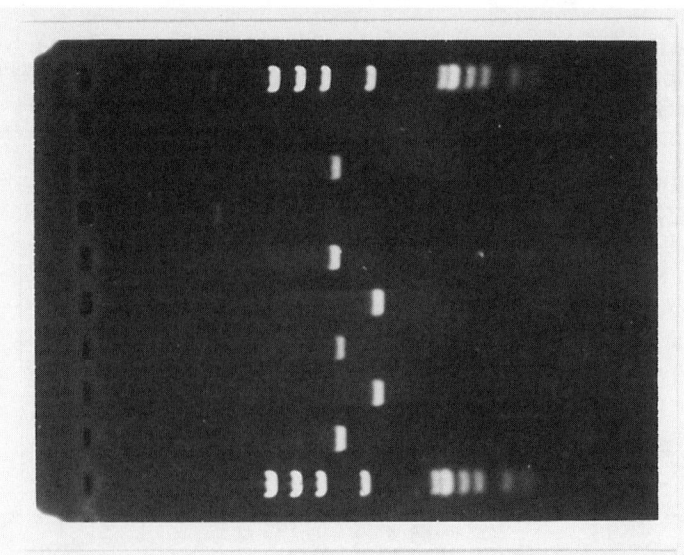

Fig. 3. Agarose gel showing the results of a PCR-based method for the rapid identification of Z. lentus. The species specific primers (pLENT1 and pLENT2) were designed from the ITS1 and ITS2 sequences of Z. lentus.

Lane order of gel: lane 1 DNA size marker; lanes 2 and 3, Z. lentus NCYC D2627^T, primers pITS1/pITS4 and pLENT1/pLENT2; lanes 4 and 5, Z. lentus NCYC 2406, primers pITS1/pITS4 and pLENT1/pLENT2; lanes 6 and 7, Z. bailii NCYC 1416 ^{T}, primers pITS1/pITS4 and pLENT1/pLENT2; lanes 8 and 9, Z. bisporus NCYC 1495 ^{T}, primers pITS1/pITS4 and pLENT1/pLENT2; lane 10, DNA size marker. *N.B.* All four strains shown generate a PCR fragment with the conserved primers pITS1/pITS4 *(8)*.

to identify the food spoilage yeast *Z. lentus*, with species-specific primers designed from its ITS1 and ITS2 sequences. As **Fig. 3** shows, these primers were used in conjunction with the PCR, and *Z. lentus* was readily distinguished from its close relatives *Z. bailii* and *Z. bisporus*, with only *Z. lentus* generating a PCR fragment with the *Z. lentus*-specific primers (all three species generated PCR fragments as expected with the pITS1:pITS4 primer combination). This test has important commercial implications as *Z. lentus* has exceptional preservative resistance and a slow rate of growth that can hinder its early detection using conventional methods.

Although direct sequence analysis of 18S rDNA, 25S rDNA, or the ITS region can take 2–3 d on average to identify a yeast isolate (compared to 3–4 wk using conventional physiological testing), PCR methods using ITS-derived primers offer the possibility of identifying specific yeast species of interest (e.g., potent food spoilage yeasts) within a single working day.

Appendix

Authenticated yeast strains can be obtained from the following culture collections:

1. National Collection of Yeast Cultures (NCYC), Dept. of Food Safety Science Genetics & Microbiology, Institute of Food Research, Norwich Research Park, Colney, Norwich NR4 7UA, UK. Tel: (0)1603-255274. Fax: (0)1603-458414. E-mail: NCYC@bbsrc.ac.uk. Web site: http://www.ifrn.bbsrc.ac.uk/ncyc/
2. Centraalbureau voor Schimmelcultures (CBS), Yeast Division, Julianalaan 67, 2628 BC Delft, The Netherlands. Tel: (0)15-2783214. Fax: (0)15-2782355. E-mail: SALES@cbs.knaw.nl. Web site: http://www.cbs.knaw.nl
3. American Type Culture Collection (ATCC), 10801 University Boulevard, Manassas, VA 20110-2209, USA. Tel: 703-365-2700. Fax: 703-365-2701. Web site: http://www.atcc.org/

References

1. Woese, C. R. (1987) Bacterial evolution. *Microbiol. Rev.* **51,** 221–271.
2. James, S. A., Collins, M. D. and Roberts, I. N. (1996) Use of an rRNA internal transcribed spacer region to distinguish phylogenetically closely related species of the genera *Zygosaccharomyces* and *Torulaspora*. *Int. J. Syst. Bacteriol.* **46,** 189–194.
3. Control of Substances Hazardous to Health Regulations (1988) Approved code of practice, Her Majesty's Stationery Office, London.
4. Feng, D. F. and Doolittle, R. F. (1987) Progressive sequence alignment as a prerequisite to correct phylogenetic trees. *J. Mol. Evol.* **35,** 351–360.
5. Genetics Computer Group (1991) Program manual for the GCG package, version 7. Genetics Computer Group, Madison, WI.
6. Felsenstein, J. (1993) PHYLIP: Phylogenetic Inference Package, version 3.5. University of Washington, Seattle.
7. Saiki, R. K., Gelfand, D. H., Stoffel, S., et al. (1988) Primer-directed enzymatic amplification of DNA with a thermostable DNA polymerase. *Science* **239,** 487–491.
8. White, T. J., Bruns, T. D., Lee, S., and Taylor, J. W. (1990) Amplification and direct sequencing of fungal ribosomal RNA genes for phylogenetics, in *PCR Protocols* (Innis, M., Gelfand, D. H., Sninsky, J. J., and White, T. J., eds.), Academic Press, San Diego, CA, pp. 315–322.
9. Pearson, W. R. and Lipman, D. J. (1988) Improved tools for biological sequence comparison. *Proc. Natl. Acad. Sci. USA* **85,** 2444–2448.
10. Saitou, N. and Nei, M. (1987) The neighbor-joining method: a new method for reconstructing phylogenetic trees. *Mol. Biol. Evol.* **4,** 406–425.
11. Murray, R. G. E., Brenner, D. J., Colwell, R. R., et al. (1990) Report of the ad hoc committee on approaches to taxonomy within the *Proteobacteria*. *Int. J. Syst. Bacteriol.* **40,** 213–215.
12. Felsenstein, J. (1985) Confidence limits on phylogenies: an approach using the bootstrap. *Evolution* **39,** 783–791.

13. Swofford, D. L., Olsen, G. J., Waddell, P. J. and Hillis, D. M. (1996) Phylogenetic inference, in *Molecular Systematics*, 2nd ed. (Hillis, D. M., Moritz, C., and Mable, B. K., eds.), Sinauer Associates Inc., Sunderland, MA, pp. 407–514.
14. De Rijk, P., Neefs, J-M., Van de Peer, Y. and De Wachter, R. (1992) Compilation of small ribosomal subunit RNA sequences. *Nucleic Acid Res.* **20**, 2075–2089.
15. James, S. A., Collins, M. D., and Roberts, I. N. (1994) Genetic interrelationship among species of the genus *Zygosaccharomyces* as revealed by small-subunit rRNA gene sequences. *Yeast* **10**, 871–881.
16. Mankin, A. S., Skryabin, K. G., and Rubstov, P. M. (1986) Identification of ten additional nucleotides in the primary structure of yeast 18S rRNA. *Gene* **44**, 143–145.
17. James, S. A., Cai, J., Roberts, I. N., and Collins, M. D. (1997) A phylogenetic analysis of the genus *Saccharomyces* based on 18S rRNA gene sequences: description of *Saccharomyces kunashirensis* sp. nov. and *Saccharomyces martiniae* sp. nov. *Int. J. Syst. Bacteriol.* **47**, 453–460.
18. James, S. A., Roberts, I. N., and Collins, M. D. (1998) Phylogenetic heterogeneity of the genus *Williopsis* as revealed by 18S rRNA gene sequences. *Int. J. Syst. Bacteriol.* **48**, 591–596.
19. Cai, J., Roberts, I. N., and Collins, M. D. (1996) Phylogenetic relationships among members of the ascomycetous yeast genera *Brettanomyces*, *Debaryomyces*, *Dekkera*, and *Kluyveromyces* deduced by small-subunit rRNA gene sequences. *Int. J. Syst. Bacteriol.* **46**, 542–549.
20. Barnett, J. A., Payne, R. W., and Yarrow, D. (1990*)* *Yeasts: Characteristics and Identification*, 2nd ed., Cambridge University Press, UK.
21. Guadet, J., Julien, J., Lafey, J. F., and Brygoo, Y. (1989) Phylogeny of some *Fusarium* species, as determined by large subunit rRNA sequence comparison. *Mol. Biol. Evol.* **6**, 227–242.
22. Kurtzman, C. P. and Robnett, C. J. (1991) Phylogenetic relationships among species of *Saccharomyces*, *Schizosaccharomyces*, *Debaryomyces* and *Schwanniomyces* determined from partial ribosomal RNA sequences. *Yeast* **7**, 61–72.
23. Peterson, S. W. and Kurtzman, C. P. (1991) Ribosomal RNA sequence divergence among sibling species of yeasts. *Syst. Appl. Microbiol.* **14**, 124–129.
24. Kurtzman, C. P. and Phaff, H. J. (1987) Molecular taxonomy, in *The Yeasts*, 2nd ed. (Rose, A. H. and Harrison, J. S., eds.), Academic Press, London, pp. 63–94.
25. Kurtzman, C. P. and Blanz, P. A. (1998) Ribosomal RNA/DNA sequence comparisons for assessing phylogenetic relationships, in *The Yeasts: a taxonomic study*, 4th ed. (Kurtzman, C. P. and Fell, J. W., eds.), Elsevier Science B.V., Amsterdam, pp. 69–74.
26. Steels, H., Bond, C. J., Collins, M. D., Roberts, I. N., Stratford, M., and James, S. A. (1999) *Zygosaccharomyces lentus* sp. nov., a new member of the yeast genus *Zygosaccharomyces* Barker. *Int. J. Syst. Bacteriol.* **49**, 319–327.
27. James, S. A., Collins, M. D., and Roberts, I. N. (1996) Use of an rRNA Internal Transcribed Spacer region to distinguish closely related species of the genera *Zygosaccharomyces* and *Torulaspora*. *Int. J. Syst. Bacteriol.* **46**, 189–194.

28. Lott, T. J., Kuykendall, R. J., and Reiss, E. (1993) Nucleotide sequence analysis of the 5.8S rDNA and adjacent ITS2 region of *Candida albicans* and related species. *Yeast* **9,** 1199–1206.
29. Liu, Z. and Kurtzman, C. P. (1991) Phylogenetic relationships among species of *Williopsis* and *Saturnus* gen. nov. as determined from partial rRNA sequences. *Antonie Leeuwenhoek* **60,** 21–30.

II

PATHOGENS

7

Conductimetric Method for Evaluating Inhibition of *Listeria monocytogenes*

Graciela Font de Valdez, Graciela Lorca, and María Pía Taranto

1. Introduction

One of the major problems in the food industry, particulary in dairy products, is the occasional presence of the pathogen *Listeria monocytogenes*, which has been associated with food-borne disease outbreaks *(1)*. The association of listeriosis with the consumption of processed foods has prompted a number of studies to determine the impact of food processing and preservation procedures on the survival of *L. monocytogenes*. The preservation methods include the addition of antimicrobial products (weak acids *[2]*, bacteriocin, etc.), the decrease in water activity, or high-temperature short time pateurization (73.9°C for 16.4 s) *(3)*.

The microbial growth can be determined by traditional methods (optical density, end-product formation, pH) or by conductimetric methods, which measure changes in the electrochemical characteristics of the culture medium (*see* **Note 1**).

The conductance of growth media depends on the number and nature of charge carriers and is changed by microbial metabolism, which converts poorly charged carriers such as carbohydrates, proteins, and lipids into more effective charged carriers, such as ionized acids and amines.

Automatic monitoring changes in the metabolic activity of microorganisms may be a useful method for quality control of dairy products, cosmetics, meat, poultry, frozen foods, pharmaceutical products, fermentation, and growth inhibitors.

2. Materials

2.1. Growth Media

Brain–heart infusion (BHI) broth *(4)*: This medium is used extensively for maintaining and culture of *Listeria*, and it is commercially available from different manufacturers (e.g., Difco, Oxoid). It contains infusion from calf brains (200 g/L), infusion from beef heart (250 g/L), proteose peptone (10 g/L), dextrose (2 g/L), sodium chloride (5 g/L), and disodium phosphate (2.5 g/L). Final pH is 7.4–0.2. Sterilize at 121°C for 15 min.

2.2. Media for Determining the Effect of Antimicrobial Compounds

Different chemical or biological compounds may be used. As example we have considered the antimicrobial effect of bile acids.

BHIB broth (*see* **Note 4**): BHI broth supplemented with the sodium salts of the conjugated bile acids, taurocholic (TCA), glycocholic (GCA), taurodeoxycholic (TDCA), glycodeoxycholic (GDCA), taurochenodeoxycholic (TCDCA), glycochenocodeoxycholic (GCDCA), and the unconjugated bile acids, cholic (CA), deoxycholic (DCA), and chenodeoxycholic (CDCA) (Sigma, St. Louis, MO). Add bile acids to a final concentration of 1, 2, 4, and 6 mM. Sterilize at 121°C for 20 min.

2.3. Materials for Determination the Effect of Bile Acids

1. An active culture grown in BHI broth for about 16 h at 37°C (overnight culture).
2. Phosphate buffer: Solution A (NaH_2PO_4 0.2 M): Weigh 27.6 g $NaH_2PO_4 \cdot H_2O$ and make up to 100 mL with distilled water (dH_2O). Solution B (Na_2HPO_4, 0.2 M): weigh 53.05 g $Na_2HPO_4 \cdot 7 H_2O$ and make to 100 mL with dH_2O. To prepare 1 L of sodium phosphate buffer 0.1 M pH 7.0, mixture 195 mL of solution A and 305 mL of solution B, and bring to 1000 mL with dH_2O. Sterilize at 121°C for 20 min.
3. BHI broth: used as control.
4. BHIB broth: BHI broth supplemented with different concentrations of bile acids, as mentioned in **Subheading 2.2.**

2.4. Materials for the Conductimetric Method

1. **Items 1–3** and **4** from **Subheading 2.3.**
2. Bactometer™ Microbial Monitoring System (BioMérieux) with BPS R03-1 software, or similar.
3. 2 mL-Capacity sterile conductance cells.

3. Methods

3.1. Optical Density Measurement

1. Grow the strain of *Listeria* under study in BHI broth at 37°C for 16 h.
2. Harvest the cells by centrifugation at 5000g for 10 min.
3. Wash twice the pellet obtained with 0.1 M phosphate buffer, pH 7.0.
4. Resuspend the cells to the original volume with the buffer by vortexing.
5. Inoculate (0.5%) (*see* **Note 3**) BHI and BHIB broth with the bacterial suspension.
6. Incubate at 37°C in a water bath.
7. Read the optical density at 560 nm (OD_{560}) against blank of medium (uninoculated broth) every hour for the first 8 h and after 24 h of incubation.
8. Plot OD values against incubation time.

3.2. Conductimetric Method

1. Prepare the culture inoculum as described in **Subheading 2.3.** (*see* **Note 2**).
2. Dispense the culture medium into the conductance cells.
3. Insert the module into the incubator. The modules can be inserted at any moment one by one or in runs, without interfering with the ongoing analyses.
4. Identify the modules (e.g., test type, product, sample number).
5. Detection times appear automatically on the computer screen. Every well in each of the module placed in the incubator is electrically measured every 6 min, enabling a growth curve to be established according to the time and percentage of electrical variation.
6. Results can be edited in report form illustrated with growth curves.

4. Notes

1. Changes in the electrical properties of culture media are related to the number of microorganisms in a sample and to the metabolic activity of those microorganisms *(4)*. These changes induce at a given moment a significant variation leading to a sudden inflection in the curve. This inflection point is known as the detection time (*see* **Fig. 1**).
2. Make 1/10 and 1/100 dilutions in the detection medium before placing the samples into the equipment for reading conductance. As the Bactometer produces detection times that are comparable to plate counts, it is possible to perform calibration curves that relate the detection time to colony-forming units. Data points must be evenly distributed over a 4–5 log range (*see* **Fig. 2**).
3. For determining the effect of bile acids (or other antimicrobial compound) on the cells, it is important to use a low inoculum (0.5%, v/v) in order to have a low initial OD_{560} value and to ensure that the culture is at an early exponential phase of growth.

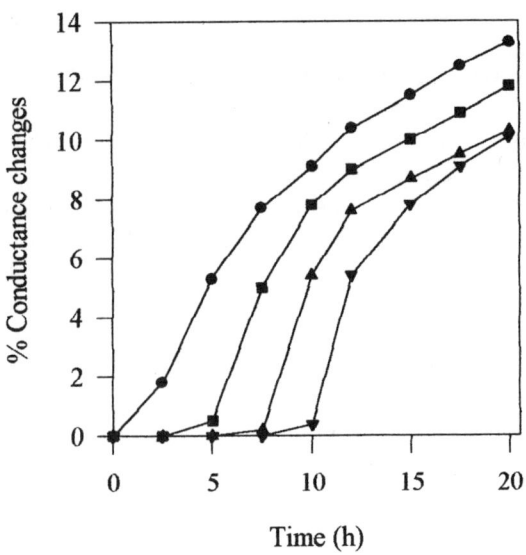

Fig. 1. Changes in the conductance of L. monocytogenes at 37°C in BHIB broth at different concentrations of bile acids. (•) control, (■) 1 mM, (▲) 2mM, (▼) 4 mM.

4. The chemical composition of the culture medium should be perfectly adapted to the conductimetric method. This fact enables the global detection of microorganisms in a product or a particular microbiota (e.g., coliforms, lactic acid bacteria, listeria). Detection media (g/L):

LM broth: This medium is used for detection and enumeration of lactic acid bacteria in fruit juice, dairy products, and foods. It contains Bacto peptone (10 g/L), tryptone (5 g/L), peptonized milk (10 g/L), dextrose (4 g/L), and yeast extract (7.5 g/L) at a final pH of 6.5. Sterilize at 121°C for 15 min.

MPCA broth: This medium is used for detection and enumeration of total aerobic flora in milk and foods. It contains yeast extract (20 g/L), dextrose (4 g/L), and tryptone (20 g/L) at a final pH 6.5–0.1. Sterilize at 121°C for 15 min.

CM broth: This medium is used for detection of coliforms in food. It contains proteose peptone (10 g/L), yeast extract (6 g/L), lactose (20 g/L), sodium lauryl sulphate (1 g/L), sodium deoxycholate (0.1 g/L), bile salts (Oxgall, Difco) (1 g/L), and bromcresol purple solution (10 ml/L). To prepare the bromcresol purple, add 0.35 g of bromcresol purple to 2 mL of 0.1 N NaOH. Bring to a final volume of 100 mL, mix, and sterilize by filtration. For enterobacteria add dextrose (20 g/L) for a final pH 6.8–0.1. Sterilize at 121°C for 15 min.

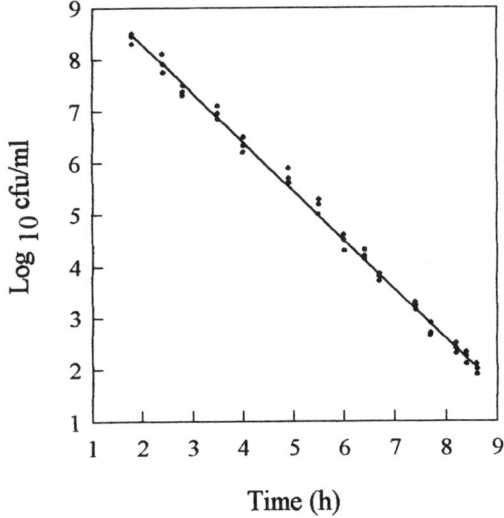

Fig. 2. Coliform calibration curve in yogurt.

The conductimetric method may also be used for monitoring changes in activity of cheese starter cultures stored at refrigeration temperature *(5)* as well as the activity of those frozen or freeze-dried *(6)* cultures in milk fermentation.

References

1. Farber, J. M. and Peternkin, P. I. (1991) *Listeria monocytogenes*, a foodborne pathogen. *Microbiol. Rev.* **55,** 476–511.
2. Eklund, T. (1983) The antimicrobial effect of dissociated and undissociated sorbic acid at different pH levels. *J. Appl. Bacteriol.* **54,** 383–389.
3. Lin, J., Smith, M. P., Chapin, K. C., et al. (1996) Mechanism of acid resistance in enterohemorragic *Escherichia coli. Appl. Environ. Microbiol.* **62,** 3094–3100.
4. Neviani, E., Muchetti, G., and Lanzanova, M. (1993) Analysis of conductance changes as the growth index of lactic acid bacteria in Milk. *J. Dairy Sci.* **76,** 2543–2548.
5. Tsai, K. and Luedecke, L. O. (1989) Impedance measurement of changes in activity of lactic cheese starter culture after storage at 4°C. *J. Dairy Sci.* **72,** 2239–2241.
6. Martos, G., Pesce de Ruiz Holgado, A., Oliver, G., and Font de Valdez, G. (1999) Use of conductimetri to evaluate *Lactobacillus delbruekii* ssp. *bulgaricus* subjected to freeze-drying. *Milchwissenschaft* **54(3),** 128–130.

8

Molecular Detection of Enterohemorrhagic *Escherichia coli* O157:H7 and Its Toxins in Beef

Kasthuri J. Venkateswaran

1. Introduction

Most *Escherichia coli* strains are harmless commensals in the human gut, but some strains are known to cause disease. The enterohemorrhagic *E. coli* (EHEC) strains of serotype O157:H7 causes hemorrhagic colitis, which may develop into life-threatening hemolytic uremic syndrome. Polymerase chain reaction (PCR) is a powerful tool to multiply a target molecule to detectable quantities. In the multiplex PCR method, two or more primer sets are used to simultaneously amplify multiple target sequences. Many researchers developed multiplex PCR for the detection of the LT (heat-labile toxin), SLT-I (Shiga-like toxin) and SLT-II producing *E. coli*. Antibody- or DNA-based assays for identifying SLTs or bacteria-carrying SLT genes will not discriminate O157:H7 isolates from the numerous other serotypes that also produce SLTs enterotoxins.

Bioassays and conventional methods are used to differentiate toxigenic *E. coli* from nontoxic strains (*see* **Notes 1–3**). In order to overcome the limitations of these existing methods, a multiplex PCR assay would be useful that simultaneously identifies isolates of O157:H7 and the types of SLT it encodes (*see* **Notes 4–7**). The first set of primers is directed to the *uid A* gene, which encodes for β-glucuronidase in *E. coli*. Although O157:H7 isolates do not exhibit β-glucuronidase activity, they carry the *uidA* gene. Exploiting the uniqueness of a 92 base change in *uidA* gene, a second set of primers is designed in a mismatch amplification mutation assay format to preferentially amplify the *uidA* allele in O157:H7 strains. The third and fourth sets of primers are directed to the conserved regions within the genes encoding for SLT-I and SLT-II genes, respectively.

From: *Methods in Biotechnology, Vol. 14: Food Microbiology Protocols*
Edited by: J. F. T. Spencer and A. L. Ragout de Spencer © Humana Press Inc., Totowa, NJ

The development of molecular methodologies to detect pathogenic microorganisms in food and clinical and environmental samples has led to improved patient diagnosis and more precise determination of the public health risk associated with food consumption and environmental exposures. Studies with the PCR have largely concentrated on the identification of bacterial strains or toxin genes with DNA extracted from pure cultures, but PCR methods for detection of bacteria in food samples have not been adequately explored. The food particles pose a major problem in the amplification of desired PCR amplicons. Here, we have overcome with a protocol that would eliminate PCR inhibitory substances originated from food.

The multiplex PCR method described here is a highly effective means for specifically detecting and characterizing EHEC organisms directly from food. The major advantage of this protocol over existing assays is that it can identify the types of SLT encoded by the strain and simultaneously discriminate other SLT-producing *E. coli* strains from O157:H7, the predominant serotype implicated in disease.

2. Materials

1. Phosphate buffered-saline (PBS) : 0.2 M NaH_2PO_4, 0.2 M Na_2HPO_4, 0.9 % NaCl; pH 7.0.
2. Microbiological media: trypticase soy broth (TSB), Trypticase soy agar, eosin methylene blue (EMB) agar, brilliant green lactose bile (BGLB) broth and *E. coli* (EC) broth are commercially available. We purchased from Nissui Pharmaceuticals, Tokyo, Japan.
3. Suitable oligonucleotide synthesizer should be used to synthesize primers. We used a Beckman oligonucleotide synthesizer (Fullerton, CA).
4. Any available DNA thermal cycler will be used for PCR amplification. We used the Perkin-Elmer system (Foster City, CA).

3. Methods

3.1. Food Homogenization

1. Weigh 25 g of beef aseptically into a sterile stomacher bag that contain 100-mm filters (Gunze, Tokyo, Japan) and add 225 mL of TSB.
2. Using a sterile food homogenizer (SH-001, Elmex, Tokyo, Japan), blend the sample for about 1–2 min. Care should be taken to avoid generation of heat in the process.
3. Aseptically siphon a out 100-µm filtered homogenized food sample and use as the inoculum.

3.2. Conventional Identification of E. coli

1. Grow the test organism in trypticase soy agar at 37°C for 18–24 h.
2. Pick a well-isolated colony and inoculate into BGLB broth and incubate at 37°C for 48 h. A gas-positive tube is considered for total coliforms.

Enterohemorrhagic E. coli

3. Transfer a portion of BGLB–gas-positive tube into EC broth and incubate at 44.5°C for 48 h. A gas-positive EC broth tube is considered positive for fecal coliforms.
4. The gas-positive EC tube is then streaked onto EMB agar. Metallic sheen colonies on EMB agar plates are presumptively identified as *E. coli*.

3.3. Molecular Identification of E. coli

3.3.1. Bacteria and Food Sample Preparation for PCR

1. Grow the bacterial strain in trypticase soy agar plates for 18 h at 37°C.
2. Pick a well-isolated colony and resuspend the bacterial cells in sterile PBS to a concentration of 10^5 colony-forming units (CFU)/ml and use as a DNA template for PCR.
3. To identify *E. coli* from food samples, inoculate the homogenized food samples as detailed in **Subheading 3.1.** in TSB and incubate at 37°C for 6 h (*see* **Notes 8–10**).
4. Remove 1 mL/of the enriched food homogenate and wash once with sterile PBS and resuspend in 1 mL/of PBS.
5. Filter 400 µL of the PBS-washed sample through 5-µm ultrafree tubes (cat. no. UFC3 0GV; Millipore, Bedford, Mass.) and centrifuge. This step will eliminate any particulate matter that might inhibit PCR reaction (*see* **Notes 11** and **12**).
6. The 5-µm filtrate should then pass through 0.2 µm ultrafree tubes (cat. no. SE3P009E4; Millipore) and centrifuge to remove bacteria only. This step will eliminate any dissolved matter that might inhibit PCR reaction.
7. Resuspend the 0.2-µm trapped materials in 400 µL of sterile PBS. Use 10-µL volumes of samples as the template for a PCR assay without extracting DNA.
8. All centifugation conditions are: $10,000g$, 4°C, for 10 min.

3.3.2. Multiplex PCR

1. Prepare DNA template as described in Subheading 3.3.1. and use 10-mL as the template.
2. Add 1-mL (final concentration, 1 mM) of various synthesized oligonucleotide primers specific for *uidA* (UAL-754 and UAR-900), O157:H7-specific *uidA* (PT-2 and PT-3), SLT-I (LP-30 and LP-31) and II (LP-43 and LP-44) genes into the PCR reaction mixture.
3. 10X PCR buffer: Chemicals (Tris-HCl, 100 m*M*; MgCl$_2$, 15 m*M*; KCl, 500 m*M*; pH 8.3) dissolved in appropriate solution provided by the manufacturer will be used.
4. Deoxynucleotide triphosphates (dNTP): Make up a single solution containing 0.2 m*M* of dATP, dGTP, dCTP, and dTTP and use appropriate concentration as described in **Subheading 3.3.2., item 7**.
5. Primer sets that are used to amplify *uidA* (UAL-754 and UAR-900), O157:H7-specific (PT-2 and PT-3), SLT-I (LP-30 and LP-31), and SLT-II (LP-43 and LP-44) gene-specific fragments are as follows:

UAL-754	5' AAA ACG GCA AGA AAA AGC AG 3'
UAR-900	5' ACG CGT GGT TAC AGT CTT GCC 3'
PT-2	5' GCG AAA ACT GTG GAA TTG GG 3'
PT-3	5' TGA TGC TCC ATC ACT TCC TG 3'
LP-30	5' CAG TTA ATG TGG TGG CGA AGG 3'
LP-31	5' CAC CAG ACA ATG TAA CCG CTC 3'
LP-43	5' ATC CTA TTC CCG GGA GTT TAC G 3'
LP-44	5' GCG TCA TCG TAT ACA CAG GAG C 3'

6. *Taq* DNA polymerase: Supplied by the manufacturer at a concentration of 5 U/µL is used.
7. 1X Reaction mix: Forty-nine microliters of the reaction mix is required for each sample. To make 49 µL of this mix, combine 5.0 µL of 10X PCR buffer, 4.0 µL of 0.2 m*M* dNTP, 1 µL each of various primers (1.0 µ*M*), 0.125 µL of *Taq* DNA polymerase (5 U/µL) with 38.875 mL of DNAse- and RNAse-free distilled water.
8. Mineral oil: Molecular biology grade mineral oil purchased from any manufacturer can be used. Recent advancements are made in various thermal cyclers where mineral oil is not necessary.

3.3.3. PCR and Electrophoresis Conditions

1. Amplification: Using a DNA thermal cycler, program for 30 cycles consisting of 30 s at 94°C, 1.5 min at 58°C, and 2.5 min at 72°C, with a final extension step at 72°C for 7 min.
2. Electrophoresis: Remove 15-µL aliquots of each PCR mix and analyze for various amplification products by submarine gel electrophoresis on 2% agarose gels.
3. Run the electrophoresis for 50 min at 100 V.
4. Stain the gel with ethidium bromide for 15 min.
5. Visualize the stained bands by UV transillumination and photograph.
6. Include a suitable molecular size marker (100 bp ladder; Gibco-BRL) in each gel.

4. Notes

1. The majority of *E. coli* isolates shows typical metallic sheen colonies on EMB agar plates. However, *S. dysenteriae* mimics *E. coli* on EMB agar plates. Similarly, gas may be produced from lactose at 44.5°C by *E. coli* O157:H7. *E. coli* O157:H7 did not exhibit β-glucuronidase activity. However, some heat labile-toxin producing *E. coli* (ATCC 43886) strains may not produce fluorescence on 4-methylumbelliferyl-β-D-glucuronide-supplemented commercial agar that contains lactose. The absence of sorbitol fermentation by O157:H7 is a characteristic phenotype used to isolate *E. coli* O157:H7 from clinical and food specimens. Although useful, confirmation with O157 and H7 antisera is required, as since other bacteria share this serotype and because there are strains of O157:H7 that can ferment sorbitol. As both *S. dysenteriae* and *S. sonneii* do not utilize sorbitol, they show green coloration and mimic *E. coli* O157:H7 in these agars. Therefore, biochemical characteristics alone will not differentiate *E. coli* O157:H7 from other toxigenic and nontoxigenic strains.

2. Antibodies to the O157 antigen are used in many assays to detect O157:H7 in clinical and food samples. Cross-reaction of somatic antigen O157 and flagellar antigen H7 between O157 and O25, O26, O78, O111 as well as between H7 and H11, H⁻ is established. These tests, however, provide no information on the toxin types produced by the isolates and are not specific, as the O157 antigen is present in other *E. coli* species. Also, anti-O157 sera often cross-reacts with *Citrobacter freundii, E. hermanii* and other bacteria. Analyses of food products with anti-O157 serum have recognized O157 isolates that neither produced SLT nor were of the H7 serotype. Furthermore, production of SLT toxins is not confined to *E. coli* O157:H7 strains and these toxins are produced in other serogroups of *E. coli*.
3. The standard bioassays used for identification of pathogenic *E. coli*, such as cytopathic effects on Y-1 adrenal cells and rabbit ileal loop, are not readily adaptable for screening large numbers of *E. coli* isolates.
4. The *uidA* gene that is responsible for β-glucuronidase activity is a good marker for the differentiation of all types of *E. coli* strains from other group of coliforms, but species of the genus *Shigella* also possesses this gene.
5. Analysis of amplification products showed that all reference strains of O157:H7 serotype were correctly identified simultaneously with the SLT type known to be produced by these strains (*see* **Fig. 1**, lanes 2–5). As anticipated, no products were amplified from wild-type *E. coli* (*see* **Fig. 1**, lane 1), whereas the expected toxin gene-specific products, but not O157:H7-specific products, were amplified from the SLT-producing non-O157: H7 serotypes examined (*see* **Fig. 1**, lanes 6–14).
6. The type of SLT identified by the multiplex PCR assay correlated well with the Vero cell toxicity data. Among non-*E. coli* strain, only *S. dysenteriae* exhibits an SLT I amplicon. The Shiga toxin of *S. dysenteriae* type 1 is almost identical to the SLT I of O157:H7; therefore, this is not unexpected. Although the multiplex PCR assay will not discriminate between *S. dysenteriae* type 1 and non-O157:H7 EHEC serotypes that produce only SLT I, the O157:H7-specific primers readily distinguish *S. dysenteriae* type 1 species from O157: H7 isolates.
7. The major advantage of this method over existing assays is that it can identify the types of SLT encoded by the strain and at the same time discriminate other SLT-producing *E. coli* from O157:H7, predominant serotype implicated in disease.
8. When whole bacterial cells are used, 10^2 CFU *E. coli* in 10-μL PCR mixture is necessary to amplify the PCR bands. However, a minimum of 10^6 CFU/g is needed to amplify specific PCR products from food.
9. A 6-h incubation of contaminated food in a normal bacteriological medium would allow proliferation of 10^2 CFU *E. coli*/g initial inoculum to a detectable level. Similarly, if the food homogenate is incubated overnight (16 h) at 37°C in shaking condition, a initial inoculum of 1 CFU *E. coli*/g slurry would attain a requisite density ($>10^9$ CFU/g), thus producing all PCR amplicons.
10. PCR amplification is possible, even when *E. coli* and other coliforms are in a ratio of 10^9:1.

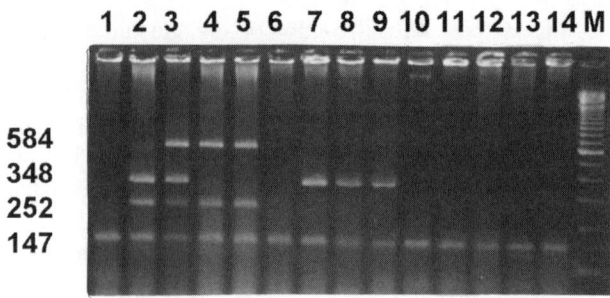

Fig. 1. Agarose gel electrophoresis of amplicons generated by multiplex PCR from *E. coli* strains isolated from various outbreaks. **Lane 1**, typical E. coli ATCC 25922; **lane 2**, O157:H7 strain producing SLT-I, **lane 3**, O157:H7 strain producing SLT-I and SLT-II; **lane 4** and **5**, O157:H7 strains producing SLT-II; **lane 6**, virulent strain not producing SLT-I or SLT-II; **lanes 7–9**, strains other than O157 serovar producing SLT-I; **lanes 10–14**, strains isolated from urinary tract and veterinary infections; **lane M**, 100-bp marker. Number to the left of the gel are molecular sizes (base pairs).

11. Food particles and other unknown metabolic by-products may be inhibitory for PCR reaction. Hence, a two-step filtration procedure is necessary to remove any PCR inhibitory substances (*see* **Subheading 3.3.1., steps 5–7**).
12. The two-step filtration procedure is successful for identifying appropriate PCR amplification products in various other food-borne pathogens directly from food enrichment culture without extracting DNA. Some examples are *Salmonella, Vibrio cholerae, V. parahaemolyticus, Bacillus cereus* groups, *Camplyobacter* spp. etc.

9

Detection of *Listeria monocytogenes* by the Nucleic Acid Sequence-Based Amplification Technique

Burton W. Blais and Geoff Turner

1. Introduction

The rapid detection of foodborne pathogens such as *Listeria monocytogenes* requires ultrasensitive techniques that give measurable responses with low numbers of the target bacteria in food samples or enrichment cultures. Although a number of approaches are possible for detection of low levels of target bacteria, including enzyme immunoassay (EIA) and nucleic acid probe hybridization, perhaps the most efficient approach from the viewpoint of detectability and specificity are the nucleic acid amplification techniques. Nucleic acid amplification targets specific nucleotide sequences within the bacterial genome and raises the number of copies of the region of interest to levels that are detectable by conventional means (e.g., agarose gel electrophoresis, DNA probes). A well-known example is the polymerase chain reaction (PCR) technique, which amplifies target DNA sequences by a mechanism involving hybridization of two oligonucleotide primers to opposite strands flanking the target region, followed by a repetitive series of cycles involving template denaturation, primer annealing, and the extension of primers by a DNA polymerase, typically the thermostable *Taq* polymerase, which enables the use of an automated thermal cycling device in this process. The amplification in a typical PCR process can result in a million-fold increase in the number of target sequence copies after 20 cycles *(1)*.

As an alternative to the PCR technique, Nucleic Acid Sequence-Based Amplification (NASBA™) was developed for the ultrasensitive detection of specific nucleotide sequences, such as viruses and pathogenic foodborne bacteria *(2,3)*. NASBA is a homogeneous, isothermal, in vitro amplification process involving the actions of three enzymes, reverse transcriptase (RT), ribonuclease H (RNase H), and bacteriophage T7 RNA polymerase (T7 RNA pol), as well as two target sequence-specific oligonucleotide primers (one of which bears a bacteriophage T7 RNA pol binding and preferred transcriptonal initiation sequence at its 5' end), acting in concert to amplify target sequences more than 10^8-fold within 90 min.

The scheme for the NASBA reaction is illustrated in **Fig. 1.** A NASBA reaction system can be designed to target either RNA or DNA in a sample. For amplification of RNA targets, the process begins when one primer (P1), containing a 3' terminal sequence complementary to the target nucleic acid, and a 5' terminal sequence corresponding to the T7 RNA polymerase promoter, anneals with the target RNA [RNA(+)], followed by extension of the annealed primer with RT. The RNA strand of the resulting RNA:DNA hybrid is then digested by RNase H, and the second primer (P2), which is complementary to the remaining DNA strand [DNA(–)], then anneals to the latter and is extended by RT to form a double-stranded DNA template with a transcriptionally active T7 promoter. This template is acted on by T7 RNA pol, which generates multiple transcripts [RNA(–)]. P2 then anneals to each transcript and is extended by RT, followed by digestion of the RNA strand by Rnase H to yield a single-stranded DNA intermediate [DNA(+)]. After annealing of P1 to this strand and extension by RT, a transcriptionally active double-stranded DNA template is created that results in the synthesis of yet more transcripts [RNA (–)] as before. This is the basis for the exponential amplification that is achieved by the NASBA system. For amplification of a sequence on a double-stranded DNA target, the process is similar to the foregoing, with the exception that the two strands must first be separated by heat denaturation in order to allow P1 to anneal to the complementary DNA strand [DNA(+)]. The annealed primer is then extended by RT, followed by a second round of denaturation, yielding the single-stranded DNA intermediate [DNA(–)] which anneals with P2 and continues in the NASBA reaction scheme as described. It should be noted that, as none of the enzymes used in the NASBA process are thermally stable, the denaturation and priming of the DNA template are carried out separately from the rest of the NASBA reaction.

The main advantages of NASBA over other amplification techniques such as PCR are (1) rapid accumulation of amplicons (predominantly RNA), (2) no need for specialized equipment (e.g., thermal cycling device), and (3) amplification of either RNA or DNA targets depending on experimental design. It will be evident from the description of the process that NASBA is particularly well suited for the amplification of RNA targets, such as rRNA and mRNA, and this is especially advantageous for food safety testing applications, where it is important to distinguish between viable and nonviable pathogenic bacteria in a

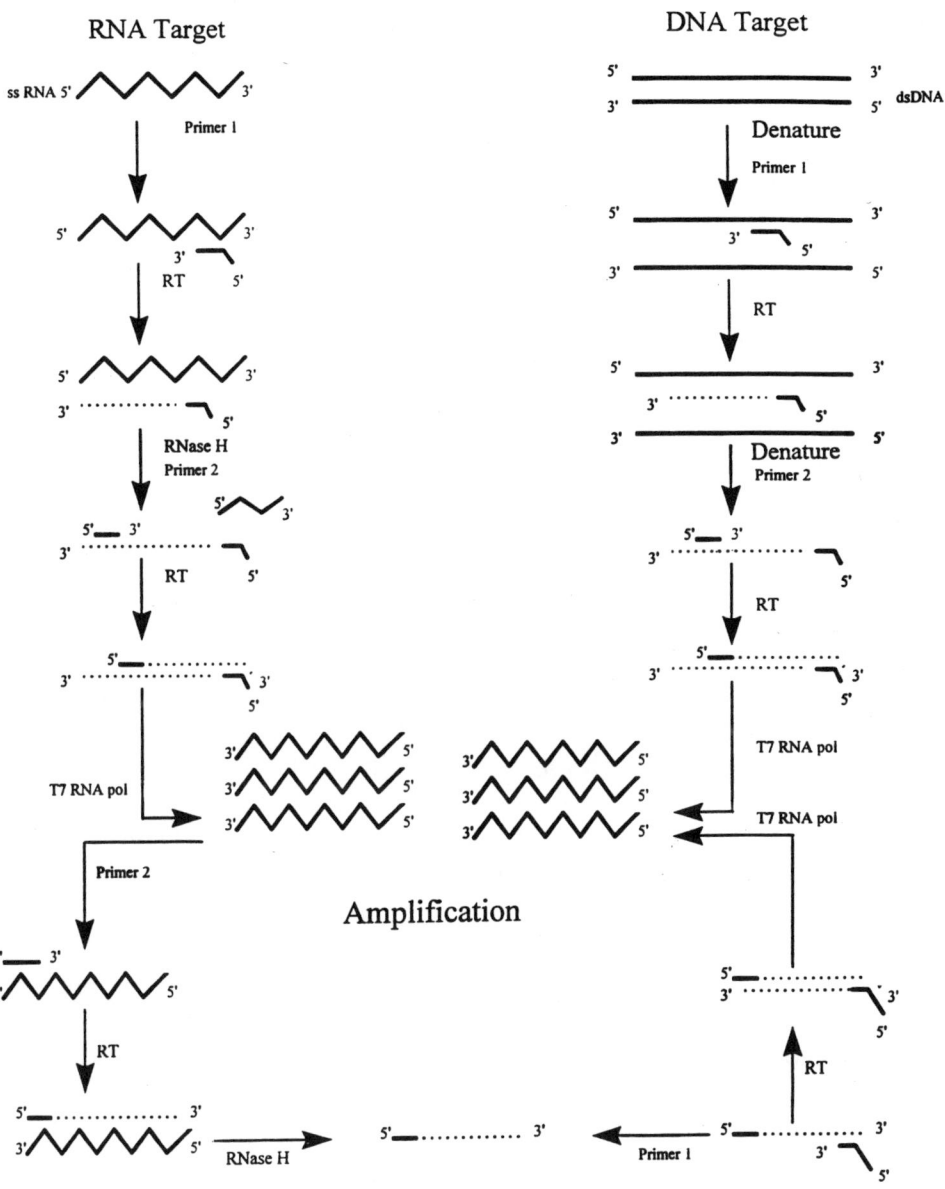

Fig. 1. Scheme for the NASBA reaction.

food sample. Because RNAs are generally unstable in the environment, only viable bacterial cells would be expected to contain significant amounts of

RNA, whether in the form of rRNA or, if gene expression systems are active, mRNA. The following sections describe two different NASBA systems developed in our laboratory for the detection of the well-known pathogen *L. monocytogenes* in foods *(3,4)*. The basis for the specificity of these systems is the amplification of hemolysin gene (*hlyA*) sequences unique to *L. monocytogenes*. In one approach, we describe a method for amplifying *hlyA* sequences in the genomic DNA of *L. monocytogenes* with incorporation of digoxigenin as a marker, followed by a simple microtiter plate-based colorimetric assay of the amplicons. The other system, which targets an *hlyA* mRNA sequence expressed in live *L. monocytogenes* cells, uses a simple dot-blot assay system for the detection of amplicons incorporating biotin as a marker. The major steps involved in the operation of a NASBA detection system comprise the following: (1) induction of mRNA synthesis (if detecting RNA targets); (2) extraction of nucleic acids from the test sample; (3) performance of the NASBA reaction; and (4) detection of amplicons using a simple colorimetric assay system (microtiter plate or dot blot format).

2. Materials
2.1. Bacterial Strains and Media

The *L. monocytogenes* American Type Culture Collection (ATCC) strain no. 43256 may be utilized as a positive control for the NASBA experiments. Alternatively, any other typical *L. monocytogenes* isolate may serve. Note that some strains of *L. monocytogenes* do not express the *hlyA* gene, and should therefore not be utilized as positive controls. The bacteria are routinely grown by inoculation into trypticase soy broth (TSB) (Difco, Detroit, MI) and shaking for 16 h at 30°C.

2.2. Chemicals, Reagents, and Supplies
2.2.1. Chemicals and Biologicals

The following is a list of chemicals and biologicals that will be required during the course of a NASBA experiment (chemicals can be purchased from any reputable supplier. However, wherever a specific supplier is indicated, it is highly recommended that the item be purchased from that source. In any case, it is advisable to purchase molecular biology grade chemicals wherever possible).

1. Ammonium acetate.
2. Antidigoxigenin antibody-peroxidase conjugate (Boehringer Mannheim, Québec, Canada).
3. Avian myeloblastosis virus reverse transcriptase (AMV-RT) (Seikagaku America, Falmouth, MA).
4. Biotin-11-UTP (Sigma Chemical Co., St. Louis, MO).

5. Boiled Rnase.
6. Blocking reagent (Boehringer Mannheim).
7. Bovine serum albumin.
8. Chloroform.
9. Deoxyribonucleotide triphosphates: dATP, dGTP, dCTP, and dTTP.
10. Digoxigenin-11-UTP (Boehringer Mannheim).
11. Dimethylsulfoxide.
12. Dithiothreitol.
13. Ethanol.
14. Ethylenediamine tetraacetic acid (EDTA).
15. Formamide.
16. Isoamyl alcohol.
17. N-lauroylsarcosine.
18. Lysozyme.
19. Mineral oil.
20. Nuclease-free, deionized distilled water (e.g., diethyl pyrocarbonate treated).
21. Phenol (Tris saturated).
22. Proteinase K.
23. Ribonucleotide triphosphates: ATP, GTP, CTP, and UTP.
24. RNase H (Pharmacia, Québec, Canada).
25. RNasin™ (Promega Corp., Madison, WI).
26. RNeasy Total RNA Kit (QIAGEN, Valencia, CA).
27. Sodium chloride.
28. Sodium citrate.
29. Sodium dodecyl sulfate.
30. Sodium phosphate, mono- and dibasic salts.
31. Sodium sarcosinate.
32. Streptavidin-peroxidase conjugate (Sigma).
33. Sucrose.
34. *Taq* DNA polymerase (Boehringer Mannheim).
35. Tetramethylbenzidine (TMB) membrane peroxidase substrate (Kirkegaard and Perry Laboratories, Gaithersburg, MD).
36. Tetramethylbenzidine (TMB) microwell peroxidase substrate (Kirkegaard and Perry Laboratories).
37. Triton X-100.
38. Trizma base (Tris).
39. T7 RNA polymerase (Pharmacia).
40. Tween-20.

2.2.2. Supplies and Equipment

(Care should be taken to ensure that all glassware and disposable plasticware are DNase- and Rnase-free.)

1. Hot plate or dry heating block.
2. Incubators.
3. Microcentrifuge.
4. 1.5 mL-capacity microcentrifuge tubes.
5. 0.6 mL-capacity microcentrifuge tubes.
6. Microtiter plates (Immunlon 1, Dynex Technologies, Chantilly, VA).
7. Microtiter plate reader (optional).
8. Nylon membrane (Boehringer Mannheim).
9. Pipetors covering volume range of 0.5–1000 µL, with appropriate tips.
10. Quartz cuvetes.
11. Spectrophotometer.
12. Thermal cycler.
13. Ultraviolet light source (254 nm).
14. Vortex mixer.
15. Water bath (2).

2.3. Oligonucleotides

All oligonucleotides can be synthesized inexpensively by a contract laboratory specializing in this field. For best results, it is recommended that oligonucleotides be purified by gel electrophoresis or high-performance liquid chromatography, a service that is usually provided at minimal additional cost by the contract laboratory. See **Subheading 4.1.** for further notes on the selection and preparation of oligonucleotides.

2.3.1. Oligonucleotides for hlyA Genomic DNA Detection

2.3.1.1. PRIMERS FOR PREPARATION OF DNA CAPTURE PROBE

A DNA capture probe for detection of NASBA amplicons using a microtiter plate format can be readily prepared by amplifying *L. monocytogenes*-specific *hlyA* sequences using the PCR technique, as described in **Subheading 3.3.1.1.** To amplify a 259-mer DNA capture probe spanning nucleotides 958–1217 of the published *hlyA* sequence *(5)*, the primer set 5'-GAGCAGTTGCAAGCG-3', and 5'-AGGTTGCCGTCGAT-3' will be required.

2.3.1.2. PRIMERS FOR NASBA REACTION

The primers for the NASBA reaction are designed to amplify the region spanning nucleotides 939–1234 of the published *hlyA* sequence, and consist of the set P1, 5'-<u>AATTCTAATACGACTCACTATAGGG</u>AGCGGCAAAGCTGTTAC-3' (the T7 RNA polymerase-binding and preferred transcriptional initiation sites are indicated by underscoring); and P2, 5'-TATCGCGTAAGTCTCCG-3'.

Detecting L. monocytogenes *with NASBA* 73

2.3.2. Oligonucleotides for hlyA mRNA Detection

2.3.2.1. OLIGONUCLEOTIDE CAPTURE PROBE

A 56-mer oligonucleotide capture probe for detection of NASBA amplicons can be prepared based on the *hlyA* sequence, 5'-CAAGGATTGGA TTACAATAAAAACAATGTATTAGTATACCACGGAGATGCAGTGAC-3'.

2.3.2.2. PRIMERS FOR NASBA REACTION

The primers for the NASBA reaction are designed to amplify a 133-mer region of *hlyA* mRNA, and consist of the set P1, 5'-<u>AATTCTAAT ACGACTCACTATAGGGAGA</u>TAACCTTTTCTTGGCGGCACA-3' (the T7 RNA polymerase-binding and preferred transcriptional initiation sites are underscored); and P2, 5'-GTCCTAAGACGCCAATCGAA-3'.

3. Methods
3.1. Sample Preparation
3.1.1. Purification of Genomic DNA

The following procedure may be used to prepare a stock of purified *L. monocytogenes* genomic DNA for amplification of DNA target sequences (*see also* **Subheading 4.2.1.**):

1. Inoculate 5 mL of TSB with one loopful of *L. monocytogenes*, and grow at 30°C to the midlog phase (approx 10^8 CFU/mL).
2. Transfer2X 1 mL portions of the cells to 1.5 mL-capacity microcentrifuge tubes. Pellet the cells by centrifugation at 10,000*g* for 15 min.
3. Resuspend the cell pellets in 100 µL each of 50 m*M* Tris HCl (pH 8.0) containing 25% (w/v) sucrose, and combine into a single tube. Freeze the suspension by placing in a –20°C freezer for 30 min.
4. Thaw the cell suspension at room temperature, and add 20 µL of a freshly prepared lysozyme solution (5 mg/mL in distilled H_2O). Incubate for 5 min at room temperature.
5. Add 40 µL of 0.25 *M* EDTA (pH 8.0). Incubate for 5 min at room temperature.
6. Add 5 µL of fresh diethyl pyrocarbonate (DEPC) solution (prepared by combining 6.5 µL of DEPC with125 µL 95% ethanol). Poke a hole in the lid of the microcentrifuge tube, and incubate for 15 min in a water bath adjusted to 65°C.
7. Add 10 µL of a boiled Rnase solution (1 mg/mL in distilled H_2O). Incubate for 15 min in a 65°C water bath. Transfer to a new microcentrifuge tube with an intact lid.
8. Add 40 µL of a proteinase K solution (2 mg/mL in distilled H_2O), followed by 80 µL of 2% (w/v) Na sarcosinate in distilled H_2O.
9. Incubate with gentle agitation overnight (approx 16 h) at 37°C.
10. Add 200 µL of TE buffer (10 m*M* Tris/1 m*M* EDTA, pH 8.0).
11. Add 400 µL chloroform:isoamyl alcohol (24:1). Gently agitate for 10 min at room temperature.

12. Centrifuge at 10,000g for 20 min. Using a pipet, remove the aqueous (top) layer, being careful not to disturb the white interface. Discard the lower layer and the interface.
13. Add 200 µL of Tris-saturated phenol and 200 µL chloroform:isoamyl alcohol (24:1) to the aqueous layer. Gently agitate for 10 min at room temperature.
14. Centrifuge at 10,000g for 20 min, and collect the aqueous layer as in **step 12**.
15. Add 400 µL of chloroform:isoamyl alcohol (24:1) to the aqueous layer. Gently agitate for 10 min at room temperature.
16. Centrifuge at 10,000g for 20 min, and collect the aqueous layer as before.
17. Transfer 400 µL of the aqueous layer to a clean microcentrifuge tube, and add 35 µL of 3 M ammonium acetate (pH 5.2), followed by 1 mL of 95% ethanol. Incubate at –20°C for at least 2 h.
18. Centrifuge at 10,000g (4°C) for 30 min. Discard the ethanol phase.
19. Rinse the pellet with ice-cold 70% ethanol, and centrifuge at 10,000g (4°C) for 20 min. Discard the ethanol phase and air dry the DNA pellet.
20. Add 20–100 µL of TE buffer to resuspend the DNA pellet.
21. Determine DNA concentration by measuring A_{260} of solution. Store at –20°C.

3.1.2. Purification of mRNA

3.1.2.1. INDUCTION OF HLYA MRNA SYNTHESIS

For detection of mRNA targets in the NASBA reaction, it will first be necessary to induce the bacterial cells to express the gene of interest (*see* **Subheading 4.2.2.**). Transcription of the *hlyA* gene in *L. monocytogenes* can generally be induced by exposure of a cell suspension to sodium azide at a slightly elevated temperature. Mix 1 mL of a cell suspension (e.g., broth culture) with 10 µL of 10% (w/v) sodium azide and incubate at 37°C for 30 min. Proceed immediately to the RNA purification step.

3.1.2.2. PURIFICATION OF TOTAL BACTERIAL RNA

In order to recover the induced *hlyA* mRNA for use in the NASBA reaction, the total RNA in the cell suspension can be purified rapidly and conveniently using a commercially available kit, the QIAGEN RNeasy Total RNA Kit. In this kit, cell suspensions are first lysed and homogenized under conditions that destroy RNases while maintaining the RNA intact. The sample is then adjusted with a high-salt buffer to allow binding of the RNA to a silica gel-based membrane, and, after washing, the bound RNA is eluted from the membrane. Following the manufacturer's instructions, elute the total RNA from the membrane with 50 µL of Rnase-free distilled H_2O. If not used immediately, the RNA should be stored at –20°C.

3.2. NASBA Reactions

The following protocols outline the procedures for amplifying either DNA (e.g., genomic DNA sequence) or RNA (e.g., mRNA) targets. For amplification of DNA targets, each set of reactions should include a positive control sample in which 50 ng of purified *L. monocytogenes* genomic DNA is introduced into the NASBA reaction. For amplification of RNA targets, a sample prepared from a pure broth culture of *L. monocytogenes* (approx 10^8–10^9 CFU/mL) should be included as a positive control. In the interest of obtaining reliable amplifications, the specified temperatures should be strictly adhered to. Refer to **Subheading 4.3.** for further considerations on the operation of a NASBA system.

3.2.1. Amplification of DNA Targets

In this particular NASBA system, the amplicons incorporate a digoxigenin label in order to facilitate the subsequent detection of the amplicons using a microtiter plate hybridization assay. The sample may be either DNA purified according to the scheme outlined in **Subheading 3.1.1.**, or cell lysates prepared by other means (*see* **Notes**). In any case, the procedure for DNA targets proceeds in two phases, template priming and amplification.

3.2.1.1. PRIMING DNA TEMPLATE

Prior to entering the NASBA cycle, the DNA template must be denatured and "primed" by annealing with primer P1, as follows.

1. In a 1.5-mL-capacity microcentrifuge tube, combine 2 µL of sample with 16 µL of NASBA reaction mixture (50 m*M* Tris-HCl (pH 8.5); 62.5 m*M* KCl; 15 m*M* MgCl$_2$; 1.25 m*M* each of dATP, dGTP, dCTP, and dTTP; 2.5 m*M* each of ATP, GTP, CTP and UTP) containing 1 pmol of primer P1 (*see* **Subheading 2.3.1.2.**).
2. Heat at 100°C for 5 min in a dry heating block or boiling water bath, then transfer to a water bath adjusted to 50°C and allow 2 min for the tube to equilibrate. Note that it may be necessary to centrifuge the tube for a few seconds after the high temperature heating step in order to spin down any condensation forming at the top of the tube.
3. Add 2 µL containing 10 m*M* dithiothreitol (DTT) and 10 U AMV-RT, and incubate further at 50°C for 15 min. Heat the tube at 100°C for 5 min in order to denature the newly synthesized complex, chill on ice, then proceed immediately to the amplification phase.

3.2.1.2. AMPLIFICATION PHASE

After initial priming, the target is then subjected to the amplification cycle as follows.

1. In a new 1.5-mL-capacity microcentrifuge tube, combine 5 µL of the P1-primed sample with 18 µL of NASBA reaction mixture (*see* **Subheading 3.2.1.1.**) containing 14 m*M* DTT, 1 pmol each of primers P1 and P2 (*see* **Subheading**

2.3.1.2.) and 0.25 m*M* digoxigenin-11-UTP. Heat the mixture at 65°C for 5 min, then transfer to a water bath adjusted to 41°C and allow 2 min to equilibrate.
2. Add 2 µL of a mixture containing 8 U AMV-RT, 40 U T7 RNA polymerase, 0.1 U RNase H, 12.5 U RNasin, and 2.6 µg bovine serum albumin. Return the tube to the 41°C water bath and incubate for 90 min.
3. Centrifuge for a few seconds to spin down any condensation, and place on ice or freeze at –20°C until ready for analysis of the amplicons.

3.2.2. Amplification of RNA Targets

The amplification of RNA targets proceeds in a manner that is similar to that for DNA targets, with the exception that no initial priming step is required, as the primer P1 can anneal directly to the single-stranded target sequence. In this particular NASBA system, the amplicons incorporate a biotin label to permit their detection using a simple dot-blot assay on membrane discs. The samples may be either total RNA purified according to **Subheading 3.1.2.**, or cell lysates prepared by other means (*see* **Notes**).

3.2.2.1. AMPLIFICATION PHASE

The sample is introduced directly into the amplification reaction as follows: In a 1.5-mL-capacity microcentrifuge tube, combine 5 µL of sample with 18 µL of NASBA reaction mixture (*see* **Subheading 3.2.1.**) containing 14 m*M* of DTT, 1 pmol each of primers P1 and P2 (*see* **Subheading 2.3.2.2.**) and 0.25 m*M* biotin-11-UTP. Heat the mixture at 65°C for 5 min, then transfer to a water bath adjusted to 41°C and allow 2 min to equilibrate. Proceed as for **steps 2** and **3** in **Subheading 3.2.1.2.**

3.3. Detection of Amplicons

3.3.1. Detection of Amplicons from DNA Targets

In **Subheading 3.2.1.**, target sequences originating from DNA molecules (e.g., genomic DNA) were amplified in the presence of digoxigenin-11-UTP, resulting in the incorporation of this label in the amplicons (predominantly minus-strand RNA). The procedure outlined in this section pertains to a microtiter plate-based assay for the detection of the digoxigenin-labeled NASBA amplicons by hybridization with an immobilized DNA probe. The hybridized amplicons are then detected by sequential reactions of the microtiter plate wells with anti-digoxigenin antibody–peroxidase conjugate and tetramethylbenzidine substrate solution. The intensity of the resulting color in the wells is proportional to the amount of amplicon produced in the NASBA reaction.

The DNA probe-coated microtiter plates must be prepared in advance. In the present NASBA system, the DNA probe is conveniently made by amplification of a portion of the target sequence using the PCR technique.

3.3.1.1. Preparation of DNA Probe

A capture DNA probe can be prepared using the PCR technique with purified *L. monocytogenes* genomic DNA as template:

1. In a 0.6-mL-capacity microcentrifuge tube, combine 10 µL of distilled H_2O containing 10 ng of *L. monocytogenes* genomic DNA (*see* **Subheading 3.1.1.**) with 89.5 µL of PCR reaction mixture (11 m*M* Tris-HCl (pH 8.3); 0.22 m*M* each of ATP, CTP, GTP, and TTP; 2.2 m*M* $MgCl_2$; 55 m*M* KCl; and 0.11% (v/v) Triton X-100) containing 1.1 µ*M* each of primers P1 and P2 (*see* **Subheading 2.3.1.1.**). Overlay the mixture with mineral oil, then place in a thermal cycler set to hold the temperature at 80°C for for 10 min.
2. Add 0.5 µL containing 1 U Taq DNA polymerase to the tube, then subject the sample to 35 cycles of the following sequence: denaturation at 94°C for 30 s, primer annealing at 55°C for 30 s and primer extension at 72°C for 90 s. Allow an additional 2 min at 72°C after the last cycle in order to ensure completion of primer extension.
3. Gently insert a pipet tip under the mineral oil layer and remove all of the PCR reaction mixture to a new 1.5-mL-capacity microcentrifuge tube. Add 0.1 vol of 3 *M* ammonium acetate (pH 5.2), followed by 2.5 vol of ethanol, and place at –20°C for at least 2 h.
4. Proceed as for **steps 18–21** in **Subheading 3.1.1.**

3.3.1.2. Immobilization of DNA Probe on a Microtiter Plate

1. Denature an aliquot of DNA probe stock by heating at 100°C for 10 min, then immediately dilute to a final concentration of 2 µg/mL in ice-cold coating buffer (0.3 *M* Tris-HCl (ph 8.0); 0.5 *M* $MgCl_2$; 1.5 *M* NaCl).
2. Transfer 100 µL of diluted, denatured DNA to each well of a microtiter plate, then seal the wells (e.g., using adhesive tape or microtiter plate sealing film pressed firmly over the wells), and incubate at 37°C for 16 h.
3. Empty the wells of their contents and allow to air dry. The DNA can be crosslinked to the plastic by exposure to ultraviolet light (254 nm) for 3 min.
4. Wash the wells three times with approx 200 µL wash buffer (0.1 M Tris-HCl (pH 8.0); 2 m*M* $MgCl_2$; 1 *M* NaCl; 0.1% (v/v) Tween-20). Note that the wells can be washed by simply filling each with wash buffer using a squirt bottle, and emptying by vigorously shaking the plate upside down. Ensure that no residual buffer remains in the wells.
5. Block the wells by incubation at 37°C for 1 h with 100 µL of hybridization solution (5X SSC [1X SSC is 0.15 *M* NaCl plus 0.015 *M* sodium citrate]; 1 % (w/v) blocking reagent; 0.1 % (w/v) N-lauroylsarcosine; 0.02 % (w/v) sodium dodecyl sulfate; 50 % (v/v) formamide. Wash the wells three times with PBST (0.01 *M* phosphate (ph 7.2), 0.15 *M* NaCl; and 0.05 % (v/v) Tween-20). At this point, the wells can be emptied and air dried for storage at 4°C for a maximum of 4–6 wk.

3.3.1.3. Assay of NASBA Amplicons

1. In a DNA probe-coated microtiter plate well, combine 25 µL of hybridization solution (*see* **Subheading 3.3.1.2.**) with 25 µL of NASBA reaction product (*see* **Subheading 3.2.1.**), and incubate at 56°C for 90 min.
2. Wash the wells with PBST (*see* **Subheading 3.3.1.2.**), and incubate at room temperature for 20 min with 100 µL of anti-digoxigenin antibody–peroxidase conjugate diluted 1:2 000 in PBST containing 0.5 % (w/v) blocking reagent (PBST-B).
3. Wash the wells with PBST as before, and incubate at room temperature for 20 min with 100 µL of tetramethylbenzidine microwell substrate solution. A blue color will develop in the wells in the presence of bound peroxidase (positive samples). Stop the reaction by the addition of 50 µL per well of 1 M H_2SO_4 (this will change the color from blue to yellow).
4. Using a microtiter plate reader, measure the absorbance in the wells at 450 nm (A_{450}). Generally, negative control samples in which no template was added to the NASBA reaction mixture should give A_{450} values below 0.10, whereas any absorbance value above this level should be considered as a positive result. However, the cutoff absorbance values for different systems will have to be determined empirically. Alternatively, if a microtiter plate reader is not available, reactions can be scored qualitatively by visual examination of the wells.

3.3.2. Detection of Amplicons from RNA Targets

In **Subheading 3.2.2.**, target sequences originating from RNA molecules (i.e., mRNA) were amplified in the presence of biotin-11-UTP, in order to incorporate a biotin label in the amplicons. This section describes a simple membrane dot-blot technique as an alternative to the microtiter plate-based method for the detection of NASBA amplicons. In this approach, an oligonucleotide probe complementary to a portion of the NASBA amplicon (minus-strand RNA) is immobilized as a spot in the center of a small nylon membrane disc. For the assay, the disc is incubated with NASBA reaction product, and the amplicon is captured by hybridization with the immobilized probe at the center of the disc. The bound amplicon is then detected by sequential reactions with a streptavidin peroxidase conjugate and tetramethylbenzidine substrate solution. In this manner, positive results are visualized on the basis of color formation (i.e., a blue spot) in the center of the disc.

3.3.2.1. Preparation of Probe-Coated Discs

1. Dissolve the 56-mer oligonucleotide probe (*see* **Subheading 2.3.2.1.**) at a final concentration of 30 ng/µL in 6 X SSC (*see* **Subheading 3.3.1.2.**), and apply 1 µL to the center of an approx 8-mm nylon membrane disc (nylon membrane discs can be punched from a sheet of membrane using a clean standard hole puncher). Allow the disc to air dry, then expose to ultraviolet light (254 nm) for 5 min to crosslink

the probe to the membrane. Mark the edge of the disc with a pencil to keep track of the side receiving the DNA spot (you can also scribe a number or letter to permit sample identification later).
2. Block the disc by incubation at 52°C for 1 h in hybridization solution (*see* **Subheading 3.3.1.2.**) containing 25% (v/v) formamide instead of 50% (v/v). Note that several discs can blocked together in a bulk volume of hybridization solution (i.e., sufficient volume to submerge all of the discs).
3. Wash disc three times for 5 min at room temperature with PBST in a 250-mL-capacity beaker (constant gentle agitation). Note that several discs can be washed together in a bulk volume (e.g., 100 mL) of PBST. At this stage, discs can be used immediately in the dot blot assay or stored dry at 4°C.

3.3.2.2. Dot-Blot Assay

1. Place a probe-coated disc in a 1.5-mL-capicity microcentrifuge tube containing 0.5 ml of hybridization solution, then add 25 µL of NASBA reaction product (*see* **Subheading 3.2.2.1.**) and mix well. Place the tube in a water bath adjusted to 52°C, and incubate for 30 min.
2. Drain the liquid (if necessary, touch lightly with the corner of a paper towel or blotting paper to draw off residual liquid) and, leaving the disc in the tube, wash three times for 5 min at room temperature with 1.5 mL of PBST (vortex briefly after addition of PBST).
3. Leaving the disc in the tube, add 0.5 mL of streptavidin–peroxidase conjugate diluted to 0.25 µg/mL in PBST containing 0.05% blocking reagent, and incubate at room temperature for 15 min.
4. Wash disc with PBST as before, then remove from tube and place on the flat surface of a Petri dish with the side bearing the DNA spot facing upward. Pipet 50 µL of tetramethylbenzidine membrane peroxidase substrate solution on the disc, and incubate at room temperature for 15 min. Reactions on the discs are scored qualitatively as follows: positive, colored spot formation at center of disc; negative, no colored spot formation.

4. Notes
4.1. Selection of Oligonucleotides
4.1.1. Selection of NASBA Primers

Primer P1 bears the T7 RNA polymerase-binding and preferred transcriptional initiation sites (indicated by underscoring in the sequences presented in **Subheading 2.3.1.2.** and **2.3.2.2.**) appended at its 5'-end, followed by a target-complementary sequence terminating at the 3'-end. The target-complementary sequence on P1 will vary depending on the target, and typically has a length of about 20 nucleotides (nt) (although this can vary from 15–30 nt). In selecting a target-complementary sequence, a G + C content of 45–60% is recommended, and care should be taken to avoid tracts of the same nucleotides (L. Malek, personal communication). Furthermore, as with PCR primers, it is important to avoid 3'-terminal complementarity between primers P1 and P2, to reduce the formation of primer dimers. With NASBA this issue is perhaps even more important than with PCR, as typical NASBA reaction conditions are relatively less stringent.

4.1.2. Selection of DNA Capture Probe

In selecting primers for synthesis of a DNA capture probe by PCR, it is important to avoid sequences complementary to the NASBA primers P1 and P2. In this manner, any primer dimers produced in the NASBA reaction will not hybridize with the immobilized DNA capture probe during the subsequent amplicon assay procedure. Thus, in the present example, the primers used for preparation of the DNA capture probe (*see* **Subheading 2.3.1.1.**) were designed to amplify *hlyA* sequences internal to the region defined by the NASBA primers P1 and P2 (*see* **Subheading 2.3.1.2.**).

4.1.3. Synthesis of Long Oligonucleotides

When synthesizing long oligonucleotides, such as the primer P1, it is important to consider that, during the synthesis procedure, oligonucleotides are synthesized one base at a time starting at the 3'-end. Unlike biological systems, the chemical couplings are not 100% efficient (typical coupling efficiencies using current synthesizer models are about 98–99.5% for each step). This means that between 0.5–2% of the oligonucleotides are not extended at each base addition. For a short oligonucleotide, the major product will still be the full-length sequence. However, as longer oligonucleotides are made, the proportion of full-length oligonucleotides in the product pool decreases exponentially [e.g., for a 61-mer involving 60 couplings, the proportion of full-length product resulting from a process having 98% coupling efficiency can be calculated as: $(0.98)E60 \times 100\% = 30\%$]. Therefore, in order to obtain a homogeneous preparation of full-length primers necessary for efficient NASBA reactions, it is highly recommended that any oligonucleotides exceeding 30 nt in length be purified by gel electrophoresis or HPLC.

4.2. Target Nucleic Acid Preparation

4.2.1. DNA Purification

A simpler, more rapid alternative would be to either lyse whole cells by boiling for 10 min in the presence of 1% (v/v) Triton X-100, and then using this crude cell lysate directly in the NASBA reaction. This approach has generally worked well for sample preparation in PCR applications. However, in some instances, such as when the sample consists of an enrichment culture, the sample matrix (e.g., food sample, enrichment broth components) may contain substances that inhibit the NASBA reaction, necessitating the use of a DNA purification step. Another alternative to the DNA purification procedure detailed in this chapter would be to utilize a genomic DNA extraction kit offered by manufacturers such as QIAGEN, Promega, Life Technologies (Gibco-BRL; Gaithersburg, MD), and Amersham Pharmacia Biotech (Uppsala, Sweden).

4.2.2. RNA Purification

Because the detection of mRNA targets depends on in vivo transcriptional activity, it is essential to understand the regulatory mechanisms involved in the expression of a gene of interest. In this manner, it may be posssible to manipulate the organism's environment in order to induce transcription of a particular gene, as is done in the present example involving expression of the *L. monocytogenes hlyA* gene by addition of sodium azide and elevation of the temperature. It should be noted that some strains of *L. monocytogenes* carry the *hlyA* gene but have lost the ability to express it (especially after prolonged storage on plating media at 4°C), in which case the NASBA approach targeting genomic DNA may be more appropriate.

4.3. Amplification Reactions

4.3.1. NASBA Reaction Conditions

It is important to recognize that the NASBA reaction system involves the concerted actions of three separate enzymes—RT, T7 RNA polymerase, and Rnase H—each with its own optimal operating conditions of temperature, pH, ionic strength, and so on. Therefore, combining the three activities in a single reaction system (e.g., NASBA) represents a compromise in terms of temperature and buffer conditions, and presents a very narrow window in which to operate the system. Any deviation from the recommended temperatures or buffer conditions can significantly affect the efficiency of the amplification. Therefore, great care must be taken in the preparation of all buffers and reagents, and in regulating the temperatures at each stage of the procedure. A properly adjusted water bath will provide better temperature control than a dry incubator, such as a heating block.

4.3.2. RNase Contamination

RNases are ubiquitous in biological systems, and frequently contaminate the laboratory environment (including glassware). These robust enzymes are very difficult to eliminate, being capable of renaturation even after autoclaving. Because RNA molecules are so central to the NASBA reaction, it is crucial to take the necessary steps to minimize the introduction of RNases in the system. The NASBA reaction mixtures described herein contain the RNase inhibitor RNasin to minimize the effects of contamination by the sample matrix. Nonetheless, certain precautions are warranted to prevent contamination from other sources. Care should be taken to avoid touching surfaces and reagents that will come in direct contact with the NASBA system (it is essential to wear gloves at all times). All glassware and water used in the preparation of buffers and reagents should be treated to eliminate RNase contamination. Solutions and glassware can be decontaminated by treatment with DEPC, but it should be borne in mind that residual DEPC can interfere with the NASBA reaction. Glassware can also be decontaminated using RNase Away (Molecular Bio-Products, Inc., San Diego, CA). Molecular biology grade reagents should be used wherever possible.

4.3.3. Contamination with Amplicons

Amplicons from previous amplifications can quickly contaminate the laboratory environment and pose a very serious problem to laboratories routinely carrying out reactions such as PCR and NASBA. NASBA is particularly prone to this problem as the levels of amplification can exceed by several orders of magnitude those achieved with PCR. It is essential that negative controls devoid of target sequences or sample matrix be run with each set of NASBA reactions in order to determine whether a contamination problem exists. If a contamination problem does exist, it is likely to involve one or more of the following laboratory items: bench and equipment surfaces, pipetters, stock buffers and solutions, reaction mixtures, reagents, and glassware. To minimize the risk of false-positive reactions due to contamination, the different stages involved in preparing the NASBA reagents and carrying out the procedure should be physically separated. For instance, NASBA reaction mixtures (which may be prepared in bulk ahead of time, then aliquoted and stored at –20°C) should not be prepared in the same physical environment where the amplification reactions are carried out. Similarly, post-amplification tubes should never be opened in the same area where samples are prepared and added to the NASBA reaction mixtures.

References

1. Sooknanan, R., Malek, L., Wang, X. I., Siebert, T., and Keating, A. (1993) Detection and direct sequence identification of BCR-ABL mRNA in Ph+ chronic myeloid leukemia. *Exp. Hematol.* **21,** 1719–1724.
2. Compton, J. (1991) Nucleic acid sequence-based amplification. *Nature (London)* **350,** 91–92.
3. Blais, B. W., Turner, G., Sooknanan, R., and Malek, L. (1997) A nucleic acid sequence-based amplification system for detection of *Listeria monocytogenes hlyA* sequences. *Appl. Environ. Microbiol.* **63,** 310–-313.

4. Blais, B. W., Turner, G., Sooknanan, R., Malek, L., and Phillippe, L. M. (1996) A nucleic acid sequence-based amplification (NASBA) system for *Listeria monocytogenes* and simple method for detection of amplimers. *Biotechnol. Tech.* **10,** 189–194.
5. Mengaud, J., Vicente, M., Chenevert, J., Pereira, J. M., Geoffroy, C., Gicquel-Sanzey, B., Baquero, F., Perez-Diaz, J., and Cossart, P. (1988) Expression in *Escherichia coli* and sequence analysis of the listeriolysin O determinant of *Listeria monocytogenes*. *Infect. Immun.* **56,** 766–772.

10

Detection of *Escherichia coli* O157:H7 by Immunomagnetic Separation and Multiplex Polymerase Chain Reaction

Ian G. Wilson

1. Introduction

Since 1983, when *Escherichia coli* O157:H7 was first recognised as a cause of hemorrhagic colitis and hemolytic uremic syndrome (HUS), verotoxin-producing strains of *E. coli* (VTEC) have been identified as the cause of increasing numbers of cases of serious human illness *(1,2)*. One of a number of large outbreaks in Scotland resulted in 20 deaths, and a massive outbreak in Japan affected more than 9000 people *(3,4)*. In England and Wales, 90 % of VTEC cases are sporadic *(5)*.

The serotype predominantly associated with serious illness is *E. coli* O157:H7, but other serotypes and coliforms have been reported to cause similar disease *(6,7)*. The morbidity and mortality are markedly higher than for other serotypes, and the sequelae of infection are both more serious and more common than most food-borne pathogens. Young children and elderly people are at particular risk of developing hemolytic uraemic syndrome or thrombotic thrombocytopenic purpura (TTP). Up to 10% of patients may develop these complications. Some die, and a small number may require hospital care for many months afterward because of renal impairment or other sequelae *(8)*.

The factors surrounding the emergence of this organism have been discussed widely. *E. coli* O157:H7 is most commonly associated with beef, in particular undercooked hamburgers *(5,9,10)*. It has also been the cause of outbreaks associated with direct animal and human contact, apple juice *(11,12)*, fermented sausage *(13)*, dairy products *(14–16)*, water *(17,18)*, plant products *(19)*, and other foods. The infectious dose is considered to be very low, perhaps less than 10 bacteria *(10,20)*. Fecal contamination from infected cattle, sheep, and wild birds *(21)* appears to be the most common ultimate source of the organism, and it can survive for long periods in feces *(22)*.

Conventional tests for *E. coli* will not detect *E. coli* O157:H7 because most strains are unable to ferment sorbitol, produce glucuronidase, or grow at 44°C. *E. coli* O157 would be detected only as a coliform in many screening tests. Given its increasing prevalence, this serotype should now be sought specifically. Although non-O157 *E. coli* serotypes which produce verotoxins and other VT-producing coliforms have been shown to cause illness, *E. coli* O157:H7 is more virulent. Non-O157 organisms are responsible only for a small number of verocytotoxin-mediated illnesses.

Immunomagnetic separation (IMS) uses a suspension of uniform polymer spheres containing oxides of iron *(23,24)*. These have superparamagnetic properties and possess no residual magnetism when removed from a magnetic field *(23)*. Conjugated with specific antisera, they provide a rapid and effective alternative to other methods of concentrating specific bacterial cells *(25,26)*.

Multiplex PCR offers a rapid and definitive test for the possession of virulence genes which have been shown to be important in strains causing illness. Both the chromosomally-encoded attaching and effacing gene (*eae*) which codes for an outer membrane protein that mediates intimate attachment to the mucosa, and the bacteriophage-encoded verocytotoxin (VT1 and VT2)/shiga-like toxin genes (*slt*I or *slt*II) appear to be necessary to cause illness *(27)*. The possession of only one of these genes may indicate lower the likelihood of causing illness. These genes are not exclusive to *E. coli* O157:H7, but may correlate with pathogenicity in other species and serotypes *(6,7,28)*. Confirmation by serological and biochemical methods is essential, as PCR alone cannot determine genus and serotype identity.

The method described in **Subheading 3.** is useful for routine food examination and for risk assessment because it enables the confirmation of *E. coli* O157:H7 culture from food and the testing of isolates for virulence genes. The principle involves sample preparation and enrichment culture, immunomagnetic separation, and detection and confirmation (selective plating, serological and biochemical diagnostic tests, and multiplex PCR).

2. Materials

1. Modified tryptone soya broth (MTSB) (Oxoid, Basingstoke, Hampshire, UK). Methods for preparing Oxoid media can be found in the Oxoid Manual. Store at 4°C.
2. Dynabeads anti-*E. coli* O157 (Dynal, Wirral, UK, techserve@dynal.u-net.com). Store at 4°C.
3. Cefixime tellurite sorbitol MacConkey (CT-SMAC) agar (Oxoid). Store at 4°C.
4. MacConkey agar (Oxoid). Store at 4°C.
5. Nutrient agar. Store at 4°C.

6. Phosphate-buffered saline (PBS) pH 7.4 with 0.05% Tween-20 (PBST) in 5-mL aliquots. Store at 4°C.
7. Saline solution (0.85% NaCl). Store at 4 °C.
8. λ-Glucuronidase broth. Store at 4°C.
9. Latex agglutination test: *E. coli* O157 (Oxoid).
10. API 20E biochemical test strips (Bio-Mérieux, Marcy l'étoile, France). Store at 4°C.
11. Magnetic particle concentrator (MPC, Dynal, UK). Dynal MPC-M holds 10 Eppendorf tubes and allows working volumes of 1.5 mL.
12. Columbia blood agar (Oxoid). Store at 4°C.
13. Tris acetate EDTA buffer (TAE).
14. 5.0 mM MgCl$_2$. Molecular biology grade.
15. 10 mM Tris hydrochloride (pH 8.3). Molecular biology grade.
16. 50 mM KCl. Molecular biology grade.
17. 0.2 mM (each) dATP, dGTP, dCTP, and dTTP (Pharmacia Biotech, St. Albans, Hertfordshire, UK).
18. Primers: 0.15 M each primer. *See* **Table 1** and **Note 1**.
19. *Thermus aquaticus* (*Taq*) DNA polymerase (Stoffel fragment) (Perkin-Elmer, South Glamorgan, UK). Store frozen.
20. Thermal cycler.
21. Agarose gel electrophoresis equipment.

3. Method

The detection of *E. coli* O157:H7 *(29)* requires four stages: selective enrichment in liquid medium, immunomagnetic separation, selective plating and colony recognition on solid media, and confirmation. PCR can be conducted on confirmed colonies (*see* **Note 2**).

3.1. Selective Enrichment Culture

1. Aseptically weigh out 25g of food into a stomacher bag with filter.
2. Add 225 mL of preincubated (42°C) modified tryptone soya broth (MTSB) to the sample and homogenize in a stomacher. If less than 25 g is available, use a 1:10 dilution with MTSB and record the weight and volume used (e.g., if 10 g of sample, add 90 mL MTSB) (*see* **Note 3**).
3. Strain through the filter and pour into a sterile screw-topped jar (*see* **Note 4**).
4. Perform enrichment culture by incubating at 42°C for 6 h to encourage the selective proliferation of *E. coli* O157:H7 (*see* **Note 5**).

3.2. Immunomagnetic Separation

Use Dynabeads anti-*E. coli* O157 (Dynal) according to the manufacturer's instructions for the isolation of *E. coli* O157:H7. The technique will vary slightly depending on the size of tubes and magnetic particle concentrators used (*see* **Notes 6** and **7**).

Table 1
Oligonucleotide Primers

Primer	Oligonucleotide sequence (5'–3')	Location within gene	Predicted size of amplified product (bp)
sltI-F	ACA CTG GAT GAT CTC AGT GG	938–957	
sltI-R	CTG AAT CCC CCT CCA TTA TG	1539–1520	601
sltII-F	CCA TGA CAA CGG ACA GCA GTT	624–644	
sltII-R	CCT GTC AAC TGA GCA CTT TG	1403–1384	780
eae-F	TCG TCA CAG TTG CAG GCC TGG T	2242–2263	
eae-R	CGA AGT CTT ATC AGC CGT AAA GT	3350–3328	1109
P11P	GAG GAA GGT GGG GAT GAC GT	16S rRNA	
P13P	AGG CCC GGG AAC GTA TTC AC	1175–1390	216

1. Add 20 µL of resuspended Dynabeads (Dynal) to an appropriate number of 1.5- mL Eppendorf tubes.
2. Add 1 mL of inoculated and incubated MTSB enrichment broth to the tubes (*see* **Note 8**).
3. Vortex briefly and incubate at room temperature for 30 min on a 360° rotating mixer (Dynal) at moderate speed. This enables the beads to encounter and specifically bind bacteria within the liquid phase.
4. Concentrate the beads using the MPC. Tubes are held in place by a perspex cover and the magnetic strip is slid into the device to begin concentration. Leave to separate for 5 min (*see* **Note 9**).
5. With the magnetic strip still in place, slowly rotate and invert the MPC three times to bring the beads to the back of the tube. Open the tubes carefully using an Eppendorf tube opener to avoid dislodging beads or creating aerosols from liquid trapped in the cap. Carefully remove the supernatant by pipetting from the side of the tube not holding the rust-colored bead–bacteria complexes (*see* **Note 10**). Discard supernatant as potentially infected waste (*see* **Note 11**).
6. Remove the magnetic strip, add 1 mL of PBS pH 7.4, containing 0.05% Tween-20 (PBST) to the tubes. Use one tube of PBST per sample to avoid cross-contamination. Close the lids and mix by inverting the MPC three times to resuspend the beads. Replace the magnetic strip and repeat steps 4–6 three times.
7. Aspirate the supernatant, remove the magnetic strip, and add 30 µL PBST to each tube. Resuspend using a vortex mixer. High numbers of a virulent pathogen with a low infectious dose may be present. Use appropriate containment and work carefully.

3.3. Selective Plating

1. Plate 50 µL of Dynabeads onto cefixime tellurite sorbitol MacConkey (CT-SMAC) agar.
2. Streak out for single colonies and incubate at 37°C for 24 h (*see* **Notes 13–15**).

3.4. Identification

Examine CT-SMAC plates for colorless non-sorbitol-fermenting colonies (sorbitol-fermenting colonies are pink/red). If present, subculture five non-sorbitol-fermenting colonies onto MacConkey agar and incubate overnight at 37°C. Serological and biochemical confirmation must be conducted.

1. Perform latex slide agglutination tests (Oxoid) on lactose-positive, Gram-negative bacilli. Prepare two saline suspensions on a glass slide using a loopful of growth from the MacConkey agar. Add a loopful of *E. coli* O157 antiserum to one of the suspensions. Rock the slide for 30–60 s. If agglutination occurs with the antiserum but not the saline, the test is positive. Lactose-fermenting cultures which autoagglutinate must be subcultured to nutrient agar, incubated until adequate growth is obtained, and retested.
2. Confirm tests that cause agglutination in this reaction biochemically using the API 20E system (bio-Mérieux) according to the manufacturer's instructions. It is also useful to inoculate a β-glucuronidase broth using a colony from the MacConkey purity culture. *E. coli* O157:H7 is usually β-glucuronidase negative and urease positive, unlike most other *E. coli* serotypes. Record as *E. coli* O157:H7 any colonies with an acceptable API profile and positive somatic O157 agglutination response.
3. Presence or absence of *E. coli* in 25g can then be reported.

3.5. DNA Extraction

1. Subculture confirmed *E. coli* O157:H7 strains on to Columbia blood agar and incubate at 37°C overnight.
2. Using a 2-mm loop, pick a colony off into 0.5 mL TAE buffer in an Eppendorf tube and place it in a boiling waterbath for 10 min to lyse the cells. This crude extraction method is satisfactory when the target organisms have been grown in pure culture. For higher purity DNA, phenol–chloroform extraction or one of many commercial systems can be used (*see* **Note 16**).

3.6. Multiplex PCR Assay

Oligonucleotide primers can be prepared by commercial companies based on previously published sequences *(30,31)* that give amplification products of sizes which could be satisfactorily separated and distinguished from each other by agarose gel electrophoresis.

The nucleotide sequence of each primer and the corresponding locations within the *slt* I, *slt*II, and *eae* genes are shown in **Table 1**. Conserved primers P11P and P13P amplify a 216-bp fragment of the V6 region of the 16S rRNA gene *(32)*. This amplification is included to verify the presence of target DNA from the sample. Negative (1µL pure H_2O, **Fig. 1**, lane 8) and positive (1 µL of extracted genomic DNA from the 3 strains listed in **Fig. 1**, lanes 5–7) controls should be included in each run. Three positive control strains of *E. coli* should be used that possess the *slt*I, *slt*II and *eae* genes. We chose an *E. coli* O128 that has the *slt*I gene only, *E. coli* O157:H7 with *slt*II and *eae*, and *E. coli* O157:H7 NCTC 12079 with *slt*I, *slt*II, and *eae* (*see* **Note 17**).

1. Make a master mix containing 5.0 mM MgCl$_2$, 10 mM Tris hydrochloride (pH 8.3); 50 mM KCl; 0.2 mM (each) dATP, dGTP, dCTP, and dTTP (Pharmacia Biotech); 0.15 M each primer; and 2.5 U of *Taq* DNA polymerase (Stoffel fragment) (Perkin-Elmer) (*see* **Note 18**).
2. Add 1 L template DNA and carry out the amplifications in 25 L vol using a DNA thermal cycler (Bio-Rad) for 1 cycle of 3 min at 96°C followed by 35 cycles of 30 s at 94°C, 30 s at 60°C, and 1 min at 72°C with a final extension at 72°C for 5 min.

3.7. Electrophoresis

1. Following PCR, electrophorese 15 L vol with tracker dye in 1% agarose (Bio-Rad) in 1X TAE buffer containing 0.25 g ethidium bromide per mL. Include a 100-bp ladder (Gibco-BRL) in each gel as a molecular-size marker. Perform electrophoresis for 40 min at 100 V; visualize by UV transillumination, and capture by photography or an image grabber attached to a computer and print.
2. Examination of the gel image shows the possession of virulence genes in the *E. coli* O157:H7 isolates. The different sizes of amplicons (*see* **Table 1**) show clearly which genes have been amplified, so the presence of virulence genes in *E. coli* isolates can be readily identified (*see* **Fig. 1**).

4. Notes

1. Many authors have published methods for *E. coli* O157:H7 using alternative primers and other variations on the methods described here.
2. This method is for the detection of the organism, not its enumeration.
3. MTSB should be prewarmed to 42°C to prevent temperature shock to the organisms which may have been damaged by environmental or processing stresses. The 42°C temperature reduces competition from nontarget organisms and ensures the highest recovery of viable *E. coli* O157:H7 *(33)*.
4. Substrates containing high proportions of fat may interfere with immunomagnetic separation. The fatty matrix of cheeses may interfere with the settling of immunomagnetic beads, allowing their aspiration during washing steps and false-negative results. Other reported causes of reduced sensitivity include nonspecific binding of non-target bacteria to beads *(25)* and reaction inhibition when PCR is used *(26,34)*. In general, IMS is very reliable and these sources of interference should not present a great problem. Nevertheless, some samples may require preparation to remove fats or particulates which reduce recovery of cells.
5. The method described above should be suitable for use with fecal samples with little modification. IMS-PCR has been reported to overcome the PCR inhibition caused by bilirubin and bile salts in feces. Normally, a 500-fold dilution is needed

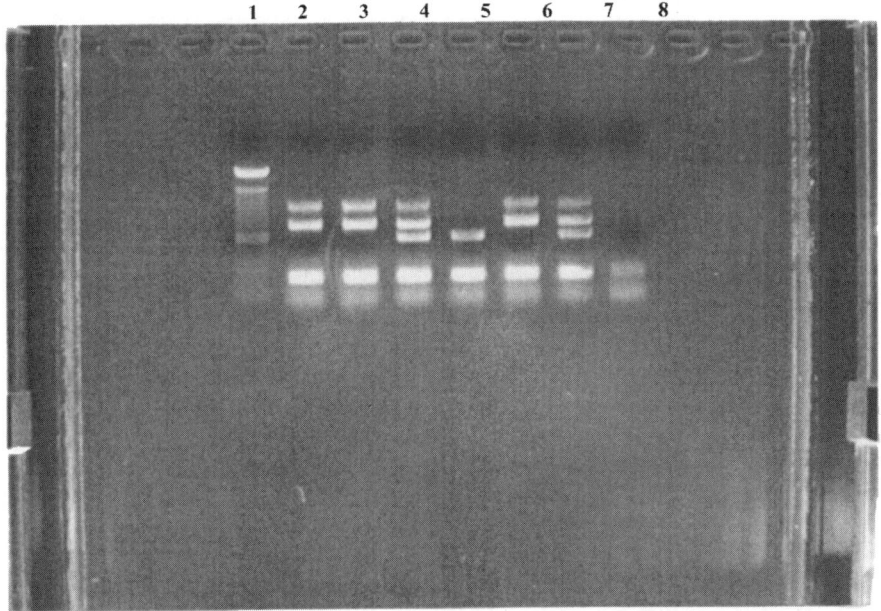

Fig. 1. Multiplex PCR of DNA from *E. coli* strains for *slt*I, *slt*II (including variants), and *eae* genes (1.0% agarose with ethidium bromide). Lanes: **1**, molecular size marker (100bp ladder); **2**, *E.coli* O157 *slt*-II, *eae*⁺ (Patient A); **3**, *E.coli* O157 *slt*-II, *eae*⁺ (Patient B [wife of A]); **4**, *E.coli* O157 *slt*-I, *slt*-II, *eae*⁺ (Patient C); **5**, *E.coli* O128 *slt*-I; **6**, *E.coli* O157 *slt*-II, *eae*⁺; **7**, *E.coli* O157 NCTC 12079 *slt*-I, *slt*-II, *eae*⁺; **8**, blank (H_2O).

to overcome inhibition, but with IMS extraction only a 10-fold dilution and the addition of T4 gene 32 protein were needed *(36)*.

6. A range of magnetic separators is available to allow various numbers and sizes of tubes from single tubes to 96-well plates to be used depending on the processing needs of the lab. The beads are visible on the wall of the tube as a rusty smear. Care should be taken not to disturb them. Pipetting should be performed slowly, and the pipet tip should be placed in the supernatant opposite the beads to avoid reducing recovery by accidental aspiration of bound cells. With the smaller volumes involved in using the 96-well concentrator, the beads are not be visible and careful pipetting is particularly important. Using a multichannel pipet and undamaged tips, aspirate every second row along the side with no beads attached, twist the MPC 180° and aspirate the remaining rows along the side with no beads attached. With this technique, the angle of the operator's hand does not need to change.

7. Nude paramagnetic beads are available and can be conjugated with specific antibodies, or with lectins by the user. This can be useful for capturing specific or

generic microorganisms or nucleic acids. Dynabeads DNA are also available for the nonsequence-specific capture of DNA from lysed cells.
8. It has been demonstrated that enrichment and selective plating using 100 µL but not 10 µL of broth is of similar sensitivity in recovering *E. coli* O157 as IMS methods, but is less selective.
9. A machine that automates the bead-washing process is expected to become available soon. This will allow larger numbers of samples to be processed easily in busy laboratories.
10. Considerable care is needed when pipetting to ensure that immunomagnetic beads are not aspirated accidentally (reducing recovery of the organism), or cross-contaminated.
11. A new pipet tip should be used for each tube to prevent cross-contamination.
12. The UK Health and Safety Executive requires that *E. coli* O157 should be worked with in Category 3 containment once identification is confirmed. Cultures which may contain VTEC may be worked with using good laboratory practice until identity is confirmed.
13. VT genes are lost from the bacterial genome very readily *(35)*. After storage at –70°C, isolates that were positive may be negative when retested. The most reliable results will be obtained if isolates are tested promptly after isolation, and with as little storage and subculturing as possible.
14. Control organisms and control DNAs are neccessary to assure the quality of results.
15. Colonies scraped from plates will give better results than those centrifuged from broths. Broth cultures tend not to amplify well.
16. Boiling to release DNA is quick and simple, but more highly purified DNA is likely to give better results, if time to purify it is available. Rapid DNA purification methods may improve the clarity and reproducibility of results
17. The presence of VT genes does not necessarily mean that they are expressed. The finding of VT genes is not positive confirmation that the isolate caused illness. Confirmation using a serological test for verocytotoxins is advisable.
18. Titration of Mg^{2+} will ensure that the most effective concentration is being used to maximize DNA amplification. Suboptimal concentrations may give rise to spurious bands or an absence of bands.

References

1. Ammon A. (1997) Surveillance of enterohaemorrhagic *E. coli* (EHEC) infections and haemolytic uraemic syndrome (HUS) in Europe. *Eurosurveillance* **2,** 91–96.
2. Armstrong, G. L., Hollingsworth, J., and Morris, J .G. (1986) Emerging foodborne pathogens: *Escherichia coli* O157:H7 as a model of entry of a new pathogen into the food supply of the developed world. *Epidemiol. Rev.* **18,** 29–51.
3. Anonymous (1996) Enterohaemorrhagic *Escherichia coli* infection. *Weekly Epidemiol. Record* **30,** 229–230.
4. Gutierrez, E. and Netley, G. (1997) Japanese *Escherichia coli* outbreak is still puzzling health officials. *Lancet* **348,** 540.

5. Parry, S. M., Salmon, R. L., Willshaw, G. A. and Cheasty, T. (1998) Risk factors for and prevention of sporadic infections with vero cytotoxin (shiga toxin) producing *Escherichia coli* O157. *Lancet* **351**, 1019–1022.
6. Law, D. (1997) The significance of verocytotoxin-producing *Escherichia coli* other than *E. coli* O157. *PHLS Microbiol. Dig.* **14**, 72–75.
7. Tarr, P. (1994) *Escherichia coli* O157:H7: overview of clinical and epidemiological issues. *J. Food Prot.* **57**, 632–636.
8. MacDonald, I. A. R., Gould, I. M. and Curnow, J. (1996) Epidemiology of infection due to *Escherichia coli* O157: a 3-year prospective study. *Epidemiol. Infect.* **116**, 279–284.
9. Gill, C. O., McGinnis, J. C., Rahn, K. and Houde, A. (1996) The hygienic condition of manufacturing beef destined for the manufacture of hamburger patties. *Food Microbiol.* **13**, 391–396.
10. Willshaw, G. A., Thirlwell, J., Jones, A. P., et al. (1994) Vero cytotoxin-producing *Escherichia coli* O157 in beefburgers linked to an outbreak of diarrhoea, haemorrhagic colitis and haemolytic uraemic syndrome in Britain. *Lett. Appl. Microbiol.* **19**, 304–307.
11. Besser, R. E, Lett, S. M., Weber, J. T., Doyle, M. P., Barrett, T., Wells J G et al. (1993) An outbreak of diarrhoea and haemolytic uremic syndrome from *Escherichia coli* O157 in fresh-pressed apple juice. *JAMA* **111**, 2217–2220.
12. Anonymous (1997) Outbreaks of *Escherichia coli* O157:H7 infection and cryptosporidiosis associated with drinking unpasteurized apple cider–Connecticut and New York, October 1996. *JAMA* **277**, 781–782.
13. Anonymous (1995) *Escherichia coli* O157:H7 outbreak linked to commercially distributed dry-cure salami. *M.M.W.R.* **44(9)**, 157–160.
14. Sharp, J. C. M., Reilly, W. J., Coia, J. E., Curnow, J., and Synge, B. A. (1995) *Escherichia coli* O157 infection in Scotland: an epidemiological overview, 1984-94 *PHLS Microbiol. Dig.* **12**, 134–140.
15. Duncan, S. E. and Hackney, C. R. (1994) Relevance of *Escherichia coli* O157:H7 to the dairy industry. *Dairy Food Environ. Sanit.* **14**, 656–660.
16. Morgan, D., Newman, C. P., Hutchinson, D. N., Walker, A. M., Rowe, B. and Majid, F. (1993) Verocytotoxin producing *Escherichia coli* O157 infections associated with the consumption of yoghurt. *Epidemiol. Infect.* **111**, 181–187.
17. Ackman, D., Marks, S., Mack, P., Caldwell, M., Root, T., and Birkhead G. (1997) Swimming-associated haemorrhagic colitis due to *Escherichia coli* O157:H7 infection: evidence of prolonged contamination of a fresh water lake. *Epidemiol. Infect.* **119**, 1–8.
18. Keene, W. E., McAnulty, J. M., Hoesly, F. C., Williams, L. P., Hedberg, K., Oxman G. L. et al. (1994) A swimming-associated outbreak of hemorrhagic colitis caused by *Escherichia coli* O157:H7 and *Shigella sonnei*. *New Engl. J. Med.* **331**, 579–584.
19. Anonymous (1997) Outbreaks of *Escherichia coli* O157:H7 infection associated with eating alfalfa sprouts—Michigan and Virginia, June–July 1997. *JAMA* **278**, 809–810.
20. Bolton, F. J., Crozier, L., and Williamson, J. K. (1996) Isolation of *Escherichia coli* O157 from raw meat products. *Lett. Appl. Microbiol.* **23**, 317–321.

21. Wallace, J. S., Cheasty, T. and Jones, K. (1997) Isolation of verocytotoxin-producing *Escherichia coli* O157 from wild birds. *J. Appl. Microbiol.* **82,** 399–404.
22. Wang, G., Zhao, T., and Doyle, M. P. (1996) Fate of enterohaemorrhagic *Escherichia coli* O157:H7 in bovine feces. *Appl. Environ. Microbiol.* **62,** 2567–2570.
23. Anonymous (1998) *Biomagnetic Techniques in Molecular Biology*, 3rd ed. Dynal, Oslo, Norway.
24. Safarik, I., Safarkov, M., and Forsythe, S. J. (1995) The application of magnetic separations in applied microbiology. *J. Appl. Bacteriol.* **78,** 575–585.
25. Tomoyasu, T. (1998) Improvement of the immunomagnetic separation method selective for *Escherichia coli* O157 strains. *Appl. Environ. Microbiol.* **64,** 376–382.
26. Chen, J., Johnson, R., and Griffiths, M. (1998) Detection of verotoxigenic *Escherichia coli* by magnetic capture-hybridization PCR. *Appl. Environ. Microbiol.* **64,** 147–152.
27. Heuvelink, A. E., Van de Kar, N. C. A. J., Meis, J. F. G. M., Monnens, L. A. H., and Melchers, W. J. G. (1995) Characterization of verocytotoxin-producing *Escherichia coli* O157 isolated from patients with haemolytic uremic syndrome in Western Europe. *Epidemiol. Infect.* **115,** 1–14.
28. Meng, J., Zhao, S., Doyle, M. P., Mitchell, S. E., and Kresovich, S. (1997) A multiplex PCR for identifying Shiga-like toxin-producing *Escherichia coli* O157:H7. *Lett. Appl. Microbiol.* **24,** 172–176.
29. Roberts, D., Hooper, W. L., and Greenwood, M. (1995) Section 6: Isolation and enrichment of micro-organisms, in *Practical Food Microbiology, Methods for the Examination of Food for Organisms of Public Health Significance*, PHLS, London.
30. Gannon, V., King, R., Kim, J., and Golsteyn Thomas, E. (1992) Rapid and sensitive method for detection of Shiga-like toxin-producing *Escherichia coli* in ground beef using the polymerase chain reaction. *Appl. Environ. Microbiol.* **58,** 3809–3815.
31. Beebakhee, G., Louie, M., de Azavedo, J., and Brunton, J. (1992) Cloning and nucleotide sequence of the *eae* gene homologue from enterohaemorrhagic *Escherichia coli* serotype O157:H7. *FEMS Microbiol. Lett.* **91,** 63–68.
32. Widjojoatmodjo, M., Fluit, A., and Verhoef, J. (1994) Rapid identification of bacteria by PCR-single-strand conformation polymorphism. *J. Clin. Microbiol.* **32,** 3002–3007.
33. Bolton, E.J., Crozier, L., and Williamson, J. K. (1995) New technical approaches to *Escherichia coli* O157. *PHLS Microbiol. Digest* **12,** 67–70.
34. Wilson, I. G. (1997) Inhibition and facilitation of nucleic acid amplification. *Appl. Environ. Microbiol.* **63,** 3741–3751.
35. Karch, H., Meyer, T., Rüssmann, H., and Heeseman, J. (1992) Frequent loss of shiga-like toxin genes in clinical isolates of *Escherichia coli* upon subcultivation. *Infect. Immunol.* **60,** 3464–3467.
36. Widjojoatmodjo, M. N., Fluit, A. C., Torensma, R., Verdonk, G. P. and Verhoef, J. (1992) The magnetic immuno polymerase chain reaction for direct detection of salmonella in fecal samples. *J. Clin. Microbiol.* **30,** 3195–3199.

11

Detection of *Campylobacter jejuni* and Thermophilic *Campylobacter* spp. from Foods by Polymerase Chain Reaction

Haiyan Wang, Lai-King Ng, and Jeff M. Farber

1. Introduction

Campylobacter spp. is one of the most commonly reported bacterial causes of acute diarrheal disease in humans throughout the world *(1–3)*. The thermophilic *Campylobacter jejuni, C. coli, C. lari,* and *C. upsaliensis* are the most important species, with *C. jejuni* accounting for more than 95% of all the human *Campylobacter* infections *(4,5)*. Poultry, raw milk, and water have been implicated as the major vehicles for *Campylobacter* infection *(6,7)*, although other foods may also become a source of infection through cross-contamination from other food types, a food handler, or a work surface during food preparation *(3)*. Because campylobacters have fastidious growth requirements and relatively inert biochemical characteristics, identification of these organisms and differentiation between species within the genus *Campylobacter* by cultural methods are time consuming and difficult *(8–10)*. The accuracy of some biochemical tests is also affected by bacterial inoculum size *(11)*, which can be difficult to control. Additionally, *Campylobacter* cells are usually present in very low numbers and may become injured in foods and environmental water, and therefore become nonculturable *(12–15)*. Because of the foregoing, nucleic acid-based detection methods became alternatives for the detection of campylobacters.

The polymerase chain reaction (PCR) is an *in situ* DNA replication process that allows for the exponential amplification of target DNA in the presence of synthetic oligonucleotide primers and a thermostable DNA polymerase. PCR has found its applications in a wide range of disciplines such as medicine, plant pathology, evolutionary biology, molecular biology, and clinical, food, and environmental microbiology. PCR-based methods have been developed for quick identification of purified cultures of a number of foodborne pathogens including *Salmonella* spp. *(16)*, *Listeria monocytogenes (17)*, *E. coli* O157:H7 *(18)*, and *Campylobacter* spp. (*see* **Table 1**) *(2,19–30)*. PCR-based methods have also been applied to detect bacteria directly from food or clinical samples with or without pre-enrichment *(4,31–36*, and *see* **Table 1**).

PCR-based methods can be made very specific for certain species, and may detect the target organisms in the presence of other organisms. Currently, a number of PCR protocols targeting different genes within the genus *Campylobacter* have been developed and applied for detection or identification of these organisms in pure cultures or directly from foods, water, and clinical samples *(4,36*, and *see* **Table 1**). *C. jejuni*-specific PCRs based on unknown genes *(4,21)* have been applied to the detection of this species from poultry, milk, and other foods *(2,21,33,36)*, whereas the *hip* gene-based *C. jejuni*-specific PCR was used to identify this species directly from clinical samples *(23)*. Linton and colleagues *(23)* also used different sets of primers to identify *C. coli*, or both *C. jejuni* and *C. coli* from clinical samples. A multiplex PCR assay combining one pair of *C. jejuni*-specific primers *(21,22)* with another pair of *C. jejuni* and *C. coli*-specific primers *(26)* in the same PCR reaction, allowed for the simultaneous detection and differentiation of *C. jejuni* and *C. coli (37)*. There are also PCR methods that can detect thermophilic *C. lari* and *C. upsaliensis*, in addition to *C. jejuni* and *C. coli (2,29,35,36)*.

PCR-based methods may not be suitable for processed foods, as amplification can be obtained from DNA originating from both viable and nonviable cells. This problem may be solved by applying a recently developed reverse transcriptase PCR (RT-PCR) targeting mRNA, which will detect only viable thermophilic campylobacters *(38)*.

Theoretically, PCR should be able to detect a single copy of a target gene in a PCR tube provided enough amplification cycles are used. However, pathogenic bacteria usually occur in food samples at very low concentrations, i.e., < 1 cell in 5–10 µL added in one PCR reaction, and therefore pre-enrichment of the samples is usually necessary. Additionally, because the PCR may be inhibited by components present in food samples, enrichment media, or DNA extraction solutions *(39)*, a sample preparation procedure is usually required before the application of PCR. Several researchers have attempted to detect campylobacters directly from foods, either with *(2)* or without *(21,22,27)* pre-enrichment. However, the sample preparation procedures used in these studies involved several steps, such as a two-step PCR (nested PCR) *(21,22,27)*, and achieved limited sensitivity. We have successfully applied a simple sample preparation procedure based on buoyant density centrifugation (BDC) *(31)* for

Table 1
Specificity and Primer Sequences of *Campylobacter* PCR

Specificity	Primer sequence	Target gene	Sample type	Reference
C. jejuni	VS15 (5'-GAATGAAATTTTAGAATGGGG-3') (L) VS16 (5'-GATATGTATGATTTTATCCTGC-3') (R)	Unknown	Pure culture	19,20
C. jejuni	C-1 (5'-CAAATAAAGTTAGAGGTAGAATGT) (L)[a] C-4 (5'-GGATAAGCACTAGCTAGCTGAT) (R)[b]	Unknown	Directly from raw/cooked poultry, vegetables, fruits	21,22
C. jejuni	HIP400F (5'-GAAGAGGGTTTGGGTGGTG-3') (L) HIP1134R (5'-AGCTAGCTTCGCATAATAACTTG-3') (R)	*hip*	Clinical	23
C. jejuni	(5'-GGATTTCGTATTAACACAAATGGTGC-3') (L) (5'-CTGTAGTAATCTTAAAACATTTTG-3') (R)	*flaA*	Clinical	24
C. coli	CC18F (5'-GGTATGATTTCTACAAGGGAG-3') (L) CC519R (5'-ATAAAAGACTATCGTCGCGTG-3') (R)	(putative) Aspartokinase	Clinical	23
C. coli	CSF (5'-ATATTTCCAAGCGCTACTCCCC-3') (L) CSR (5'-CAGGCAGTGTGATAGTCATGGG-3') (R)	Unknown	Pure culture	20
C. jejuni *C. coli*	CCCJ609F (5'-AATCTAATGGCTTAACCATTA-3') (L) CCCJ1442R (5'-GTAACTAGTTTAGTATTCCGG-3') (R)	16SrRNA	Clinical	23
C. jejuni *C. coli*	pg50 (5'-ATGGGATTTCGTATTAAC-3') (L) pg3 (5'-GAACTTGAACCGATTTG-3') (R)	*flaA*	Water	25,26
C. jejuni *C. coli* *C. lari*	CF03 (5'-GCTCAAAGTGGTTCTTATGCNATGG-3') (L) CF02 (5'-AAGCAAGAAGTGTTCCAAGTTT-3') (R) CF04 (5'-GCTGCGGAGTTCATTCTAAGACC-3') (R)	*flaA*, *flaB*	Directly from milk, dairy products, water	27,28
C. jejuni *C. coli* *C. lari*	C442 (5'-GGAGGATGACACTTTTCGGAGC-3') (L) C490 (5'-ATTACTGAGATGACTAGCACCCC-3') (R)	16S rRNA	Directly from enriched chicken products	2
C. jejuni/ *C. coli*/ *C. lari*/ *C. upsaliensis*	GTP 1.1 (5'-GCCAAATGTTGGiAARTC) (L)[c] GTP 2.1 (ATCAAGCCCTCCiCTRTC) (R)[d] GTP 2.2 (ATCiAGiCCTSSiCTRTC) (R)[d]	(putative) GTPase	Pure cultures	29
C. jejuni/ *C. coli*/ *C. lari*/ *C. upsaliensis*	BO4263 (5'-AGAACACGCGGACCTATATA-3') BO4264 (5'-CGATGCATCCAGGTAATGTAT-3')	(putative) Oxidase	Water, milk	30

[a] L- left primer (sense); [b] R- right primer (antisense); [c] i - inosine; [d] S, G, or C; R, G, or A.

the PCR detection of less than 0.3 CFU/mL of *C. jejuni* from naturally contaminated chicken rinses (<0.1 CFU/g of chicken) *(33)*, and of around 16 CFU/mL inoculated into 2% milk. Another simple sample preparation technique called immunomagnetic separation (IMS) was used for PCR detection of thermophilic campylobacters from spiked poultry and milk samples *(36)*. In this chapter, we describe a *C. jejuni*-specific PCR and a thermophilic campylobacter-specific PCR, combined with BDC and IMS, respectively, for sample preparation from chicken rinse and milk.

2. Materials and Equipment

2.1. Primer Sequences

2.1.1. For PCR Specific for C. jejuni *(4) (see* **Note 1***)*

CL2 (5'-TGACGCTAGTGTTGTAGGAG-3') (L)
CR3 (5'-CCATCATCGCTAAGTGCAAC-3') (R)

2.1.2. For PCR Specific for Thermophilic Campylobacter *spp. (35)*

6-1 (5'-GTCGAACGATGAAGCTTCTA-3') (L)
18-1 (5'-TTCCTTAGGTACCGTCAGAA-3') (R)

2.2. Buffers and Reagents for PCR

1. PCR buffer (10X stock): 100 mM Tris-HCl, 15 mM MgCl$_2$, 500 mM KCl, pH 8.3. Final concentration in mixture is 1X *(see* **Note 2***)*.
2. Nucleotide stock: Solution of the sodium salts of dATP, dCTP, dGTP, and dTTP, each at a concentration of 10 mM in water. Final concentration in reaction mixture is 0.2 mM.
3. Primers: Can be synthesized in house or by commercial companies (e.g., Sigma-Genosys, The Woodlands, TX) with cartridge (reverse-phase) or high performance liquid chromatography (HPLC) purification. Stock is made in water at a concentration of 0.1 mM. Final concentration in reaction mixture is 0.1–1.0 µM.
4. *Taq* DNA polymerase: 5 U/µL, available from Roche Diagnostics (Laval, Quebec, Canada) or other suppliers (e.g., Perkin Elmer).
5. PCR-grade water *(see* **Note 3***)*.
6. DNA molecular marker VIII: From Roche Diagnostics in a concentration of 0.25 µg/µL, or appropriate DNA markers for amplicon sizes around 400 bp.
7. TBE (5X stock): 0.45 M Tris-borate, 0.01 M EDTA. For 1 L of solution, add 54 g Tris base, 27.5 g boric acid, and 3.72 g EDTA into water, pH 8.0. Use 0.5X TBE to make agarose gel and run electrophoresis *(40)*.
8. Gel-loading buffer: 0.25% Bromophenol blue, 0.25% xylene cyanol FF, and 15% Ficoll in water *(40)*.
9. Ethidium bromide: Stock solution at 10 mg/mL in water is commercially available from Cedarlane Laboratories Ltd. (Hornby, Ontario, Canada): add 1 drop (2.5 µL) to

50 mL of water to make a final concentration of 0.5 µg/mL. It can also be made from powder *(40)*. Store stock solutions in a light-tight container (e.g., a bottle completely wrapped in aluminum foil). *Caution: Always wear gloves when handling ethidium bromide, as it is a powerful mutagen and a possible carcinogen.*

2.3. Reagents and Materials for Sample Preparation
2.3.1. For Buoyant Density Centrifugation (BDC)

1. Preston broth: The powder for base medium and the supplement are commercially available from Oxoid (Nepean, Ontario, Canada), and the broth can be made according to the instructions on the bottle, or from the formula provided by Bolton and Robertson *(41)* for enrichment of campylobacters from chicken rinses and milk.
2. Standard isotonic medium (SIM): 0.1% Peptone and 0.85% NaCl in Percoll® (Pharmacia Biotech, Uppsala, Sweden) *(31)*.
3. Peptone water: 0.1 g Peptone in 100 mL water.
4. 40% SIM: Dilute SIM in peptone water.

2.3.2. For Immunomagnetic Separation (IMS)

1. Exeter enrichment broth *(13)* or modified Rosef broth *(42)* for enrichment of campylobacters from milk or chicken samples (*see* **Note 4**).
2. BHI–YE broth: BHI broth containing 1% yeast extract.
3. Biomag Protein G magnetic particles (Metachem Diagnostics, Northampton, UK) (*see* **Note 5**): coated in antibody solution at either 1×10^8 or 5×10^8 particles/mL (*see* **Subheading 3.** for coating) and used at 10^6 particles/mL sample for bacterial capturing.
4. Polyclonal anti-*Campylobacter* antibody (Kirkgaard and Perry Labs, Inc., Gaithersburg, MD), diluted at 1:100 in PBS containing 0.5% Bovine Serum Albumin (BSA) for coating of magnetic particles.
5. PBST: PBS containing 0.5% Tween-20.
6. PCR-grade water.

2.4. Equipment (4,36)

1. Microcentrifuge with speed up to 20,000*g*.
2. Shaking incubator.
3. Anaerobic jars.
4. Magnetic particle concentrator (Dynal UK, Ltd).
5. Spira-mixer (Denley, UK).
6. Multimixer.
7. Block heater, to heat samples before PCR.
8. PE 9600 or PE 2400 thermal cycler (PE Biosystems, Ltd., Mississauga, Ontario, Canada), or Progene thermal cycler (Techne, Princeton, NJ) or

other equivalent models made for small and thin-wall PCR tubes (200 µL). These models allow for rapid heating, have heated lids, and do not require oil overlay in the tubes.
9. Water bath: to cool agarose solution and keep it at 60°C before casting gel.
10. Horizontal gel electrophoresis system for agarose gel electrophoresis of PCR products.
11. Gel Print 2000i (Bio/Can Scientific, Missisauga, Ontario, Canada) or other gel documentation systems with UV illumination for DNA visualization.

2.5. Disposables

1. PCR tubes: In size of 200 µL for PCR, available from DiaMed Lab Supplies Inc. (Missisauga, Ontario, Canada) or other suppliers.
2. Graduated microcentrifuge tubes, in 1.5 mL sizes for BDC, available from DiaMed or other suppliers.
3. Microcentrifuge tubes: With screw top, pretreated with PBS, pH 7.4, containing 5% BSA/PBS (pH 7.4) for IMS.

3. Methods
3.1. Master Mixture Preparation

1. Prepare a master reaction mixture by adding the following reagents to a sterile test tube as listed in **Table 2**.
2. Aliquot into 45 µL/PCR tube, and keep them at –20°C until use.

3.2. Sample Preparation

3.2.1. Cell Lysis from Pure Cultures

1. Suspend bacterial growth from agar plate into PBS buffer or water to make a cloudy solution (10^7–10^8 CFU/mL), heat at 105°C for 10 min to release DNA from bacterial cells (*see* **Notes 6** and **7**). It is ready for use after brief cooling on ice.
2. For bacterial growth in broth, heat directly as above (*see* **Notes 6** and **7**).
3. Briefly spin down (2000g) the condensed liquid on the inside wall of the tube.

3.2.2. Whole-Chicken Rinse Preparation and Enrichment for BDC (33)

1. Rinse the whole fresh chickens in 200 or 500 mL peptone water by shaking in an automated paint shaker for 1 min.
2. Add 25 mL of chicken rinse in 100 mL of Preston broth in an Erlenmeyer flask (250 mL) and place in an anaerobic jar.
3. Degas the jar, fill the jar with gas mixture (5% O_2, 10% CO_2, 85% N_2), and incubate at 37°C for 3–4 h, followed by 42°C for 16–20 h, with shaking at 60–80 rpm.

Table 2
Preparing Master Reaction Mixture

Reagent	Per reaction	100 reactions	Final concentration
dH$_2$O	37.7 µL	3.8 mL	—
10X reaction buffer	5 µL	0.5 mL	1X
dNTPs	1 µL	100 µL	0.2 mM
Primer (CL2)	0.5 µL	50 µL	1.0 µM
Primer (CR3)	0.5 µL	50 µL	1.0 µM
Taq polymerase	0.3 µL	30 µL	1.5 U/reaction

3.2.3. Milk Spiking and Enrichment for BDC

1. Make *C. jejuni* dilutions in PBS and inoculate into 25 mL of 2% milk.
2. Add the spiked milk into 100 mL of Preston broth and enrich as above (*see* **Subheading 3.2.2.**).

3.2.4. PCR Sample Preparation by BDC from Enriched Chicken Rinses and Milk (33)

1. Layer 0.9 mL of sample gently over the top of 0.6 mL of 40% SIM in a microcentrifuge tube and centrifuge at 16,000g for 1 min in a microcentrifuge.
2. Remove supernatant carefully down to 0.1 mL, and resuspend in 1.0 mL of 1X PCR reaction buffer.
3. Centrifuge the suspension at 10,000g for 5 min.
4. Remove supernatant down to a final volume of 10–20 µL and resuspend the pellet in this small volume.
5. Heat in a block heater at 105°C for 10 min and cool on ice.
6. Briefly spin down (2000g) the liquid on the inside wall of the tube.
7. Add 5 µL of cell lysate into 45 µL of PCR master mix (*see* preparation in **Subheading 3.1.**).

3.2.5. Milk Spiking and Enrichment for IMS (36)

1. Make 10-fold serial dilutions of *C. jejuni* culture (from broth or plate) in 2% milk.
2. Add 1 mL of above dilutions to 9 mL of enrichment broth.
3. Incubate microareobically at 37°C for 3–4 h, then at 42°C for 24–48 h.

3.2.6. Chicken Rinse Preparation, Spiking, and Enrichment for IMS (36)

1. Add 9 mL of BHI–YE broth per gram of chicken skin and homogenize in a stomacher for 3 min.

2. Make 10-fold serial dilutions of *C. jejuni* culture (from broth or plate) as in **Subheading 3.2.5**.
3. Add 1 mL of the foregoing dilutions to 9 mL of enrichment medium.
4. Incubate microareobically at 37°C for 3–4 h, then at 42°C for 24–48 h.

3.2.7. IMS of Campylobacter Cells from Enriched Milk and Chicken Samples (36)

1. Coat Biomag Protein G magnetic particles with anti-*Campylobacter* antibody by incubating the particles at a concentration of 1×10^8 particles per mL of antibody solution at 4°C for 3 h with gentle rotation on a Spira-mix.
2. Wash the particles twice for 5 min each time with PBST by gentle rotation as above to remove any unbound antibody. Recover the particles using the magnetic particle concentrator.
3. Add 10 μL of antibody-coated Biomag magnetic particles to 1 mL of enriched milk or chicken samples and incubate at room temperature for 1 h with gentle rotation on a multimixer (*see* **Note 8**).
4. Recover the bound bacterial cells using the magnetic particle concentrator (MPC).
5. Wash the particles once with PBST, and then once with PCR-grade water.
6. Resuspend the particles in 50 μL of PCR-grade water and lyse the bacteria cells as above (*see* **Subheadings 3.2.1.** and **3.2.4.**).
7. Add lysate directly to the PCR master reaction mix and start the PCR program.

3.3. Controls

Each PCR run should include positive and negative controls to ensure that the PCR assay is working (positive control) and that DNA contamination is not occurring (negative control). The positive control can be a pure culture of *C. jejuni* at 10^7–10^8 CFU/mL in PBS. The negative control can be a negative chicken rinse that has gone through all the enrichment, sample preparation, and other steps the same way as the real samples.

3.4. PCR Cycle Programs

3.4.1. C. jejuni Specific

Start the PCR by denaturation for 10 min at 95°C, followed by 25 cycles of denaturation at 95°C for 15 s, annealing at 48°C for 15 s, and extension at 72°C for 30 s. Allow a final 10 min at 72°C for the completion of primer extension after the last cycle. Hold the PCR products at 4°C or store at –20°C until the amplicon is ready to be analyzed by electrophoresis.

3.4.2. Thermophilic Campylobacter Specific

Start the PCR by denaturation for 2–10 min at 94°C, then followed by 35 cycles of 94°C for 1 min, 65°C for 1 min, and 75°C for 0.5 min. Hold the PCR products at 4°C or store at –20°C until ready to be analyzed by electrophoresis.

3.5. Agarose Gel Electrophoresis

1. Prepare 1.5% agarose gel by adding 0.75 g agarose (Sigma) into 50 mL of 0.5X TBE and 5 mL water (to make up the volume lost during microwaving), and microwave for 2–3 min or until the agarose particles dissolve completely. Cool the solution to 60°C in a water bath before casting the gel.
2. Mix 10 µL of PCR product with 2 µL of gel-loading buffer and load in well carefully. Add 5 µL of molecular weight marker III mixed with 1 µL gel-loading buffer to the first well.
3. Electrophorese at 10 V/cm for 1–1.5 h.
4. Stain the gel in 0.5 µg/mL of ethidium bromide for 30 min (*see* **Note 9**).
5. Destain the gel in water for 20 min to remove the ethidium bromide background in agarose.
6. View and photograph the gel under UV light.
7. Determine the amplicon sizes based on molecular size marker.

3.6. Expected Sizes of Amplicons

The *C. jejuni*-specific PCR should produce a single band of 402 bp for *C. jejuni*, whereas the PCR specific for thermophilic campylobacters will produce a single band of 409 bp for *C. jejuni, C. coli, C. lari,* and some strains of *C. upsaliensis*. The positive control should show the expected size(s), and the negative control should not have any bands.

4. Notes

1. Our primers cannot distinguish subspecies of *C. jejuni* (*4*).
2. The MgCl$_2$ concentration affects the specificity of PCR; gelatin or BSA or other chemicals may be incorporated into the PCR mix to enhance PCR sensitivity. For the thermophilic *Campylobacter*-specific PCR, the MgCl$_2$ concentration in the 10X stock is 3.5 mM, and 0.1% gelatin is included in the 10X PCR buffer.
3. The impurities in some water supplies may affect PCR performance. In our lab, we use autoclaved double-distilled water for PCR. Reverse osmosis purified water can also be used for PCR (*36*).
4. Desirable features of an enrichment broth for use in IMS include (a) high selectivity for *Campylobacter*, (b) capability of resuscitating injured *Campylobacter* cells, and (c) noninterference with bacterial binding to the beads. Because the components of Preston broth are very similar to that of Exeter broth, either one could be used for the IMS or BDC method.
5. Because the magnetic beads (especially the antibody-coated ones) are very expensive, the IMS procedure may not be cost effective (also *see* **Note 8**).
6. When performing PCR from isolated colonies, bacterial growth (not too heavy) can also be picked up with the sharp end of a toothpick and added directly into the PCR mix. Because too much DNA may deplete primers and thus affect the amplification,

overloading with cells should be avoided. Sometimes, PCR inhibitors present in foods or selective enrichment media may lead to false-negative reactions. Diluting such samples may yield positive results.
7. Mohran and coworkers *(43)* found that some of the *C. jejuni* and *C. coli* strains that they isolated in Egypt were resistant to boiling and did not release PCR-detectable DNA. If this happens in your system, alternative DNA extraction methods should be sought.
8. Sometimes, it is necessary to centrifuge the enriched samples before the addition of magnetic beads to eliminate the inhibitors to immunobinding. This again adds more steps and costs to the IMS method.
9. Ethidium bromide can be added to the agarose solution before casting, and therefore staining would not be needed after electrophoresis *(40)*.

References

1. Altekruse, S. F., Stern, N. J., Fields, P. I., and Swerdlow, D. L. (1999) *Campylobacter jejuni*—an emerging foodborne pathogen. *Emerg. Infect. Dis.* **5 (1),** 28–35.
2. Giesendorf, B. A. J., Quint, W. G. V., Henkens, M. H. C., et al. (1992) Rapid and sensitive detection of *Campylobacter* spp. in chicken products by using the polymerase chain reaction. *Appl. Environ. Microbiol.* **58,** 3804–3808.
3. Roels, T. H., Wickus, B., Bostrom, H. H., et al. (1998) A foodborne outbreak of *Campylobacter jejuni* (O:33) infection associated with tuna salad: a rare strain in an unusual vehicle. *Epidemiol. Infect.* **121,** 281–287.
4. Ng, L.-K., Kingombe, C. I. B., Yan, W., et al. (1997) Specific detection and confirmation of *Campylobacter jejuni* by DNA hybridization and PCR. *Appl. Environ. Microbiol.* **63,** 4558–4563.
5. Van Doorn, L.-J., Verschuuren-van Haperen, A., van Belkum, A., et al. (1998) Rapid identification of diverse *Campylobacter lari* strains isolated from mussels and oysters using a reverse hybridization line probe assay. *J. Appl. Microbiol.* **84,** 545–550.
6. Tauxe, R. V. (1992) Epidemiology of *Campylobacter jejuni* infections in the United States and other industrialized nations, in *Campylobacter jejuni: Current Status and Future Trends* (Nachamkin, I., Blaser, M.J., and Tompkins, L.S., eds.), American Society for Microbiology, Washington, DC, pp. 9–19.
7. Tauxe, R. V., Hargrett-Bean, N., Patton, C. M., and Wachsmuth, I. K. (1988) *Campylobacter* isolates in The United States, 1982–1986. *M.M.W.R.*, CDC Surveillance, **37 (No. SS-2),** 1–13.
8. On, S. L. W. and Holmes, B. (1992) Assessment of enzyme detection tests useful in identification of campylobacteria. *J. Clin. Microbiol.* **30,** 746–749.
9. Penner, J. L. (1988) The genus *Campylobacter*: a decade of progress. *Clin. Microbiol. Rev.* **1,** 157–172.
10. Sanders, G. (1998) Isolation of *Campylobacter* from food, in *Compendium of Analytical Methods* (Warburton, D., ed.), vol 3. HPB laboratory procedure MFLP-46. Polyscience Publications, Laval, Québec, Canada.
11. On, S. L. W. and Holmes, B. (1991) Effect of inoculum size on the phenotypic characterization of *Campylobacter* species. *J. Clin. Microbiol.* **29,** 923–926.

12. Beumer, R. R., de Vries, J., and Rombouts, F. M. (1992) *Campylobacter jejuni* non-culturable coccoid cells. *Int. J. Food Microbiol.* **15,** 153–163.
13. Humphrey, T.J. (1986) Techniques for the optimum recovery of cold injured *Campylobacter jejuni* from milk or water. *J. Appl. Bacteriol.* **61,** 125–132.
14. Medema, G. J., Schets, F. M., van de Giessen, A. W., and Havelaar, A.H. (1992) Lack of colonization of 1 day old chicks by viable, non-culturable *Campylobacter jejuni. J. Appl. Bact.* **72,** 512-516.
15. Rollins, D. M. and Colwell, R. R. (1986) Viable but nonculturable stage of *Campylobcter jejuni* and its role in survival in the natural aquatic environment. *Appl. Environ. Microbiol.* **52,** 531–538.
16. Rahn, K., De Grandis, S. A., Clarke, R. C., et al. (1992) Amplification of an *invA* gene sequence of *Salmonella typhimurium* by polymerase chain reaction as a specific method of detection of *Salmonella. Mol. Cell. Probes* **6,** 271–279.
17. Border, P. M., Howard, J. J., Plastow, G. S., and Siggens, K. W. (1990) Detection of *Listeria* species and *Listeria monocytogenes* using polymerase chain reaction. *Lett. Appl. Microbiol.* **11,** 158–162.
18. Pollard, D. R., Johnson, M. W., Lior, H., et al. (1990) Differentiation of Shiga toxin and verocytotoxin type I genes by the polymerase chain reation. *J. Infect. Dis.* **162,** 1195–1198.
19. Stonnet, V. and Guesdon, J.-L. (1993) *Campylobacter jejuni*: specific oligonucleotides and DNA probes for use in polymerase chain reaction-based diagnosis. *FEMS Immunol. Med. Microbiol.* **7,** 337–344.
20. Stonnet, V., Sicinschi, L., Mégraud, F., and Guesdon, J. L. (1995) Rapid detection of *Campylobacter jejuni* and *Campylobacter coli* isolated from clinical specimens using the polymerase chain reaction. *Eur. J. Clin. Microbiol. Infect. Dis.* **14,** 355–359.
21. Winters, D. K. and Slavik, M. F. (1995) Evaluation of a PCR based assay for specific detection of *Campylobacter jejuni* in chicken washes. *Mol. Cell. Probes* **9,** 307–310.
22. Winters, D. K., O'Leary, A. E., and Slavik, M. F. (1998) Polymerase chain reaction for rapid detection of *Campylobacter jejuni* in artificially contaminated foods. *Lett. Appl. Microbiol.* **27,** 163–167.
23. Linton, D., Lawson, A. J., Owen, R. J., and Stanley J. (1997) PCR detection, identification to species level, and fingerprinting of *Campylobacter jejuni* and *Campylobacter coli* direct from diarrheic samples. *J. Clin. Microbiol.* **35,** 2568–2572.
24. Nachamkin, I., Bohachick, K., and Patton, C. M. (1993) Flagellin gene typing of *Campylobacter jejuni* by restriction fragment length polymorphism analysis. *J. Clin. Microbiol.* **31,** 1531–1536.
25. Oyofo, B. A. and Rollins, D. M. (1993) Efficacy of filter types for detecting *Campylobacter jejuni* and *Campylobacter coli* in environmental water samples by polymerase chain reaction. *Appl. Environ. Microbiol.* **59,** 4090–4095.
26. Oyofo, B. A., Thornton, S. A., Burr, D. H., et al. (1992) Specific detection of *Campylobacter jejuni* and *Campylobacter coli* by using polymerase chain reaction. *J. Clin. Microbiol.* **30,** 2613–2619.
27. Wegmüller, B., Lüthy, J., and Candrian, U. (1993) Direct polymerase chain reaction of *Campylobacter jejuni* and *Campylobacter coli* in raw milk and dairy products. *Appl. Environ. Microbiol.* **59,** 2161–2165.

28. Kirk, R. and Rowe, M. T. (1994) A PCR assay for the detection of *Campylobacter jejuni* and *Campylobacter coli* in water. *Lett. Appl. Microbiol.* **19,** 301–303.
29. Van Doorn, L.-J., Giesendorf, B. A. J., Bax, R., et al. (1997) Molecular discrimination between *Campylobacter jejuni, Campylobacter coli, Campylobacter lari* and *Campylobacter upsaliensis* by polymerase chain reaction based on a novel putative GTPase gene. *Mol. Cell. Probes* **11,** 177–185.
30. Jackson, C. J., Fox, A. J., and Jones, D. M. (1996) A novel polymerase chain reaction assay for the detection and speciation of thermophilic *Campylobacter* spp. *J. Appl. Bacteriol.* **81,** 467–473.
31. Lindqvist, R. (1997) Preparation of PCR samples from food by a rapid and simple centrifugation technique evaluated by detection of *Escherichia coli* O157:H7. *Int. J. Food Microbiol.* **37,** 73–82.
32. Wang, H., Blais, B. W., and Yamazaki, H. (1995) Rapid confirmation of polymyxin-cloth enzyme immunoassay for group D salmonellae including *Salmonella enteritidis* in eggs by polymerase chain reaction. *Food Control* **6,** 205–209.
33. Wang, H., Farber, J. M., Malik, N., and Sanders, G. (1999) Improved PCR detection of *Campylobacter jejuni* from chicken rinses by a simple sample preparation procedure. *Int. J. Food Microbiol.* **52,** 39–45.
34. Wernars, K., Heuvelman, C. J., Chakraborty, T., and Notermans, S. H. W. (1991) Use of the polymerase chain reaction for direct detection of *Listeria monocytogenes* in soft cheese. *J. Appl. Bacteriol.* **70,** 121–126.
35. Van Camp, G., Fierens, H., Vandamme, P., et al. (1993) Identification of enteropathogenic *Campylobacter* species by oligonucleotide probes and polymerase chain reaction based on 16s rRNA genes. *Syst. Appl. Microbiol.* **16,** 30–36.
36. Docherty, L., Adams, M. R., Patel, P., and McFadden, J. (1996) The magnetic immuno-polymerase chain reaction assay for the detection of *Campylobacter* in milk and poultry. *Lett. Appl. Microbiol.* **22,** 288–292.
37. Harmon, K. M., Ransom, G. M., and Wesley, I. V. (1997) Differentiation of *Campylobacter jejuni* and *Campylobacter coli* by polymerase chain reaction. *Mol. Cell. Probes* **11,** 195–200.
38. Sails, A. D., Bolton, F. J., Fox, A. J., et al. (1998) A reverse transcriptase polymerase chain reaction assay for the detection of thermophilic *Campylobacter* spp. *Mol. Cell. Probes* **12,** 317–322.
39. Rossen, L., Norskov, P., Holmstrom, K. and Rasmussen, O. F. (1992) Inhibition of PCR by components of food samples, microbial diagnostic assays and DNA-extraction solutions. *Int. J. Food Microbiol.* **17,** 37–45.
40. Sambrook, J., Fritsch, E. F., and Maniatis, T. (1989) *Molecular Cloning, A Laboratory Manual,* 2nd ed. Cold Spring Harbor Laboratory Press, Cold Spring Harbor, NY, pp. 6.15.
41. Bolton, F. J. and Robertson, L. (1982) A selective medium for isolating *Campylobacter jejuni/coli. J. Clin. Pathol.* **35,** 462–467.
42. Lammerding, A. M., Garcia, M. M., Mann, E. D., et al. (1988) Prevalence of *Salmonella* and thermophilic *Campylobacter* in fresh pork, beef, veal and poultry in Canada. *J. Food Proteins* **51,** 47–52.
43. Mohran, Z. S., Arthur, R. R., Oyofo, B. A., et al. (1998) Differentiation of *Campylobacter* isolates on the basis of sensitivity to boiling in water as measured by PCR-detectable DNA. *Appl. Environ. Microbiol.* **64,** 363–365.

12

Magnetic Capture Hybridization Polymerase Chain Reaction

Jinru Chen and Mansel W. Griffiths

1. Introduction

Polymerase chain reaction (PCR) is a novel DNA amplification technique that has brought fundamental change to clinical diagnosis and rapid detection of food-borne pathogens. The technique needs little technical time and has a quick turnover. The results of PCR are accurate, sensitive, and specific. However, the application of PCR for the detection of pathogens directly from food is limited because natural food components often affect the activity of *Taq* DNA polymerase, the enzyme catalyzing DNA amplification.

Magnetic capture hybridization polymerase chain reaction (MCH-PCR) uses a capture probe to separate DNA template from food by DNA hybridization and biotin-streptavidin based magnetic separation. Captured-template DNA is used subsequently in PCR amplification. MCH-PCR was initially used to overcome the inhibitory effect of humic acid present in soil samples during PCR amplification *(1)*. Chen and colleagues *(2)* adapted the technique for the detection of verotoxigenic *Escherichia coli* (VTEC) from ground beef. The technique has also been applied to the detection of *Bacillus cereus* and *Bacillus thuringiensis* in soil samples (Damgaard and co-workers *[3]*); *Mycobacterium paratuberculosis* and *Mycobacterium avium* subsp. *silvaticum* in tissue and fecal samples *(4)*; and *Staphylococcus* spp. from clinical samples *(5)*.

In MCH-PCR, bacterial cells are lysed by heat treatment. Template DNA present in the supernatant of lysed bacterial suspension is separated from cell debris by centrifugation, and captured by a DNA probe that is internal to the PCR product to be amplified and has a biotin label on the 5' end. After hybridization, the DNA hybrid of capture probe and template is isolated through chemical binding between biotin on the hybrid and streptavidin on paramagnetic particles. The hybrid on paramagnetic particles can be used directly in subsequent PCR amplification.

2. Materials and Equipment
2.1. Equipment

1. Hot plate.
2. Bench-top centrifuge.
3. Hybridization oven.
4. Magnetic separator.
5. Rotator.
6. DNA thermal cycler.
7. DNA gel electrophoresis apparatus.
8. Gel document system.

2.2. Reagents and Buffers

1. Sterile dH_2O.
2. Streptavidin-coated paramagnetic particles (*see* **Note 1**).
   10 mg/mL
3. Binding buffer.
   a. 10 mM Tris-HCl
   b. 1 mM EDTA
   c. 100 mM NaCl
   d. pH 8.0
4. Hybridization buffer.
   a. 50% (vol/vol) formamide
   b. 5% standard saline citrate (SSC)
   c. 2% (w/v) blocking agent
   d. 0.1% (w/v) N-lauroylsarcosine
   e. 0.02% (w/v) sodium dodecyl sulfate (SDS)
5. PCR reagents.
   a. PCR reaction buffer (5%)
   b. dNTP (10 mM)
   c. *Taq* DNA polymerase (1 U/µΛ)
   d. Oligonucletide primers (0.1 µg/µΛ)
6. TBE buffer.
   a. 0.089 M Tris-base
   b. 0.089 M boric acid
   c. 0.002M EDTA
   d pH 8.0
7. Ethidium bromide.
   10 mg/mL

Magnetic Capture Hybridization

3. Procedures

3.1. Synthesis of Capture Probe

A capture probe should be designed based on the internal sequence of MCH-PCR product to be amplified. It can be synthesized in conventional PCR by the use of biotin-labeled oligonucleotide primers. Consequently, biotin labels are located on the 5' end after the probe is made. If a shorter probe is desired, a synthesized oligonuceotide with a biotin label can be used directly as a capture probe. Synthesized probe can be stored at refrigeration temperature for future use.

3.2. Preparation of DNA Template

Tested bacteria are grown in brain–heart infusion (BHI) agar at 37°C. Overnight cultures are transferred into BHI broth and incubated at 37°C for about 6 h. If food samples are tested, enrichment is sometimes required depending on initial load of tested pathogen in a sample. One milliliter of enriched broth or liquid bacterial culture is centrifuged at 12,000g for 2 min using a benchtop centrifuge. Bacterial pellets are washed twice with 1 mL of sterile dH_2O and resuspended subsequently in 100 µL of dH_2O. The bacterial suspensions are heated in a water bath for 10 min. After heat treatment, the cultures are placed on ice immediately to prevent DNA from annealing. Heat-treated cells are centrifuged at 12,000g for 2 min. Supernatants were taken and used in the hybridization assay.

3.3. Hybridization

Biotin-labeled DNA probe is heated at 100°C for 10 min and cooled down rapidly on ice (*see* **Note 1**). Denatured probe is mixed with single-stranded DNA template and hybridization buffer (300 µL). Hybridization is carried out at 42°C for 4–16 h with rotating in a hybridization oven.

3.4. Magnetic Capture

Streptavidin-coated paramagnetic particles (30 µg per sample) are taken from storage and washed three times with, and subsequently resuspended in, binding buffer (*see* **Note 2**). Washed paramagnetic particles are then added to hybridization mix and incubated at room temperature for about 1 h on a rotator. The particles are collected at the end of the incubation and washed three times with sterile H_2O and used directly in PCR amplification. A magnetic separator is used to collect paramagnetic particles between washing.

3.5. PCR Amplification

Fifty microliters of PCR mix contains the following.

1. X µL of template DNA on paramagnetic particles.
2. 5 µL of buffer (10%).
3. 5 µL of dNTP (10 mM).

4. 1 µL of each of the two primers (0.1 µg/µL).
5. 0.4 µL of *Taq* DNA polymerase (1 U/µL).

The total volume of the mix is adjusted to 50 µL using dH_2O.

The amplification conditions include one cycle at 94°C for 5 min, 58°C for 1 min and 72°C for 6 min, followed by 30–50 cycles of 94°C for 2 min, 60°C for 1 min, and 72°C for 1 min. The amplification is followed by a holding period at 72°C for 10 min.

Amplified products are analyzed using DNA electrophoresis. Alternatively, the method can be automated by using an alkaline phosphatase-labeled DNA probe and detecting hybridization by flow injection with a chemiluminescent substrate *(2)*. In this way the detection cycle was less than 30 min and the sensitivity was at the femtomole level.

4. Notes

1. Depending on individual pathogen, the ratio of DNA template and capture probe in hybridization needs to be adjusted. Magnetic particles collected from hybridization mix should be washed properly with dH_2O to remove the residue of hybridization buffer.
2. Commercial paramagnetic particles contain preservatives. Therefore, proper washing is needed before use. The amount of paramagnetic particles used in PCR amplification is critical. An excessive amount of particles present in PCR mix during MCH-PCR could lead to the failure of DNA amplification.

References

1. Jacobsen, C. S. (1995) Microscale detection of specific bacterial DNA in soil with magnetic capture—hybridization and PCR amplification assay. *Appl. Environ. Microbiol.* **61,** 3347–3352.
2. Chen, J., Johnson, R., and Griffiths, M. W. (1998) Detection of verotoxigenc *Escherichia coli* by magnetic capture hybridization PCR. *Appl. Environ. Microbiol.* **1,** 147–152.
3. Damgaard, P. H., Jacobsen, C. S., and Sorenson, J. (1996) Development and application of a primer set for specific detection of *Bacillus thuringiensis* and *Bacillus cereus* in soil using magnetic capture hybridization and PCR amplification. *Syst. Appl. Microbiol.* **19,** 436–441.
4. Millar, D. S., Withey, S. J., Tizad, M. L. V., et al. (1995) Solid-phase hybridization capture of low-abundance target DNA sequences: Application to the polymerase chain reaction detection of *Mycobacterium paratuberculosis* and *Mycobacterium avium* subsp. *silvaticum. Anal. Biochem.* **226,** 325–330.
5. Kolbert, C. P., Connolly, J. E., Lee, M. J., and Persing, D. H. (1995) Detection of the Staphylococcal *mecA* gene by chemiluminescent DNA hybridization. *J. Clin. Microbiol.* **33,** 2179–2182.
6. Chen, X., Zhang, X. E., Chai,Y. Q., et al. (1998) DNA optical sensor: A rapid method for the detection of DNA hybridization. *Biosens. Bioelect.* **13,** 451–458.

13

Enterococci

Anavella Gaitán Herrera

1. Introduction

Lancefield's group D is composed of *Streptococcus equinus, S. faecalis, S. bovis,* and *S. faecium. S. avium* (Q and D antigens). Group D is found in the intestinal tract of warm-blooded animals. The term "enterococcus" refers only to the species *S. faecalis* and *S. faecium*. The medium employed for enumeration of coliforms in foods is KF agar.

2. Materials
2.1. Presumptive Test

1. Glass Petri dishes.
2. 1 mL Bacteriological pipets.
3. Incubator at 35–37°C.
4. KF streptococcus agar.
5. 3% Hydrogen peroxide.

2.2. Confirmation Test

1. Brain–heart infusion broth and BHI with added 6.0% sodium chloride.
2. Hydrogen peroxide, 3% aqueous solution.

3. Methods
3.1. Presumptive Test

1. Prepare samples by methods described in previous chapters.
2. To Petri dishes pipet 1 mL of each dilution of the food.

3. Add to each dish 15 mL of KF agar tempered at 45°C.
4. Mix the dishes by rotating and incubate at 35–37°C for 48–72 h.
5. Count all colonies on the KF agar deep red and light pink (*S.faecalis-S. faecium*) on the plates containing 30–300 colonies and multiply by the dilution factor.
6. Compute the number of presumptive enterococci per gram of food.

3.2. Confirmation Test

1. Select 5–10 light pink colonies from KF plates, pick individually into tubes of BHI, incubate at 35–37°C for 18–24 h, until turbidity appears.
2. Observe for typical Gram-positive oval cocci in short chains or pairs.
3. Mix 3 mL of each culture in another tube with about 0.5 mL of 3% hydrogen peroxide. Failure of bubbles to appear indicates mixture is catalase negative. Confirms that the culture is a *Streptococcus*.
4. Inoculate one tube each of BHI with the confirmed *Streptococcus* isolates. Incubate at 44–46°C for 48 h. Look for growth.
5. Inoculate one tube each of BHI containing 6.5% NaCl with the confirmed *Streptococcus* isolates, incubate at 35–37°C for 72 h and look for growth.
6. Catalase negative *Streptococcus* that grows at 44–46°C and in the BHI plus 6.5% NaCl confirms that the culture is an enterococcus.
7. To identify the species, the cultures belonging to serological group D can be obtained by testing with specific serum.

References

1. Speck, M. L. (1984) *Compendium of Methods for the Microbiological Examination of Foods*, 2nd ed. Compiled by the APHA Technical Commitee on Microbiological Methods for Foods, American Public Health Association, Washington, DC.
2. Elliot, R. P., Clark, D. S., et al. (1978) *Microorganisms in Foods 1. Significance and Methods of Enumeration*. 2nd ed. A publication of the International Comission on Microbiological Specifications for Foods (ICMSF) of the International Association of Microbiological Societies, University of Toronto Press, Canada.

14

Salmonella

Anavella Gaitán Herrera

1. Introduction

Salmonellosis is one of the most common, if not the most common infectious diseases transmitted by contaminated poultry foods *(1,2)*. A critical goal in food processing plants and governmental control agencies is to prevent *Salmonella* contamination of food products and this prevention depends to a great exent on an adequate quality control program. *Salmonella* detection is still highly dependent on employing appropriate culture media. The cells may be stressed during processing of the food. Standard bodies and media they specify for *Salmonella* detection are named in **Table 1** *(3,4)*.

2. Methods

2.1. Resuscitation and Preenrichment

Salmonella occurring in dried, processed foods are usually present in low numbers and in injured cells are frequently accompanied by competing organisms that are present in numbers thousands of times greater. Satisfactory resuscitation and preenrichment generally requires a nutritious nonselective medium. Buffered peptone water and lactose broth are commonly used, but other nutrient media such as tryptone soya and nutrient broths may also be used *(5)*. The isolation procedure should provide the following: nutrients for multiplication to favor the ratio of *Salmonella* to non-*Salmonella* microoranisms, repair of cell damage, rehydratation, and dilution of toxic or inhibitory substances *(2,4)*.

2.2. Selective Enrichment

Selective enrichment broths are employed for the purpose of increasing the *Salmonella* population while at the same time inhibiting multiplication of other

Table 1
Regulatory Agencies That Specify Detection Procedures for *Salmonella* and the Culture Media to Be Used

Agency	Culture media		
	Pre-enrichment	Enrichment	Plating
ISO	Buffered peptone water	Rappaport–Vassiladis (RV) broth, Selenite cysteine broth	Brillant green agar
APHA	Lactose broth	Selenite cysteine broth, tetrathionate broth (USP)	SS agar, bismuth sulphite agar, Hektoen agar
AOAC/FDA	Lactose broth, tryptone soya broth, nutrient broth	Selenite cysteine broth, tetrathionate broth (USP)	Brillant green agar, Hektoen agar, XLD agar, bismuth sulphite agar
IDF	Buffered peptone water, distilled water plus, brillant green 0.002%	Muller–Kauffman tetrathionate broth, Selenite cysteine broth	Brillant green agar Bismuth sulphite agar
BSI	Buffered peptone water	Rappaport–Vassiliadis (RV) broth, selenite cysteine broth	Brillant green agar, any other solid selective medium

organisms in the food sample. A variety of inhibitors is in use, the most widely used of which are bile, tetrathionate, sodium selenite, and either brillant green or malachite green dyes. Additionally, media may be supplemented with antibiotics, commonly novobiocin. Activity of inhibitory agents may be futher enhanced by incubation of the enrichment culture at higher temperatures, usually between 41 and 43°C.

2.3. Selective Plating Media

These are formulated so that *Salmonella* may appear in the form of discrete colonies, whereas the growth of competing non-*Salmonella* microorganisms is suppressed. Non-*Salmonella* colonies are generaly distinguished by their ability to produce hydrogen sulfide, and to utilize one or more carbohydrates incorporated in the media. Selective media that have been used for the isolation of *Salmonella* include brillant green (BG), bismuth sulfite (BS), *Salmonella shigella*, MacConkey's, desoxycholate citrate, Hektoen enteric (HE), xylose lysine desoxycholate (XLD), and xylose lysine brillant green.

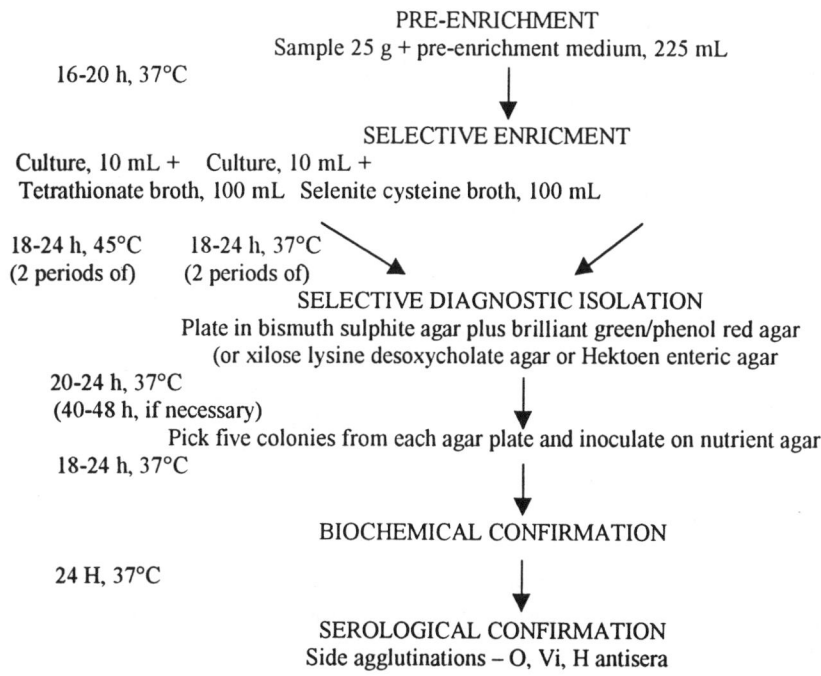

Fig. 1. Isolation procedure.

2.4. Differential Media

Differential media are ussually tubed. The more commonly used differential agar are triple sugar iron (TSI) agar (production of hydrogen sulfide and utilization of glucose, lactose, and sucrose) and lysine iron (LI) agar (production of hydrogen sulfide and decarboxlilation of lysine).

2.5. Confirmatory Serological Tests

The genus *Salmonella* is characterized serologically by especific antigenic components. The antigens are divided into somatic (O), flagelLar (H), and capsular (K). A tube of the antigen antiserum mixture is incubated for 1 h in a 50°C water bath and the culture usually is tested with a polyvalent flagellar (H) antiserum.

Isolation procedures as typified by BSI/ISO and FDA/AOAC BAM are shown in **Figs. 1** and **2**.

3. Notes

1. Selenite broth base (lactose): discard the prepared medium if large amounts of reduced selenite can be seen as a red precipitate in the bottom of the bottles.

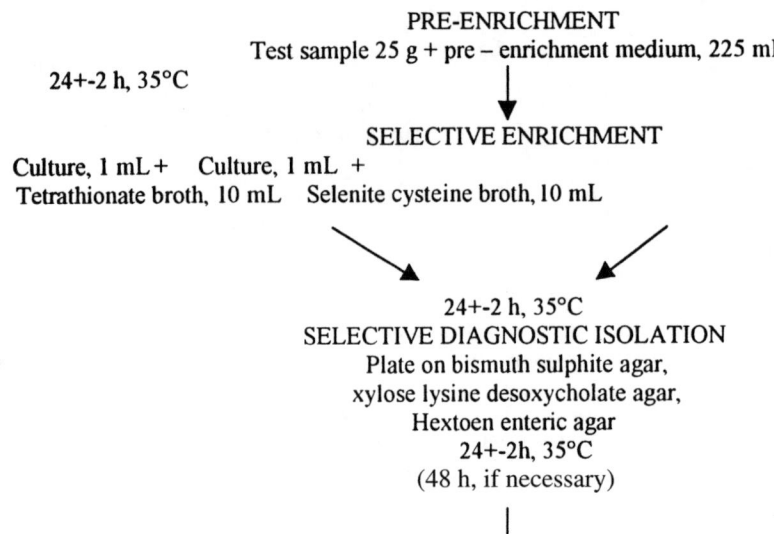

Fig. 2. Isolation procedure.

2. Do not incubate longer than 24 h because the inhibitory effect of selenite is reduced after 6–12 h incubation. Mannitol fermentation by *Salmonella* helps correct the alkaline pH swing, which can occur during incubation.
3. Take subcultures from the upper third of the broth column, which should be at least 5 cm deep.
4. Occasionally *Salmonella* cultures showing atypical biochemical results (H_2S negative, lactose positive, or dulcitol-negative reactions) may be isolated. It should be realized that the classification of an isolate as *Salmonella* depends ultimately on the antigenic structure of the organism, and not unqualifiedly on its biochemical characteristics.
5. Some foods contain microbial inhibitors that may affect the efficiency of the analytical method.
6. Culture media and reagents, including antisera, should be subjected to quality control procedures.

References

1. Post, D. E. (1997) Food-borne pathogens, in *Monograph Number I: Salmonella.* Oxoid Setting Standards.
2. Unipath Ltd. (1990) *The Oxoid Manual,* 6th ed., Basingstoke, UK.
3. American Public Health Association (1976) *Compendium of Methods for the Microbiological Examination of Foods,* APHA Inc., Washingon, DC.
4. Association of Official Analytical Chemists (1989) *F.D.A. Bacteriological Analytical Manual,* 6th ed. AOAC, Arlington, VA.
5. American Public Health Association (1980) *Standard Methods for the Examination of Water and Wastewater,* 15th ed. APHA Inc., Washington, DC.

15

Campylobacter

Anavella Gaitán Herrera

1. Introduction

The association of infection with consumption of contaminated water and foods, particularly poultry, is clearly established *(1)*. *Campylobacter* food poisoning outbreaks occur either sporadically, affecting individuals and small groups suchs as families, or larger community outbreaks. *Campylobacter jejuni* commonly occurs in undercooked chicken *(2)*. Cross-contamination from raw poultry to foods that are not cooked before eating is also a cause. *Campylobacter* spp. are readily destroyed by temperatures used in Pasteurization and cooking *(3,4)*. They may survive for several weeks in a moist environment but quickly die in dry conditions, particularly at room temperature. Acidic conditions rapidly destroy them and they show no unusual resistance to disinfectants. **Table 1** presents characteristicsof catalase-positive *Campylobacters*.

2. Materials

1. *Campylobacter* enrichment broths: Hunt and Radle enrichment broth (BAM).
2. *Campylobacter* plating media: *Campylobacter* agar (CAT) and Campy-Cefex agar.
3. Antibiotic supplement Amphotericin B 10 mg/L, cefoperazone 8 mg/L, teicoplanin 4 mg/L, vancomycin 10mg/L, trimethopim lactate 12.5 mg/L, sodium cefoperazone 33 mg/L (15 mg/L) and cycloheximide 200 mg/L.
4. *Campylobacter* growth supplement (FBP) sodium pyruvate 0.125 g, sodium metabisulfite 0.125 g, ferrous sulfate (hydrated salt) 0.125 g. Aseptically add 2 mL of sterile distilled water and add to 500 mL of culture medium.
5. Jar with *Campylobacter* gas-generating kits.
6. CampyGen creates a suitable atmosphere for growth of *Campylobacter* spp. by decreasing the carbon dioxide content in the air.

Table 1
Characteristics of Catalase-Positive *Campylobacters*

		Catalase-positive *Campylobacter*			
		C. jejuni	*C. coli*	*C. fetus*	*C. fetus* ss. *venerealis*
	Oxidase	+	+	+	+
	Ferment sugars	–	–	–	–
	NO$_3$ reduction	+	+	+	+
	NO$_2$ reduction	–	–	–	–
	H$_2$S (SIM)	–	–	–	–
	H$_2$S (Strip)	+	+	+	(V)
	Hippurate hydrolysis	+	–	–	–
	1% glycine	+	+	+	–
T	3.5% NaCl	–	–	–	–
O	25°C	–	–	+	+
L	30.5°C	–(*Vr*)	+	+	+
E	37°C	+	+	+	+
R	42°C	+	+	–(V)	–(V)
A	Aerobic (plate)	–	–	–	–
N	5% O$_2$ (plate)	+	+	+	+
C	Nalidixic acid[a]	S	S	R	R
E	Cephalotin[b]	R	R	S	S
	Falta TTC[c]	S(V)	R(V)	S	S

Vr = results are variable; + = positive; – = negative; S = sensible; R = resistant.
[a]High concentration nalidixic acid sensitivity disc (30 µg). Any zone of inhibition on agar plates is regarded as a positive or a sensitive strain.
[b]High concentration cephalothin sensitivity disc (30 µg).
[c]TTC = 2,3,5-triphenyltetrazolium chloride. More than 6 mm zone of inhibition is regarded as a positive or sensitive strain.

3. Methods

Cells of *C. jejuni* are sensitive to air and they usually survive for only 1–2 d in solids, 2–4 d in liquids, and 10–20 d in semisolid media at room temperature. Their survival twofold when held under refrigeration al 4°C (*see* **Note 1**).

3.1. Campylobacter *Enrichment Broths*

A large size is desirable for isolation, to obtain a cell concentrate from relatively large (1–2 kg) samples of poultry products, a rinse technique may be used. However, the taking of swabs is recommended for sampling carcases of large animals. When a large sample is not available, pieces of food (10–25 g) can be added directly to an enrichment broth without the preparation of a cell concentrate. In this case, the sample may be blended at low speed (*see* **Note 2**).

1. Surface rinse technique: Rinse the surface of the sample (1–2 kg) by shaking or massaging it with 250 mL of nutrient broth (without agar) in a sterile plastic bag. Filter the washing through two layers of cheesecloth. Centrifuge the filtrate at 16,000g for 20 min at 4°C. Discard the supernatant fluid and suspend the pellet in a minimum (2–5 mL) volume of an enrichment broth.
2. Swab-sampling technique: Dip a sterile swab into an enrichment broth and press the swab against the container wall to remove excess moisture. When swab samples are to be transported, use a transport medium. Wipe the surface of the sample (25–100 cm^2) with the moist swab.

Hunt and Radle Enrichment Broth (BAM/FDA)

Base	g/L
Nutrient broth	10.0
Yeast extract	6.0
Water distilled	950 mL
Supplement A	
Ferrous sulfate	0.25 g
Sodium metabisulfite	0.25 g
Sodium pyruvate	0.25 g
Lysed horse blood	50.0 mL
Supplement B	
Vancomycin	10.0 mg
Trimethoprim lactate	12.5 mg
Sodium cefoperazone	15.0 mg
Amphotericin B	2.0 mg
Supplement C	
Sodium cefoperazone	15.0 mg

3.2. Campylobacter *Plating Media Enrichment*

1. Transfer the cell concentrate or food sample into the enrichment broth.
2. Place a weighted ring on the neck of the flask, and insert a sterile Teflon tube, 0.317 cm or 1/8 in. in diameter, from the gas cylinder between the cotton plug and the neck of the flasks so that the tube is below the level of he broth. Allow the gas mixture to flow at a rate of 5–10 mL/ min.
3. Wrap the cotton plug of the flask with two layers of parafilm and let stand at room temperature for 30 min.
4. Incubate the flasks in a water bath, preset at 42°C for 48 h under a constant flow of the gas mixture (5% O_2, 10% CO_2, 85% N_2).
5. Filter the enrichment culture (5–7 mL) through a 4.7-cm-diameter membrane filter (0.65 µm pore size) in a sterile Millipore filter unit connected to a vacuum pump, or filter through a Swinny adapter fitted to a 10-mL syringe. Use a low vacuum, 12.7 cm or 5 in. of Mercury.

Humphrey developed Exeter agar made by adding agar to Exeter broth. *C. jejuni* strains produce gray/moist flat spreading colonies. Some strains may have a green hue or a dry appearance with or without a metallic sheen. *C. coli* strains tend to be creamy-gray color, moist, slightly raised, and often producing discrete colonies.

Prepare *Campylobacter* blood-free agar base, sterilize by autoclaving at 121°C for 15 min. Cool to 50°C. Aseptically add one vial of CAT supplement reconstituted with 4 mL of sterile distilled water. Mix well and pour into Petri dishes. Store plates in the dark and preferably in wrapped or sealed containers.

3.2.1. Campylobacter Agar (CAT)

Basal medium blood-free *Campylobacter* agar base:	mg/L
Nutrient broth	25.0
Bacteriological charcoal	4.0
Casein hydrolysate	3.0
Sodium deoxycholate	1.0
Ferrous sulphate	0.25
Sodium pyruvate	0.25
Agar	12.0
pH 7.4 +− 0.2	

3.2.2. Campylobacter Selective Supplement (CAT)

Base	mg/L
Cefoperazone	8.0
Teicoplanin	4.0
Amphotericin B	10.0

3.2.3. Campy-Cefex Agar

Campy-Cefex agar was formulated by Stern, Wojton and Kwiatek as a selective-differential medium for the isolation of *C. jejuni* from chicken carcasses. Campy-Cefex was found to be as productive and selective as the other media. The high concentration of cycloheximide in Campy-Cefex medium enables it more effectively inhibit the growth of molds and yeast, which is frequently associated with poultry samples (*see* **Note 3**).

Base	g/L
Brucella agar	44.0
Ferrous sulfate	0.5
Sodium bisulfite	0.2
Sodium pyruvate	0.5

Distilled water   950 mL
Lysed horse blood   50 mL

3.2.4. Antibiotic Supplement

Base	mg/L
Sodium cefoperazone	33.0
Cycloheximide	200

3.3. Test for Identification

Prepare a fresh culture in 5 mL of Brucella broth contained in a 25-mL flask. The culture is used for inoculation of the following tubes for biochemical and grouth tests, and one agar plate for sensitivity to antimicrobial compounds. Incubate all tubes aerobically or microaerobically at 42°C for 2–3 d (tubes for 25 and 30.5°C incubate for 5–7 d). Place a disc each of nalidixic acid (30 µg) and cephalothin (30 µg) on each plate. Incubate the plates at 42°C for 2–3 d under a microaerobic atmosphere. After the incubation period, measure the diameter of the transparent zone around the disc (5).

4. Notes

1. Colonies tend to swarm when initially isolated. However, reduction in moisture content of culture media can markedly alter the appearance to round, entire, sometime butyrous, colonies. The extent of this change in colony appearance is variable. The effect may explain differences seen in different laboratories and on different culture media. It cannot be reversed by increasing the moisture in the atmosphere during incubation.
2. If plates are first examined after 24 h incubation, read them immediately and quickly return them to a microaerobic atmosphere to ensure continued viability.
3. *Campylobacter*-selective supplement contains cycloheximide and is toxic swallowed, inhaled, or by skin contact. As a precaution when handling, wear gloves and eyes/face protection.

References

1. Bryan, F. L. and Doyle, M. P. (1995) Health risks and consequences of *Salmonella* and *Campylobacter jejuni* in raw poultry. *J. Food Protein* **58,** 326–344.
2. Griffiths, P. L. and Park, R. W. A. (1990) Campylobacters associated with human diarrhoeal disease. *J. Appl. Bact.* **69,** 281–301.
3. Jones, R. G. and Skinner, F. A. (1992) *Identification Methods in Applied and Environmental Microbiology.* Society for Applied Bacteriology Technical Series No. 29, Board, Blackwell Scientific Publishers, Oxford, UK.
4. Vandrzant, C. and Splittstoesser, D. F. (1990) *Compendium of Methods for the Microbiological Examination of Foods,* 3rd ed., American Public Health Association, Washington, DC.
5. Unipath Ltd. (1990) *The Oxoid Manual*, 6th ed., Basingstoke, UK.

16

Listeria monocytogenes

Anavella Gaitán Herrera

1. Introduction

Listeria monocytogenes was discovered as a pathogen of animals and humans in the 1930s. As far as humans are concerned the organism was initially identified as a cause of abortion in early pregnancy, stillbirth, or of septicemia after an uneventful birth. Ecological surveys have demobstrated that *Listeria* in general, and *L. monocytogenes* in particular, are naturally occurring in a wide variety of domestic animals, particularly sheep and chickens. *L. monocytogenes* has four attributes: the alleged elevated heat resistance, the ability for relatively rapid growth at refrigeration temperatures, a marked tolerance of reduced pH values, and growth in the presence of over 5% sodium chloride.

L. monocytogenes have been solated from raw staple foods including chicken, red meat, seafood, and, of course, raw milk (*see* **Note 1** and **refs.** *1* and *2*).

2. Materials

1. Cultures of *Staphylococcus aureus* NCTC 1803 and *Rhodococcus equi* NCTC 1621.
2. Sheep blood agar plate.
3. Henry's oblique illumination.
4. Oxford and PALCAM agar.
5. FDA *Listeria* enrichment broth (LEB).

2.1. Enrichment Media

Cold preenrichment cultures in which samples are added to a nonselective nutritious medium and refrigerated was the standard procedure until recent years (*see* **Note 2**). Cold enrichment takes advantage of the psychrophilic property of *Listeria* spp. *Listeria* will multiply at the low temperature, which inhibits multiplication of the accompanying medophilic flora *(3)*. Detection of *Listeria* colonies is generally assisted by viewing with a plate microscope with the plate illuminated at an oblique angle as described by Henry (*see* **Fig. 1**). Colonies of *Listeria* spp. are blue or blue/gray making them easily distinguishable to enable suitable colonies to be picked for futher testing *(4)*. The explanation of the phenomenon is that the cells in a *Listeria* colony tend to lie the same plane, thus forming a crude diffraction grating, which changes the wave length of light transmitted through the colony.

2.1.1. FDA Listeria Enrichment Broth (LEB)

Formula	(g/L)
Tryptone soya	30.0
Yeast extract	6.0
pH 7.2 ± 0.2	

2.1.2. Listeria Selective Enrichment Supplement

Formula[a]	
Nalidixic acid	20.0 mg (equivalent to 40 mg/L)
Cycloheximide	25.0 mg (equivalent to 50 mg/L)
Acriflavine hydrochloride	7.5 mg (equivalent to 15 mg/L)

[a]Each vial is sufficient for 500 mL of medium.

Suspend the supplement in 2 mL of sterile distilled water. Sterilize by autoclaving at 121°C for 15 min. Cool to 50°C and aseptically add the *Listeria*-selective enrichment supplement, mix well, and distribute into sterile containers.

2.1.3. Listeria Selective Agar Base

Formula	(g/L)
Columbia blood agar base	39.0
Aesculin	1.0
Ferric ammonium citrate	0.5
Lithium chloride	15.0
pH 7.0 ± 0.2	

Bring gently to boil to dissolve. Sterilize by autoclaving at 121°C for 15 min. Cool to 50°C and aseptically add the contents of one vial of *Listeria*-selective supplement (Oxford formulation).

Listeria monocytogenes

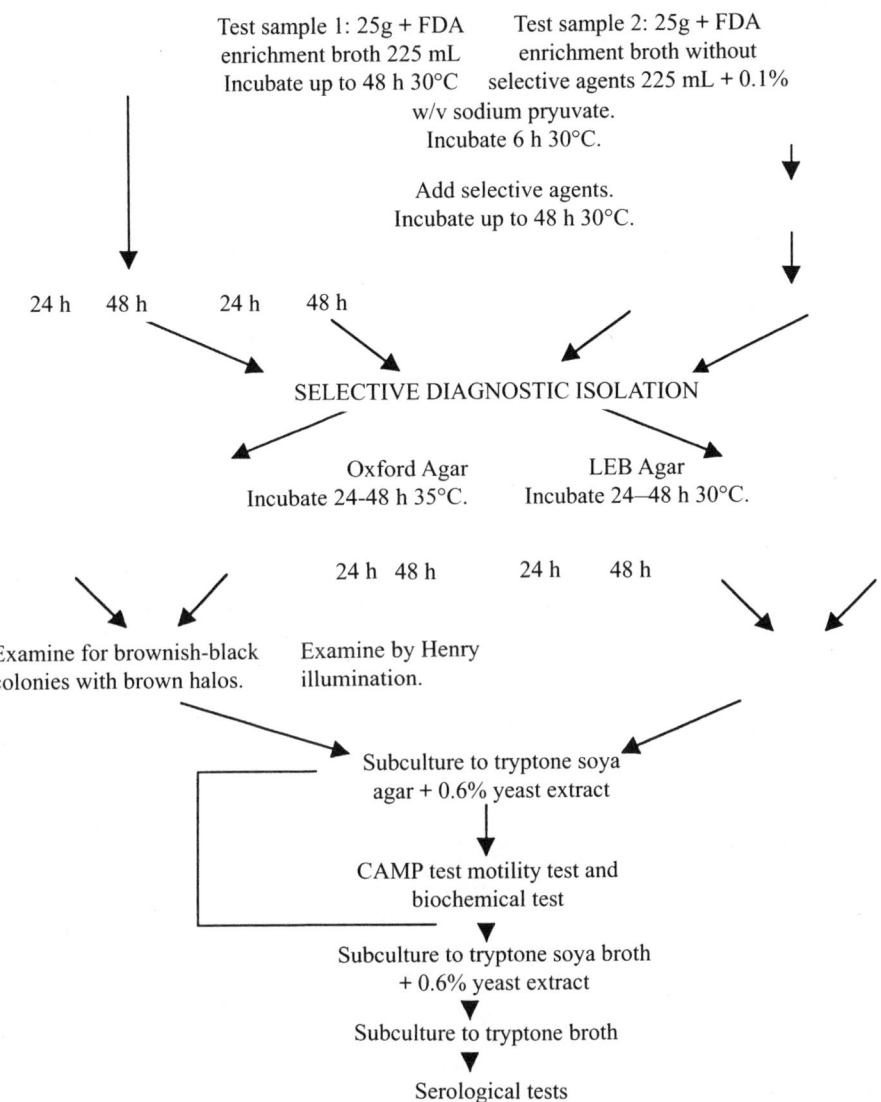

Fig. 1. *Listeria monocytogenes* flowsheet.

2.1.4. Listeria *Selective Supplement (Oxford Formulation)*

Formula[a]

Cycloheximide	200.0 mg (eqivalent to 400 mg/L)
Colistin sulphate	10.0 mg (equivalent to 20 mg/L)

Acriflavine	2.5 mg (equivalent to 5 mg/L)
Cefotetan	1.0 mg (equivalent to 2 mL/L)
Fosfomycin	5.0 mg (equivalent to 10 mg/L)

^aEach vial is sufficient for 500 mL of medium.

Reconstituted with 5 mL of ethanol/sterile distilled water (1:1). Mix well and pour into sterile Petri dishes.

2.1.5. PALCAM Agar

Selective and differential diagnostic medium for the detection of *L. monocytogenes*.

Formula	(g/L)
Columbia blood agar base	39.0
Yeast extract	3.0
Glucose	0.5
Aesculin	0.8
Ferric ammonium citrate	0.5
Mannitol	10.0
Phenol red	0.08
Lithium chloride	15.0
pH 7.2 ± 0.2	

Suspend in 500 mL of distilled water. Bring gently to boil to dissolve completely. Sterilize by autoclaving at 121°C for 15 min. Cool to 50°C and aseptically add the contents of one vial of PALCAM selective supplement, reconstituted.

2.1.6. PALCAM Selective Supplement

Formula[a]	mg
Polymixin B	5.0
Acriflavine hydrochloride	2.5
Ceftazidime	10.0

[a]Each vial is sufficient for 500 mL of medium.

Vial is reconstituted with 2 mL of sterile distilled water. Mix and pour into sterile Petri dishes. The addition of 2.5% (v/v) egg yolk emulsion to the medium may aid the recovery of damaged *Listeria* spp.

L. monocytogenes hydrolyzes aesculin resulting in the formation of a black halo around colonies, not fermented mannitol, thus easy differentiation from contaminants such as enterococci and staphylococci can be made (these will ferment mannitol and produce a change from red to yellow in the pH indicator phenol red). Incubation under microaerophilic conditions serves to inhibit strict

Listeria monocytogenes

aerobes such as *Bacillus* spp. and *Pseudomonas* spp. that might otherwise appear on the medium.

After 48 h incubation, typical *Listeria* spp. from colonies that are approximately 2 mm in diameter, gray-green in color with a black sunken center and a black halo against a cherry-red medium background.

3. Methods
3.1. Plating Media (see Note 3)

Oxford and PALCAM agars have employed as highly effective plating media that are widely specified in official methodology (FDA/BAM) (*see* **Fig. 2**). The medium utilizes the selective inhibitory components lithium chloride, acriflavine, colistin sulphate, cefotetan, cycloheximide, and fosfomycin and the indicator system aesculin and ferrous iron for the isolation and differentiation of *L. monocytogenes*. *L. monocytogenes* hydrolyzes aesculin, producing black zones around the colonies due to the formation of black iron phenolic compounds derived from the aglucon. Gram-negative bacteria are completely inhibited. Typical *L. monocytogenes* colonies are almost always visible after 24 h, but incubation should be continued for a futher 24 h to detect slow-growing strains (*see* **Note 4** and **ref. 4**).

3.2. Identification Tests

Colonies that appear to be *Listeria* spp. on plating media can be confirmed using four simple tests (*see* **Notes 5** and **6**).

1. Motility: Heavily inoculate brain–heart infusion or nutrient broth and incubate at room temperature. Examine microscopically at 4–6 h. If tumbling, roating motility is not seen, reexamine at 18 h before discarding as negative. *Listeria* spp. do not form flagellae above 30–33°C and motility may not occur if cultures are incubated above 30°C.
2. Catalase: Emulsify a colony in a drop of hydrogen peroxide on a glass slide. Immediate bubbling indicates a positive catalase test. *Listeria* spp. are catalase positive. False positive catalase reactions may occur if a colony is taken from a medium containing blood.
3. Microscopy: Examine a Gram-stain of growth from a suspected colony. *Listeria* cells have a distinctive appearance and disposition. They are short Gram-positive rods that ocur as straight pairs, pairs arranged in V formation and pairs adjacent to each other. Most non-*Listeria* spp. can be eliminated by these three screaning tests.
4. The CAMP test (Christie-Atkins-Munch-Peterson): Prepare sheep blood agar plates by pouring a thin layer of 5% v/v blood agar made with washed sheep cells. Streak cultures of *Staphylococcus aureus* NCTC 1803 and *Rhodococcus equi* NCTC 1621 across the sheep blood agar plate. Then streak the test strains at right angles to the *S. aureus* and *R. equi* leaving a minimum of 12 mm between cultures. Incubate at 37°C overnight (*see* **Tables 1** and **2** and **ref. 5**).

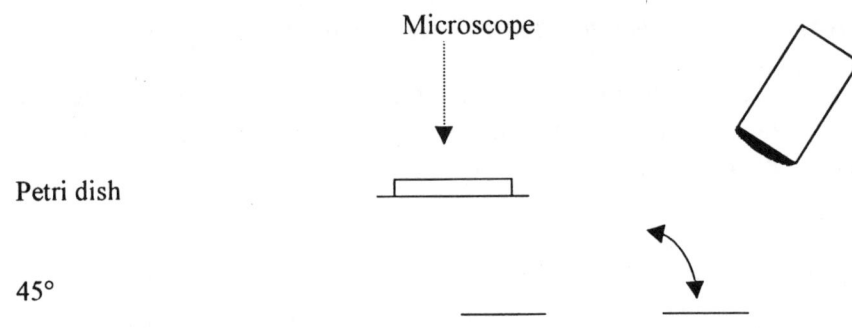

Fig. 2. Plating methodology.

Table 1
CAMP Reactions for the Hemolytic Species of *Listeria*

	Staphylococcus aureus	Rhodococcus equi
L. monocytogenes	+	−
L. seeligeri	+	−
L. ianovii	−	+

Table 2
Tests to Be Done and the Results Shown by *L. monocytogenes*

Test	L. monocytogenes
Beta hemolysis (horse blood)	+
Camp test	
Staphylococcus aureus	+
Rhodococcus equi	−
Catalase	+
Oxidase	−
Nitrate reduction	−
Methyl red (MR)	+
Voges–Proskauer (VP)	+
H_2S	−
Urea	−
Acid from:	
Glucose	+
Xylose	−
Rhamnose	+
Mannitol	−
Aesculin	+

4. Notes

1. Do not use beyond the expiration date or if the product is caked, discolored, or shows any signs of deterioration.
2. *Listeria* enrichment supplement medium contains cycloheximide and is toxic if swallowed, inhaled, or skin contact. When handing, wear gloves and eye/face protection.
3. Store the prepared medium at 2–8°C, tightly capped, in the dark, and use as soon as possible.
4. Store prepared medium away from light. Acriflavine can photooxidize to form compounds inhibitory to *Listeria* spp.
5. Broth cultures are more dangerous than colonies on agar plates.
6. Store the supplement at 2–8°C away from light and use before the expiration date on the label.

References

1. Dever, F. P., Shaffner, D. W., and Slade, P. J. (1993), Methods for the detection of foodborne *Listeria monocytogenes* in the US. *J. Food Safety* **13**, 263–292.
2. Post, D. E. (1989) The detection of *Listeria*, in *Reducting the Risk of Listeria*. I.B.C. Technical Services Ltd. Symposium, London, UK.
3. Farber, J. M. (1993) Current research on *Listeria monocytogenes* in foods: an overview. *J. Food Protect.* **56**, 640–643.
4. Farber, J. M. and Peterkin, P. I. (1991) *Listeria monocytogenes*, a foodborne pathogen. *Microbiol. Rev.* **55**, 476–511.
5. Varnham, A. H. and Evans, M. G. (1991) *Foodborne Pathogens: An Illustrated Text*. Wolfe Publishing Ltd., London.
6. British Standard (1993) Microbiological examination for dairy purposes, B5 4285, Part 3. *Methods for Detection and/or Enumeration of Specific Groups of Microorganisms*. Section 3.15. Detection of *Listeria monocytogenes*, ISO 10560.

III

FERMENTED FOODS

17

Methods for Plasmid and Genomic DNA Isolation from Lactobacilli

M. Andrea Azcárate-Peril and Raúl R. Raya

1. Introduction

Lactobacillus represents a major genus of the lactic acid bacteria that have widespread use in fermented food production. They are also found in the mouth, intestinal tract, and vagina of many animals. They produce bioantagonists compounds such as lactic acid and bacteriocins and hence are used as probiotics to treat gastroenteric disorders *(1)*. In view of their commercial importance, a complete characterization of their genome and further genetic manipulation may have great potential for the improvement of these microorganisms. However, compared with that of *Lactococcus lactis*, the genetic knowledge of lactobacilli is sparse and the availability of genetic systems for its study is limited. Natural plasmids of lactobacilli may play a significant role in vector development techniques that are the basis for applied genetics. Resident plasmids are abundant in *Lactobacillus* strains. Although some industrially significant characteristics such as metabolic functions, restriction/modification systems, and drug resistance are plasmid DNA encoded, most of these plasmids remain cryptic *(2)*. Several protocols have been described for plasmid DNA isolation in this genus *(3,4)*, mainly based in the Birnboim and Doly method *(5)*. In this chapter, we describe two methods used in our laboratory for the isolation of plasmid and genomic DNA. They are based on standard protocols developed for other lactic acid bacteria and render DNA suitable for polymerase chain reaction (PCR) and restriction digestion experiments.

2. Materials

2.1. Culture Media

1. MRS broth *(6)*: 10 g/L peptone, 10 g/L meat extract, 5 g/L yeast extract, 20 g/L glucose, 5 g/L sodium acetate, 1 g/L Tween-80, 2 g/L ammonium citrate, 2 g/L K_2HPO_4, 0.2 g/L $MgSO_4 \cdot 7H_2O$, and 0.05 g/L $MnSO_4 \cdot 4H_2O$; pH 6.5.
2. MB broth *(7)*: 10 g/L yeast extract, 10 g/L glucose, 1 g/L Tween-80, 0.05 g/L $(NH_4)_2HPO_4$, and 0.005 g/L $MgSO_4 \cdot 7H_2O$; pH 6.5.

Autoclave for 17 min at 121°C. Media can be stored at room temperature for months.

2.2. Total DNA Isolation

1. Lysis buffer: 75 mM sodium chloride, 25 mM EDTA, 20 mM Tris–HCl, pH 7.5, containing 10 mg/mL of lysozyme. The solution should be prepared just prior to use.
2. 10% sodium dodecyl sulfate (SDS).
3. 25 mg/mL proteinase K, in distilled water.
4. 5 M sodium chloride.
5. A 24 : 1 (v/v) mixture of chloroform and isoamyl alcohol.
6. 100% isopropanol.
7. 70% ethanol chilled to –20°C.
8. TE buffer: Tris–HCl 10 mM, and EDTA 1 mM, pH 8.0. Autoclave for 20 min at 121°C.
9. TE-buffer-saturated phenol *(8)*.
10. 3 M sodium acetate, adjusted to pH 5.2 with glacial acetic acid.
11. 37°C water bath.
12. 55°C water bath.
13. Refrigerated centrifuge.
14. pH Meter.

2.3. Plasmid DNA Isolation

1. Resuspension buffer: 6.5% sucrose, 50 mM Tris–HCl, and 1mM EDTA; pH 8.0.
2. Lysozyme solution: 10 mg/mL lysozyme, dissolved in resuspension buffer. The solution should be prepared fresh just prior to use.
3. Lysis solution: 3% SDS, 50 mM Tris–HCl, and 5 mM EDTA. Adjust the pH to 12.25–12.4 with 5 N sodium hidroxide just prior to use.
4. Fresh 5N sodium hidroxide.
5. 3N potassium acetate, adjusted to pH 4.8 with glacial acetic acid.
6. 100% isopropanol.
7. TE buffer.
8. 10 mg/mL DNase-free RNase (Sigma, St. Louis, MO) dissolved in sterile distilled water.

9. TE-buffer-saturated phenol.
10. A 24:1(v/v) mixture of chloroform and isoamyl alcohol.
11. Optional: 8 N LiCl dissolved in sterile distilled water.
12. 37°C water bath.
13. Refrigerated microcentrifuge.
14. pH Meter.

2.4. Agarose Gel Electrophoresis

1. 0.8% agarose dissolved in TAE or TBE buffer *(8)*.
2. TAE (1X: 0.04 M Tris–acetate, 0.001 M EDTA, pH 8.0) or TBE (0.5X: 0.045 M Tris–borate, 0.001 M EDTA; pH 8.0) buffer *(8)*.
3. 10 mg/mL ethidium bromide.
4. Power supply and accessories for electrophoresis.

3. Methods
3.1. Total DNA Isolation (see Note 1)

1. Inoculate 30 mL of MRS broth with 300 µL of an overnight culture. Incubate 16 h at 37°C (*see* **Note 2**).
2. Centrifuge the culture at 7000g for 10 min and resuspend the cells in 5 mL of lysis buffer. Incubate in a 37°C water bath for 2 h.
3. Add 500 µL of 10% SDS and 100 µL of 25 mg/mL proteinase K. Incubate in a 55°C water bath for 2 h. Shake occasionally.
4. Add 2 mL of 5 M sodium chloride and 6 mL of chloroform–isoamyl alcohol. Incubate at room temperature for 30 min.
5. Centrifuge at 19,000g for 10 min to eliminate cellular debris. Transfer the upper phase to fresh tube.
6. Precipitate the nucleic acids by adding 1 volume of 100% isopropanol. Take the DNA filament with a glass stick and wash it with ice-cold 70% ethanol.
7. Air-dry the DNA and resuspend it in 600 µL of TE buffer.
8. Extract the sample by adding 1 volume of TE-buffer-saturated phenol. Mix by inverting the tube and centrifuge at 19,000g for 5 min. Transfer the aqueous upper phase to a fresh tube and repeat this step.
9. Treat the upper phase with 1 volume of the chloroform–isoamyl alcohol mixture. Centrifuge at maximum speed for 5 min and carefully transfer the upper phase to a new tube.
10. Add 50 µL of 3M sodium acetate pH 5.2, and 2 volumes of chilled 100% ethanol. Incubate at –70°C for 1 h or at –20°C overnight. Centrifuge at 19,000g for 10 min.
11. Wash the pellet twice with 70% ethanol chilled at –20°C and air-dry the pellet.
12. Resuspend de DNA in 50–100 µL of TE buffer and store at 4°C (*see* **Notes 3 and 4**).

3.2. Plasmid DNA Isolation (see Note 1)

1. Inoculate 10 mL of broth with a single colony. Incubate the culture for 16 h at 37°C.
2. Centrifuge 2–10 mL of the culture at 16,000g for 2 min.
3. Wash the cells with 1 mL of resuspension buffer and resuspend the pellet in 200 µL of lysozyme solution (*see* **Notes 5** and **7**).
4. Incubate on ice for 1 h.
5. Add 400 µL of the lysis solution (pH 12.25–12.4; *see* **Note 6**) and incubate on ice 5 min to allow a complete cell lysis.
6. Neutralize with 400 µL of 3M potassium acetate and mix by inverting the Eppendorf tube several times (*see* **Note 7**). Do not vortex.
7. Keep the sample on ice (at least 10 min), mix by inversion, and centrifuge at 16,000g for 10 min.
8. Carefully transfer the supernatant to a fresh tube and recentrifuge if necessary.
9. Precipitate the clear supernatant with 1 volume of isopropyl alcohol. Incubate on ice for 30 min.
10. Centrifuge at room temperature at 16,000g for 20 min.
11. Carefully pour off the supernatants and allow pellet to drain completely (*see* **Note 8**).
12. Resuspend the pellet in 500 µL of TE-buffer and add 50 U of DNase-free RNase. Incubate at 37°C for 45 min.
13. Add 500 µL of TE-buffer-saturated phenol and invert several times to mix. Centrifuge at 16,000g for 5 min and transfer the upper phase to a fresh tube (*see* **Note 9**).
14. Treat the upper phase with 500 µL of chloroform-isoamyl alcohol. Invert several times to ensure the complete elimination of any trace of phenol and centrifuge at 16,000g for 5 min.
15. Precipitate the upper phase by adding 1 mL of ice-cold ethanol and incubate at –70°C for at least 30 min. Centrifuge at 16,000g for 15 min at 4°C.
16. Wash the pellet twice with ice-cold 70% ethanol, air-dry, and resuspend the pellet in 20 µL of TE buffer.
17. Examine 10 µL in an 0.8–0.9% agarose gel and stain with ethidium bromide (*see* **Note 4**).

4. Notes

1. Wear gloves all the time to minimize the risk of contamination with DNases. Other sources of DNases are water and solutions, so use sterilized water and do not store solutions for more than 30 d. Use disposable gloves when handling dangerous solutions like sodium hydroxide, TE-buffer-saturated phenol, chloroform–isoamyl alcohol and ethidium bromide. Fresh-prepared solutions, especially sodium hydroxide and lysozyme, have to be used to optimize the procedure.
2. The conditions to achieve a complete cell lysis have to be optimized: Although overnight cultures can be used for almost all lactobacilli, some strains might need to be processed in the exponential growth phase. For exopolisaccharide-producing strains, the use of the low-salt-medium MB is recommended.

3. The pellet can also be resuspended in sterile distilled water instead TE if DNA will be treated with restriction enzymes. If water is used instead, TE plasmid DNA should be conserved at −20°C. Repeated thaw and freeze of the DNA can cause damage in large genomic and plasmid DNA.
4. Measure the absorbance of the final solution at 260 and 280 nm to check DNA concentration and purity *(8)*.
5. In some cases, the incubation with lysozyme solution can be made at 37°C. However, it should not be done for more than 30 min because plasmid DNA in some strains can be damaged.
6. Carefully control the pH of the lysis solution during the plasmid isolation. A pH of at least 12.25 is needed to denature chromosomal DNA, and a pH over 12.4 can cause damage to the plasmid DNA.
7. Sucrose can be replaced by 50 m*M* glucose in the resuspension buffer for plasmid DNA isolation. Also, potassium acetate can be replaced by 3 *M* sodium acetate, pH 4.8.
8. After precipitation with isopropanol, do not overdry the pellet, it can make the subsequent resuspension difficult and shear the DNA.
9. The TE-buffer-saturated phenol extraction can be avoided in plasmid DNA isolation if there are problems with subsequent restriction enzymes treatments. Add 1 volume of 8*N* lithium chloride and incubate at −20°C for 30 min. Centrifuge at 16,000*g* for 15 min and precipitate the supernatant with ice-cold ethanol.

References

1. Davidson, P. M. and Hoover, D. G. (1993) Antimicrobial components from lactic acid bacteria, in *Lactic Acid Bacteria* (Salminen, S. and von Wright, A., eds.), Marcel Dekker, New York, pp. 127–159.
2. Wang, T. and Lee, B. (1997) Plasmids in *Lactobacillus. Crit. Rev. Biotechnol.* **17,** 227–272.
3. Klaenhammer, T. R. (1984) A general method for plasmid isolation in lactobacilli. *Curr. Microbiol.* **10,** 23–28.
4. Muriana, P. and Klaenhammer, T. (1991) Cloning, phenotypic expression, and DNA sequence of the gene for lactacin F, an antimicrobial peptide produced by *Lactobacillus* ssp. *J. Bacteriol.* **173,** 1779–1788.
5. Birnboim, H. and Doly, J. (1979) A rapid alkaline extraction procedure for screening recombinant plasmid DNA. *Nucleic Acids Res.* **7,** 1513–1523.
6. De Man, J., Rogosa, M., and Sharpe, M. (1960) A medium for the cultivation of lactobacilli. *J. Appl. Bacteriol.* **23,** 130–135.
7. Bruno-Bárcena, J. M., Azcárate-Peril, M. A., Ragout, A., Font de Valdez, G., Raya, R., and Siñeriz, F. (1998) Fragile cells of *Lactobacillus casei* suitable for plasmid DNA isolation. *Biotechnol. Techniques* **12,** 97–99.
8. Sambrook, J., Fritsch, E. F., and Maniatis, T. (eds.) (1989) *Molecular Cloning. A Laboratory Manual*, Cold Spring Harbor Laboratory Press, Cold Spring Harbor, NY.

18

Methods for the Detection and Concentration of Bacteriocins Produced by Lactic Acid Bacteria

Sergio A. Cuozzo, Fernando J. M. Sesma, Aída A. Pesce de R. Holgado, and Raúl R. Raya

1. Introduction

Lactic acid bacteria (LAB), used for centuries by man to preserve food, produce a wide variety of antagonistic compounds, including lactic acid, hydrogen peroxide, and bacteriocins. Bacteriocins are antimicrobial peptides that are bactericidal toward bacteria taxonomically close to the producer *(1)*. The bacteriocins produced by LAB have been extensively studied *(2)* and classified into three main groups *(3)*: (I) lantibiotics, small peptides (<5 kDa), which are characterized by the presence of lanthionine and/or β-methyl-lanthionine residues in the polypeptide; (II) nonlantibiotic, low-molecular-weight (<10 kDa), heat-stable peptides; and (III) nonlantibiotic, large (>30 kDa), heat-labile peptides. Class II bacteriocins can be subdivided into (IIA) *Listeria*-active peptides, (IIB) two peptide bacteriocins, (IIC) Sec-dependent bacteriocins, and (IID) class II bacteriocins that do not belong to the other subgroups.

Nisin, produced by *Lactococcus lactis*, is a well-characterized lantibiotic extensively used as a food preservative, particulary in cheese and other milk products. Furthermore, the specific actions of some bacteriocins toward foodborne pathogenic bacteria and undesirable flora like *Listeria* spp. and *Clostridium* spp. have increased the interest in these compounds. The search for new bacteriocins is therefore of great significance because of their potential use in fermented food and feed.

Several direct and deferred procedures to detect bacteriocin activity have been described *(4)*. In this chaper, we describe several methods used in our laboratory for the detection, titration, and partial purification of bacteriocins produced by LAB.

2. Materials

2.1. Culture Media

1. MRS broth *(5)*: 10 g/L peptone, 10 g/L meat extract, 5 g/L yeast extract, 20 g/L glucose, 5 g/L sodium acetate, 1 g/L Tween-80, 2 g/L ammonium citrate, 2 g/L K_2HPO_4, 0.2 g/L $MgSO_4 \cdot 7H_2O$, and 0.05 g/L $MnSO_4 \cdot 4H_2O$, pH 6.5.
2. MRS base agar: MRS with agar 1.5%.
3. MRS top agar: MRS with agar 0.7%.

2.2. Detection of Bacteriocin Activity In Situ

1. Tubes with 10 mL MRS base agar.
2. Tubes with 5 mL MRS top agar.
3. Petri dishes containing 10 mL MRS base agar.
4. Stove at 30°C.
5. Water bath at 45°C.

2.3. Titration by Diffusion Zone

1. Tubes with 5 mL MRS top agar.
2. Petri dishes containing 10 mL MRS base agar.
3. MRS broth.
4. Membrane filter of 0.45 µm of pore size.
5. Sterile tubes.
6. Hollow punch, 0.5 cm in diameter.
7. Water bath at 45°C.

2.4. Microtiter Plate Assay

1. Membrane filter of 0.45 µm of pore size.
2. Microtiter plates.
3. Sterile tubes.
4. Microplate reader.

2.5. Bacteriocin Purification

1. $(NH_4)_2SO_4$ powder.
2. Vacuum evaporator.
3. Membrane filter of 0.45 µm of pore size.
4. Columns C18 or C8 from SUPELCO (Supelclean SPE).
5. Isopropanol.
6. Methanol.
7. Deionized water.

8. 2-Propanol or acetonitrile.
9. Peristaltic pump.
10. Refrigerated centrifuge.

3. Methods

3.1 Detection of Bacteriocin Activity In Situ

1. Add 100 µL of a 1:100,000 dilution of an overnight culture of the bacteriocin producer strain to a tube containing 10 mL of molten MRS base agar at 45°C (*see* **Note 1**).
2. Mix gently and pour the mixture into a Petri dish containing MRS base agar. Let solidify.
3. Add 5 mL of molten MRS top agar (45°C) over the second layer. Slide the plate in circles on the bench top immediately to spread the top agar over the plate. Once the top agar has solidified (about 10 min), incubate overnight at 30°C (*see* **Note 2**).
4. Mix gently 70 µL of an overnight culture of the sensitive strain with 5 mL of molten MRS top agar (45°C) and pour onto the three-layer Petri dish. Slide the plate in circles.
5. Incubate the plates (with the agar side up) overnight at 30°C.
6. Bacteriocin activity is detected by the presence of growth inhibition zones (halos) of the indicator strain (*see* **Note 3**).

3.2 Diffusion Zone

1. Mix gently 70 µL of an overnight culture of the sensitive strain with 5 mL of molten MRS top agar (45°C) and pour the content into a Petri dish containing 10 mL of solidified MRS base agar (1.5%).
2. Make wells 0.5 cm in diameter with a hollow punch.
3. In each well, add 30 µL of undiluted and serial twofold (1/2, 1/4, 1/8, 1/n) dilutions of the membrane-filtered supernatant containing the bacteriocin (*see* **Note 4**). Use sterile MRS broth to prepare the dilutions.
4. Incubate the plates overnight at 30°C.
5. Bacteriocin activity is expressed in arbitrary units (AU) per milliliter, as follows: AU/mL = 1/0.03 mL × maximal dilution, which inhibits growth of the indicator strain (*see* **Note 5**).

3.3. Microtiter Plate Assay (7)

1. Add 100 µL of the bacteriocin dilutions (twofold dilutions) into the wells of a microtiter plate (96 wells).
2. Add 20 µL of an exponentially growing indicator culture (A_{600} = 0.1–0.4) and 80 µL of MRS broth to each well. Mix gently.

3. After 3 h of incubation at 30°C, evaluate growth inhibition by spectrophotometrically measuring the optical density at 600 nm with a MR 700 microplate reader.
4. One bacteriocin unit (B.U.) was defined as the amount of bacteriocin that inhibited growth of the indicator microorganism by 50%, when comparing with control culture without bacteriocin.

3.4. Concentration and Partial Purification of Bacteriocin

3.4.1. Precipitation by Ammonium Sulfate

1. Centrifuge the culture at 13,000g for 10 min and transfer the supernatant to a separate vessel.
2. Precipitate the bacteriocin by adding $(NH_4)_2SO_4$ slowly and mixing continuously (most of the bacteriocins are precipitable with at concentration of 40–80%). This step can be perfomed at room temperature.
3. Centrifuge at 10,000g and resuspend the pellet in deionized water or in 5% isopropanol.
4. Apply the bacteriocin extract directly to a solid-phase extraction column (SPE). Alternatively, the extract can be dialyzed in a benzoylated nitrocellulose sack (which must retain proteins with molecular weights greater than 2000) against deionized water at 4°C overnight, with gently stirring .

3.4.2. SPE Column

1. Apply the extract to a C18 or C8 solid-phase extraction column (*see* **Note 6**). Depending on the supernatant volume, it is possible to use columns with different sorbent capacities. SPE columns are previously washed with two volumes of isopropanol or methanol and one volume of deionized water. Once applied, the bacteriocin samples are applied to the top of the column and are drawn through the packing bed (flow must be 1–5 mL/min). A syringe is used or, alternatively, a peristaltic pump coupled with a fraction collector can also be used.
2. Columns are washed with deionized water (or other polar solvent) (*see* **Note 7**).
3. Bacteriocins are eluted with a nonpolar solvent (usually 2-propanol or acetonitrile). Elution is more efficient using a gradient of the nonpolar solvent (*see* **Note 7**).
4. Determine bacteriocin activity in the collected fractions by the diffusion zone method (*see* **Subheading 3.2.**).

3.4.3. Adsorption and Desorption

Alternatively the bacteriocin can be concentrated by using the adsorption/desorption properties at different pHs *(6)*.

1. Grow the producer strain in 1 L of MRS broth at 30°C for 18–20 h (without pH control).

2. Heat the culture to 70°C for 30 min (to inactivate proteases and to kill cells).
3. Adjust the pH to 6.5 with 4 N NaOH (to adsorb bacteriocin to the cells).
4. Collect by centrifugation (15,000g, 15 min) and wash the pellet twice in 5 mM sodium phosphate buffer (pH 6.5).
5. Resuspend in 10 mL of 100 mM NaCl at pH 2.0 (adjusted with 5% phosphoric acid) and mix with a magnetic stirrer for 2 h at 4°C.
6. Centrifuge the cell suspension at 29,000g, 20 min at 4°C, and evaluate the supernatant for inhibitory activity and stored at –20°C for future assays.

4. Notes

1. The optimal dilution of the producer culture should permit the growth of 150–200 colonies by plate.
2. Addition of the third layer of top agar over the bacteriocin producer strain is very important to avoid dissemination of colonies.
3. The colony size should be 0.5–1 mm in diameter. Larger colonies will produce a spread of the halo of inhibition.
4. The supernatant containing the bacteriocin are obtained by centrifugation of the culture at 13,000g for 10 min. They are sterilized by filtration through a membrane of 0.45 µm, without detectable loss of activity. If the bacteriocin is sufficiently thermostable to withstand temperatures that kill the producer cells, it can be sterilized by autoclaving.
5. A spot of undiluted supernatant might show a confluent clear area of inhibition whether it contains a phage or a bacteriocin. In the case of a phage, the dilutions would show a decreasing number of discrete phage plaques; a bacteriocin, on the other hand, would show a diffuse thinning of growth getting less marked with increasing dilution of the supernatant. In general, no bacteriocin activity was detectable in dilutions beyond 1:1000, whereas phage activity was still present at much higher dilutions.
6. Although C8 or C18 columns can be prepared in the lab, it is recommended to use the ready-to-use commercial columns available from several suppliers.
7. The bacteriocin recovered from C8 or C18 columns can be further purified from high-performance liquid chromatography or sodium dodecyl sulfate–polyacrylamide gel electrophoresis.

5. References

1. Tagg, J. R., Dajani, A. S., and Wannamaker, L. W. (1976) Bacteriocins of gram-positive bacteria. *Bacteriol. Rev.* **40,** 722–756.
2. De Vuyst, L. and Vandamme, E. J. (1994) Bacteriocins of lactic acid bacteria, in *Microbiology, Genetics and Applications* (De Vuyst, L. and Vandamme, E. J., eds.), Blackie Academic & Professional, London.
3. 3. Nes, I. F., Diep, D. B., Håverstein, L. S., Brurberg, M. B., Eijsink, V., and Holo, H (1996) Biosynthesis of bacteriocins in lactic acid bacteria. *Antonie van Leeuwenhoek* **70,** 113–128.

4. Klaenhammer, T. R. (1988). Bacteriocins of lactic acid bacteria. *Biochimie* **70,** 337–349.
5. De Man, J., Rogosa, M. and Sharpe, M. (1960) A medium for the cultivation of lactobacilli. *J. Appl. Bacteriol.* **23,** 130–135.
6. Yang, R., Johnson, M. C., and Ray, B. (1992) Novel method to extract large amount of bacteriocin from lactic acid bacteria. *Appl. Environ. Microbiol.* **58,** 3355–3359.
7. Toba, T., Samant, S. K., and Itoh, T. (1991) Assay system for detecting bacteriocin in mucrodilution wells. *Lett. Appl. Microbiol.* **13,** 102–104.

19

Meat Protein Degradation by Tissue and Lactic Acid Bacteria Enzymes

Silvina Fadda, Graciela Vignolo, and Guillermo Oliver

1. Introduction

During the fermentation and ripening of dry fermented sausage, a large number of biological reactions occur in the sausage mince. Proteolysis is considered to be one of the major processes involved in texture and flavor development. Moreover, small peptides and free amino acids, thus originated, are essential constituents of the nonvolatile compounds with flavor properties (1). Free amino acids are also the origin of other aroma volatile compounds that are involved in further enzymatic and chemical reactions (2,3). This is the result of the proteolytic activity of both endogenous and microbial enzymes. In recent years, the proteolytic system of lactobacilli involved in meat fermentation is becoming the focus of an increasing number of studies because of the technological roles of these organisms (5–7). The study of the enzymology of dry fermented sausages is quite complex because of the coexistence of endogenous and microbial enzymes, which prevent the determination of the independent contribution of each enzymatic system in the whole proteolytic event. For this reason, it was necesary to establish aseptic conditions to evaluate the real function of endogenous enzymes and lactic acid bacteria (LAB).

The techniques used to analyze meat protein degradation will be considered under the following headings:

1. Experimental meat systems: Aseptic meat system involving muscular protein extraction: (sarcoplasmic and myofibrillar); modification of the experimental meat system to optimize the growth of lactobacilli and beaker sausage to evaluate proteolysis by approaching the real sausage conditions.

2. Meat protein degradation: The roles of endogenous enzymes and LAB were determined by complementary methods: electrophoresis in denaturing conditions (sodium dodecyl sulfate–polyacrylamide gel electrophoresis [SDS-PAGE]) to evaluate the large protein degradation; fluorometric method (orthophthalaldehyde) and reverse phase high pressure liquid chromatography (rp-HPLC) to analyze the peptidase activity and hydrolysis products (peptides and amino acids).

2. Materials

2.1. Sarcoplasmic and Myofibrillar Systems

1. Stomacher 400 blender (London, UK).
2. Bovine *Semimembranosus* muscle and porcine *Longissimus dorsi* after 24 h of exsanguinization.
3. Distilled water (dH_2O).
4. Phosphate buffer 20 mM, pH 7.4.
5. Whatman paper No. 5; sterifilter system (Millipore); 0.22-µm membrane filter.
6. Glucose.
7. Phosphate buffer 0.1 N with 0.7 M KI, pH 6.5, sodium azide (0.02%, w/v).
8. Plate count agar (PCA).

2.2. Proteolytic Activity in Aseptic Conditions; Effect of Curing Additives

1. 250 mL of each experimental system (sarcoplasmic and myofibrillar).
2. Stock solution of $NaNO_2$, 10,000 ppm. Weigh 1 mg of $NaNO_2$ and added dH_2O to reach 100 mL in a volumetric flask. Sterilize by filtration.
3. Sodium chloride.
4. Stock solution of ascorbic acid 1 mg/mL (approx 50 mL). Weigh 10 mg of ascorbic acid and adjust to 10 mL with dH_2O. Sterilize by filtration.
5. Glucose and sucrose.
6. Lactic acid solution (20%, w/v).

2.3. Proteolytic Activity of Lactic Acid Bacteria on Muscle Proteins

2.3.1. Ability of Lactic Acid Bacteria to Growth in Sarcoplasmic System

1. LAB isolated from dry fermented sausages.
2. MRS broth.
3. An active lactobacilli culture grown in MRS broth for about 16 h at 30°C.
4. 20 mM phosphate buffer, pH 7.0.
5. 30 mL of the sarcoplasmic modified medium.
6. 0.1% Peptone for decimal dilutions.
7. MRS agar for bacterial enumeration.

2.3.2. Proteolytic Activity of Different LAB Enzyme Sources on Sarcoplasmic and Myofibrillar Proteins

1. 100–150 mL of an overnight culture.
2. 20 mM phosphate buffer, pH 7.0, as washing and resuspending solutions.
3. 100–150 mL of each muscle experimental system (sarcoplasmic and myofibrillar).
4. Sucrose solutions (0.4, 0.45, and 0.6 M).
5. 10 mM MgCl$_2$ stock solution.
6. Lysozyme.

2.3.3. Curing Conditions Effect on LAB Proteolytic Activity

1. 100–150 mL of an overnight culture.
2. 150 mL of muscle experimental systems (sarcoplasmic and myofibrillar).
3. Stock solution of NaNO$_2$, 10,000 ppm (**step 2** of **Subheading 2.2.**).
4. Sodium chloride.
5. 1 mg/mL stock solution of ascorbic acid (**step 4** of item **Subheading 2.2.**).

2.3.4. Proteolytic Activity of LAB in a Sausage Experimental System

1. Porcine and bovine meat.
2. 10,000 ppm of a stock solution of NaNO$_2$: Weigh 1 mg of NaNO$_2$ and add dH$_2$O to reach 100 mL in a volumetric flask. Sterilize by filtration.
3. Sodium chloride.
4. 1 mg/mL stock solution of ascorbic acid.
5. Glucose and sucrose.
6. Poly (vinyl chloride) (PVC) films.
7. Sodium azide.
8. 150 mL of an overnight culture of each microorganism.

2.4. To develop the Proteolytic Activity

2.4.1. SDS-PAGE

1. Resolving acrylamide stock solution 30% T–0.5% C. Weigh 29.85 g of acrylamide and 0.15 gr of N-N,methylene *bis*-acrylamide into a beaker, add about 50 mL of dH$_2$O, stir until dissolved and dilute to 100 mL. Filter (0.45-µm filter) and store in cold (4°C) and protected from light. Avoid skin contact.
2. Stacking acrylamide stock solution 10% T–15% C. Weight 4.25 g acrylamide and 0.75 g of *bis*-acrylamide into a beaker, add 25 mL of dH$_2$O, stir until dissolved and dilute to 50 mL. Filter (0.45-µm filter), store in brown bottle at 4°C.
3. 3 M Tris, pH 8.8. Dissolve 32,2 g of Trizma base (desiccated) and 5.4 g Tris–HCl (desiccated) in dH$_2$O; dilute to 100 mL. Store at 4°C (shelf life about 1 mo). If desiccated solutions are used, no pH adjustment is necessary.

4. 0.5 M Tris, pH 6.8. Dissolve 0.075 g Trizma base (desiccated) and 3.85 g Tris–HCl (desiccated) in water; dilute to 50 mL. Store at 4°C.
5. SDS (20% w/v). Dissolve and filter through 0.45-µm filter. Store at room temperature.
6. Reservoir buffer concentrate (5X). 0.25 M Tris (Trizma base), 1.92 M glycine, 0.5% SDS. Store in cold and dilute to 1X before use.
7. Amonium persulfate. Prepare a 100-mg/mL solution in water just before use.
8. $N'N'N,N'$-tetramethylethylenediamine (TEMED).
9. Sample buffer: 8 M urea, 2 M thiourea, 0.05M Tris (pH 6.8), 75 mM dithiothreitol (DTT), 3% SDS, 0.05% bromophenol blue. For 100 mL 1: 1 weight: 48 g urea (deionized), 15.2 g thiourea (deionized), 0.605 g Trizma base, 3 g SDS, 5 mg bromophenol blue, and 1.155 g DTT. Transfer the solids to 150-mL glass beaker, add a stir bar and 40 mL of water (**Note 1**) and stir gently until all the solids are dissolved (**Note 2**). Adjust the pH to 7, adding 100 µL of 12 M HCl. Now, add 1.155 g of solid DTT while stirring until dissolved; continue adjusting the pH down to 6.8 using a 20-µL aliquot of 2M HCl (*see* **Note 3**). Finally, add water to 100 mL and mix. Aliquote and store at –75°C until used.

Note: For urea/thiourea deionization, proceed as follows:
- Prepare 200 mL of 6 M urea.
- Add 30 g of mixed-bed ion-exchange resin (i.e., AG = 501 × 8 [Bio-Rad, Richmond, CA]) and stir for 1–2 h.
- Monitor the conductivity of the solution (*see* **Note 4**).
- Filter the urea.
- Prepare 150 mL of 2 M thiourea.
- Add 5 g of the resin; stir for 1–2 h.
- Check the conductivity.
- Filter the deionized thiourea
- Add the other reagent as stated in this step.

10. Glycerol (50% v/v). Stable at room temperature.
11. Staining solution. Weigh out 0.5 g of Coomassie brillant blue R-250 and dissolve in 250 mL isopropanol and 100 mL of acetic acid and dilute to 1000 mL with dH_2O.
12. Destaining solution. Measure 50 mL of acetic acid and 150 mL of ethanol; mix both reagents and dilute to 500 mL with dH_2O.
13. The proteins used as standards were myosin (200.0 kDa), β-galactosidase (116.3 kDa), phosphorylase B (97.4 kDa), serum albumin (66.2 kDa), ovalbumin (45.0 kDa), carbonic anhydrase (31.0 kDa), trypsin inhibitor (21.5 kDa), lysozyme (14.4 kDa), and aprotinin (6.5 kDa) from Bio-Rad.

2.4.2. Peptide Analyses

1. Ultraviolet (UV) detector (214 nm).
2. Waters Symmetry C18 (4.6 mm inside diameter by 250 mm) column (Waters Corp. Milford, MA).
3. Eluent A: 0.1% (v/v) trifluoroacetic acid (TFA) in MilliQ water (*see* **Note 5**).
4. Eluent B: acetonitrile–water–TFA 60:40:0.085% (v/v).

2.4.3. Ortho-Phthaldialdehyde method (OPA)

1. 0.1 M Phosphate buffer, pH 7.0. Solution A (0.2 M NaH_2PO_4): Weigh 27.6 g $NaH_2PO_4 \cdot H_2O$ and make to 100 mL with distilled water (dH_2O). Solution B 0.2 M Na_2HPO_4. Weight 53.05 g of $Na_2HPO_4 \cdot 7H_2O$ and make to 100 mL with dH_2O. To prepare 1 L of 0.1 M phosphate buffer, pH 7.0, mix 195 mL solution A and 305 mL solution B and bring to 1000 mL with dH_2O.
2. SDS (20% w/v). Dissolve 20 g SDS in 100 mL dH_2O with gentle stirring. Store at room temperature. Solution may become cloudy at temperatures below 20°C but may be restored by warming to 30°C and mixing.
3. 100 mM sodium tetraborate (borax). Weight 3.81 g of $Na_2B_4O_7 \cdot 10H_2O$ and make up to 100 mL with dH_2O. This solution is stable at least 1 mo at 4°C.
4. Working reagent (OPA solution). Mix 25 mL of 100 mM borax, 2.5 mL 20% SDS, 40 mg OPA (dissolved in 1 mL methanol), and 100 µL β-mercaptoethanol. Dilute to a final 50 mL with dH_2O. This reagent must be prepared daily.
5. 0.75 N trichloroacetic acid (TCA). This solution is prepared by weighting 61.24 g TCA and diluting to 500 mL with dH_2O. The solution is stable for at least 2 wk when stored at 4°C.

2.4.4. Amino Acid Analysis

1. Pico TagWork Station™.
2. 50 mm × 6 mm derivatization glass tubes.
3. Amino acids standards (Sigma).
4. Drying solution: methanol, sodium acetate trielthilamine TEA (2:2:1).
5. Derivatization solution: prepare daily and consisted of methanol, H_2O/TEA/phenyl isothiocyanate (PICT) (7:1:1:1).
6. UV detector (240 nm).
7. A Waters Symmetry C18 (4.6 mm inside diameter by 250 mm) column.
8. Acetonitrile.
9. High-purity water (MilliQ).
10. Sample buffer: 0.005 M phosphate buffer, pH 7.4, containing 5% acetonitrile.
11. Eluent A: 70 mM acetate buffer, pH 6.55, containing 2.5% acetonitrile.
12. Eluent B: acetonitrile/H_2O/metanol (45:40:15).

3. Methods
3.1. Sarcoplasmic and Myofibrillar Experimental Systems

These systems consist of meat proteins extracted from bovine (*Semimembranosus*) or porcine (*Longissimus dorsi*) muscles. After removing fat and connective tissue in aseptic conditions, cut the lean muscle into small cubes (3 cm) and keep at –70°C until use.

3.1.1. Sarcoplasmic System

1. Weigh 20 g of lean muscle in a stomacher bag.
2. Add 200 mL of 20 mM phosphate buffer, pH 7.4 (**Note 6**).
3. Homogenize this mix in a Stomacher 400 blender for 8 min.
4. Centrifuge the protein solution at 11,180g for 20 min at 4°C, to precipitate the insolubilized protein and connective tissue.
5. Filter the supernatant containing the sarcoplasmic proteins through Whatman paper.
6. Filter-sterilize this solution by using a 250-mL capacity filter (Bio-Rad). This experimental system contain approximatly 1.95 mg/mL of protein.

3.1.2. Modified Sarcoplasmic System

To ensure the development of lactic acid bacteria in the sarcoplasmic system, it is necesary to adjust the pH at 6.5 with 1 N NaOH in **step 5**; complement the protein extract with 1% glucose (w/v), as the carbon source, add it before filter-sterilize (**step 7**) and just before use; supplement with 0.01% of Tween-80, as the growth factor, previously sterilized.

3.1.3. Myofibrillar System (**Note 7**)

1. Weigh 10 g of the pellet resulting from the sarcoplasmic protein extraction
2. Add 100 mL of 20 mM phosphate buffer, pH 7.4, previously sterilized.
3. Process this mixture for 4 min in a Stomacher 400 blender.
4. Centrifuge for 20 min (11,180g at 4°C).
5. Repeat **steps 2** and **3** twice (**Note 8**).
6. Weigh the resulting pellet and resuspend in 9 volumes of 0.1 N phosphate buffer with 0.7 M KI, pH 6.5, containing 0.02 % sodium azide (*see* **Note 9**).
7. Homogenize for 8 min in the Stomacher.
8. After centrifugation (11,180g for 20 min at 4°C), dilute the supernatant 10 times in the same buffer for enzymatic assays. The protein content of this myofibrillar extract is 0.75 mg/mL (*see* **Note 10**).

3.1.4. Sterility Control

1. Inoculate two Petri dishes with 0.5 mL of each protein extract.
2. Add 30 mL of the melted plate count agar (PCA) (*see* **Note 11**) homogenizing properly.
3. Once the agar has solidified, invert the plates and incubate for 48 h at 37°C.
4. The growth of microorganisms must be negligible with colony-forming units (CFU) below 1×10^2 CFU/mL.

3.2. Proteolytic Activity in Aseptic Conditions; Effect of Curing Conditions (see Note 12)

1. $NaNO_2$: Add 200 µL and 400 µL of a 10,000-ppm stock solution of $NaNO_2$ to 20 mL of each experimental system (sarcoplasmic and myofibrillar) to evaluate the effect of 100 ppm and 200 ppm of $NaNO_2$ on muscle proteolysis (tubes 1 and 2).
2. NaCl: Put 0.4 g and 1 g of NaCl, previously sterilized (15 min at 121°C) in each glass tube and add 20 mL of the experimental system to evaluate the effect of 3% and 5% of NaCl on proteolysis (tubes 3 and 4).
3. Ascorbic acid: Add 2 mL of 1-mg/mL stock solution of ascorbic acid to 20 mL of the sarcoplasmic or myofibrillar experimental system (final concentration 0.1 mg/mL) (tube 5).
4. Glucose+sucrose: Add 0.15 g of each sugar (0.75% w/v) to 20 mL of experimental medium and filter sterilize once more, to study the sugars effect on the muscle proteolysis (tube 6).
5. Control: 20 mL of experimental muscle extract (tube 8).
6. pH: This variable is adjust with lactic acid (20% v/v solution) in aseptic conditions to reach pH 4.2 and pH 5.0 (tubes 9 and 10).
7. To study the combination effect, all agents (3%NaCl, 200 ppm $NaNO_2$, and 0.1 mg/mL ascorbic acid) were supplemented at the same time into 20 mL of the sarcoplasmic and myofibrillar experimental systems (tube 11).
8. Incubate a set of tubes at the required temperature.
9. Take the samples (1 mL) at 0, 24, 48, 72, and 96 h stored at −70°C until proteolytic analyses.

3.3. Proteolytic Activity of LAB on Muscle Proteins

3.3.1. Culture Conditions

Lactobacillus is routinely propagated in MRS broth and incubated at 30°C. They were harvested at logarithmic phase (an overnight culture), washed twice with 20 mM phosphate buffer (pH 7.0), and inoculated in the required concentrations.

3.3.2. Ability of LAB to Grow in the Sarcoplasmic System

1. Inoculate 30 mL of the sarcoplasmic modified medium with an overnight culture grown in MRS broth to yield an initial number of 10^5 CFU/mL corresponding to an optical density (680 nm) of 0.15 (*see* **Note 13**).
2. Incubate this culture for 96 h at 30°C.
3. Take samples every 24 h for pH, bacterial development, and proteolytic analysis. Take 1 mL for each analysis.
4. The pH values are monitored potentiometrically using a pH meter. Take 2 mL of the sample to carry out this measure.
5. Determine bacterial growth in MRS agar using 100 µL of sample + 900 µL of peptone 0.1% to make decimal dilutions.

- Inoculate 0.5 mL of each dilution in a Petri dish and add the melted MRS agar by homogeneizing properly.
- Once solidified, add 10 mL more of MRS agar in a second layer (*see* **Note 14**).
- Once solidified, invert the plates an incubate 48 h at 30°C.

3.3.3. Proteolytic Activity of Different LAB Enzyme Sources on Sarcoplasmic and Myofibrillar Proteins

To determine the role of intracelullar enzymes (peptidases and aminopeptidases) in muscle subtrates three independent assays are carried out for each type of protein (sarcoplasmic and myofibrillar), using as enzymatic sources either whole cell suspensions (WC), cell-free extracts (CFE), or a combination of WC+CFE (1:1).

3.3.3.1. OBTENTION OF WHOLE CELLS

1. Harvest whole cells (WC) from of an overnight culture in MRS broth (30 mL) by centrifugation (9060g, 4°C) and wash twice with phosphate buffer solution (20 mM, pH 7.0).
2. Resuspend the washed cells in 6 mL of the same buffer solution (20% of initial volume).
3. Aseptically add the 6 mL of WC to 30 mL of the sarcoplasmic or myofibrillar system.

3.3.3.2. OBTENTION OF CELL-FREE EXTRACTS

1. Harvest cells from 30 mL of an overnight culture grown in MRS broth by centrifugation and wash twice with phosphate buffer solution (20 mM; pH 7.0).
2. Resuspend the WC in the same buffer (10% of initial volume) supplemented with sucrose and 1 mg/mL lysozyme (*see* **Note 15**).
3. After incubation at 30°C for 1 h, remove the cell-wall fraction by centrifugation (25,160g for 20 min at 4°C).
4. Wash the pellet in 20 mM phosphate buffer, pH 7.0, and resuspend in the same buffer without sucrose (*see* **Note 16**).
5. Sonicate the cell suspension for 15 min in three cycles of 5 min each with intermediate rest of 2 min (*see* **Note 17**).
6. Remove cell debris by centrifugation (44,740g for 20 min at 4°C). The supernatant constitute the cell-free extract (CFE).
7. Aseptically add 6 mL of CFE to 30 mL of sarcoplasmic or myofibrillar extract.

3.3.3.3. WHOLE CELLS+CELL-FREE EXTRACTS

1. In the case of the combination of the WC and CFE, combine 3 mL of each enzymatic sample, twice concentrated (*see* **Note 18**).
2. Incubate each reaction mixture at 37°C for 96 h under shaking conditions (*see* **Note 19**).
3. Take the samples at 0 and 96 h for further analysis of pH, bacterial count, and proteolytic events.

Meat Protein Degradation

3.3.4. Curing Conditions Effects on LAB Proteolytic Activity

1. Inoculate the sarcoplasmic and myofibrillar systems (30 mL) supplemented with NaCl (3%), $NaNO_2$ (200 ppm), and ascorbic acid (0.1 mg/mL) with WC (6 mL) as described in **Subheading 3.3.2.** with the selected *Lactobacillus* strain (*see* **Note 20**).
2. Incubate the reaction mixture in shaking conditions for 96 h at the required temperature.
3. Take the samples at 0 and 96 h and analyze for protein degradation and peptide and amino acid content.

3.3.5. Proteolytic Activity of LAB in a Sausage Experimental System

1. Aseptically process 50 g of porcine and 50 g of bovine meat (1:1) in a home processor.
2. Supplement this mix with 3% of NaCl, 200 ppm $NaNO_2$, 0.1 mg/mL of ascorbic acid, and 1.5% (w/v) of glucose and sucrose (0.75% of each sugar); mix vigorously. Divide the system into four batches.
3. This sausage-like system is divided in the number of batches according to the LAB strains to be assayed. Cells are collected from a MRS broth overnight culture and added to reach a final concentration of approx 2×10^8 CFU/mL. A mixed culture of the strains and a control (without bacterial inocula) is also included (*see* **Note 21**).
4. Properly homogenize each batch to ensure the efficient distribution of the microorganisms and additives.
5. Stuff into artificial casings of PVC (*see* **Note 22**). Heat seal properly.
6. Incubate at 25°C for 4 d.
7. Take the samples every 24 h and analyze for protein degradation, protein, peptide, and free amino acids contents, pH, and bacterial development.

3.4. Methods Employed to Measure the Proteolytic Events

3.4.1. SDS-PAGE

The hydrolysis of muscle proteins is monitored by SDS-PAGE analysis *(8)* using 12% and 10% polyacrylamide gels for sarcoplasmic and myofibrillar proteins, respectively.

Table 1 lists the components necessary for making 10% and 12% polyacrylamide resolving gels and stacking 3% acrylamide gel. The quantities listed are those necessary to prepare two Mighty Slab gels.

1. Prepare the resolving gel mixture in 25 mL Erlenmeyer flasks containing everything except the ammonium persulfate and the TEMED, which should be added immediatly prior to pouring the gel. Homogenize gently, avoiding bubbles (*see* **Note 23**).
2. Pour the gel gently into the previously assemble slab gel casting unit, leaving space for the sample well comb. Take care not to introduce any air bubbles. Over-

Table 1
Composition of Polyacrylamide Resolving and Stacking Gels

Components	Final percentage (w/v) acrylamide		
	Resolving gels 10%	12%	3% Stacking gel
Acrylamide stock + *bis*-acrylamide	4 mL	4.8 mL	—
Stacking stock acrylamide + *bis*-acrylamide	—	—	1.5 mL
3M Tris pH 8.8	3 mL	3 mL	—
0.5 MTris, pH 6.8	—	—	1.25 mL
20% (w/v) SDS	60 µL	120 µL	25 µL
H$_2$O	2.46 mL	1.6 mL	1.17 mL
50% (v/v) Glycerol	2.4 mL	2.4 mL	1 mL
10% (w/v) Persulfate	75 µL	75 µL	30 µL
TEMED	10 µL	10 µL	10 µL

lay the gel with isopropanol/H$_2$O solution (1:3) and allow to polymerize for aproximately 20 min at room temperature (*see* **Note 24**).
3. After polymerization, rinse off the propanol solution and the unpolymerized upper layer with water.
4. Prepare the stacking gel mixing the reagents as in the resolving gel preparation and pour onto the top of the resolving gel. Use a sample well comb to form wells for sample loading. Polymerization should be complete in 13–30 min at room temperature.
5. Remove the sample well comb and wash the wells extensively with distilled water to remove any unpolymerized acrylamide. Load the gel into the electrophoresis tank and fill the lower and upper tanks with 1X reservoir buffer. Gently remove the gel combs after the upper reservoir is filled. You are now ready to load the samples.
6. Mix samples with an equal volume of sample buffer: Boil in a water bath for 4–5 min at 100°C to denature the sample (*see* **Note 25**).
7. Apply 20 µL of sarcoplasmic or 25 µL of myofibrillar proteins onto the gels using a Hamilton syringe and store at –70°C until required (*see* **Note 25**).
8. Electrophorese at 50 mA at room temperature until the front marker reaches the botom of the gel (approx 50 min).
9. Separate the gel plates, discard the stacking gel, and place the resolving gel gently into staining solution. Stain the gel for 2 h with constant agitation and then transfer to destain (*see* **Note 26**).
10. Destain for 2 h by replacing the destaining solution periodically (*see* **Note 27**).

3.4.2. rp-HPLC for Peptide Analyses

1. Deproteinize 2 mL of the sample with 5 mL of acetonitrile.
2. Vortex vigorously and allow to stand for 15 min at 4°C.

3. Centrifuge the solution at 11,180g for 15 min and transfer the supernatant to a clean 25-mL round-bottom flask.
4. Concentrate the supernatant to dryness by evaporation and resuspend in 200 µL of solvent A (*see* **Note 28**).
5. Apply 15 µL of resuspended sample into a Waters Symmetry column previously equilibrated under basal conditions.
6. Elute 1% solvent B in an isocratic phase for 5 min, followed by a linear gradient from 1% to 100 % solvent B for 20 min, at a flow rate of 0.9 mL/min and at 40°C.
7. Peptides are detected at 214 nm.
8. Analyze the peptide peaks qualitatively by comparison between 0 and 96 h and with the control samples (*see* **Note 29**).
9. The chromatogram is concluded in 30 min.

3.4.3. OPA Method

The OPA reagent reacts with primary amines in the presence of a thiol group (i.e., β-mercaptoethanol) and is enhanced at basic pH. Under these conditions, 1-thioalkyl-2-alkylisoindoles are formed, which absorb strongly at 340 nm. This method was developed to study milk protein hydrolysis *(9)*. The modification described here is to be used for meat proteins allowing the obtention of satisfactory preliminary results:

1. Add 500 µL of 0.75 *N* trichloride acetic acid (TCA) to 250 µL of homogeinized sample.
2. Vortex vigorously and allow to stand for 15 min at 4°C.
3. Centrifuge the solution at 13,000g for 15 min and transfer the supernatant to a clean Eppendorf with a micropipet (*see* **Note 30**).
4. Add 1 mL of OPA reagent to 50 µL supernatant in a 1-mL quartz cuvet.
5. Mix briefly by inversion and incubate 2 min at room temperature.
6. Measure the absorbance at 340 nm against the blank (sterile muscle protein extract).
7. Calculate the millimoles per liter of α-amino released (m*M*) from the following relationship:

$$mM = \varepsilon \Delta A_{340} F$$

where ΔA_{340} is the experimentally observed change of absorbance at 340 nm using a 1-cm light path, F is the dilution factor corresponding to the assay procedure, and ε is the molar absorption coefficient (6000 *M*/cm).

3.4.4. rp-HPLC for Amino Acid and Natural Dipeptide Analyses

The derivatized amino acids were analyzed by reverse-phase HPLC according to the method of Aristoy and Toldrá *(10)*.

1. Deproteinize samples of 500 µL plus 50 µL of an internal standard (0.325 mg/mL hydroxyproline) with 1375 µL of acetonitrile.

2. Vortex vigorously, allow to stand for 15 min at 4°C, and centrifuge at 11,180g.
3. The supernatant is derivatized to their phenylthiocarbamyl derivatives according to the method of Bidlingmeyer et al. *(11)*:
   - Dry under vacuum 250 µL of supernatant in the Pico Tag Work Station.
   - Add 10 µL of the drying solution (methanol/sodium acetate/TEA) and dry under vacuum again.
   - Add 20 µL of the derivatization reagent (methanol/H_2O/TEA/PICT).
   - Vortex vigorously and incubate at room temperature for 20 min in darkness.
   - Dry under vaccum.
   - Resuspend the sample in 250 µL of 0.005M phosphate buffer (pH 7.4) containing 5% acetonitrile.
   - Centrifuge at 11,180g at 4°C and the supernatant constitute the derivatized amino acid sample.
4. Inject 15 µL of each derivatized sample including the amino acid standard mixture into the column previously equilibrated in basal conditions.
5. rp-HPLC is performed at 40°C during 72 min with a flow rate of 1 mL/min.
6. The separation gradient is as follows:

0–13 min:	100% of solution A
13–16.5 min:	3–31% solution B
16.5–30 min:	3.1–9% solution B
30–50 min	9–34% soultion B
50–60 min:	34% solution B
60–63 min:	34–56% solution B
65–70 min:	100% solution B

7. Calculate the amino acid concentration by the relationship of the peak area and estándar amino acid concentration as follows:

$$Mg\% \; AA = AA \; factor \times Aa \; area \times \frac{ProOH \; weight \times 100}{ProOH \; area \quad 0.5}$$

4. Notes

1. When preparing the deionized urea for the sample buffer, do not add too much water, the urea and thiourea takes up over half the final volume. Use latex gloves to avoid contamination of the sample buffer with skin proteins.
2. Avoid temperatures above 40°C. Heated urea speeds cyanate formation.
3. The buffer capacity rapidly declines as the pH is lowered. Do not overshoot.
4. The conductivity should be less than 1 µΩ when the deonizaton is complete. Another parameter to take in account when the deionization is finished is the change of resin color. If not, more resin may need to be added.
5. Acetonitrile, methanol, and trifluoroacetic acid used for HPLC analysis must be HPLC grade.
6. The sarcoplasmic proteins are soluble in low-ionic-strength solutions. Eight minutes of stomacher process is sufficient to ensure the soluble protein extraction.

7. Myofibrillar proteins are soluble only in high-ionic-strength solutions. It is neccesary to increase the ionic strength by adding KI, but not higher than $0.7M$ to protect microbial amino peptidases.
8. In the obtention the myofibrillar system, a complete elimination of sarcoplasmic proteins must be done by properly washing the meat extract.
9. The myofibrillar extract cannot be filter sterilized because its gel nature. Sodium azide is added to prevent bacterial development.
10. The myofibrillar extract must be diluted 10-fold just before use to avoid an excessive inhibition of bacterial proteinases by KI.
11. PCA media is used to quantify the total aerobic organisms.
12. Curing additives concentrations are related to the technological parameters used in industry. As regard as pH values, they are chosen to evaluate the proteolytic phenomena in the middle (pH 5.0) and final (pH 4.2) stages of dry sausage fermentation.
13. To avoid interference with the sarcoplasmic system, an optical density of 680 nm was selected.
14. To ensure microaerophilic conditions, a second layer of MRS agar is applied.
15. It is necessary to supplement the buffer solution with sucrose and $MgCl_2$ at different concentrations according to the strain to ensure cell integrity (i.e., $0.4\ M$ sucrose for *L. plantarum*, $0.6\ M$ sucrose and 5 mM $MgCl_2$ for *L. curvatus* and *L. casei*, and $0.45\ M$ sucrose for *L. sake*). Lysozyme is used to digest the cellular wall.
16. To support osmotic shock, resuspension buffer (20 mM phosphate, pH 7.0) must be free of sugars!
17. To prevent thermal enzyme inactivation, it is necessary to pause for 2 min between each sonication cycle.
18. Whole cells and cell-free extract are concentrated when added together, to keep the same enzyme content and to prevent dilution of the reaction mixture.
19. To ensure enzyme–substrate interaction of muscle proteins during the incubation time at 30–37°C, shaking must be used.
20. The additive concentrations employed when combination effect were analyzed are such that best proteolytic changes were observed. When BAL are added no sugar were added to avoid a pH decrease.
21. The strains selected for use as mixed cultures must be compatible.
22. Avoid air bubbles when stuffing the meat mixture. It is known that the air bubbles produce defects (i.e., mold contamination) in the final product
23. The dissolved oxygen in the gel solutions retards the acrylamide–bis polymerization. Use disposable latex gloves and avoid polyacrylamide skin contact!
24. Optimal time to gel polymerization solution is 10–20 min. If the gel sets more quickly, decrease the amount of ammonium persulfate; if the gel takes longer, increase the ammonium persulfate. There is a considerable batch-to-batch variation in the acrylamide and amonium persulfate so the levels need to be readjusted with your own reagents. The acrylamide is gelled when a clear line can be observed 1–2 mm below the isopropanol solution.

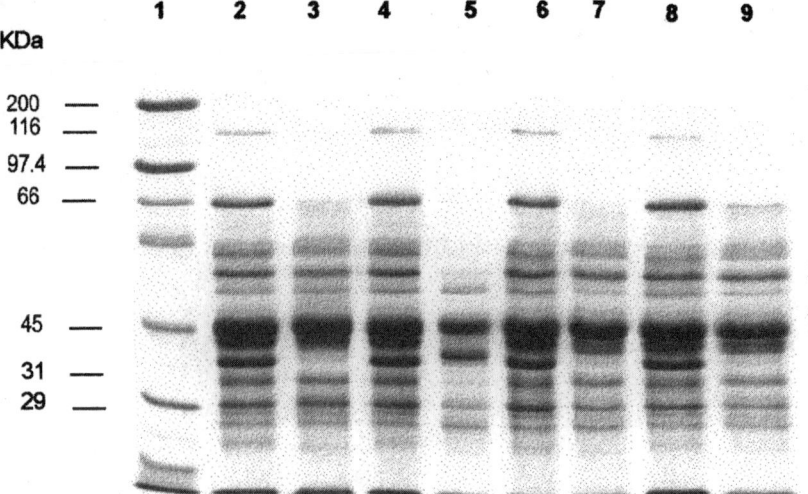

Fig. 1. Proteoplytic activity of *Lactobacillus* on sarcoplasmic proteins after 96 h at 37°C, whole cells. **1.** Molecular weight standards, **2.** *L. plantarum* CRL681 0 h, **3.** *L. plantarum* CRL681 96 h, **4.** *L. casei* CRL705 0 h, **5.** *L. casei* CRL705 96 h, **6.** *L. curvatus* NCDO904 0 h, **7.** *L. curvatus* NCDO904 96 h, **8.** *L. sake* CECT4808 0 h, **9.** *L. sake* CECT4808 96 h.

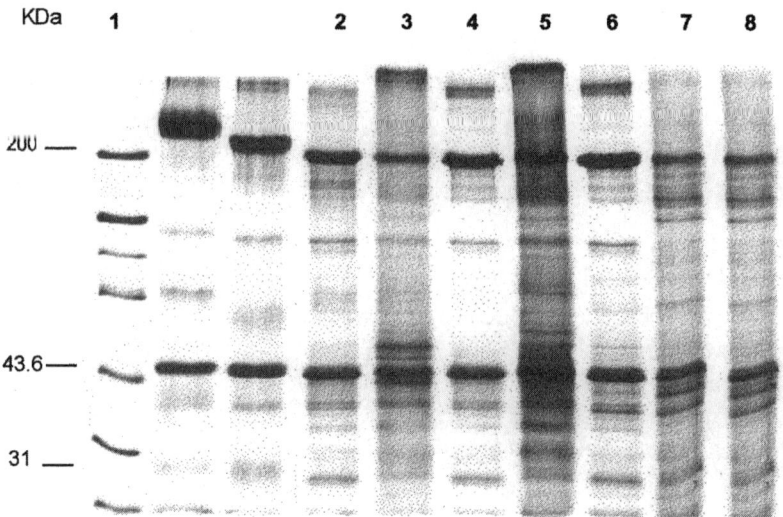

Fig. 2. Proteolytic activity of *Lactobacillus* on myofirbrillar proteins after 96 h at 37°C, Whole cells + cell-free extracts. **1.** Molecular weight standards, **2.** *L. plantarum* CRL681 0 h, **3.** *L. plantarum* CRL681 96 h, **4.** *L. casei* CRL705 0 h, **5.** *L. casei* CRL705 96 h, **6.** *L. curvatus* NCDO904 0 h, **7.** *L. curvatus* NCDO904 96 h, **8.** *L. sake* CECT4808 96 h.

25. Sample preparation: protein concentration can be readily determined by the method of Bradford. Band resolution is improved if small volumes of sample are applied to the gels. However, with all kinds of samples, avoid protein concentrations exceeding 5 mg/L. The sample volume applied onto the gel depends on the protein concentration. It is necessary to have at least 20 µg of protein in the total volume: sample + sample buffer. Do not boil more than 5 min because the urea will form cyanate, which reacts with proteins and alters the migration rates.
26. Staining should be done immediatly to avoid band broadening by diffusion. Alternatively, soak gel in 50% methanol–10% acetic acid–40% dH_2O until ready to stain.
27. The most important proteolytic changes could be observed when lactic acid bacteria WC and WC+CFE were present in the reaction mixture on both sarcoplasmic and myofibrillar proteins, respectively (*see* **Figs. 1** and **2**).
28. The samples must be concentrated before HPLC analysis because of the low concentration of free peptides in the muscle.
29. Hydrophilic peptides (desirable flavor products) are elueted first, whereas hydrophobic ones (bitter peptides) are eluted in the last 10 min.
30. The samples after the TCA treatment can be frozen at –20°C until further use.

References

1. Verplaetse, A. (1994) Influence of raw meat properties and processing technology on aroma quality of raw fermented meat products in *Proceedings of the 40th International Congress on Meat and Technology,* The Hague, pp. 45–65.
2. Flores, M., Aristoy, M. C., Spanier, A., and Toldrá, F. (1997) Non-volatile components effects on quality of Serrano dry cured ham as related to processing time. *J. Food Sci.* **62,** 1235–1239.
3. Maga, J. A. (1982) Pirazines in foods: an update. *Crit. Rev. Food Sci. Nutr.* **16,** 1–18.
4. Montel, M. C., Seronine, M. P., Talon, R., and Hebraud, M. (1995) Purification and characterization of a dipeptidase from Lactobacillus sake. *Appl. Environ. Microbiol.* **61,** 837–839.
5. Sanz, Y. and Toldrá, F. (1997) Purification and characterization of an aminopeptidase from Lactobacillus sake. *J. Agric. Food Chem.* **45,** 1552–1558.
6. Sanz, Y., Mulholland, F., and Toldrá, F. (1998) Purification and characterization of a tripeptidase from Lactobacillus sake. *J. Agric. Food Chem.* **46,** 349–353.
7. Sanz, Y. and Toldrá, F. (1998) Aminopeptidases from Lactobacillus sake affected by amines in dry sausages. *J. Food Sci.* **63,** 894–896.
8. Fritz, J., Swartz, D. R., and Greaser, M. L. (1989) Factors affecting polyacrilamide gel electrophoresis and electroblotting of high molecular weight myofibrillar proteins. *Anal. Biochem.* **180,** 205–209.
9. Church F. C., Swaisgood H. E., Porter H. D., and Catignani, G. L. (1983) Spectrophotometric assay using *o*-phtaldialdehyde for determination of proteolysis in milk and isolated milk proteins. *J. Dairy Sci.* **66,** 1219–1227.
10. Aristoy, M. C. and Toldrá, F. (1991) Deproteinization techniques for HPLC amino acid analysis in *J. Agric. Food Chem.* **39,** 1792–1795.

11. Bidlingmeyer, B. A., Cohen, S. A., Tarvin, T. L., and Forst, B. A. (1987) A new, rapid, high sensitivity analysis of amino acids in food type samples. *J. Assoc. Off. Anal. Chem.* **70,** 241–247.
12. Bradford, M. M. (1976) A rapid and sensitive method for quantification of microgram quantities of protein utilizing the principle of protein-dye binding. *Anal. Biochem.* **72,** 248–254.

20

Maintenance of Lactic Acid Bacteria

Graciela Font de Valdez

1. Introduction

Freeze-drying is commonly used for the long-term preservation and storage of microorganisms in stock collections as well as for the production of starter cultures for the food industry. The choice of an appropriate suspending medium is of primary importance to increase the survival rate of the lactic acid bacteria (LAB) during and after freeze-drying although the success of the process also depends on several factors such as growth phase, extent of drying, rehydration, suspension medium, cryoprotectors, and so forth *(1–3)*. During freezing or freeze-drying, cellular damage may occur, resulting in a mixed population containing unharmed cells and dead cells as well as those sublethally injured. Damage may not lead directly to death since in a suitable environment the injured cells may repair and regain normal functions *(4,5)*. Information on the requirements for recovery from sublethal injury is important from the standpoint of food microbiology and culture collections.

LAB can also be preserved for short-term storage. The techniques used will be considered under the following headings:

1. Short-term maintenance for daily or weekly use. Rich, undefined media such as MRS, LAPTg, M17, or Elliker broth are commonly used (*see* **Subheading 2**).
2. Long-term preservation, as in a culture collection, where immediate access is less important, but maintenance of the characteristics of the species and strains is the primary objective.

2. Materials
2.1. Growth Media

1. MRS broth (6). For 1 L:
   a. Nitrogen source: polypeptone (10 g), meat extract (10 g), and yeast extract (5 g).
   b. Carbon source: glucose (20 g).
   c. Salts: K_2HPO_4 (2 g), ammonium citrate (2 g), sodium acetate (5 g), $MgSO_4 \cdot 7H_2O$ (0.2 g), and $MnSO_4 \cdot 4H_2O$ (0.05 g).
   d. Growth factor: Tween-80 (1 mL).
   The pH is adjusted to 6.4 ± 0.2 before autoclaving at 121°C for 15 min.
2. LAPTg broth (7). For 1 L:
   a. Nitrogen source: yeast extract (10 g), universal peptone (10 g), and tryptone (16 g).
   b. Carbon source: glucose (10 g).
   c. Growth factor: Tween-80 (1 mL).
   The pH is adjusted to 6.6 before autoclaving at 121°C for 15 min.
3. M17 medium for lactococci (8). For 1 L:
   a. Nitrogen source: phytone peptone (5 g), polypeptone (5 g), yeast extract (5 g), and beef extract (2.5 g).
   b. Carbon source: lactose (5 g).
   c. Ascorbic acid (0.5 g), β-disodium glycerophosphate (19 g), 1.0 M $MgSO_4 \cdot 7H_2O$ (1 mL).
   The pH is adjusted to 7.1 before autoclaving at 121°C for 15 min.
4. Elliker medium for lactococci (9). For 1 L:
   a. Nitrogen source: tryptone (20 g), yeast extract (5 g), and gelatin (2.5 g).
   b. Carbon source: dextrose (5 g), lactose (5 g), and sucrose (5 g).
   c. Sodium chloride (4 g), sodium acetate (1.5 g), and ascorbic acid (0.5 g)
   The pH is adjusted to 6.8 before autoclaving at 121°C for 15 min. All media mentioned can be agarized by addition of 15 g/L agar (*see* **Notes 1–5**).
5. Nonfat milk powder (NFM) (100 g). Sterilize by autoclaving at 121°C for 15 min.

2.2. Short-Term Storage

1. Storage on liquid medium: tubes of any of the broth media, as described previously; pipet with sterile tips.
2. Inoculum: bacterial cells, grown for 16 h in any of the media described to approx 10^8–10^9 colony-forming units (CFU)/mL.

2.3. Long-Term Storage
2.3.1. Lyophilization

1. Cultures grown in any of the culture media described, for 16 h (overnight) at 37°C. In the case of thermophilic species (e.g., *Streptococcus salivarius* ssp. *thermophilus*), the optimum incubation temperature may be in the range 39–41°C.

Maintenance of Lactic Acid Bacteria

2. Suspending media: NFM (20%), sodium glutamate (10%). For preparing NFM, reconstituted skim milk powder may be used.
3. Freeze-drying apparatus, if available.
4. Acetone/dry-ice freezing bath, if a commercial freeze-drying apparatus is not at hand.
5. Ampules, standard.
6. Labels, permanent markers.
7. Sterile Pasteur pipets.
8. Small cotton plugs for ampules.
9. Torch, for sealing ampules under vacuum.
10. When using double vials:
    a. Tubes with flat bottom (outer vials).
    b. Vials, 2 mL, flat bottom (inner vials).
    c. Silica gel, to confirm the dryness of the lyophilized samples.
11. When a commercial drier is not available:
    a. Vacuum pump in good condition.
    b. Acetone/dry-ice bath, for freezing.
    c. Manifold for attaching ampules for vacuum-drying.

2.3.2. Freezing

1. Storage in liquid medium or in NFM; pipet with sterile tips.
2. Inoculum: washed bacterial cells obtained by centrifugation of cultures grown for 16 h in any of the media described, and taken to half of the initial volume (approx 10^8–10^9 CFU/mL) with sterile distilled water.
3. Cryoprotectant: glycerol, sterile.
4. Small cultures tubes or cryovials, 3–4 mm in diameter, sterile.
5. Freezer, capable of operation at –70°C.
6. Waterproof markers.

2.3.3. Storage Under Liquid Nitrogen

1. Plastic ampules, 2 mL; screw cap, sterile.
2. Inoculum: bacterial cells, grown at the selected temperature for 16 h in any of the liquid media as described, to a cell density of 10^8–10^9 CFU/mL.
3. 10% or 20% cryoprotectant, sterile glycerol solution, 10% v/v dimethyl sulfoxide (DMSO).
4. Liquid-nitrogen refrigerator, and also a domestic freezer (–30°C) if available, or a cooling bath.
5. Dewar flask for transporting frozen ampules.
6. Racks for holding the ampules in the freezer, and other ancillary equipment.
7. Permanent markers for ultralow temperatures (*see* **Note 6**).

3. Methods
3.1. Short-Term Storage

1. Label the tubes containing 5–10 mL of the selected liquid medium carefully with a waterproof marker (preferently black or blue).
2. Inoculate (4%, v/v) with an overnight active culture. Keep under refrigeration (4–6°C) without previous incubation. Cells remain viable up to 10 d.
3. When needed, place the tubes at room temperatures for 15–20 min and incubate at the proper temperature for 16 h. Make at least two or three transfers in fresh medium before using.

3.2. Long-Term Storage
3.2.1. Freezing

1. Carefully label 1 to 2-mL screwcap cryovials with a waterproof marker (black or blue preferred).
2. Inoculation into NFM: Harvest and wash once by centrifugation the cells from a 10-mL overnight active culture. Resuspend the cell pellet into 1–2 mL 10% NFM supplemented with 1% (w/v) glucose, 0.5% (w/v) yeast extract, and 10% (v/v) glycerol (final concentration) and store in a domestic freezer (-20°C to –30°C) or even better, at –60 to –70°C.
3. Inoculation into glycerol solution: Take an aliquot of the washed pellet and make up to a glycerol concentration of 15–50%. Different workers prefer different concentrations of glycerol in the final mixture.
   a. Transfer the mixture to the sterile cryovials, freeze, and store as described in **step 2**.
   b. Routine transfers are made by scraping a little of the culture from the surface of the frozen medium and transferring to fresh medium.
4. If the freezer should fail, the culture can be transferred and a fresh subculture frozen later.
5. Survival (shelf life) is for several years, cultures stored at –70°C surviving longer than those kept at –20°C. Several tubes of each strain should be stored.
6. For thawing, place the cryovials at room temperature or in a water bath at 37°C and inoculate tubes containing 5–10 mL of the proper liquid medium. Incubate the tubes at the selected temperature for 16–18 h. Make at least two or three transfers in fresh medium before using (*see* **Note 7**).

3.2.2. Storage Under Liquid Nitrogen

1. Mix equal quantities of inoculum (washed) and the glycerol solution (or other cryoprotectant) in a sterile tube, so that the final concentration of glycerol is 10% (v/v). Transfer 1 mL of the mixture to each of the ampules.
2. Freeze the preparations in a domestic freezer or cooling bath, to –30°C, at a rate of about 5°C/min and allow to dehydrate for 2 h.

Maintenance of Lactic Acid Bacteria

3. Transfer the frozen ampules, without thawing, to the liquid-nitrogen refrigerator. Use the chilled Dewar flask if necessary.
4. Maintain the level of liquid nitrogen to where the ampules are completely submerged. Shelf life is for many years.
5. Cultures are revived by rapid thawing in a water bath at 37°C.

3.2.3. Lyophilization

3.2.3.1. DOUBLE-VIAL SYSTEM

1. Label the inner vials carefully. Labels may be printed by machine to reduce the possibility of mislabeling or written with waterproof markers.
2. Prepare outer vials by placing a small amount of silica gel granules (6–16 mesh) in the vial to cover about half of the bottom. Add a small cotton wad to cushion the inner vial and heat at 100°C overnight. The silica gel should be dark blue after heating; this serves as a moisture indicator during storage. Place vials in a dry box (<10% relative humidity) to cool.
3. Aseptically, mix equal amounts of inoculum (washed) and suspending medium in a sterile tube or bottle.
4. Inoculation of the inner vial: Six drops of the mixture (0.2 mL) are transferred to the bottom of each vial with a sterile Pasteur pipet. Do not touch the side of the vial.
5. Replace the cotton plug and trim it so the cotton is even with the rim of the vial. Place the inner vial in a pan, in racks, or in boxes in a freezer at –60 to –70°C and let the samples freeze for 1–2 h. In the case of a domestic freezer (–20°C to –30°C), let the samples freeze overnight.
6. Chamber-type freeze-dryer: The plates of the freeze dryer should be frozen as well. Let the condenser cool at –60°C to –70°C (about 30–45 min) and then place the frozen inner vials on the plates. Evacuate the system to below 30 µmHg (4 Pa).
7. Start the process in the afternoon and allow to run about 18 h. The system is monitored by a thermistor vacuum gage. When the vacuum sensor is placed between the product and the condenser, it will show an increase in pressure as drying occurs. However, when drying is complete, the pressure should return to below 30 µmHg.
8. When the cycle is complete, close the vacuum line between the chamber (containing the plates with the dried samples) and the condenser. Open the valve on the inlet port to admit air, allowing pressure in the cabinet to reach atmospheric.
9. Insert the inner vials into the outer vials. Tamp a 1/4-in. plug of glass fiber paper above the cotton-plugged inner vial. Heat the outer vial in an air/gas torch, rotating the vial and keeping the flame just above the glass fiber paper until the glass begins to constrict. Pull the top of the vial slowly with forceps until the constriction is a narrow capillarly tube. Cool the vials in a dry cabinet.
10. Attach each vial to a port of a manifold. Each port has a single-holed rubber stopper that fits the open end of the vial. Evacuate the system to less than 50 µmHg (7 Pa). Seal the vials at the capillarly using a double-flame air/gas torch.

11. Store vials at 2–8°C. To open the vials, heat the tip of the outer vial in a flame, then squirt a few drops of water on the hot tip to crack the glass. Strike with a file or pencil to remove the tip. Remove the fiber paper insulation and the inner vial. Use forceps to gently remove the cotton plug and rehydrate with 0.3–0.4 mL of appropriate broth medium. When resuspended, transfer the content to 5–6 mL of broth and incubate at the selected temperature for 16–18 h (*see* **Notes 8–12**).

3.2.3.2. Manifold System

1. There are two types of ampules used in the manifold method. A 1-mL bulb-shaped ampule (8.0 mm outside diameter) facilitates shell freezing and is generally used when cryoprotectants are added. A 1-mL tubular ampule (8.0 mm outside diameter) is used when the culture is suspended in skim milk without a cryoprotector; the product will freeze as a pellet.
2. Plug the ampules lightly with cotton. Autoclave at 121°C for 60 min. Label the ampules before or after sterilization as described in **step 1** of **Subheading 3.2.3.1.**
3. Prepare the cell suspension as described in **step 3** of **Subheading 3.2.3.1.**
4. Dispense 0.2 mL of the material into each of the sterile ampules. Depress the cotton plug approximately 1/2 in. below the rim of the ampule using a sterile probe and flame the rim to remove any residual cotton fibers that would interfere with the integrity of the vacuum system.
5. Attach a 1-in. piece of nonpowdered amber Latex IV tubing to the rim of each ampule using a tubing stretcher.
6. If tubular ampules are used, freeze the material by direct immersion in a dry ice/ethylene glycol bath or by a controlled freezer (**step 5** of **Subheading 3.2.3.1.**), after which ampules are immersed in the bath for further processing.
7. For material in bulb-shaped ampules, immerse in a dry-ice/ethylene glycol bath and rotate the ampule to effect shell freezing. Some lyophilizers contain an alcohol bath that may be cooled to temperatures below –70°C, according to the equipment.
8. Let the condenser cool at –60°C to –70°C (about 30 min) and then attach ampules to the ports on the manifold while they are immersed in the dry-ice slush. This is particularly important when the material is frozen as a pellet.
9. In shell-frozen samples or when ampules cannot be maintained in the dry-ice bath while drying, they can be attached one by one to the manifold port, turning on the vacuum after each connection of the ampule.
10. Attach a thermistor vacuum gage between the condenser and the vacuum pump. An additional gage may be placed between the manifold and the condenser.
11. Start the cycle early in the afternoon and allow to run overnight. Ambient temperature (not over 25°C) would be the heat source for drying. At the end of the drying cycle, use a double-flame air/gas torch to seal the ampules, moving the flame up and down the ampule within a 1-in. area below the cotton plug. After the ampules have cooled, a cellulose sleeve may be applied to protect the label and the ampule tip.
12. Store the freeze-dried material at 2–8°C.

13. To open single-vial preparations, first remove the cellulose film with a sharp blade or by soaking briefly in water. Then, score the ampule once with a file 1 in. from the tip. Desinfect the ampule with alcohol-dampened gauze. Wrap gauze around the ampule and break at the scored area. Rehydrate in appropriate broth at once (*see* **Notes 13** and **14**).

4. Notes

1. MRS broth has become the standard culture medium for lactobacilli. It is available from Merck in the dehydrated form, but it can be made up in the laboratory if desired. The reagents must be of very high quality. For some purposes, such as in fermentation tests, the meat extract and glucose are omitted.
2. LAPTg broth is not as rich as MRS broth, but it can be used for both lactobacilli and streptococci (including lactococci).
3. M17 broth has become the standard for genetic investigations because lactococcal bacteriophages can be efficiently demonstrated and distinguished on M17 agar. Plaques larger than 6 mm in diameter could be observed as well as turbid plaques, indicating lysogeny. *Streptococcus salivarius* ssp. *thermophilus* and enterococci also grow well in this medium.
4. Elliker broth is probably the most cited for the isolation and growth of lactococci, although it is unbuffered. This disadvantage can be overcome by the addition of suitable buffer substances. Addition of 0.4% (w/v) of diammonium phosphate improves the enumeration of lactic streptococci on Elliker agar.
5. Maintenance of the cultures at refrigeration temperatures is useful for routine work in the laboratory. However, it is not recommended as the conservation method because subculturing is susceptible to contamination and errors during execution.
6. Freezing in liquid nitrogen is technically simple but requires preliminary experimentation to establish the optimum freezing conditions, because strains vary widely in their levels of survival at different freezing rates. On the whole, a two-step system is employed, as microbial cells should not freeze directly to the temperature of liquid nitrogen. The cells are first cooled to a temperature between $-20°C$ and $-40°C$ and allowed to dehydrate at this temperature in order to remove sufficient water from the cells to allow subsequent rapid cooling to $-196°C$.
7. Removal of too much water from the cells during freezing leads to a concentration of solutes within the cells that may be harmful (osmotic stress). However, too much water remaining inside the cell can result in intracellular damage from ice-crystal formation (mechanical damage). The delicate balance between these two events is maintained by carefully controlling the cooling rate when freezing cells, and warming the frozen product as quickly as possible during the thawing process.
8. The freeze-drying process consists of three steps: prefreezing of the sample to ensure a solidly frozen starting structure, primary drying during which most of the water is removed, and secondary drying to remove bound water. Prefrezing of the cell suspension is a key step. Rapid cooling results in small ice crystals,

useful in preserving the cell structures, but the product is more difficult to freeze-dried. Slower cooling results in larger ice crystals and less restrictive channels in the matrix during the drying process.
9. During the prefreezing process, most products form *eutectics*. Eutectics are solutes that freeze at lower temperatures than the surrounding water. The higher the concentration of solutes in the suspending medium, the lower the freezing temperature to reach the eutetic temperature. For effective freeze-drying, the product must be cooled until all of the eutetic mixtures are frozen (eutectic temperature). Small pockets of unfrozen material remaining in the product may not be visible; however, under high vacuum, the melted pockets will expand, causing the product to bubble. These facts will affect the structural stability of the the freeze-dried product.
10. Sublimation of the frozen water requires very careful control of the two parameters involved in lyophilization, temperature and pressure. No matter what type of freeze-drying system is used, conditions must be created to encourage the free flow of water molecules from the sample, which depends on the vapor pressure differential between the product and the condenser. Therefore, the condenser temperature must be significantly lower than the product temperature. Sublimation of water is enhanced by heat, which can be applied by several means: directly, through a thermal conductor shelf as is used in tray drying, or by using ambient heat as in manifold drying.
11. After all ice crystals have been sublimed, bound moisture is still present in the product, which appears dry but the residual mosisture may be as high as 7–8%. The further storage stability of the dried cells depends on the reduction of humidity to optimal values (about 1–2%). In manifold systems and tray dryers with external condensers, the drying end point can be determined by valving off the path to the condenser and measuring the pressure above the product with a vacuum gage. If drying is still occurring, the pressure in the system will increase.
12. Two of the most important factors that can affect the stability of freeze-dried cells are moisture and oxygen. The amount of moisture remaining in the material depends on the nature of the product (suspending medium, cryoprotectants) and the length of secondary drying. The freeze-dried samples should be stored at refrigeration temperatures. The higher the storage temperature, the faster the product will degrade.
13. When opening single-vial preparations, care should be taken not to have the gauze too wet, or alcohol could be sucked into the culture when the vacuum is broken.
14. After addition of the liquid for rehydration, let the sample settle for 5–10 min before transferring to the appropriate culture medium for growing.

References

1. De Valdez, G. F., De Giori, G. S., Ruiz Holgado, A. P., and Oliver, G. (1985). Effect of the drying medium on the residual moisture content and viability of freeze-dried lactic acid bacteria. *Appl. Environ. Microbiol.* **49,** 413–415.
2. De Valdez, G. F., De Giori, G. S., Ruiz Holgado, A. P., and Oliver, G. (1985). Effect of the rehydration medium on the recovery of freeze-dried lactic acid bacteria. *Appl. Environ. Microbiol.* **50(5),** 1339–1341.

3. De Valdez, G. F. and Diekmann, H. (1993). Freeze-drying conditions of starter cultures for sourdoughs. *Cryobiology* **30,** 185–190.
4. De Valdez, G. F. and de Giori, G.S. (1993). Effect of freezing and thawing on the viability and amino acids uptake by *L. delbrueckii* ssp. *bulgaricus*. *Cryobiology* **30,** 329–334.
5. Fernandez Murga, M. L., de Ruiz Holgado, A. P., and de Valdez, G.F. (1998). Survival rate and enzyme activities of *Lactobacillus acidophilus* following frozen storage. *Cryobiology* **36,** 315–319.
6. De Man, J. C., Rogosa, M., and Sharpe, M. E. (1960). A medium for the cultivation of lactobacilli. *J. Appl. Bacteriol.* **23,** 130–135.
7. Raibaud, P., Coulet, M., Galpin, J. V., and Mocquot, G. (1961). Studies on the bacterial flora of the alimentary tract of pigs. II Streptococci: selective enumeration and differentation of the dominant group. *J. Appl. Bacteriol.* **24,** 285–291.
8. Terzaghi, B. E., and Sandine, W. E. (1981). Bacteriophage production following exposure of lactic streptococci to ultraviolet radiation. *J. Gen. Microbiol.* **122,** 305–311.
9. Elliker, P. R., Anderson, A.W., and Hannesson, G. (1956). An agar medium for lactic acid streptococci and lactobacilli. *J. Dairy Sci.* **39,** 1611–1612.

21

Probiotic Properties of Lactobacilli

Cholesterol Reduction and Bile Salt Hydrolase Activity

Graciela Font de Valdez and María Pía Taranto

1. Introduction

Among the probiotic effects attributed to lactic acid bacteria (LAB), the assimilation of cholesterol *(1)* would be of particular interest for reducing the absorption of dietary cholesterol from the digestive system into the blood. Several studies have indicated that the cholesterol removal would be related to the ability of the cultures to deconjugate bile salts *(2)*.

Bile tolerance and gastric juice resistance *(3)* are another important characteristics of probiotic lactic acid bacteria used as adjuncts because they enable them to survive, to grow, and to perform their beneficial action in the gastrointestinal tract (GIT). Although the degree of tolerance required for maximum growth in the GIT is not known, it seems reasonable that the most bile- and acid-resistant species should be selected.

The techniques used to selecting LAB for probiotic purposes will be considered under the following headings:

1. Bile tolerance, using differents bile salts concentrations.
2. Resistance to gastric juice.
3. In vitro cholesterol reduction on the presence of bile salts.
4. Bile salt hydrolase activity, determined by three complementary methods: plate assay, colorimetric method, and high-performance liquid chromatography (HPLC).

2. Materials

2.1. Growth Media for 1 L

MRS broth *(4)* is used extensively for maintaining and culturing lactobacilli.

1. Nitrogen source: polypeptone (10 g), meat extract (10 g), and yeast extract (5 g).
2. Carbon source: Glucose (20 g).
3. Salts: Sodium acetate (5 g), ammonium citrate (2 g), KH_2PO_4 (2 g), $MgSO_4 \cdot 7H_2O$ (0.25 g), $MnSO_4 \cdot 4H_2O$ (0.058 g).
4. Growth factor: Tween-80 (1.08 mL) (*see* **Fig. 1**).
5. Final pH 6.4 ± 0.2. Sterilize at 121°C for 20 min.

2.2. Media for Determination of Bile Tolerance for 1 L

MRSO broth: MRS broth supplemented with 0.5, 1, 1.5, and 3 g bile Oxgall (Difco Laboratories) to obtain a final concentration of 0.05%, 0.1%, 0.15%, and 0.3 %, respectively. (*See* **Note 1**). Sterilize at 121°C for 20 min.

2.3. Media for Determination of Acid Resistance for 1 L

Artificial gastric juice: NaCl (2 g), pepsine (3,2 g), adjusted at a final pH 2–2.3 with HCl without dilution (approx 7 mL), and take to 1 L with distilled water. As a control, artificial gastric juice adjusted at a final pH 6.5–7.0 with 5 *N* NaOH is used. Sterilize by filtration (filter membrane 0.22 µm).

2.4. Media for Determination of Cholesterol Reduction for 1 L

MRSOCH broth: Add 3 g bile oxgall to MRS broth and sterilize at 121°C for 20 min. Supplement the culture medium with 1% (v/v) of Lipids Cholesterol Rich (Sigma, L-4646) stored at 4°C. The Lipids Cholesterol Rich is sterile and must be added at the moment of using.

2.5. Media for Bile Salt Hydrolase Activity for 1 L

2.5.1. Plate Assay

1. MRS agar: Add 12 g granular agar to MRS broth and sterilize the medium at 121°C for 20 min.
2. MRSBA agar: Supplement MRS broth with 12 g granular agar, 5 g thyoglycholate (*see* **Note 2**) and 5 g of the sodium salt of one of the following compounds: taurocholic acid (TCA), glycocholic acid (GCA), taurodeoxycholic acid (TDCA), or glycodeoxycholic acid (GDCA). In all cases, the final concentration of the conjugated acid is approx 4 m*M*. Sterilize at 121°C for 20 min.

2.5.2. Colorimetric Method

MRSTT broth: MRS broth supplemented with 5 g thioglycholate and 5 g TCA (final concentration, 4 m*M*). Sterilize at 121°C for 20 min.

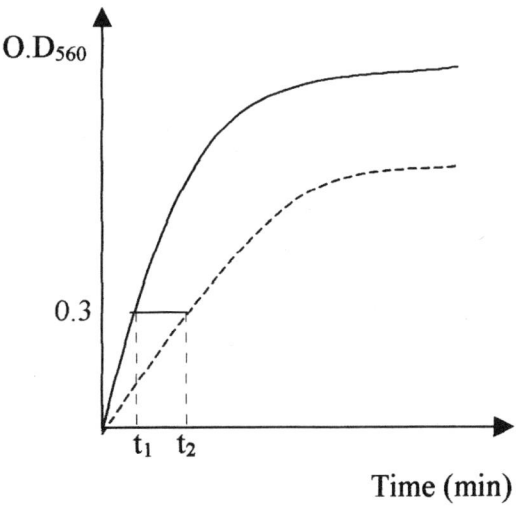

Fig. 1. Example for calculating growth delay. $D = t_2 - t_1$.

2.5.3. HPLC

MRSTC broth: MRS broth supplemented with 0.5 g cysteine chlorhydrate (*see* **Note 2**) and 5 g TCA (final concentration, 4 mM). Sterilize at 121°C for 20 min.

2.6. Materials for Bile Tolerance

1. An active culture grown in MRS broth for about 16 h at 37°C (overnight culture).
2. Phosphate buffer: Solution A (0.2 M NaH$_2$PO$_4$): Weigh 27.6 g NaH$_2$PO$_4$·H$_2$O and make to 100 mL with distilled water (dH$_2$O). Solution B (0.2 M Na$_2$HPO$_4$): Weigh 53.05 g Na$_2$HPO$_4$·7H$_2$O and make to 100 mL with dH$_2$O. To prepare 1 L of 0.1 M sodium phosphate buffer (pH 7.0), mix 195 mL of solution A and 305 mL of solution B and bring to 1000 mL with dH$_2$O. Sterilize at 121°C for 20 min.
3. MRSO broth: MRS broth supplemented with different concentrations of bile salts, as mentioned in **Subheading 2.2.**

2.7. Materials for Gastric Juice Resistance

1. An overnight culture in MRS broth.
2. 0.1 M phosphate buffer, pH 7.0, sterile.
3. Artificial gastric juice, pH 2–2.2 and pH 6.5–7.0.
4. 0.1% peptone water: To prepare 1 L, weigh 1 g peptone and bring it to 1000 mL with dH$_2$O. Sterilize at 121°C for 20 min.
5. MRS agar.

2.8. Materials for Cholesterol Reduction

1. An overnight culture in MRSOCH broth.
2. Ethanol (95%), prepared at the moment of using or stored at 4°C.
3. Potassium hydroxide (50%), stable at room temperature.
4. Hexane pure, stored at 4°C.
5. Distilled water.
6. Nitrogen gas.
7. *o*-Phthalaldehyde reagent. This reagent contains 0.5 mg *o*-phthalaldehyde (Sigma) per mL of glacial acetic acid. It is prepared at the moment of using.
8. Concentrated sulfuric acid.

2.9. Materials for Bile Salt Hydrolase Activity

2.9.1 Plate Assay

1. An overnight culture in MRS.
2. MRS agar.
3. MRSBA agar (*see* **Fig. 2**).

2.9.2. Colorimetric Assay

1. An overnight culture in MRSTT broth.
2. NaOH (1 N), stored at room temperature.
3. Distilled water.
4. HCl (10 N), stable at room temperature.
5. Ethyl acetate pure, stored at 4°C.
6. Nitrogen gas.
7. NaOH (0.01 N), stable at room temperature.
8. Sulfuric acid (16 N), stable at room temperature.
9. 1% Furfuraldehyde. This reagent is particularly toxic and must be prepared at the moment of using.
10. Glacial acetic acid.

2.9.3. HPLC Analysis

1. An overnight culture in MRSTC broth.
2. Membrane filter (0.22 µm).
3. C18 Spherisorb 5-µm column (250 × 4.6 mm).
4. Programmable solvent module (126 M).
5. 106 Dioxide array detector.
6. Ultraviolet detector (210 nm).
7. Eluent A: 65% methanol in 0.03 M sodium acetate, pH 4.3.
8. Eluent B: 90% methanol in 0.07 M sodium acetate, pH 4.3 (*see* **Note 3**).

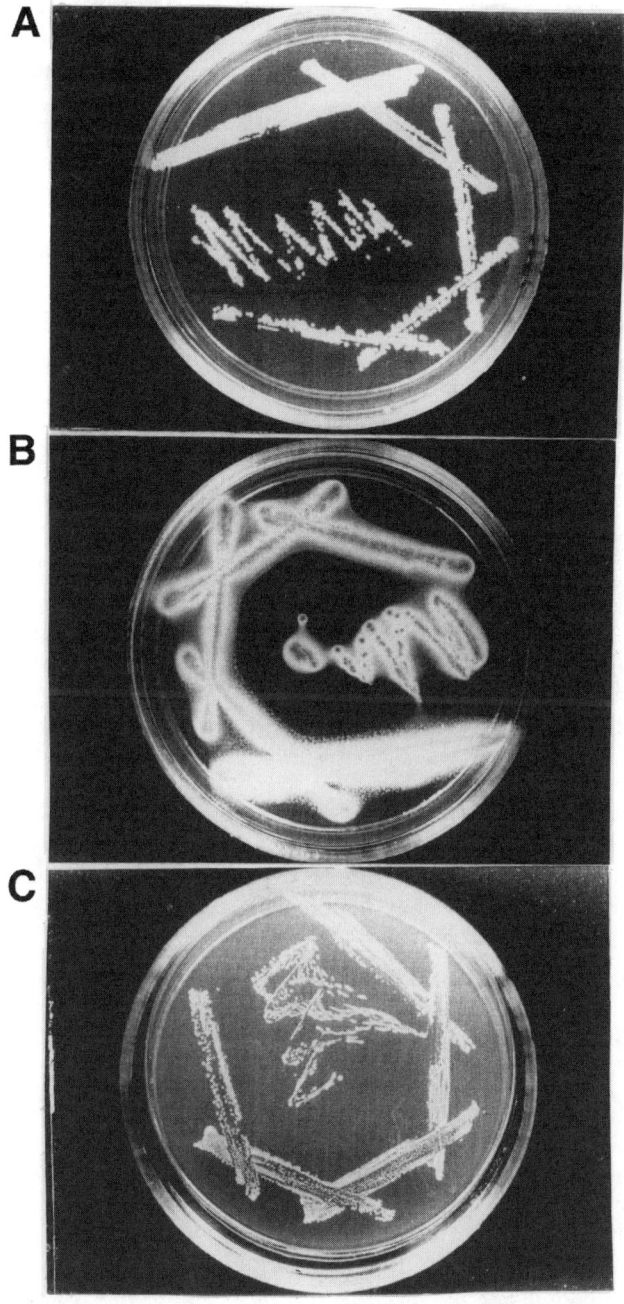

Fig. 2. Bile salt hydrolase activity in LAB on MRS agar: (**A**) plate with TDCA; (**B**) plate with GDCA; (**C**) control.

9. Standard solution of bile acids: taurocholic (0.2 mg/mL), glycocholic (0.2 mg/mL), and cholic acids (0.4 mg/mL). Filter the solution before injection by using a 0.22-μm filter membrane.

3. Methods
3.1. Bile Tolerance

1. Grow the lactobacilli under study in MRS broth at 37°C for 16 h.
2. Harvest the cells by centrifugation at 5000g for 10 min.
3. Wash twice the pellet obtained with 0.1 M phosphate buffer, pH 7.0.
4. Resuspend the cells to the original volume with the buffer by vortexing.
5. Inoculate (0.5%) (*see* **Note 4**) MRS and MRSO broth with the bacterial suspension.
6. Incubate at 37°C in water bath.
7. Read the optical density at 560 nm (OD_{560}) against the blank (uninoculated broth) every hour for the first 8 h and after 24 h of incubation.
8. Plot optical density values against incubation time (*see* **Note 5**).

3.2. Gastric Juice Resistance

1. Repeat **steps 1–4** of **Subheading 3.1.**
2. Inoculate (2%) the artificial gastric juice pH 2–2.3 and pH 6.5–7.0 with the bacterial suspension.
3. Incubate both media at 37°C and take samples at 0, 1, 2, 3, and 4 h and after 24 h for cell viability.
4. Plate in MRS agar (in mass) proper dilutions from 10-fold serial dilutions prepared in 0.1% peptone water.
5. Incubate the plates at 37°C for 24 h, and count the resulting colonies after that time. Results are expressed as colony-forming units (CFU) per milliliter (CFU/mL).

3.3. Cholesterol Reduction (5)

1. Inoculate (1%) 20 mL of MRSCHO broth an overnight culture in MRS broth and incubate at 37°C for 16 h. Uninoculated MRSCHO broth (control) is processed in the same way.
2. Remove the cells by centrifugation at 8000g for 5 min. Place the sample (0.5 mL supernatant) into a clean glass tube.
3. Add 3 mL of 95% ethanol to each tube, followed by 2 mL of 50% potassium hydroxide. Mix after the addition of each component.
4. Heat the tubes in water bath at 60°C for 10 min. Cool at room temperature (20°C).
5. Carefully add 5 mL of hexane. Mix vigorously with a vortex for 20 s. Add 3 mL dH_2O and repeat the mixing with the vortex.
6. Let the tubes settle at room temperature for 15 min or until complete phase separation (aqueous and organic phase).

Probiotic Properties of Lactobacilli

7. Transfer 2.5 mL of the hexane layer (upper phase) into a clean tube. Evaporate hexane to dryness at 60°C under nitrogen gas flow.
8. Resuspend the residue formed in 4 mL of o-phthalaldehyde reagent. Keep the tubes at room temperature for 10 min and then pipet 2 mL of concentrated sulfuric acid slowly down the inside of each tube. Mix thoroughly as described previously.
9. After standing at room temperature for an additional 10 min, read the absorbance at 550 nm (A_{550}) against the reagent blank (*see* **Note 6**).
10. The results are expressed as micrograms (µg) of cholesterol per milliliter.

3.4. Bile Salt Hydrolase Activity

3.4.1. Plate Assay (6)

1. Melt the agar media: MRS agar and MRSBA agar (described in **Subheading 2.5.1.**) in boiling water. Pour each melted medium separately into sterile Petri dishes (60 × 15 mm).
2. Once solidified, invert the plates and place in an anaerobic chamber for at least 48 h before using.
3. Inoculate each plate on surface with an overnight culture grown in MRS broth by using a 10-µL loop.
4. Incubate the plates at 37°C in anaerobic jars (Systen Oxoid) for 72 h.
5. The bile salt hydrolase activity of the cultures is evidenced by the formation of a white precipitate around the colonies grown in MRSBA agar. This precipitate is not observed in MRS agar (control) without bile salts, where colonies are translucent (*see* **Note 7**).

3.4.2. Colorimetric Assay (7)

1. Inoculate (1%) 20 mL of MRSTT broth with an overnight culture grown in MRS broth and incubate at 37°C for 16 h.
2. Adjust the pH to 7.0 with 1 *N* NaOH and take to 25 mL with distilled water (*see* **Note 8**).
3. Remove the cells by centrifugation at 12,000*g* at 4°C for 10 min.
4. Adjust 15 mL of the resulting supernatant fluid to pH 1.0 by using 1 *N* HCl and take to 24 mL with distilled water. (*See* **Note 8**).
5. Transfer 3 mL of each sample to glass test tubes and add 9 mL of ethyl acetate.
6. Mix the content of each tube and let settle to allow the complete phase separation.
7. Transfer 3 mL of the ethyl acetate layer (upper phase) to a clean test tube and evaporate to dryness at 60°C under nitrogen gas flow.
8. Dissolve the residue formed with 1 mL of 0.01 *N* NaOH.
9. Carefully add 6 mL of 16*N* H_2SO_4 to each tube, followed by the addition of 1 mL of 1% furfuraldehyde.
10. Mix the tubes, heat at 65°C for 13 min in water bath, and cool at room temperature.

11. Add 5 mL of glacial acetic acid to each tube, mix the content vigorously, and read the absorbance at 660 nm (A_{660}) against the reagent blank (*see* **Note 9**).
12. The results are expressed as micromoles (µmol) cholic acid per milliliter.

3.4.3. HPLC Analysis (8)

1. Inoculate (1%) MRSTC broth with an overnight culture grown in MRS broth and incubate at 37°C for 16 h.
2. Remove the cells by centrifugation at 8000g for 5 min and filter the supernatant by using a of 0.22-µm filter membrane.
3. Inject 50 µL of each standard solution, previously filtrated and diluted (1:2).
4. Perform the analysis on a 250 × 4.6-mm C18 Spherisorb 5-µm column.
5. Use the following elution program: isocratic elution with 15% of eluent B, 85% of eluent A for 10 min, then a 25-min linear gradient to 90% of solvent B.
6. Maintain the mobile-phase composition at 90% of solvent B. The flow rate is 1 mL/min.
7. Perform the detection at 210 nm, and process the data with System Gold™ software.
8. After running the standard, follow **steps 3–7** for the samples obtained in **step 2**, which are also filtered and diluted (1:2) (*see* **Note 10**).

4. Notes

1. The range of bile salt concentration used (0.05–0.3%) corresponded to that found in the human intestinal tract; 0.3% bile is the maximun concentration that is present in healthy men.
2. Lactic acid bacteria are anaerobic or microaerophile. Culture growth in liquid and agarized media for evaluation of both cholesterol reduction and bile salt hydrolase activity were performed under anaerobiosis. Thioglycholate and cysteine chlorhydrate are used to mantain a low redox potential.
3. Methanol and sodium acetate used for HPLC analysis must be of HPLC grade.
4. For determining the bile tolerance of the cells, it is important to use a low inoculum (0.5%, v/v) in order to have a low initial OD_{560} value and to assure that the culture is at an early exponential phase of growth.
5. The comparison of the cultures is based on the time required for each of them to increase the OD_{560} by 0.3 units (generally, this value is found in the early exponential phase of growth) in both MRS and MRSO broths; the difference in time (min) between the culture media is considered as the growth delay *(D)* (*see* **Fig. 1**).
6. The A_{550} is compared with a standard curve to determine the concentration of cholesterol. The same procedure for the samples is used for the standard curve, except that the following amounts of cholesterol (Sigma L-4646) are assayed in place of the samples: 0, 10, 20, 30, 40, 50, 60, 70, 80, 90, and 100 µg. The A_{550} values are plotted against microgram of cholesterol.

7. TDCA and GDCA produce the most sharply defined halos (white dense precipitate and diffused halos around colonies, respectively (*see* **Fig. 2**), whereas TCA and GCA are slightly less effective. Strains with a high bile salt hydrolase activity for TDCA and GDCA will also release high concentrations of deoxycholic acid, which may inhibit the growth in the plate. In this case, it is suggested to use a lower TDCA or GDCA concentration (2 m*M*).
8. The pH must be carefully adjusted to 7.0 and 1.0 in each case. In this step, the complete separation of the two forms of the bile acid (i.e., conjugated and unconjugated taurocholic acid) as well as the further extraction of the cholic acid released is performed.
9. The A_{660} is compared with a standard curve to determine the concentration of cholic acid. The same procedure used for the samples is applied for the standard curve, except that the following amounts of cholic acid are assayed in place of the samples: 0, 0.5, 0.7, 1.0, 1.5, 2.0, 2.5, 3.0, 3.5, and 4.0 μmol. The A_{660} values are plotted against micrograms of cholic acid per milliliter.
10. The early eluting peak (about 0.1 U absorbancy in the elution time 1.6–4 min), corresponds to residues of MRS broth. The elution time for taurocholic, glycocholic, and cholic acids are 4.75, 6.22, and 17.11 min, respectively. Therefore, the first peak does not interfere with the measurements of interest.

References

1. Hepner, G., Fried, R., Jeor, S., Fusetti, L., and Morin R. (1979) Hypocholesterolemic effect of yogurt and milk. *Am. J. Clin. Nutr.* **32,** 19–24.
2. Walker, D. K. and Gilliland, S. E. (1983) Relationship among bile tolerance, bile salt deconjugation, and assimilation of cholesterol by *Lactobacillus acidophilus*. *J. Dairy Sci.* **76,** 956–961.
3. Kilara, A. (1982) Influence of *in vitro* gastric digestion on survival of some lactic cultures. *Milchwissenschaft* **37,** 129–132.
4. De Man, J. C., Rogosa, M., and Sharpe, M. E. (1960) A medium for the cultivation of lactobacilli. *J. Appl. Bacteriol.* **23,** 130–135.
5. Rudel, L. L. and Morris, M. D. (1973). Determination of cholesterol using *o*-phthalaldehyde. *J. Lipid Res.* **14,** 364.
6. Dashkevicz, M. P. and Feighner, S. D. (1988) Development of a differential medium for bile salt hydrolase-active *Lactobacillus* sp. *Appl. Environ. Microbiol.* **55,** 11–16.
7. Irvin, J. L., Johnson, C. G., and Kopalo, J. (1944) A photometric method of determination of cholates in bile and blood. *J. Biol. Chem.* 439–457.
8. Scalia, S. (1988) Simultaneus determination of free and conjugated bile acids in human gastric juice by high performance liquid chromatography. *J. Chromatogr.* **431,** 259–269.

22

Identification of Exopolysaccharide-Producing Lactic Acid Bacteria

A Method for the Isolation of Polysaccharides in Milk Cultures

Fernanda Mozzi, María Inés Torino, and Graciela Font de Valdez

1. Introduction

Exopolysaccharides (EPS) are exocellular polysaccharides that can be found either attached to the cell wall in the form of capsules or secreted into the extracellular environment in the form of slime.

Among the wide variety of EPS-producing microorganisms, lactic acid bacteria (LAB) have gained a lot of attention because of the interesting properties of these polymers and the GRAS (Generally Recognized as Safe) status of the group. LAB capable of synthesizing EPS are often used in the manufacture of fermented milks *(1)* and more recently, EPS$^+$ starter cultures have been used in the elaboration of low-fat cheeses (mozzarella) *(2)*. The presence of these kind of polymers improves the texture of the fermented milks, prevents the syneresis, and can increase moisture retention in low-fat mozzarella.

Milk has been commonly used as a culture medium for EPS production by LAB. The techniques for the identification of EPS-producing strains and for the isolation of these polymers from milk cultures will be considered under the following headings:

1. Mucoid colonies formation on agar medium.
2. Ropy appearance in milk cultures.
3. Capsular polysaccharides; India ink negative staining.
4. Isolation of EPS.

2. Materials

2.1. Growth Media

Liquid Non-Fat Skim Milk (NFSM): NFSM (100 g/L). Sterilize at 115°C for 20 min.

2.2. Media for Detecting Mucoid Colonies

Milk agar: Liquid NFSM with 1.5% agar (*see* **Note 1**). Sterilize at 115°C for 20 min.

2.3. Media for Determining Ropy Appearance in Milk Cultures and for EPS Isolation

Liquid NFSM: NFSM (100 g/L). Sterilize at 115°C for 20 min.

2.4. Materials for Detecting Mucoid Colonies on Milk Agar

1. An active culture grown in liquid NFSM for about 16–24 h at the usual growth temperature.
2. Plates of milk agar medium.
3. Peptone water 0.1% as dilution medium. To prepare 1 L, weigh 1 g peptone and bring it to 1000 mL with distilled water. Sterilize at 121°C for 20 min.
4. Sterile toothpicks.

2.5. Materials for Observing Ropiness in Milk Culture

1. An active culture grown in liquid NFSM for about 16–24 h at the usual growth temperature.
2. Pipets.
3. Meter scale.

2.6. Materials for Detecting Capsular EPS

1. An active culture grown in liquid NFSM for about 16–24 h at the usual growth temperature.
2. India ink.
3. Distilled water
4. Very clean microscope slides
5. Cover slips

2.7. Materials for Isolating EPS

1. An active culture grown in liquid NFSM for about 16–24 h at the usual growth temperature.
2. Sterilized distilled water.

Identification of Exopolysaccharide-Producing LAB

3. Sterilized NaOH (2 N), stable at room temperature.
4. Pronase E Type XIV from *Streptomyces griseus* (Sigma Chemical Co.). Add 10 mg of the enzyme to 100 mL of milk culture (*see* **Note 2**). Employ sterile material to weigh.
5. Merthiolate, add to reach 0.1% final concentration.
6. 95° Ethanol. Store at 4°C before use.
7. Distilled water.
8. Dialysis sacks that retain compounds with molecular weight greater than 12,000 (Sigma Diagnostics) (*see* **Note 3**).

2.8. Materials for the EPS Quantification by the Phenol–Sulfuric Method

1. Very clean glass tubes (*see* **Note 4**).
2. Distilled water.
3. Phenol reagent (80% w/v). To prepare 10 mL, weigh 8 g phenol using a 10-mL test tube, dissolve at 37°C, and then add distilled water to the final volume. Phenol is toxic and must be carefully prepared.
4. Concentrated sulfuric acid (*see* **Note 5**).

3. Methods
3.1. Detecting Mucoid Colonies on Milk Agar (3)

1. Grow the LAB under study in liquid NFSM at the usual temperature for 16–24 h.
2. Melt the milk agar medium in boiling water and pour into sterile Petri dishes.
3. Prepare appropriate 10-fold dilutions of the NFSM culture using water peptone as diluent.
4. Transfer and spread 0.5 mL of the desire dilution on surface of milk agar plates.
5. Allow the plates to set, then invert and incubate for about 48–72 h at the appropriate temperature (*see* **Note 6**).
6. Touch the colonies formed with sterile toothpicks, look for typical threadlike structure.

3.2. Determining Ropy Appearance in Milk Cultures (3)

1. Grow the LAB under study in liquid NFSM at the usual temperature for 16–24 h.
2. Touch the milk cultures with sterile pipets and let the coagulated milk drop.
3. Measure the length of the string formed with a meter scale. Measurements between 0 and 5 mm are recorded as nonropy (–) and those higher than 6 mm as ropy (+).

3.3. Determination of Capsular EPS (4)

1. Grow the LAB under study in liquid NFSM at the usual temperature for 16–24 h.
2. Prepare a dilution of the coagulated LAB culture with water peptone or distilled water.

3. Place a loop of India ink on a very clean, grease-free microscope slide.
4. Mix into the India ink a little of the bacterial suspension.
5. Place a cover slip on the mixture, avoiding air bubbles, and press firmly with blotting paper until the film of liquid is very thin.
6. Place one drop of immersion oil on the microscope slide. Examine with the oil-immersion objective. The capsule will be seen as a clear area around the bacterium.

3.4. Isolation of EPS from Milk Cultures (5)

1. Grow the LAB under study in 50–100 mL (*see* **Note 7**) liquid NFSM at the usual temperature for 16–24 h. Perform the following steps in sterile conditions (*see* **Note 8**).
2. Dilute the coagulated cultures twice with sterile distilled water. Mix with a sterile pipet to break the coagulum.
3. Adjust the pH to 7.5 (*see* **Note 9**) with sterile NaOH (2 *N*). Mix with a sterile pipet.
4. Add the Pronase E and merthiolate (*see* **Note 10**). Mix with a sterile pipet.
5. Incubate at 37°C (*see* **Note 11**) for about 24 h.
6. Remove the cells by centrifugation (16,000g, 30 min at 4°C) (*see* **Note 12**). The EPS are present in the supernatant. For the calculation of EPS production, the volume of sterile water added in **step 2** must be subtracted from the supernatant volume obtained after centrifugation.
7. Concentrate the clarified supernatant five times by evaporation using a rotavap evaporator (*see* **Note 13**).
8. Recover the EPS from the concentrated supernatant by precipitation at 4°C for 24 h with 3 vol of cold 95° ethanol.
9. Dissolve the precipitated EPS with 2–3 mL of distilled water.
10. Dialyze under stirring against distilled water at 4°C during 48–72 h (*see* **Note 14**). The volume of liquid retained inside the sacks (containing the EPS) must be considered for calculations.
11. Store at –20°C or lyophilize and store at 4°C.

3.5. Quantification of EPS Production (8)

1. In very clean glass tubes, add the following:
   a. 800 µL of sample (dialyzed EPS solution) (*see* **Note 15**)
   b. 40 µL of phenol reagent; mix by vortexing.
   c. 2 mL of sulfuric acid (*see* **Note 16**); mix by vortexing.
2. Prepare a reagent blank using 800 µL of distilled water instead of sample.
3. Read the absorbance at 490 nm (A_{490}) against the reagent blank (*see* **Note 17**).
4. The determinations must be performed in triplicate (*see* **Note 18**).
5. The results are expressed as milligrams (mg) of EPS per liter (*see* **Note 19**).

4. Notes

1. LAB are very exacting microorganisms and usually it is necessary to add yeast extract (5 g/L) and glucose (10 g/L) to enhance the growth on solid media.

2. To simplify the procedure when many samples are manipulated, prepare a solution of the enzyme to an appropriate concentration (i.e., for five samples of 100 mL each, prepare 5 mL of a 10-mg/mL solution and add 1 mL). The solution must be freshly prepared with a sterile tube and sterile distilled water.
3. The cutoff of the dialysis sacks depends of the molecular weight of the EPS isolated. LAB usually produce EPS of about 10^4–10^6.
4. The phenol sulfuric method is very sensitive and the employed materials must be very clean. The new glass tubes must be carefully washed with acidulated water and rinsed three times with distilled water. Once used, keep them with the reaction mixture. When needed, discard the reaction mixture and rinse several times with water and three times with distilled water.
5. Sulfuric acid used must be high quality.
6. LAB are anaerobic or microaerophile and many species do not grow well on the surface of solid media incubated aerobically. When necessary, incubate the Petri dishes in an oxygen-free atmosphere. The incubation period will be longer than 48 h.
7. The amount of EPS produced by LAB is scarce (usually in the order of milligrams per liter) and the isolation from a small volume of culture is very erroneous.
8. The sterile conditions must be maintained because of the further incubation period with the Pronase.
9. The pH 7.5 is the optimal one for the Pronase activity.
10. Pronase is added to digest the milk proteins and merthiolate to prevent further cell growth during the additional incubation period.
11. The incubation temperature of 37°C is the optimal for the Pronase activity.
12. The presence of EPS as capsule usually decreases the adhesion of the cell pellet and the centrifugation parameters must be extended. When a high amount of EPS remains attached to the cell wall, it must be removed from the pellet to avoid lower EPS values. In this cases, the pellet is washed with sterile 0.9% NaCl and centrifuged again. The supernatant fluid is decanted with ethanol and the pellet resuspended in 1 mL 5% trichloroacetic acid. It is then vortexed at maximum velocity for 1 min and transferred to an Eppendorf tube. The debris are centrifuged at 1000*g* for 5 min at room temperature, washed twice in Pronase buffer *(7)*, and resuspended in 1 mL of Pronase buffer. Pronase (20 mg/mL) is added and the mixture is incubated at 40°C for 20 h. The digest is dissolved in 5 mL of distilled water and dialyzed *(8)*.
13. The temperature used during the concentration may be about 60–65°C.
14. Dialysis is one of the critical steps in the EPS isolation because the residual sugar from culture medium is removed, avoiding inflated values of EPS quantification. The water for dialysis must be changed twice each day.
15. The amount of sample added depends of the EPS concentration. Usually 50 µL (plus 750 µL of distilled water to reach the final volume of 800 µL) gives appropriate A_{490} values.
16. The sulfuric acid must be quickly added in the middle of the reaction mixture.

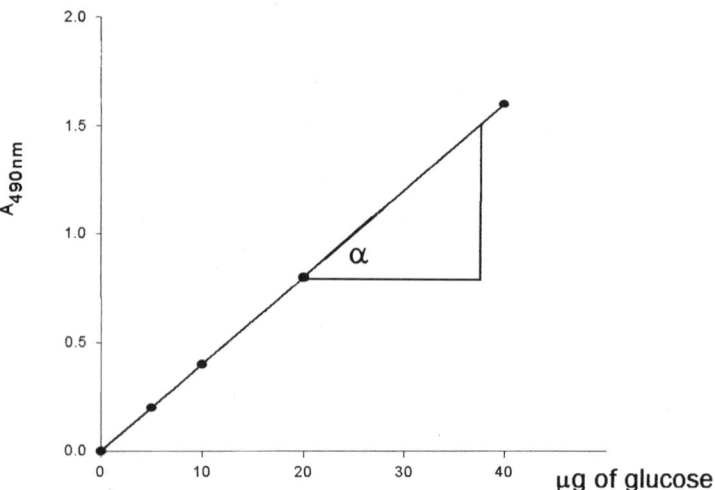

Fig. 1. Plot of A_{490} values versus micrograms of glucose.

17. The A_{490} of the blank reagent must be less than 0.1. The measured A_{490} difference (sample and blank) should be at least 0.2 absorbance units to achieve sufficiently accurate results. If lower values are obtained, the amount of sample employed in the reaction must be increased.
18. Because of the high sensitivity of the phenol sulfuric method, a single determination is unacceptable. Those A_{490} values for each sample that differ in 0.1 or less absorbance units are averaged.
19. The amount of EPS is calculated with the following formulas:

$$\text{EPS (mL/mg)} = \frac{\overline{A_{490}} \times F}{\text{mL}}$$

where A_{490} is the averaged absorbance values, F is the factor calculated from the standard curve (see below), and mL is the amount of sample used in the phenol sulfuric reaction.

EPS mg (in mL of dialyzed) = EPS (mg/mL) × mL of dialyzed

where mL of dialyzed is the amount of liquid retained into the dialysis sack.

$$\text{EPS production (mg/L)} = \frac{\text{EPS (mL dialyzed)} \times 1000}{\text{mL of supernatant} - \text{mL of sterile water (\textbf{step 2})}}$$

Standard curve for calculating the factor F. The same procedure used for the phenol sulfuric method in samples is applied for the standard curve, except that glucose (1 g/L) is used as standard reagent in place of the samples. The amount of

Table 1
Example of Calculating Factor F from the Standard Curve

	X_i (µg of standard glucose)	Y_i (A$_{490}$)	X_iY_i	X_i^2
	0	0.04	0	0
	5	0.2	1	25
	10	0.4	4	100
	20	0.8	16	400
	25	0.9	22.5	625
	30	1	30	900
	40	1.6	64	1600
$(\Sigma X_i)^2 = 16,900$ Total	$\Sigma 130$	4.94	137.5	3650

$a = (7 \times 137.5 - 130 \times 4.94)/(7 \times 3650 \times 16900)$
$a = 0.037$
$F = 1/a = 27$

glucose standard added are: 0, 5, 10, 20, 25, 30, and 40 mL (or micrograms because the concentration of the standard is 1 g/L). The A$_{490}$ values are plotted against µg glucose in **Fig. 1**.
The mathematical expression of a line is: $y = ax + b$, where

$$a = \text{tg } \alpha = \frac{A_{490}}{\mu g \text{ glucose}}; \quad \mu g \text{ glucose} = \frac{A_{490} 1}{(-) a}; \quad F = \frac{1}{a}$$

In order to choose the best line, the principle of least squares is generally used. By this method, the a value is obtained and F can be calculated:

$$a = \frac{n\Sigma X_iY_i - \Sigma X_i \Sigma Y_i}{n\Sigma X_i^2 - (\Sigma X_i)2}$$

where n is the number of trials. In this case, $n = 7$ (0, 5, 10, 20, 25, 30, and 40). See **Table 1** for values used in the calculation.

References

1. Sutherland, I. W. (1990) Food usage of polysaccharides, in *Biotechnology of Microbial Polysaccharides* (Sutherland, I. W., ed.,), Cambridge University Press, Cambridge, pp. 117–125.
2. Perry, D. B., McMahon, D. J., and Oberg, C. J. (1997) Effect of exopolysaccharide producing cultures on moisture retention in low-fat mozzarella cheese. *J. Dairy Sci.* **80**, 799–805.
3. Mozzi, F. (1995) Ph.D. thesis. Universidad Nacional de Tucumán, Argentina.
4. Harrigan, W. F. and McCance, M. E. (1976) *Laboratory Methods in Food and Dairy Microbiology*, Academic, London.
5. Cerning, J., Bouillanne, C., and Desmazeaud, M. J. (1986) Isolation and characterization of exocellular polysaccharide produced by *Lactobacillus bulgaricus*. *Biotechnol. Lett.* **8(9)**, 625–628.

6. Dubois, M., Gilles, K. A., Hamilton, J. K., Rebers, P. A. and Smith, F. (1956) Colorimetric method for determination of sugars and related substances. *Anal. Chem.* **28,** 350–356.
7. Sambrook, J., Fritsch, E. F., and Maniatis, T. (1989) *Molecular Cloning: A Laboratory Manual,* 2nd ed., vol. 3, Cold Spring Harbor Laboratory Press, Cold Spring Harbor, NY, p. B16.
8. Escalante, A., Wacher-Rodante, C., García-Garibay, M., and Farrés, A. (1998) Enzymes involved in carbohydrate metabolism and their role on exopolysaccharide production in *Streptococcus thermophilus. J. Appl. Microbiol.* **84,** 108–114.

23

Differentiation of Lactobacilli Strains by Electrophoretic Protein Profiles

Graciela Savoy de Giori, Elvira Maria Hébert, and Raúl R. Raya

1. Introduction

Lactic acid bacteria (LAB) comprise a diverse group of Gram-positive, non-spore-forming microorganisms *(1)*. Fermentable carbohydrates are used as energy sources and are degraded to lactate (homofermentatives) or to lactate and additional products such as acetate, ethanol, carbon dioxide, formate, or succinate (heterofermentatives). Lactobacilli species are widely used in food technology. The manufacture of high-quality products requieres close attention to characterization, differentiation, and maintenance of lactobacilli starter culture strains. The species identification of LAB depends mainly on physiological and biochemical criteria. These conventional methods are time-comsuming and difficult and the results are often ambiguous. Therefore, for microbiological quality control, it is necessary to develop more reliable and quicker identification methods. The electrophoretic separation of cellular proteins is a sensitive technique, applied in bacterial systematics, that mainly provides information on the similarity of strains within the same species or subspecies. The present chapter deals with the most common technique of polyacrylamide gel electrophoresis (PAGE) in the presence of denaturing agents (sodium dodecyl sulfate [SDS]) of whole and wall–cell-associated proteins extracts for strain typing of lactobacilli. These methods will be useful for culture maintenance by giving each particular strain a fingerprint.

2. Materials

1. MRS broth *(2)*: polypeptone (10 g/L), meat extract (10 g/L), yeast extract (5 g/L), glucose (20 g/L), sodium acetate (5 g/L), Tween-80 (1.08 g/L), ammonium citrate (2 g/L), K_2HPO_4 (2 g/L), $MgSO_4 \cdot 7H_2O$ (0.2 g/L), and $MnSO_4 \cdot 4H_2O$ (0.05 g/L), pH 6.4 ± 0.2. Sterilize in an autoclave at 121°C for 15 min.
2. Phosphate buffer: Solution A (0.2 M NaH_2PO_4): weigh 27.6 g $NaH_2PO_4 \cdot H_2O$ and make to 100 mL with distilled water (dH_2O). Solution B (0.2 M Na_2HPO_4): weigh 53.05 g of $Na_2HPO_4 \cdot 7H_2O$ and make to 100 mL with dH_2O. To prepare 1 L of 0.1 M phosphate buffer, (pH 7.0) mix 195 mL of solution A and 305 mL of solution B and bring to 1000 mL with dH_2O.
3. Acrylamide solution: Weigh 29.2 g acrylamide and 0.8 g *N,N'*-methylene-*bis*-acrylamide, add about 50 mL of dH_2O, stir until disolved, and dilute to 100 mL. Remove insoluble material by filtration. Store at 4°C in a dark bottle (30 d maximum to avoid hydrolysis of acrylamide to acrylic acid). **Caution:** Acrylamide is neurotoxic; repeated skin contact, inhalation, or swallowing may cause a nervous system disorder.
4. Separation gel buffer: Dissolve 18.15 g Tris in 80 mL dH_2O, adjust to pH 8.8 with HCl and make up to 100 mL with dH_2O. The solution is stable for at least 2 wk when stored at 4°C.
5. Stacking gel buffer: Dissolve 6 g Tris in 80 mL dH_2O, adjust to pH 6.8 with HCl and make up to 100 mL with dH_2O. The solution is stable for at least 2 wk when stored at 4°C.
6. 10% (w/v) sodium dodecyl sulfate (SDS): Dissolve 10 g SDS in 100 mL dH_2O with gentle stirring. Store at room temperature. Solution may become cloudy at temperatures below 20°C, but clarity may be restored by warming to 30°C and mixing.
7. 10% (w/v) Ammonium persulfate (APS): Dissolve 100 mg APS in 1 mL dH_2O. Always use a freshly prepared solution.
8. Reservoir buffer: Prepare solution by combining 6.05 g Tris, 28.8 g glycine, and 2 g SDS. Dissolve with dH_2O to a final volume of 2 L. Final pH should be about 8.3. Store at 4°C; warm to 37°C before use if precipitation occurs.
9. Sample buffer: Prepare solution by combining dH_2O (4 mL), stacking gel buffer (1 mL), glycerol (0.8 mL), 10% SDS (1.6 mL), β-mercaptoethanol (0.4 mL), and bromophenol blue (0.002%). Dilute the sample 1:1 with sample buffer and heat at 100°C for 5 min.
10. Staining solution: Weigh 0.5 g of Coomassie brilliant blue R-250 and dissolve in 500 mL of methanol; add 400 mL dH_2O and 100 mL acetic acid. Store at room temperature. This reagent is stable for several months.
11. Destaining solution: 25% (v/v) methanol and 10% (v/v) acetic acid.
12. *N,N,N',N'*-Tetramethylethylenediamine (TEMED).
13. Glass beads (0.1 mm in diameter, Sigma)
14. SDS molecular weight markers (Sigma).
15. Bio-Rad protein assay (Bio-Rad).
16. Vertical gel electrophoresis unit (Sigma).

3. Methods

3.1. Cultivation of Bacteria

1. Inoculate 1% of the microorganisms in 5 mL MRS broth (*see* **Notes 1** and **2**) and incubate overnight at 30°C (mesophilic bacteria) or 37°C (thermophilic bacteria).
2. Collect cells by centrifugation at 10,000g.
3. Wash the pellet twice with 5 mL of 0.1 M sodium phosphate buffer pH 7.0.
4. Resuspend washed cells in 200 µL of the above-mentioned buffer by vortexing. Be sure that the pellet is completely resuspended before proceeding.

3.2. Preparation of Protein Samples

3.2.1. Whole-Cell Protein Extract

1. Vortex the resuspended cells eight times for 1 min with 200 mg glass beads with 1-min intervals on ice (*see* **Note 3**).
2. Centrifuge the tubes at 12,000g for 7 min to remove glass beads and unbroken cells.
3. Transfer the supernatant to a clean tube by carefully decanting from the pellet. Store at –20°C. For long term storage the samples should preferably be kept at –60°C.

3.2.2. Wall–Cell Associated Proteins Extract

1. Add 40 µL of 10% SDS to 200 µL of resuspended cells.
2. Heat the cell suspension at 100°C for 7 min.

3.3. Electrophoresis of Protein Samples

The SDS-PAGE procedure considered here is based on the technique described by Laemmli *(3)*. All percentages are expressed as weight per volume (w/v), except when indicated otherwise.

3.3.1. Preparation of Separation Gel (12% Acrylamide)

1. Place a comb into the assembled gel sandwich (*see* **Note 4**). With a pen, place a mark on the glass plate at 2 cm below the teeth of the comb. This will be the level to which the separating gel is poured.
2. To a clean flask, add 16.75 mL of dH$_2$O, 12.5 mL separation gel buffer, 20 mL acrylamide solution, 0.5 mL of 10% SDS, 0.25 mL of 10% APS, and 25 µL TEMED (*see* **Notes 5** and **6**).
3. Mix thoroughly by stirring. Holding the flask by the neck and swirl it 8–10 cycles. Pour the solution smoothly to the mark and avoid the inclusion of air bubbles.
4. Immediately overlay the solution with 1.5 mL water-saturated isobutanol to obtain a flat surface (*see* **Note 7**).
5. Allow the gel to polymerize for 45 min to 1 h at 20°C. The acrylamide is gelled when a clear line can be observed 1–2 mm below the water-saturated–isobutanol (*see* **Note 8**).

6. Rinse off the overlaying solution four times with dH$_2$O. This is especially important with alcohol overlays (*see* **Note 9**).
7. The gel can be used after 12 h or stored for 48 h at 4°C in a closed plastic bag (to avoid evaporation; *see* **Note 10**).

3.3.2. Preparation of Stacking Gel (5% Acrylamide)

1. Dry the area above the separating gel with filter paper before pouring the stacking gel.
2. Mix the following in a clean flask: 7.0 mL dH$_2$O, 1.3 mL stacking gel buffer, 1.6 mL acrylamide solution, 0.05 mL of 10% SDS, 0.05 mL of 10% APS, and 10 µL TEMED. Swirl the solution gently but thoroughly (*see* **Note 5**).
3. Rinse the top of the gel with approximately 1 mL of stacking gel solution, pour off, and fill the cassettes inmediately with stacking gel solution.
4. Insert the comb into cassettes; avoid trapping air bubbles.
5. Allow the gel to polymerize at least 1 h at 20°C.
6. Remove the comb and rinse the wells with dH$_2$O.

3.3.3. Loading the Samples

1. After polymerization, fill the upper and lower reservoirs with reservoir buffer. Remove any air bubbles from the bottom of the gel so that good electrical contact is achieved. This can be done by swirling the lower buffer with a pipet until the bubbles clear.
2. Insert the samples through the reservoir buffer and into the wells with a Hamilton syringe or a GE Loader tip (Bio-Rad) attached to a 20-µL pipetman. The volume depends on the concentration of proteins in the extract (*see* **Note 11**). Apply approximately 50 µg of protein (determined with Bio-Rad Protein Assay) from each extract. Adjust the volume to 15 µL with phosphate buffer and mix 1 with sample buffer. Load 20 µL of these samples in each well. Include one reference pattern and a mixture of molecular-weight markers on each slab gel (*see* **Note 12**).

3.3.4. Running the Gel

1. Attach the electrical leads to a suitable power supply with the proper polarity. Be sure to connect the positive lead to the lower chamber.
2. Run the gel at constant current of 10 mA per gel until the tracking dye reaches 1 cm of the botttom of the gel.
3. Turn off the power supply and disassemble the glass plates.

3.3.5. Staining and Destaining

1. Immerse the gel in a container and cover it with staining solution (*see* **Notes 13** and **14**).
2. Shake gently for 1–2 h.

3. Remove stain from the container.
4. Rinse the gel and the gel container with water to remove excess staining solution.
5. Gels may be destained by soaking in several changes of destaining solution until the background is clear (*see* **Note 15**).

3.3.6. Storage

Stained gels may be stored between cellophane sheets or dried onto Mylar Sheets using a gel dryer (Bio-Rad).

3.3.7. Interpretation of Bacterial Protein Electrophoretic Profiles

Visual comparison is the most frequently used method for the interpretation of bacterial protein electrophoretic patterns (*see* **Notes 16** and **17**). However, computer programs have been developed to allow the standardization, normalization, and comparison of data *(4,5)*. Pay special attention to the bands present in the range 20–66 kDa.

4. Notes

1. The cultivation medium chosen should support good growth of the microorganisms. Rich media are recommended.
2. To identify the species it is necessary to include one reference pattern of a neotype strain and compare the electrophoretic profiles.
3. Other methods to lyse cells can be used, such as lysozyme treatment *(6)*, sonication, or pressure techniques.
4. When assembling cassettes, always use clean glass plates and label the bottom right corner of each gel cassette.
5. The electrophoretic reagents must be high quality.
6. The high-molecular-weight proteins (>100 kDa) are better resolved on 7–8% acrylamide gels. Reduce the volume of acrylamide in the separating gel and replace it with water. The stacking gel composition can remain the same.
7. Isobutanol chemically attack the acrylic plastic of the sandwich clamp.
8. If the gel sets more quickly than 1 h, decrease the amount of TEMED; if the gel takes longer, increase the amount of TEMED.
9. Do not allow alcohols to remain on the gel more than 1 h because the top of the gel will dehydrate.
10. It is sometimes convenient to cast the separating gel the day before running the gel. If the stacking gel is to be cast the following day, place about 10 mL of a 1:4 dilution of separating buffer on top of separating gel after rinsing with dH_2O. This will prevent dehydration of the separating gel during overnight storage.
11. Band resolution is improved if small volumes of sample are applied to the gels.
12. Molecular-weight markers or a reference bacterial protein extract must be used as standards to evaluate the reproducibility of the electrophoretic system.
13. Use disposable plastic gloves to handle the gel.

14. Staining should be done inmediately to avoid band broadening by diffusion.
15. If a gel becomes overdestained or fades, the staining can be repeated.
16. One advantage of electrophoretic methods over conventional physiologic test is that once the bacteria are isolated and identified to the genus level, the protein can be prepared and SDS-PAGE results can be determined in 1 d. In contrast, the physiologic tests required for species identification can require an additional incubation of at least 7 d.
17. The electrophoretic protein patterns provide a reliable way to differentiate species, whereas differentiation at the strain level is not always possible. We recommend using both methods (whole cells and wall–cell-associated protein extracts) to identify species.

References

1. Kandler, O. and Weiss, N. (1986) Regular, nonsporing Gram-positive rods, in *Bergey's Manual of Systematic Bacteriology* (Sneath, P. H. A., Mair, N., Sharpe, E., and Holt, J. G., eds.), Williams & Wilkins, Baltimore, pp. 1208–1260.
2. De Man, J. C., Rogosa, M., and Sharpe, M. E. (1960) A medium for the cultivation of lactobacilli. *J. Appl. Bacteriol.* **23,** 130–135.
3. Laemmli, U. K. (1970) Cleavage of structural proteins during the assembly of the head of bacteriophage T4. *Nature* **227,** 680–685.
4. Pot, B., Gillis, M., Hoste, B., Van de Velde, A., Bekaert, F., Kersters, K. , et al. (1989) Intra- and intergeneric relationships of the genus *Oceanospirillum. Int. J. Syst. Bacteriol.* **39,** 23–34.
5. Vauterin, L. and Vauterin, P. (1992) Computer-aided objective comparison of electrophoresis patterns for grouping and identification of microorganisms. *Eur. Microbiol.* **1,** 37–41.
6. Pot, B., Vandamme, P. and Kersters, K. (1994) Analysis of electrophoretic whole-organism protein fingerprints, in *Chemical Methods in Prokariotic Systematics* (Goodfellow, M. and O'Donnell, A. G., eds.), Wiley, Chichester, pp. 493–522.

24

Methods to Determine Proteolytic Activity of Lactic Acid Bacteria

Graciela Savoy de Giori and Elvira María Hébert

1. Introduction

Lactic acid bacteria (LAB) are considered weakly proleolytic when compared with many other groups of bacteria (e.g., *Bacillus, Proteus, Pseudomonas*). However, most strains of LAB rely on a complex proteolytic system that allow them to liberate essential and growth-stimulatory amino acids and small peptides from the protein-rich substrates such as milk, meat, and vegetables in which they are primarily found.

The proteolytic system of dairy LAB is composed of an extracellular, cell-envelope proteinase, various intracellular peptidases, and amino acid and peptide transport systems for uptake of the products of proteolysis. The action of the cell-envelope proteinase is not only crucial to the growth of the organism in milk but also to secondary proteolysis and flavor development in cheese *(1)*.

Many procedures have been used to detect proteolysis of LAB. Traditionally, the method most widely used is that described by Hull *(2)*. The Hull method relies on the release of tyrosine- and tryptophan-containing peptides that react with Folin–Ciocalteau reagent; consequently, this method lacks sensitivity.

In this chapter, we describe different procedures, more sensitive and rapid, that are routinely used in our laboratory to determine the proteolytic activity of LAB. These techniques include the use of *o*-phthaldialdehyde (OPA), chromogenic peptide (S-Ala), or fluorescein isothiocyanate (FITC)-casein as substrate.

2. Materials

2.1. Growth Media

1. MRS broth *(3)*: polypeptone (10 g/L), meat extract (10 g/L), yeast extract (5 g/L), glucose (20 g/L), sodium acetate (5 g/L), Tween-80 (1.08 g/L), ammonium citrate (2 g/L), K_2HPO_4 (2 g/L), $MgSO_4 \cdot 7H_2O$ (0.2 g/L), and $MnSO_4 \cdot 4H_2O$ (0.05 g/L); pH 6.4 ± 0.2. Sterilize in an autoclave at 121°C for 15 min.
2. 10% Reconstituted skim milk (RSM): Dissolve 10 g of skim milk in 100 mL of distilled water (dH_2O). Sterilize by autoclaving at 115°C for 15 min.

2.2. OPA Method

1. Phosphate buffer (0.1 *M*, pH 7.0). Solution A (0.2 *M* NaH_2PO_4): Weigh 27.6 g $NaH_2PO_4 \cdot H_2O$ and make to 100 mL with distilled water (dH_2O). Solution B (0.2 *M* Na_2HPO_4), weight 53.05 g of $Na_2HPO_4 \cdot 7H_2O$ and make to 100 mL with dH_2O. To prepare 1 L 0.1 *M* phosphate buffer (pH 7.0), mix 195 mL of solution A and 305 mL of solution B and bring to 1000 mL with dH_2O.
2. 20% (w/v) Sodium dodecyl sulfate (SDS): Dissolve 20 g SDS in 100 mL dH_2O with gentle stirring. Store at room temperature. Solution may become cloudy at temperatures below 20°C, but clarity may be restored by warming to 30°C and mixing.
3. 100 m*M* sodium tetraborate (borax): Weigh 3.81 g of $Na_2B_4O_7 \cdot 10H_2O$ and make up 100 mL with dH_2O. This solution is stable at least 1 mo at 4°C.
4. The OPA solution is prepared combining the following reagents: 25 mL of 100 m*M* borax, 2.5 mL of 20% SDS, 40 mg OPA (dissolved in 1 mL methanol), and 100 µL β-mercaptoethanol. Dilute to a final volume of 50 mL with dH_2O. This reagent must be prepared daily.
5. 0.75 *N* Trichloroacetic acid (TCA): This solution is prepared by weighing 61.24 g TCA and diluting to 500 mL with dH_2O. The solution is stable for at least 2 wk when stored at 4°C.

2.3. Chromogenic Method

1. MRS–Ca: MRS medium supplemented with 1.47 g/L $CaCl_2 \cdot 2H_2O$ (10 m*M*).
2. 10 m*M* $CaCl_2$–saline solution: Dissolve 4 g NaCl and 0.73 g $CaCl_2 \cdot 2H_2O$ with 500 mL dH_2O.
3. Tris–hydroxymethyl–aminomethane (Tris) buffer (50 m*M*, pH 7.8). Dissolve 1.21 g Tris in 50 mL dH_2O, adjust to pH 7.8 with HCl, and make up to 200 mL with dH_2O. The solution is stable for at least 1 mo when stored at 4°C.
4. 5*M* NaCl: Dissolve 146.25 g NaCl with 500 mL dH_2O.
5. Phosphate buffer (0.2 *M*, pH 7.0): Weigh 27.6 g $NaH_2PO_4 \cdot H_2O$ and 53.05 g of $Na_2HPO_4 \cdot 7H_2O$. Make up to 200 mL with dH_2O.
6. 80% (v/v) Acetic acid: Add 20 mL dH_2O to 80 mL acetic acid.
7. Substrate *N*-succinyl-Ala-Ala-Pro-Phe-*p*-nitroanilide (S-Ala, Sigma). Dissolve the peptide in dimetyl sulfoxide at a concentration of 20 m*M*. Store the solution at –20°C.

2.4. FITC–Casein Assay

1. Tris buffer (50 mM, pH 7.0): Dissolve 1.21 g Tris in 50 mL dH$_2$O, adjust to pH 7.0 with HCl, and make up to 200 mL with dH$_2$O. The solution is stable for at least 1 mo when stored at 4°C.
2. FITC–casein solution (0.4%): Dissolve 4 mg of FITC–casein (Sigma) in 10 mL of 50 mM Tris–HCl buffer, pH 7.0. Store the solution at –20°C.
3. 5% TCA: Dissolve 5 g TCA in 100 mL dH$_2$O. The solution is stable for at least 2 wk when stored at 4°C.
4. Tris buffer (0.5 M, pH 8.5): Dissolve 6.057 g Tris in 50 mL dH$_2$O, adjust to pH 8.5 with HCl, and make up to 100 mL with dH$_2$O. The solution is stable for at least 1 mo when stored at 4°C.
5. The CaCl$_2$–saline solution is prepared as described in **Subheading 2.3**.
6. 10% Sodium citrate: Dissolve 50 g sodium citrate with 500 mL dH$_2$O.

3. Methods
3.1. OPA Method

The reaction of OPA with primary amines occurs only in the presence of a thiol, typically β-mercaptoethanol, and is enhanced at basic pH. Under these conditions, 1-thioalkyl-2-alkylisoindoles are formed, which absorb strongly at 340 nm *(3)*. The advantages of the OPA assay are the following: (1) the assay is simple and rapid, (2) the method is sensitive (detecting approx 7 μM primary amines) because OPA forms adducts having similar absorptivities with 18 of the 20 common amino acids (in the Hull method only tyrosine and tryptophane are detected), and (3) if proteins of known concentration are used as substrate, a percent hydrolysis can be obtained. The disadvantages of this technique are the following: (1) The OPA give a weak reaction with cysteine and none with proline and (2) the measure at 340 nm is the net result of proteolysis and consumption of amino acids during growth.

1. Inoculate 1% MRS and grow the strains overnight at 30°C (mesophilic bacteria) or 37°C (thermophilic bacteria).
2. Collect cells by centrifugation at 10,000g for 10 min at 4°C.
3. Wash the pellet twice with phosphate buffer (*see* **Notes 1** and **2**).
4. Resuspend the pellet to the original volume in the same buffer.
5. Inoculate (1%) the bacterial suspension into 10 mL RSM.
6. Mix the cultures by vortexing and incubate them for 12 h at the optimum temperature (*see* **Note 3**). Also incubate uninoculated milk as control (*see* **Note 4**).
7. To 2.5 mL of the sample homogenized by vortexing, add 0.5 mL dH$_2$O and 5 mL 0.75 N TCA while agitating the test tube to mix thoroughly the coagulated milk.
8. Vortex vigorously and allow to stand for 15 min.
9. Centrifuge the solution at 13,000g for 15 min and transfer the supernatant to a clean tube with a pipet (*see* **Note 5**).

10. Add 1 mL of OPA reagent to 50 µL supernatant in a 1-mL quartz cuvet.
11. Mix briefly by inversion and incubate 2 min at room temperature.
12. Measure the absorbance at 340 nm against the blank (uninoculated milk).
13. Calculate the mmoles per liter of α-amino acid released (mM) from the following relationship:

$$mM = \varepsilon \, \Delta A_{340} \, F \quad \text{(1 cm)}$$

where ΔA_{340} is the experimentally observed change of absorbance at 340 nm using a 1-cm light path, F is the dilution factor corresponding to the assay procedure, and ε is the molar absorption coefficient ($6000/M^-/cm$).

3.2. Chromogenic Method

In this method, the amount of p-nitroanilide (pNA) released from the chromogenic peptide by the action of LAB proteinase is measured at 410 nm. The method is as follows:

1. Grow the microorganism in 100 mL MRS–Ca at 30°C (mesophilic bacteria) or 37°C (thermophilic bacteria) at the optical density at 600 nm of 1.5 (*see* **Notes 3** and **6**).
2. Harvest cells by centrifugation at 10,000g for 10 min at 4°C (*see* **Note 7**).
3. Wash twice with $CaCl_2$_saline solution (*see* **Note 8**).
4. Resusped in 5 mL Tris buffer (50 mM, pH 7.8).
5. To 200 µL resuspended cells (enzyme solution), add 287.5 µL phosphate buffer (0.2M, pH 7.0), 225 µL 5M NaCl, and 37.5 µL S-Ala (*see* **Notes 9–11**).
6. Mix gently and incubate at 30°C (mesophilic bacteria) or 37°C (thermophilic bacteria) for 30 min.
7. Stop the reaction by the addition of 175 µL of 80% (v/v) acetic acid.
8. Centrifuge for 5 min at 13,000g.
9. Measure the release of pNA at 410 nm.
10. The concentration of pNA released can be calculated from the derived value of molar absorption coefficient ($\varepsilon = 8.800/M/cm$):

$$\mu M \, pNA = \varepsilon \Delta A_{410} \, F \times 10^3$$

where ΔA_{410} is the experimentally observed change of absorbance at 410 nm using a 1-cm light path and F is the dilution factor corresponding to the assay procedure.

One unit of enzyme is defined as the amount of enzyme required to release 1 µmol pNA per minute under the conditions of the assay.

3.3.3. FITC–Casein Assay

The procedure described here is based on the techniques described by Twining *(4)* with some modifications.

1. Grow the microrganism overnight in RSM at 30°C (mesophilic bacteria) or 37°C (thermophilic bacteria).
2. Inoculate 2 mL of this preculture in 100 mL RSM and incubate at optimum temperature for 8 h (*see* **Notes 3** and **12**).
3. Collect cells by centrifugation at 10,000g for 10 min at 4°C.
4. Adjust pH to approx 7.0 with NaOH.
5. Add 10 mL sodium citrate and let stand about 30 min at room temperature to clear.
6. Harvest cells by centrifugation at 10,000g for 10 min at 4°C (*see* **Note 7**).
7. Wash the pellet twice with $CaCl_2$ saline solution (*see* **Note 8**).
8. Resuspend the cells in Tris buffer (50 mM, pH 7.0) at optical density at 590 nm of 1.0.
9. Mix 30 µL cells and 20 µL FITC–casein.
10. Incubate at optimum temperature for 1h.
11. Stop the enzyme reaction by adding 120 µL TCA.
12. Centrifuge the mixture and neutralize 60 µL of the supernatant by diluting with 3 mL of 0.5 M Tris buffer, pH 8.5 (*see* **Note 13**).
13. Measure the fluorescein with an excitation of 490 nm and an emission of 525 nm.
14. One unit of the enzyme is defined as the amount of enzyme yielding 1% of the total initial casein fluorescence as TCA soluble fluorescence after 60 min of hydrolysis.

4. Notes

1. Saline solution, instead of phosphate buffer, can be use to wash cells.
2. Cell washes are necessary to minimize carryover of free amino acids during inoculation.
3. Proteinase activity changes during cell growth.
4. A control consisting of uninoculated milk must be sampled at the same time.
5. The samples after the TCA treatment can be frozen at –20°C until they are used.
6. It is always good practice to include positive and negative controls.
7. Keep everything on ice to preserve the enzyme activity.
8. Ca^{2+} is added to prevent the release of proteinase to the culture medium.
9. Enzyme solution may be diluted with Tris buffer.
10. Peptide solution of S-Ala and the buffered enzyme solution must be prewarmed separately before the reaction is started.
11. Other chromogenic substrates such as Suc-Ala-Glu-Pro-Phe-*p*-nitroanilide (S-Glu) or MeOsuc-Arg-Pro-Tyr- *p*-nitroanilide (MS-Arg) can be used *(5)*.
12. Culturing the cells in skim milk medium is important to prevent the repression of the enzyme activity by peptides.
13. Thorough mixing is required at each step for this assay. If the TCA supernatant fraction is not completely neutralized by the pH 8.5 buffer, low-fluorescence yields are obtained, because FITC is colorless below pH 4.0.

References

1. Exterkate, F. A. (1990) Differences in short peptide–substrate cleavage by two cell–envelope-located serine proteinases of *Lactococcus lactis* subsp. *cremoris* are related to secondary binding specificity. *Appl. Microbiol. Biotechnol.* **33,** 401–406.
2. Hull, M. E. (1947) Studies on milk proteins. II. Colorimetric determination of the partial hydrolysis of the proteins in milk. *J. Dairy Sci.* **30,** 881–884.
3. Church, F. C., Swaisgood, H. E., Porter, D. H. and Catignani, G. L. (1983) Spectrophotometric assay using *o*-phthaldialdehyde for determination of proteolysis in milk and isolated milk proteins. *J. Dairy Sci.* **66,** 1219–1227.
4. Twining, S. S. (1984) Fluorescein isothiocyanate-labeled casein assay for proteolytic enzymes. *Anal. Biochem.* **143,** 30–34.
5. Exterkate, F. A., Alting, A. C., and Bruinenberg, P. G. (1993) Diversity of cell envelope proteinase specificity among strains of *Lactococcus lactis* and its relationship to charge characteristics of the substrate-binding region. *Appl. Environ. Microbiol.* **59,** 3640–3647.

25

Methods for Isolation and Titration of Bacteriophages from *Lactobacillus*

Lucía Auad and Raúl R. Raya

1. Introduction

Lactic acid bacteria (LAB), mainly *Lactobacillus* and *Lactococcus* species, are useful microorganisms in many biotechnological processes in the food and feed industries. Bacteriophage contaminations of this important group of Gram-positive bacteria have been reported since the 1930s *(1)* and they are known to be one of the main causes of fermentation failures in the dairy industry. Phages can enter the processes from outside, survive the pasteurization of milk, and remain in factory environments for prolonged periods. Other possible sources of virulent phages are lysogenic bacteria and the prophages they harbor. Lysogeny of LAB has been recognized widely. On the analogy of the *Escherichia coli* Lambda system *(2)*, temperate phages of *Lactobacillus* have been regarded as valuable genetic tools of gene transfer (transduction) and cloning in this industrially important group of bacteria.

In this chapter, we describe the methods optimized in our laboratory for the induction, detection, and propagation of phages infecting *Lactobacillus* species. They are based on standard protocols developed for phages from Gram-negative bacteria, in particular for the coliphage Lambda *(3)*, and have been adapted for bacteriophage attacking lactic acid bacteria.

Mitomycin C is the standard drug used to activate the lytic life cycle of temperate phages *(4,5)*. Although the double-layer plaque assay *(6)* is a quantitative procedure to measure the phage concentration of a lysate solution, the propagation of phages in liquid and solid media allows the amplification the titer of the lysate *(7)*.

2. Materials
2.1. Culture Media

MRS broth *(8)*: 10 g/L peptone, 10 g/L meat extract, 5 g/L yeast extract, 20 g/L glucose, 5 g/L sodium acetate, 1 g/L Tween-80, 2 g/L ammonium citrate, 2 g/L K_2HPO_4, 0.2 g/L $MgSO_4 \cdot 7H_2O$, and 0.05 g/L $MnSO_4 \cdot 4H_2O$; pH 6.5. Autoclave for 20 min at 121°C. Media can be stored at room temperature for months.

MRS–Ca^{2+} soft agar: To MRS, add agar at 0.7% (w/v) and $CaCl_2 \cdot 6H_2O$ to a final concentration equal to 10 mM. Autoclave for 20 min at 121°C. Media can be stored at room temperature for months.

MRS–Ca^{2+} agar: To MRS, add agar at 1.5% (w/v) and $CaClI_2 \cdot 6H_2O$ to a final concentration equal to 10 mM. Autoclave for 20 min at 121°C. Media can be stored at room temperature for months.

2.2. Bacteriophage Induction from Lactobacillus Strains

1. A 37°C water bath.
2. Spectrophotometer.
3. Mitomycin C with a concentration of the stock solution equal to 0.1 mg/mL.
4. Refrigerated centrifuge.
5. Membrane filter of 0.45 µm pore size.
6. Sterile tubes.

2.3. Double-Layer Plaque Assay to Enumerate Bacteriophages

1. 1 M $CaCl_2 \cdot 6H_2O$.
2. 37°C water bath.
3. 45°C water bath.
4. Petri dishes containing MRS agar (1.5%) with $CaCl_2 \cdot 6H_2O$ to a final concentration of 10 mM.
5. Stove at 37°C.

2.4. Propagation of Phages in Liquid Medium

1. 1 M $CaCl_2 \cdot 6H_2O$.
2. 37°C water bath.
3. Refrigerated centrifuge
4. Membrane filter of 0.45 µm pore size.
5. Sterile tubes.

2.5. Propagation of Phages in Solid Medium

1. 1 M $CaCl_2 \cdot 6H_2O$.
2. 37°C water bath.
3. 45°C water bath.
4. Petri dishes containing MRS agar (1.5%) with $CaCl_2 \cdot 6H_2O$ to a final concentration of 10 mM.
5. Incubator at 37°C.
6. TE buffer: 10 mM Tris-HCl, 1 mM EDTA, pH 8.0. Autoclave for 20 min at 121°C.

3. Methods
3.1. Bacteriophage Induction from Lactobacillus Strains

1. Inoculate 5 mL of fresh MRS broth with 50 µL of an overnight culture of the lysogenic *Lactobacillus* strain. Measure the initial absorbance at 600 nm. Incubate in a water bath for 30 min at 37°C (*see* **Notes 2–4**).
2. Add mitomycin C to a final concentration of 0.1–0.5 µg/mL. Measure the absorbance at 600 nm each hour for 6–8 h (until a decrease of the optical density is observed).
3. Centrifuge the culture at 5000g for 15 min at 4°C (*see* **Note 5**).
4. Filter the supernatants containing the phage particles through a membrane filter of 0.45 µm of pore size.
5. Store the sterile supernatant at 4°C for several months (*see* **Note 6**).

3.2. Double-Layer Plaque Assay to Enumerate Bacteriophages

1. Mix 100 µL of an overnight culture of the indicator strain grown in MRS broth with 100 µL of the decimal dilution of the phage lysate suspension (10^0, 10^{-1}, 10^{-2}, 10^{-n}). Use sterile MRS broth to prepare the phage dilutions (*see* **Note 7**).
2. Add $CaCl_2 \cdot 6H_2O$ to a final concentration equal to 10 mM.
3. Preincubate the microtubes 30 min at 37°C in a water bath to allow adsorption of the phages.
4. Add the bacteria–phage mixture (200 µL) to a tube containing 3 mL of molten MRS-Ca^{2+} soft agar at 45°C. Gently mix the tube and pour the contents on Petri dishes containing 15 mL of solidified MRS–Ca^{2+} 1.5% agar. Swirl the plate in circles in the bench top immediately after pouring to spread the sample and the soft agar over the plate (*see* **Notes 8–10**).
5. Leave plates on the bench until the soft agar has solidified (about 15 min).
6. Incubate the plates upside up (with the agar side down) overnight at 37°C.
7. The following day, count the number of plaques and calculate the titer by multiplying the number of plaques by the inverse of the dilution by 10 to express the titer (T) in PFU/mL (plaque-forming units per milliliter) (*see* **Notes 11** and **12**).

3.3. Propagation of Phages in Liquid Medium

1. Inoculate at 2% a fresh, ovenight host bacteria culture in MRS broth. Add $CaCl_2 \cdot 6H_2O$ to a final concentration equal to 10 mM.
2. Incubate at 37°C for 30 min in a water bath.
3. Add at 2% a suspension of the phage lysate with at least 1.0×10^4 PFU/mL.
4. Incubate at 37°C for 6–8 h in a water bath until lysis occurred.
6. Centrifuge at 5000g for 15 min at 4°C.
7. Filter the supernatants containing the phage lysates through a membrane filter of 0.45 µm pore size.
8. Store the sterile supernatant at 4°C for several months.

3.4. Propagation of Phages in Solid Medium

1. Mix 100 µL of an overnight culture of the indicator strain grown in MRS broth with 100 µL of the phage lysate suspension (with at least 1.0×10^2 PFU/mL).
2. Add $CaCl_2 \cdot 6H_2O$ to a final concentration equal to 10 mM.
3. Incubate the tube 30 min at 37°C in a water bath to allow adsorption of the phages.
4. Add the bacteria–phages mixture (200 µL) to a tube containing 3 mL of molten MRS–Ca^{2+} soft agar at 45°C. Gently mix the tube and pour the content on Petri dishes containing 15 mL of solidified MRS–Ca^{2+} 1.5% agar. Swirl the plate in circles on the bench top immediately after pouring to spread the sample and the soft agar over the plate.
5. Leave plates on the bench until the soft agar has solidified (about 15 min).
6. Incubate the plate without inversion overnight at 37°C.
7. At the time of harvesting the phage, the plaques should touch one another, and the only bacteria growth visible should be a gauzy webbing that marks the junction between adjacent plaques (semiconfluent lysis).
8. Remove the plate of the incubator, and add 1 mL of TE buffer over the soft agar.
9. With a micropipet harvest as much of the TE buffer as possible and place in a sterile microtube.
10. Repeat **steps 8** and **9**.
11. Centrifuge at 5000g for 15 min at 4°C (*see* **Note 5**).
12. Recover the supernatant and store at 4°C (*see* **Note 6**).

4. Notes

1. For *Lactobacillus delbrueckii* ssp. *bulgaricus* strains, we observed that 42°C is the optimal temperature of induction *(11)*.
2. Optimal doses of mitomycin C ranged from 0.1–0.2 µg/mL for most of *L.* strains; however, some strains of *L. helveticus* and *L. casei* tolerate 0.5 µg/mL of the inducer well *(9,10)*. Higher concentration of mitomycin C can be toxic for the cells.
3. Wear disposable gloves and suitable protecting clothes when handling mutagenic solutions like mitomycin C.
4. Store mitomycin C solution at 4°C and protect the drug from light and high temperatures.
5. The centrifugation described in **Subheading 3.** can be achieved at room temperature if a refrigerated centrifuge is not available.
6. Lysates of phages that do not contain lipids in their structure can be stored at 4°C for several years with the addition of 0.3% chloroform.
7. Quantification of phage titers by the double-layer procedure should be run in duplicate or triplicate.
8. The phage and bacteria should be exposed to the warm agar (45°C) for as short a period of time as possible.
9. One alternative of the double-layer method is to add the phage to the tube containing molten MRS–Ca^{2+} soft agar at 45°C, then add the bacteria and pour

onto the the plate without preincubation at 37°C. However, we recommend preincubation of the phage–host bacteria mixture to obtain more uniform plaques.
10. Vigorous mixing of the tube containing the phage–host bacteria mixture may damage the phage particles and introduce air bubbles into the soft agar that could look like plaques, especially to the inexperienced eye.
11. There are some bacteriophages, particularly temperate phages, that replicate poorly, resulting in incomplete lysis during propagation in liquid medium. This problem can be solved by increasing the ratio of bacteriophage to bacteria in the infection.
12. In general, the propagation of phages in solid medium is more efficient than in liquid culture.

References

1. Whitehead, H. R. and Cox, G. A. (1935) The occurrence of bacteriophages in starter cultures of lactic streptococci. *MZJ Sci. Technol.* **16,** 319–320.
2. Campbell A. (1994) Comparative molecular biology of lambdoid phages. *Ann. Rev. Microbiol.* **48,** 193–222.
3. Sambrook, J., Fritsch, E. F., and Maniatis, T. (1989) *Molecular Cloning. A Laboratory Manual.* Cold Spring Harbor Laboratory Press, Cold Spring Harbor, NY.
4. Accolas, J. P. and Spillmann, H. (1979) Morphology of bacteriophages of *Lactobacillus bulgaricus, L. lactis* and *L. helveticus. J. Appl. Microbiol.* **47,** 309–319.
5. Sechaud, L., Cluzel, P. J., Rousseau, M., Baumgartner, A., and Accolas, J. P. (1988) Bacteriophages of lactobacilli. *Biochimie* **70,** 401–410.
6. Mata, M., Trautwetter, A., Luthaud, G., and Rikenthaler, P. (1986) Thirteen virulent and temperate bacteriophages of *Lactobacillus bulgaricus* and *Lactobacillus lactis* belong to a single DNA homology group. *Appl. Environ. Microbiol.* **52,** 812–818.
7. Leder, P., Tiemeier, D., and Enquist, L. (1977) EK2 derivates of bacteriophage lambda useful in the cloning of DNA from higher organite: the λgtWES system. *Science* **196,** 175–179.
8. De Man, J., Rogosa, M., and Sharpe, M. (1960) A medium for the cultivation of lactobacilli. *J. Appl. Bacteriol.* **23,** 130–135.
9. Sechaud, L., Rousseau, M., Fayard, B., Callegari, M. L., Quenee, P., and Accolas, J. P. (1992) Comparative study of 35 bacteriophages of *Lactobacillus helveticus:* morphology and host range. *AppL Environ. MicrobioL* **58,** 1011–1018.
10. Herrero, M., de los Reyes-Gavilan, C. G., Caso, J. L., and Suarez, J. E. (1994) Characterization of π393-A2, a bacteriophage that infects *Lactobacillus casei. Microbiology* **140,** 2585–2590.
11. Auad, L., Fortean, P., Alatossava ,T., de Ruiz Holgado, A. P., and Raya, R. R. (1997) Isolation and characterization of a new Lactobacillus delbrueckii subsp. bulgaricus temperate bacteriophage. *J. Dairy Sci.* **80,** 2706–2712.

26

Identification of Yeasts Present in Sour Fermented Foods and Fodders

Wouter J. Middelhoven

1. Introduction

Lactic acid fermentation is commonly used for food conservation. The main products of this bacterial process are lactic and acetic acids, which are toxic to many microorganisms, most yeasts included. The low pH achieved by the lactic acid fermentation, together with anaerobiosis, provides conditions adverse to spoiling and pathogenic microorganisms. Hence, the fermented commodities are stable and can be stored for a long time without loss of quality. The production of sauerkraut is a good example of this practice. In the dairy industry, buttermilk and yogurt are well-known products. This application of the lactic acid fermentation originated from central Asia in times immemorial and has spread from there to Europe and the Orient. It was unknown in other continents until these were colonized from Europe.

Only a limited number of the approximately 700 yeast species known at present are tolerant to lactic and acetic acids at low pH. Many of these inhabit fermented foods and fodders. Some of these are mild pathogens (e.g., *Candida parapsilosis*); others are very harmful when present in silage because of their rapid degradation of lactic and acetic acids under aerobic conditions, which results in loss of nutritive value. *Candida milleri, Issatchenkia orientalis,* and *Saccharomyces exiguus* are the main causative agents of aerobic spoilage of maize silage *(1)*. A review of yeast species isolated from various silages has been given by Middelhoven *(2)*.

In this chapter, easy methods for rapid yeast identification are described. They include simple light microscopy, physiological growth tests, and some additional characteristics. The yeast species to be identified are those found in various silages *(2)* and species mentioned in both recent yeast monography *(3,4)* as inhabitants of commodities like sauerkraut, buttermilk, cucumber brine, and pickles. If identification by the methods proposed in this chapter is unsuccessful, the methods prescribed in both yeast monographs should be applied. It must be kept in mind that only part of the yeasts present in nature has been described yet. Unidentifiable yeast strains may represent unknown species. They are welcomed by yeast culture collections, of which the yeast collection of the Centraalbureau voor Schimmelcultures (CBS) at Delft, The Netherlands, is the most prominent.

Before carrying out any of the tests described, it must have been proven that the isolated yeast cultures are pure and indeed are yeasts. This can be ascertained by plating the cultures on 1% yeast extract, 1% glucose, and 2% agar and by microscopical examination. Only colonies of one type should develop on the plates and light microscopy (magnification ×1000) should reveal true budding yeast cells which are considerably larger than bacteria (usually 3 μm wide or more). However, some species do not propagate by budding but by splitting cells or fragmenting mycelium (e.g., *Dipodascus, Galactomyces* sp.). Some of these yeastlike fungi that do not show budding are frequently isolated from sour foods and fodders and are treated here like they are in the most authorative yeast monographs *(3,4)*.

The colonies of most species treated in this identification key are white or cream; some species are red or orange, but never black. Black yeastlike fungi are not treated here. For their identification, a culture collection (e.g., the Centraalbureau voor Schimmelcultures [CBS] at Baarn, The Netherlands, should be consulted. Slant cultures of isolated strains should be incubated at 25°C for at least 2 wk, to be sure that no black pigment will develop.

In this chapter, a dichotomous identification key to the yeast species listed in **Table 1** is provided (**Table 2**). Identification is valid only if the isolated strain fits the morphological and physiological properties of the species that are given in **Table 3**. These were taken from recent yeast monographs *(3,4)* and from the CBS Yeast Data Base (http.www.cbs.knaw.nl). The nomenclature of the species is according to Kurtzman and Fell *(4)*. The most current synonyms are also presented (**Table 1**).

2. Materials

1. Yeast identification system ID 32 C of BioMérieux (Marcy-l'Etoile, France or Hazelwood, MO, USA).
2. Soluble starch (Merck).
3. Growth media from Difco Laboratories (Detroit, MI): YM agar, yeast nitrogen base, yeast carbon base, yeast extract, potato dextrose agar, bacto vitamin-free yeast base.

Table 1
Names and Current Synonyms of the Yeast Species Treated

1. *Arxula adeninivorans*, synonym: *Trichosporon adeninovorans*
2. *Candida apicola*, synonyms: *Torulopsis apicola, Torulopsis bacillaris*
3. *Candida boidinii*, assimilates methanol
4. *Candida glabrata*, synonym: *Torulopsis glabrata*
5. *Candida holmii*, anamorph of *Saccharomyces exiguus*
6. *Candida lactis-condensi*, synonym: *Torulopsis lactis-condensi*
7. *Candida milleri*
8. *Candida parapsilosis*
9. *Candida pseudolambica*
10. *Candida sake*, synonym: *Torulopsis sake*
11. *Candida tenuis*
12. *Candida tropicalis*
13. *Candida versatilis*, synonym: *Torulopsis versatilis*
14. *Candida wickerhamii*, synonym: *Torulopsis wickerhamii*
15. *Debaryomyces etchellsii*, synonyms: *Pichia etchellsii, Torulaspora etchellsii*
16. *Debaryomyces hansenii*, synonyms: *Candida famata, Torulopsis candida*
17. *Dipodascus capitatus*, synonyms: *Trichosporon capitatum, Geotrichum capitatum*
18. *Galactomyces geotrichum*, synonym: *Geotrichum candidum*
19. *Hanseniaspora uvarum*, synonym: *Kloeckera apiculata*
20. *Hanseniaspora valbyensis*, synonym: *Kloeckera japonica*
21. *Issatchenkia orientalis*, synonym: *Candida krusei*
22. *Kluyveromyces lactis*, synonyms: *Candida sphaerica, Kluyveromyces marxianus* var. *lactis*
23. *Kluyveromyces marxianus*, synonym: *Candida kefyr*
24. *Pichia anomala*, synonyms: *Hansenula anomala, Candida pelliculosa*
25. *Pichia burtonii*, synonyms: *Hyphopichia burtonii, Endomycopsis burtonii, Candida variabilis*
26. *Pichia canadensis*, synonyms: *Hansenula canadensis, Hansenula wingei, Candida melinii*
27. *Pichia fermentans*, synonyms: *Candida lambica*
28. *Pichia holstii*, synonyms: *Hansenula holstii, Candida silvicola*
29. *Pichia membranifaciens*, synonyms: *Pichia membranaefaciens, Candida valida*
30. *Pichia ohmeri*
31. *Pichia pijperi*, synonym: *Hanseniaspora pijperi*
32. *Pichia subpelliculosa*, synonym: *Hansenula subpelliculosa*
33. *Rhodotorula minuta*
34. *Rhodotorula mucilaginosa*, synonym: *Rhodotorula rubra*
35. *Saccharomyces barnettii*
36. *Saccharomyces cerevisiae*, synonyms: *S. carlsbergenis, S. chevalieri, S. ellipsoideus, S. italicus, S. lindneri, S. uvarum*
37. *Saccharomyces dairenensis*, synonym: *Saccharomyces dairensis* or *S. castellii*; species can only be distinguished from each other and from *S. rosinii* with certainty by molecular techniques

Table 1 (cont.)
Names and Current Synonyms of the Yeast Species Treated

38	*Saccharomyces exiguus*, synonyms: *Candida holmii, Torulopsis holmii*
39	*Saccharomyces rosinii*, can be distinguished from *S. dairenensis* and *S. castellii* only by molecular methods
40	*Saccharomyces spencerorum*
41	*Saccharomyces unisporus*
42	*Sacharomycopsis fibuligera*, synonym: *Endomycopsis fibuliger*
43	*Saccharomycopsis selenospora*, synonyms: *Guillermondella selenospora, Endomycopsis selenospora*
44	*Stephanoascus ciferrii*, synonym: *Candida ciferrii*
45	*Torulaspora delbrueckii*, synonyms: *Saccharomyces delbrueckii, Candida colliculosa*
46	*Trichosporon gracile*, fragmenting mycelium, urease positive
47	*Zygosaccharomyces bailii*
48	*Zygosaccharomyces bisporus*, synonym: *Saccharomyces bisporus*
49	*Zygosaccharomyces mrakii*, synonyms: *Saccharomyces mrakii, Torulaspora mrakii*. inositol (20 mg/L) stimulates growth.
50	*Zygosaccharomyces rouxii*, synonyms: *Saccharomyces bailii* var. *osmophilus*

3. Methods
3.1. Morphology

Cells taken from a young pure slant culture are examined microscopically (magnification ×1000) for the presence of budding yeast cells and filaments (mycelium or pseudomycelium). Several species fail to produce mycelium in slant cultures. They should be examined in slide cultures. For this purpose, a Petri dish containing a U-shaped glass rod supporting a glass microscope slide is sterilized by dry heat at 160–180°C for 2 h. A suitable agar (e.g., maize [corn] meal agar or potato dextrose agar [both commercialy available]) is melted and poured into a second Petri dish. The glass slide is quickly removed from the glass rod with a flame-sterilized pair of tweezers and dipped into the molten agar, after which it is replaced on the glass rod. After the surface of the agar has solidified, the yeast is lightly inoculated in either one or two lines along the slide and a sterile cover slip is placed over part of it. A little sterile water is poured into the Petri dish to prevent the agar from drying out. The culture is then incubated at 25°C. After 3 d the slide is examined microscopically (magnification ×400) for the formation of filaments along the edges of the streak, both under and around the cover slip. Some genera (e.g., *Arxula, Dipodascus, Galactomyces, Trichosporon*) are notable for fragmenting of the mycelium into arthroconidia, which often lie in a characteristic zigzag way.

Table 2
Identification Key

1	Nitrate assimilated	2
	Nitrate not assimilated	10
2	Ethylamine asssimilated	3
	Ethylamine not assimilated	*Candida lactis-condensi*
3	Inositol assimilated	*Arxula adeninivorans*
	Inositol not assimilated	4
4	Maltose assimilated	5
	Maltose not assimilated	9
5	No vitamins required	*Pichia anomala*
	Vitamins required	6
6	No growth at 35°C	*Candida versatilis*
	Growth at 35°C positive	7
7	No fermentation of glucose	*Pichia canadensis*
	Gas from glucose	8
8	Raffinose assimilated	*Pichia subpelliculosa*
	Raffinose not assimilated	*Pichia holstii*
9	Erythritol assimilated	*Candida boidinii*
	Erythritol not assimilated	*Candida wickerhamii*
10 (1)	Ethylamine assimilated	11
	Ethylamine not assimilated	59
11	Colonies red or pink	12
	Colonies white or cream	13
12	*N*-Acetyl-D-glucosamine assimilated	*Rhodotorula minuta*
	N-Acetyl-D-glucosamine not assimilated	*Rhodotorula mucilaginosa*
13	No budding yeast cells, fragmenting mycelium	14
	Budding yeast cells, with or without filaments	15
14	Xylose assimilated	*Galactomyces geotrichum*
	Xylose not assimilated	*Dipodascus capitatus*
15	Inositol assimilated	16
	Inositol not assimilated	18
16	Xylose assimilated	17
	Xylose not assimilated	*Saccharomycopsis fibuligera*
17	Rhamnose and raffinose assimilated	*Stephanoascus ciferrii*
	Rhamnose and raffinose not assimilated	*Trichosporon gracile*
18	2-Keto-D-gluconate assimilated	19
	2-Keto-D-gluconate not assimilated	40
19	Maltose not assimilated	20
	Maltose assimilated	23
20	Cellobiose assimilated	*Hanseniaspora uvarum*
	Cellobiose not assimilated	21
21	Growth at 30°C absent or weak	*Zygosaccharomyces mrakii*
	Growth at 30°C positive	22

Table 2
Identification Key (cont.)

22	Tolerates 1% acetic acid	*Zygosaccharomyces bailii*
		Zygosaccharomyces bisporus
	No growth in presence of 1% acetic acid	23
23	Mannitol assimilated	24
	Mannitol not assimilated	*Saccharomyces spencerorum*
24	N-Acetyl-D-glucosamine assimilated	25
	N-Acetyl-D-glucosamine not assimilated	28
25	True mycelium formed	26
	No hyphae or primitive pseudomycelium	*Debaryomyces hansenii*
26	Xylose assimilated	*Candida sake*
	Xylose not assimilated	*Pichia ohmeri*
27	Cellobiose assimilated	28
	Cellobiose not assimilated	34
28	Rhamnose assimilated	29
	Rhamnose not assimilated	30
29	Raffinose assimilated	*Debaryomyces hansenii*
	Raffinose not assimilated	*Candida tenuis*
30	Raffinose assimilated	31
	Raffinose not assimilated	32
31	Fragmenting mycelium present	*Pichia burtonii*
	No hyphae or primitive pseudomycelium	*Debaryomyces hansenii*
32	L-Arabinose assimilated	*Debaryomyces etchellsii*
	L-Arabinose not assimilated	33
33	Maximum growth temperature >40°C	*Candida tropicalis*
	Maximum growth temperature approx 30°C	*Candida sake*
34	L-Arabinose assimilated	35
	L-Arabinose not assimilated	36
35	Galactose assimilated	*Candida parapsilosis*
	Galactose not assimilated	*Candida boidinii*
36	Erythritol assimilated	*Candida boidinii*
	Erythritol not assimilated	37
37	N-Acetylglucosamine assimilated	38
	N-Acetylglucosamine not assimilated	*Torulaspora delbrueckii*
		Candida sake
38	Raffinose assimilated	*Candida apicola*
	Raffinose not assimilated	39
39	Maximum growth temperature >40°C	*Candida tropicalis*
	Maximum growth temperature about 30°C	*Candida sake*
40	(18) Soluble starch not assimilated	41
	Soluble starch assimilated	*Saccharomycopsis fibuligera*
41	Sucrose assimilated	42
	Sucrose not assimilated	49

Table 2
Identification Key (cont.)

42	Cellobiose assimilated	43
	Cellobiose not assimilated	45
43	Melezitose assimilated	44
	Melezitose not assimilated	*Kluyveromyces marxianus*
44	Lactose assimilated	*.Kluyveromyces lactis*
	Lactose not assimilated	*Pichia canadensis*
45	Raffinose assimilated	46
	Raffinose not assimilated	47
46	N-Acetylglucosamine assimilated	*Candida apicola*
	N-Acetylglucosamine not assimilated	*Kluyveromyces marxianus*
47	Glucitol assimilated	48
	Glucitol not assimilated	*Saccharomyces spencerorum*
48	1% Acetic acid tolerated (growth maybe weak)	*Zygosaccharomyces bailii*
	1% Acetic acid not tolerated	*Zygosaccharomyces rouxii*
49	(41) N-Acetylglucosamine assimilated	50
	N-Acetylglucosamine not assimilated	54
50	Glucitol assimilated	*Candida boidinii*
	Glucitol not assimilated	51
51	Cellobiose assimilated	*Hanseniaspora valbyensis*
	Cellobiose not assimilated	52
52	Fermentation of glucose positive	53
	No or weak fermentation of glucose	*Pichia membranifaciens*
53	Xylose assimilated	*Pichia fermentans*
	Xylose not assimilated	*Issatchenkia orientalis*
54	Cellobiose assimilated	55
	Cellobiose not assimilated	56
55	Glucitol assimilated	*Pichia pijperi*
	Glucitol not assimilated	*Hanseniaspora valbyensis*
56	Xylose assimilated, may be delayed	*Candida pseudolambica*
	Xylose not assimilated	57
57	Glucitol assimilated, may be delayed	58
	Glucitol not assimilated	*Saccharomyces unisporus*
58	1% Acetic acid tolerated (growth may be weak)	*Zygosaccharomyces bailii*
		Zygosaccharomyces bisporus
	1%Acetic acid not tolerated	*Zygosaccharomyces rouxii*
59 (10)	Red colonies	*Rhodotorula minuta*
	Colonies white or cream	60
60	Sucrose assimiated	61
	Sucrose not assimilated	70
61	Maltose assimilated	62
	Maltose not assimilated	66
62	2-Ketogluconate assimilated	63

Table 2 (cont.)
Identification Key

	2-Ketogluconate not assimilated	64
63	Hyphae formed	*Candida sake*
	No hyphae or some pseudomycelium	*Torulaspora delbrueckii*
64	Cellobiose assimilated	*Saccharomycopsis fibuligera*
	Cellobiose not assimilated	65
65	Mannitol assimilated	*Torulaspora delbrueckii*
	Mannitol not assimilated	*Saccharomyces cerevisiae*
66	(61) Mannitol assimilated	*Torulaspora delbrueckii*
	Mannitol not assimilated	67
67	No vitamins required or biotin only	68
	Other vitamins required	*Candida milleri*
68	Grows at 30°C	69
	No growth at 30°C	*Saccharomyces barnettii*
69	Ascospores formed	*Saccharomyces exiguus*
	No ascospores formed	*Candida holmii*
70	Mannitol assimilated	*Torulaspora delbrueckii*
	Mannitol not assimilated	71
71	True mycelium present	*Saccharomycopsis selenospora*
	No hyphae or some pseudomycelium	72
72	Grows at 40°C	*Candida sake*
	Maximum growth temperature 37°C or lower	*Saccharomyces dairenensis*
		Saccharomyces castellii
		Saccharomyces rosinii

A flask culture in 2% glucose, 0.5% yeast extract, 1% peptone (GYEP) broth is recommended. In a 100-mL conical flask, 50 mL of the broth is put and sterilized at 120°C for 20 min. The yeast is inoculated to the glass wall at the liquid surface. After incubation for 3 d at 25°C without shaking the culture is examined for the formation of a sediment and a pellicle that may creep onto the glass wall.

3.2. Assimilation of Carbon Compounds

The easiest way to study the pattern of carbon compound utilization, which in many cases is species-specific, is by using the yeast identification system ID 32 C of BioMérieux. For inoculation the manufacturer's instructions should be followed. The test strips are inspected for growth daily, up to 7 d. The test kits should be prevented from drying out. The test kits must be stored at 4°C and should not be used after the expiration date. The following carbon compounds are included in the system:

Table 3
Characteristics of Individual Yeast Species Inhabiting Sour Foods

Yeast species (nr)	Arxula adeninivorans	Candida apicola	Candida boidinii	Candida glabrata	Candida holmii	Candida lactis-condensi	Candida milleri	Candida parapsilosis	Candida pseudolambica	Candida sake
	1	2	3	4	5	6	7	8	9	10
Budding cells	+	+	+	+	+	+	+	+	+	+
Filaments	+	–	+	W	–	–	–	+	+W	+
Fragmenting	+	–	–	–	–	–	–	–	–	–
Pellicle	+	V	+	–	–	–	–	–	?	W
D-Glucose	+	+	+	+	+	+	+	+	+	+
D-Galactose	+	–	–	–	+	–	+	+	–	+
L-Sorbose	+	+D	V	–	–	–	–	+D	–	+
D-Ribose	+	V	V	–	–	–	–	V	–	V
D-Xylose	+	V	+	–	–	–	–	+	+D	+
L-Arabinose	+	–	V	–	–	–	–	+	–	–
Rhamnose	D	–	V	–	–	–	–	–	–	–
α-Methylglucoside	+	–	–	–	–	–	–	+	–	V
Sucrose	+	+	–	–	+	+	+	+	–	+
Maltose	+	–	–	–	–	–	–	+	–	+
Trehalose	+	–	–	V	+	–	+	+	–	+
Cellobiose	+	–	–	–	–	–	–	–	–	V
Melibiose	+	–	–	–	–	–	–	–	–	–
Lactose	D	–	–	–	–	–	–	–	–	–
Raffinose	+	+	–	–	+	+D	+	–	–	–
Melezitose	+	–	–	–	–	–	–	+	–	+
D-Glucosamine	+	–	V	–	–	–	–	V	+D	V
Acetyl-D-glucosamine	+	+	+	–	–	?	–	+	–	V
Soluble starch	+	–	–	–	–	–	–	–	–	–
Glycerol	+	+	+D	+D	–	–	V	+	–	+
Erythritol	+	–	+	–	–	–	–	–	–	–
Glucitol	+	+	+	–	–	–	–	+	–	+
Mannitol	+	+	+	–	–	–	–	+	–	+
Inositol	+	–	–	–	–	–	–	–	–	–
DL-Lactate	–	–	+	V	–D	–	V	–	+	V
D-Gluconate	+	V	–	+	–	–	–	+D	–	V
D-Glucuronate	+	–	–	–	–	–	–	–	–	–
2-Keto-D-gluconate	+	–D	V	V	–	–	–	+	–	+
Nitrate	+	–	V	–	–	+	–	–	–	–
Fermentation of glucose	+D	D	+	+	+	+	+	+	+	+
Ethylamine (N)	+	+	+	–	–	–	–	+	+	V
Vitamin requirement	OT	BT	B(T)	M	0B	BT	BM	B	0	0B
Urease	–	–	–	–	–	–	–	–	–	–
Max. growth T (°C)	45	<35	35	40	<35	<35	<35	>37	35	V
Cycloheximide (100 ppm)	+	–	+	–	V	–D	V	V	–	–

Table 3 (continued)

Yeast species (nr)	Candida tenuis 11	Candida tropicalis 12	Candida versatilis 13	Candida wickerhamii 14	Debaryomyces etchellsii 15	Debaryomyces hansenii 16	Dipodascus capitatus 17	Galactomyces geotrichum 18	Hanseniaspora uvarum 19	Hanseniaspora valbyensis 20
Budding cells	+	+	+	+	+	+	−	−	+	+
Filaments	+	+	−	−	+	−W	+	+	V	V
Fragmenting	−	−	−	−	−	−	+	+	−	−
Pellicle	−	W	−	−	−	V	+	+	−	−
d-Glucose	+	+	+	+	+	+	+	+	+	+
d-Galactose	+	+	+	+	+	+	+	+	−	−
L-Sorbose	V	V	−	−	+	V	V	+	−	−
d-Ribose	+	−D	V	+	−	V	V	V	−	−
d-Xylose	+	+	−D	+	+	+	−	+	−	−
L-Arabinose	V	−	−D	+	+	+W	−	−	−	−
Rhamnose	+	−	−	+	−	V	−	−	−	−
α-Methylglucoside	+	V	−D	−	+	+	−	−	−	−
Sucrose	+	V	V	−	+	+	−	−	−	−
Maltose	+	+	V	−	+	+	−	−	−	−
Trehalose	+	+	+	V	+	+	−	−	−	−
Cellobiose	+	+D	+	+	+	+	−	−	+	+
Melibiose	−	−	V	−	−	V	−	−	−	−
Lactose	+D	−	−D	−	−	V	−	−	−	−
Raffinose	−D	−	V	−	−	+	−	−	−	−
Melezitose	+D	V	−	−	+	V	−	−	−	−
D-Glucosamine	−D	V	−	−D	+	V	−	−	−	−
Acetyl-D-glucosamine	+	+	+	+	+	V	?	?	?	?
Soluble starch	V	+	−	−	−	V	−	−	−	−
Glycerol	+	V	+	+	+	+	+	+	−	−
Erythritol	V	−	−	−	−	V	−	−	−	−
Glucitol	+D	+	−	+	+	+W	−	+	V	−
Mannitol	+	+	+	+	+	+	−	V	−	−
Inositol	−	−	−	−	−	−	−	−	−	−
DL-Lactate	V	V	V	−	V	V	+	V	−	−
D-Gluconate	+D	V	−D	V	−	+W	−	−	V	−
D-Glucuronate	−	−	−	−	−	V	−	?	−	−
2-Keto-D-gluconate	+	+	+	V	+	+	−	−	+	−
Nitrate	−	−	+	+	−	−	−	−	−	−
Fermentation of glucose	V	+	D	+	+D	−W	−	V	+	+
Ethylamine (N)	+	+	+	+	+	+	+	+	+	+
Vitamin requirement	BT	0B	BT	BT	B	0B	?	0	M	M
Urease	−	−	−	−	−	−	−	−	−	−
Max. growth T (°C)	30	>40	30	<35	37	V	>37	V	<35	<30
Cycloheximide (100 ppm)	V	+	V	+	−	V	+	?	+	+

Table 3 (continued)

Yeast species (nr)	Issatchenkia orientalis 21	Kluyveromyces lactis 22	Kluyveromyces marxianus 23	Pichia anomala 24	Pichia burtonii 25	Pichia canadensis 26	Pichia fermentans 27	Pichia holstii 28	Pichia membranifaciens 29	Pichia ohmeri 30
Budding cells	+	+	+	+	+	+	+	+	+	+
Filaments	+	+	V	V	+	+	+	+	V	+
Fragmenting	−	−	−	−	+	−	−	−	−	−
Pellicle	+	+	V	V	V	−W	+	V	+	+
D-Glucose	+	+	+	+	+	+	+	+	+	+
D-Galactose	−	+	D	V	+	−	−	+	−	+
D-Sorbose	−	V	V	−	V	−	−	+	V	+
D-Ribose	−	−	V	V	+	−	−	+	−	V
D-Xylose	−	V	D	V	+	+	+	+	V	−
L-Arabinose	−	−	V	V	V	−	−	+	−	−
Rhamnose	−	−	−	−	−	+W	−	+	−	−
α-Methylglucoside	−	V	−	+	+	V	−	+	−	+
Sucrose	−	+	+	+	+	+	−	+	−	+
Maltose	−	V	−	−	+	+	−	+	−	+
Trehalose	−	+	−	+	+	+D	−	+	−	+
Cellobiose	−	+	V	+	+	+	−	+	−	+
Melibiose	−	−	−	−	−	−	−	−	−	−
Lactose	−	+	V	−	−	−	−	−	−	−
Raffinose	−	V	+	+	+	−	−	−	−	+
Melezitose	−	+	−	+	V	+	−	+	−	−
D-Glucosamine	+	−	−	−	+W	−	+	+	V	+
Acetyl-D-glucosamine	+	−	−	−	+	−	+	+	+	+
Soluble starch	−	−	−	+	+	−	−	+	−	−
Glycerol	+	V	D	+	+	+	+	+	V	+
Erythritol	−	−	−	+	+	−	−	V	−	−
Glucitol	−	+	V	+	+	+D	−	+	−	+
Mannitol	−	+	V	+	+	+D	−	+	−	+
Inositol	−	−	−	−	−	−	−	−	−	−
DL-Lactate	+	V	+	+	−	+	+	−	V	V
D-Gluconate	−	−	−	V	V	+	−	V	−	V
D-Glucuronate	−	−	−	−	−	−	−	−	−	−
2-Keto-D-gluconate	−	−	−	−	+	−	−	V	−	+
Nitrate	−	−	−	+	−	V	−	+	−	−
Fermentation of glucose	+	+	+	+	+	−	+	+	−W	+
Ethylamine (N)	+	+	+	+	+	+	+	+	+	+
Vitamin requirement	0	BM	M	0	0	+	T	BT	BT	B
Urease	−	−	−	−	−	−	−	−	−	−
Max. growth T (°C)	<40	37V	>40	<37	<37	V	<40	<40	<37	40
Cycloheximide (100 ppm)	−	+	+	−	−	−	−	+	−	−D

Table 3 (continued)

Yeast species (nr)	Pichia pijperi 31	Pichia subpelliculosa 32	Rhodotorula minuta 33	Rhodotorula mucilaginosa 34	Saccharomyces barnettii 35	Saccharomyces cerevisiae 36	Saccharomyces dairenensis 37	Saccharomyces exiguus 38	Saccharomyces rosinii 39	Saccharomyces spenceronum 40
Budding cells	+	+	+	+	+	+	+	+	+	+
Filaments	–W	+	–	–W	–	–	–	–	–	–
Fragmenting	–	–	–	–	–	–	–	–	–	–
Pellicle	+	V	–	–	–	–	–	–	–	–
D-Glucose	+	+	+	+	+	+	+	+	+	+
D-Galactose	–	V	V	V	+	V	+	+	+	+
L-Sorbose	+	–	V	V	–	–	–	–	–	–
D-Ribose	–	V	V	V	–	–	V	–	–	–
D-Xylose	+	V	+	+	–	–	–	–	–	–
L-Arabinose	–	V	+	V	–	–	–	–	–	–
Rhamnose	–	–	–	V	–	–	–	–	–	–
α-Methylglucoside	–	+	–	V	–	V	–	–	–	–
Sucrose	–	+	+	+	+	+	–	+	–	+
Maltose	–	+	–	V	–	+	–	–	–	–
Trehalose	–	+	+	+	+	+	+D	+	–	+
Cellobiose	+	V	V	V	–	–	–	–	–	–
Melibiose	–	–	–	–	–	V	–	–	–	–
Lactose	–	–	V	–	–	–	–	–	–	–
Raffinose	–	+	–	+	+	+	–	+	–	–
Melezitose	–	V	+	V	–	V	–	–	–	–
D-Glucosamine	–	–	–	V	–	–	–	–	–	–
Acetyl-D-glucosamine	–	–	+	–	–	–	–	–	–	–
Soluble starch	–	V	–	–	–	–	–	–	–	–
Glycerol	+	+	+	V	–	–	V	–	–	+
Erythritol	–	+	–	–	–	–	–	–	–	–
Glucitol	+	+	V	V	–	–	–	–	–	–
Mannitol	+	+	V	V	–	–	–	–	–	–
Inositol	–	–	–	–	–	–	–	–	–	–
DL-Lactate	+	+	V	V	+	V	–	–	–	+
D-Gluconate	–	+	+	+	+	V	V	–	–	+
D-Glucuronate	–	–	V	–	–	–	–	–	–	?
2-Keto-D-gluconate	–	–	+	–	–	–	–	–	–	–
Nitrate	–	+	–	–	–	–	–	–	–	–
Fermentation of glucose	+	+	–	–	+	+	+	+	+	+
Ethylamine (N)	+	+	V	+	–	–	–	–	–	+
Vitamin requirement	M	BT	T	T	B	M	M	–B	M	T
Urease	–	–	+	+	–	–	–	–	–	–
Max. growth T (°C)	V	V	V	<40	31	V	V	V	<37	>37
Cycloheximide (100 ppm)	–	–	+D	V	–	–	V	+D	–	–

Table 3 (continued)

Yeast species (nr)	Saccharomyces unisporus 41	Sacharomycopsis fibuligera 42	Saccharomycopsis selenospora 43	Stephanoascus ciferrii 44	Torulaspora delbrueckii 45	Trichosporon gracile 46	Zygosaccharomyces bailii 47	Zygosaccharomyces bisporus 48	Zygosaccharomyces mrakii 49	Zygosaccharomyces rouxii 50
Budding cells	+	+	+	V	+	+	+	+	+	+
Filaments	−	+	+	V	−W	+	−W	−	−W	−W
Fragmenting	−	−	−	−	−	+	−	−	−	−
Pellicle	−	V	−	+	−	+	−	−	−	−
D-Glucose	+	+	+	+	+	+	+	+	+	+
D-Galactose	+	−	+	+	V	−	V	V	+	V
L-Sorbose	−	−	−	+	V	V	V	+	−	−
D-Ribose	−	−	−	V	−	V	−	−	−	−
D-Xylose	−	−	+D	+	V	+	−	−	−	−
L-Arabinose	−	−	+D	+	−	+	−	−	−	−
Rhamnose	−	−	−	+	−	−	−	−	−	−
α-Methylglucoside	−	+	−	V	V	−	−	−	−	−
Sucrose	−	+	−	+	V	V	V	−	+	−D
Maltose	−	+	−	+	V	V	−	−	−	V
Trehalose	V	V	−	+	D	+	V	−	−	+W
Cellobiose	−	+	−	V	−	+	−	−	−	−
Melibiose	−	−	−	V	−	−	−	−	+	−
Lactose	−	−	−	−	−	V	−	−	−	−
Raffinose	−	V	−	+	V	−	−	−	+	−
Melezitose	−	V	−	−	V	−	−	−	−	−
D-Glucosamine	−	−	−	−	−	V	−	−	−	−
Acetyl-D-glucosamine	−	−	−	?	−	−	−	−	−	−
Soluble starch	−	+	V	V	−	V	−	−	−	−
Glycerol	−	+	V	+	V	+	V	+	D	+W
Erythritol	−	V	−	+	−	−	−	−	−	−
Glucitol	−	V	−	+	V	V	+	D	+	+
Mannitol	−	V	−	+	+	+	+	D	+	+
Inositol	−	+W	−	+	−	+	−	−	−	−
DL-Lactate	−	+W	+	+	V	+	−	−	−	−
D-Gluconate	−	+W	V	+	V	+	−	V	−	−W
D-Glucuronate	−	+	−	+	−	+	−	−	−	−
2-Keto-D-gluconate	−	−	−	+	+	+	V	V	+	−
Nitrate	−	−	−	−	−	−	−	−	−	−
Fermentation of glucose	+	+D	+D	−	+	−	+	+	+	+
Ethylamine (N)	+	V	−	+	V	+	+	+	+	+
Vitamin requirement	M	M	M	TB	0B	T	B	BT	0	BP
Urease	−	−	−	−	−	+	−	−	−	−
Max. growth T (°C)	<37	<40	<37	>40	V	<35	V	<35	<30	V
Cycloheximide (100 ppm)	+	+	+	+	−	+	−	−	+	−

+, positive; −, negative; W, weak response; D, delayed positive (7 days or more); V, variable results; ?, no data known; 0, no vitamins required; B, biotin required; T, thiamine required; P, pantothenate required; M, more or other vitamins required.

Monosaccharides: glucose, galactose, L-sorbose, D-ribose, D-xylose, L-arabinose, Rhamnose, α-methylglucoside
Disaccharides and trisaccharides: sucrose (saccharose), maltose, trehalose, cellobiose, melibiose, lactose, raffinose, melezitose
Amino sugars: D-glucosamine, acetyl-D-glucosamine
Polyols: glycerol, erythritol, glucitol (sorbitol), mannitol, inositol
Organic acids: DL-lactate, D-gluconate, Δ-glucuronate, 2-keto-D-gluconate

3.3. Assimilation of Nitrogen Compounds

Potassium nitrate (40 mM) is dissolved in Difco yeast carbon base. The broth is dispensed (2.5 mL) in culture tubes (15 cm high, 16 mm in diameter) and is sterilized at 120°C for 20 min. Ethylamine hydrochloride should not be sterilized in the presence of glucose. A separately sterilized concentrated solution is added aseptically to culture tubes with 2.5 mL sterile yeast carbon base, to a concentration of 40 mM. The culture tubes are inoculated with a drop of a young culture in yeast carbon base with 40 mM ammonium chloride (sterilized separately). For comparison, a culture tube with yeast carbon base without a nitrogen source is inoculated. All tubes are incubated at 25°C in a rotary shaker up to 2 wk and are inspected for growth daily. Positive growth responses should be confirmed by transfer of a loopful of the culture to a second culture tube with the same growth medium. This should also show growth.

3.4. Fermentation of Glucose

Several yeasts present in foods are able to carry out an alcoholic fermentation. This appears from the production of gas (i.e., carbon dioxide, in Durham tubes. The latter are test tubes with a small inverted tube inserted to collect any gas that may be produced. These tubes contain 10 mL of 2% glucose, and 1% yeast extract. They are inoculated with a loopful after sterilization for 20 min at 120°C. The tubes are incubated at 25°C until gas is visible in the insert, or up to 28 d if no gas is produced. The tubes are shaken at intervals of several days.

3.5. Additional Characteristics

3.5.1. Urease

In a 10 mM potassium phosphate buffer (pH 6.0), Phenol Red is dissolved (20–50 mg/L). This solution can be stored indefinitely in the dark at room temperature. Aliquots of 0.5 mL are dispensed in test tubes. Immediately before use a freshly prepared concentrated urea solution is added to a final concentration of 20 g/L. A loopful of a 1- to 2- d-old slant culture is added to the solution. The test tubes are incubated at 37°C irrespective of the yeast's optimum growth temperature. A dark red color appearing within 5 h demonstrates a pH rise due to hydrolysis of urea to ammonia and carbon dioxide. Comparison

Identification of Yeasts in Fermented Foods and Fodder

with an uninoculated blank is recommended. The reaction is characteristic of basidiomycetous yeasts (in this study the genera *Rhodotorula* and *Trichosporon*). Most ascomycetous yeasts are urease-negative.

3.5.2. Maximum Growth Temperature

Slants of appropriate growth media (e.g., malt extract agar, Difco YM agar) are inoculated and incubated at constant temperature, preferably in a thermostated water bath. The slants are inspected for growth after 1, 2, or 3 d.

3.5.3. Tolerance of 1% Acetic Acid

This test is only used to discriminate *Zygosaccharomyces* spp. It is carried out by streaking a young preculture (the same as used in the assimilation tests) on agar plates of the composition 10% glucose, 1% tryptone, 1% yeast extract, and 2% agar is sterilized for 20 min at 120°C and cooled down to approx 45–50°C. Glacial acetic acid (1 mL/100 mL) is then added and quickly mixed, and the agar is poured in Petri dishes.

3.5.4. Resistance to 100 ppm Cycloheximide

This test is included in the ID 32 C test system for assimilation of carbon compounds.

3.5.5. Assimilation of Starch and Methanol

Merck soluble starch is dissolved (5 g/L) in Difco yeast nitrogen base. Aliquots of 2.5 mL are dispensed in culture tubes (15 cm high, 16 mm in diameter) and sterilized for 20 min at 120°C. A concentrated solution of methanol in sterile water is added to culture tubes containing 2.5 mL sterile Difco yeast nitrogen base. The final methanol concentration should not exceed 5 g/L. Inoculation and incubation are as described in **Subheading 3.3**.

3.5.6. Vitamin Requirement

Two procedures are recommended. A 10-fold concentration of bacto vitamin-free yeast base is prepared by dissolving 16.7 g/100 mL. This concentrated broth should be filter-sterilized. Aliquots of 0.25 mL are added to culture tubes containing 2.25 mL sterile water. Alternatively, the vitamin-free medium can be prepared (for the composition, see **ref. 4**, p. 99). If ammonium chloride and glucose are kept separately, the growth medium can safely be sterilized at 120°C without browning and with less risk of airborne infections than filter sterilization. Add concentrated sterile vitamin solutions after sterilization, at final concentrations: biotin (20 µg/L) and/or thiamine (400 µg/L). More complex vitamin requirements are not specified in this study. For inoculation and incubation, see above. Results should be confirmed by transfer of a loopful to a second culture tube with medium of the same composition. This should also show growth.

4. Notes

1. Yeast species treated. This chapter deals with yeast species which according to **refs. 2–4** have never been isolated from foods and fodders that underwent a lactic acid fermentation (**Table 1**). However, products derived from olives (alpechin, olive brine) are very rich in yeasts not found elsewhere in commodities. These species were not included. Neither were yeast species characteristic of fruit juices and alcoholic beverages.
2. Identification Key. A dichotomous identification key to the treated species is shown in **Table 2**.
3. Characteristics of individual yeast species. In **Table 3** characteristics of the treated yeast species observed in this study are listed.

References

1. Middelhoven, W. J. and van Baalen, A. J. M. (1988) Development of the yeast flora of whole-crop maize during ensiling and during subsequent aerobiosis. *J. Sci Food Agric.* **42,** 199–207.
2. Middelhoven, W. J. (1998) The yeast flora of maize silage. *Food Technol. Biotechnol.* **36,** 7–11.
3. Barnett, J. A., Payne, R. W., and Yarrow, D. (1990) *Yeasts: Characteristics and Identification.* Cambridge University Press, Cambridge, UK (available on disk).
4. Kurtzman, C. P. and Fell, J. W. (eds.) (1998) *The Yeasts, A Taxonomic Study.* Elsevier, Amsterdam.

IV

Organisms in the Manufacture of Other Foods and Beverages

27

Protein Hydrolysis

Isolation and Characterization of Microbial Proteases

Marcela A. Ferrero

1. Introduction

Microbial proteases play an important role in industrial processes. The proteases of *Bacillus* spp., *Mucor* spp., and *Aspergillus oryzae* account for the bulk of enzyme production and represent aproximately 40% of the total worldwide enzyme sales *(1)*.

The large success of microbial proteases in food and other biotechnological systems can be attributed to the broad biochemical diversity of the microorganisms, to the genetic manipulation of the organisms, and to the improved techniques for the enzyme production and purification.

The screening of microorganism-producing proteases from nature is the first step to obtain microbial proteases with industrial purposes. Although there are many microorganisms that produce proteases in nature, for industrial purposes, it should be convenient to find extracellular protease-producing strains because of easier recovery and the stability of the enzyme.

Bacterial neutral and alkaline proteases are produced mostly by organisms belonging to the genus *Bacillus*. Neutral proteases, called also metalloproteases, generate less bitterness in hydrolyzed food proteins than do the animal proteases and, hence, are valuable for use in food industry. They are insensitive to natural plant protease inhibitors and are therefore useful in the brewing industry. Their low thermotolerance is advantageous for controlling their reactivity during the production of food hydrolysates with a low degree of hydrolysis *(2)*.

Fungi contain a wider variety of enzymes than do bacteria. For example, *A. oryzae* is used to modify wheat gluten by limited proteolysis. The addition of proteases reduces the mixing time and results in increased loaf volume. Bacterial proteases are used to improve the extensibility and strength of the dough.

The alkaline and neutral protease of fungal origin play an important role in the processing of soy sauce. Soybeans serve as a rich source of food and the hydrolysate is used in protein-fortified soft drinks and in the formulation of dietetic feeds. Treatment of soy proteins with fungal proteases results in soluble hydrolysates with high solubility, good protein yield, and little bitterness *(2)*.

Acid proteases are typically produced by fungi. They belong to the cystein and aspartic proteases groups and have commercial applications in cheese manufacture.

The isolation of protease-producing microorganisms from natural environments is relatively easy and a general procedure is carried out, utilizing, in this case, a proteinaceous substrate in the screening procedure.

As a general procedure for microorganism-producing protease screening, I will describe the isolation and enzyme characterization of neutral proteases produced by *Bacillus* strains, useful in food industries.

Methods for determinating alkaline and acid protease activities will be described in this chapter because they are very useful in food industries too.

2. Materials

2.1. Media

1. Skim milk medium (SM): Autoclave skim milk agar, composed of skim milk (10 g/L) and purified agar (15 g/L). Cool the medium to about 60°C and add Rose Bengal (2 µg/mL) or Fungizone (10 µg/mL) to avoid development of fungi. Final pH obtained is about 6.5–7.5 (*see* **Note 1**).
2. Minimal synthetic medium (ZM), with the following composition: 1.0 g/L $(NH_4)_2SO_4$, 6.0 g/L K_2HPO_4, 3.0 g/L KH_2PO_4, 0.01 g/L $MgSO_4 \cdot 7H_2O$, 0.05 g/L $CaCl_2 \cdot 2H_2O$, 0.01 g/L $MnSO_4 \cdot 2H_2O$, 0.001 g/L $FeSO_4 \cdot 7H_2O$, 0.001 g/L $ZnSO_4 \cdot 7H_2O$, 1.0 g/L trisodium citrate, and 10.0 g/L casein *(3)*.
3. Supplemented medium: The same composition described in **item 2**, but adding 2.5 g/L yeast extract. This medium is utilized when poor growth is observed in the minimal medium. Vitamins and growth factors are provided by yeast extract.

2.2. Sampling

1. Collecting vessels and tools. These can be sterile jars, flasks, or tubes for liquid samples, or small plastic bags for solids.
2. Flasks or vessels with sterile water to dissolve or disperse the sample.
3. Optical microscope for examining bacteria and kit for Gram staining.
4. Biochemical tests for identification of *Bacillus* strains.

2.3. Solutions for Determination of Protease Activity

1. Casein (10 mg/mL) dissolved in 200 mM sodium phosphate buffer, pH 7,4 and 5 mM PMSF (phenylmethylsulfonyl fluoride), for determination of neutral protease.
2. Casein (10 mg/mL) dissolved in 50 mM borate buffer, pH 9, and 5 mM EDTA, for determination of alkaline protease.
3. Bovine serum albumin (BSA) (10 mg/mL) dissolved in 0.1 M citrate buffer, pH 3.0, for determination of acid protease.
4. 10.0% TCA (trichloroacetic acid).
5. Folin–Ciocalteau reagent (Merck Co.), threefold diluted in distilled water.
6. 0.5 M Na_2CO_3 solution.
7. Standard solution of tyrosine, 1.0 mg/mL.
8. Spectrophotometer and plastic cuvets.
9. Centrifuge for Eppendorf tubes.

3. Methods

3.1. Sampling

From different substrates obtained from natural habitats, collect samples in sterile containers. Upon returning to the lab, use the following procedure:

1. Weigh out samples into 250-mL flasks or vessels of aproximately 100 mL of sterile distilled water with 0.85% NaCl added.
2. After shaking for approximately 1 h, set aside to allow solids to settle. Aliquots of supernatants are subjected to thermal shock treatment to obtain spore-formers: 5 mL of supernatant is put in several tubes and incubated at 80°C for 10 min. This procedure is used to isolate spore-forming microorganisms (4).

3.2. Screening

1. Inoculate with 4 mL of the 250-mL Erlenmeyer flasks containing 80 mL of SM. Incubate this flasks for 2–4 d in a rotary shaker at approximately 37–45°C.
2. After a good cellular density was obtained, 50 µL of proper dilution are spread on skim milk plates and incubated at 37°C until colonies are easily visible. If possible, determine the total count of proteolytic spore-former colonies. Follow the isolation scheme described in **Fig. 1** if necessary (5) (*see* **Note 2**).
3. Place at least one of each colony type obtained on skim milk agar as punctures in the middle of squares formed by a grid.
4. The productivity of the colonies is measured in these plates as a ratio of the halo formed by the casein hydrolyzed to the diameter of the producer bacteria.

3.3. Identification of the Strains

1. Individual colonies are observed under a microscope, and Gram-stained preparations are used as part of the identification.

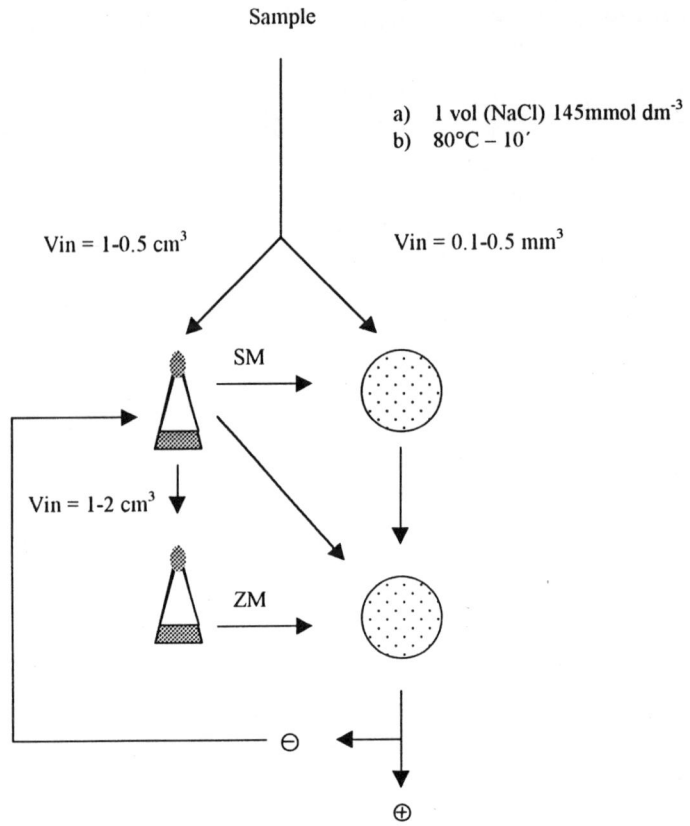
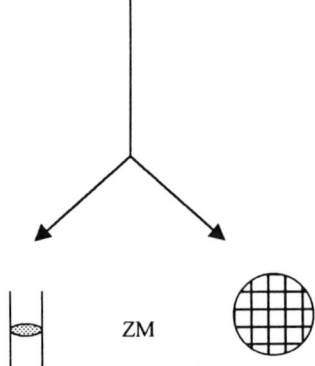

Fig. 1. Isolation scheme for extracellular enzymes producing spore-forming bacteria.

Protein Hydrolysis 231

2. Identification of isolated strains by biochemical tests described in *Bergey's Manual of Systematic Bacteriology (6)* and the API Identifiaction System *(7)*.

3.4. Determination of Protease Activity

The determination of proteolytic activity is carried out in the supernatants from the cultures with casein of the isolated strains. Supernatants were obtained by centrifugation of 1 mL of culture in Eppendorf tubes at 8000g for 20 min.

Protease activities are measured using a modified Anson method *(8)*. As a general procedure for all cases, 500 µL of substrate (casein or BSA) in proper buffer is mixed with 100 µL of supernatant of the culture in Eppendorf tubes and incubate at 37°C for 30 min in a thermostated water bath. The reaction is stoped with 100 µL of 10% TCA in an icebath for 15 min to precipite the undegraded protein. The precipitate is separated by centrifugation in a Eppendorf centrifuge at 8000g for 5 min. After that, the amino acids produced from the hydrolysis are determined by the Folin–Ciocalteau reagent, mixing 500 µL of supernatant obtained previously with 2.5 µL of Na_2CO_3 in glass tubes. The colorimetric reaction is produced by the add of 500 µL of Folin–Ciocalteau reagent and then incubation for 30 min at room temperature in the dark. The color produced is read at 660 nm against the reaction blank and the absorbance values are compared to calibration values of the tyrosine standard (*see* **Note 3**).

The samples are determined by duplicates and two controls are necessary. One of them is using water instead of a supernatant (enzymatic activity control) and the other is using a supernatant but coagulating the casein before the incubation. This last control is to measure the internal casein present in the sample.

Depending of the type of enzyme to be identified, substrate and inhibitors should be selected *(9)* (*see* **Note 4**). Some characteristics of the protease enzymes to be considered are as follow:

> Neutral proteases, in general, works at pH 5–9 and they are sensitive to metal-chelating reagents, such as EDTA *(9)*.
> Alkaline proteases work at pH 7–9 and depending of the enzyme into the general classification, they can be inhibited by PMSF, DFP, TLCK or TPCK. The two lastest are specific trypsin inhibitors *(9)*.
> Acid proteases belong to cysteine and aspartic protease groups. They exhibe optimal activity in pH 3–7. Cysteine proteases are sensitive to sulfydryl reagents, iodoacetic acid, heavy metals, and iodoacetamide. Aspartic proteases, the most representative of these enzymes, are sensitive to epoxy and diazo-ketone compounds in the presence of cooper cations and pepstatine *(9)*.

The unit of enzymatic activity unit was defined as the amount of the enzyme needed to produce 1 mg of tyrosine at 30 min and 37°C.

4. Notes

1. The screening of bacteria, yeast, or fungi is not very different. Antibiotics or fungicides are convenient to use in selecting desirable microorganims.
2. In some case, it is very important to know the specific substrates and characteristics of the enzyme to be isolated, especially for screening in agar plates. This provides the most convenient medium for obtaining good zones clear of hydrolysis.
3. The solution of tyrosine is not easy to stabilize. Tyrosine should be dissolved in a thermosttated water bath at 60°C with agitation.
4. The determination of the proteolytic activity in liquid medium must be according to the desired enzyme in order, to offer the optimal conditions for enzyme activity.

References

1. Godfrey, T. and West, S. (1996) *Industrial Enzymology*, 2nd ed., Macmillan, New York, p. 3.
2. Rao, M., Tanksale, A. M., Ghatge, M. S., and Deshpande, V. V. (1998) Molecular and biotechnological aspects of microbial proteases. *Microbiol. Mol. Biol. Rev.* **62 (3),** 597–635.
3. Ferrero, M. A. (1995) Ph.D. Thesis.
4. Castro, G. R, Ferrero, M. A., Mendez, B. S., and Siñeriz, F. (1993) Screening and selection of bacteria with amylolytic activity. *Acta Biotechnol.*
5. Ferrero, M. A., Castro, G. R., Abate, C. M., Baigorí, M. D., and Siñeriz, F. (1996) Thermostable alkaline proteases of *Bacillus licheniformis* MIR 29: isolation, production and characterization. *Appl. Microbiol. Biotechnol.* **45,** 327–332.
6. Sneath, P. H. A. (1986) Endospore-forming gram-positives rods and cocci, in *Bergey's Manual of Systematic Bacteriology*, vol. 2. (Sneath, P. A. H., Mc Nair, N. S., and Sharpe, M. E., eds.) William & Wilkins, Baltimore, pp. 1114–1207.
7. Logan, N. A. and Berkeley, C. W. (1984) Identification of *Bacillus* strains using API System. *J. Gen. Microbiol.*, **130(7),** 1871–1882.
8. Anson, M. L. (1938) Estimation of pepsin, papain and cathepsin with hemoglobin. *J. Gen. Physiol.*, **22,** 79–89.
9. Kalisz, H. M. (1988) Microbial proteinases, in *Advances in Biochemical Engineering/Biotechnology*, Vol. 36 (Fietcher, A., ed.), Springer-Verlag, Berlin, pp. 1–61.

28
Production of Polyols by Osmotolerant Yeasts

Lucía I. C. de Figueroa and María E. Lucca

1. Introduction

The growth of microbial cells is often inhibited in environments where the water activity is much reduced. However, some microorganisms have developed various strategies in order to resist the stresses to which they are exposed. The ability to adapt to fluctuations in external osmolarity and the mechanisms of osmoregulation have been elucidated in some of them *(1)*. Thus, both prokaryotic and eukaryotic microorganisms include species that can tolerate a wide range of salt concentrations and/or sugars in the culture medium during growth; they are called osmotolerant *(2)*.

In the cells of most actively metabolizing organisms, the intracellular medium must remain relatively constant in ionic strength, pH, and levels of metabolites. Thus, in media of low osmolality, there are homeostatic mechanisms that maintain these parameters within the required limits, especially in maintaining the intracellular osmolality. After a transfer to a hyperosmotic medium, adaptation to the change is required. For most organisms, the adaptive response to this change is the intracellular accumulation of organic compounds, which act as compatible solutes that are not toxic to the cells, even at the high concentrations required for stabilization of the osmotic equilibrium. In yeasts, compatible solutes are polyhydroxy alcohols, such as glycerol, arabitol, erythritol, and xylitol *(4)*. Polyols are usually used as additives in the food industry.

Glycerol is a simple amphipatic three-carbon alcohol molecule with multiple commercial applications. The lack of color and odor and the high viscosity of glycerol make it suitable as an adjunct to ointments and cosmetics. Glycerol is part of many antifreezing agents, and it is used to stabilize enzyme solutions. Glycerol originating from yeast fermentation contributes to the consistency of beer, wine, and bakery products, and the control of its level is of interest to these industries. Yeast stress tolerance in the yeast-producing industry is strictly related to the overproduction of glycerol *(6)*. Improvement of wine yeast strains for glycerol production would be advantageous in the case of wines that are lacking in body. Moreover, glycerol contributes to the taste of wine by providing sweetness. Sugar-tolerant yeasts isolated from flowers, fermenting honey, and dry fruits, mostly classified as strains of *Zygosaccharomyces rouxii*, *Candida magnoliae* and *Torulaspora delbrueckii* produce glycerol and arabitol *(6,9)*. Salt-tolerant yeasts isolated from salty environments, such as *Pichia miso*, *P. farinosa*, and *Debaryomyces hansenii*, are able to produce high yields of glycerol, arabitol, and erythritol *(5,10)*.

Xylitol is a naturally occurring five-carbon polyhydroxy alcohol with a high sweetening power. It is increasingly used as a food sweetener, dental caries reducer, and sugar substitute in the treatment of diabetics. It is a normal intermediate of carbohydrate metabolism in humans and animals. It is also widely distributed in the plant kingdom, particularly in certain fruits and vegetables *(11)*.

Xylitol is currently produced by the nonspecific chemical reduction of D-xylose. Therefore, the biological production of xylitol could be of economical interest because it does not require a pure xylose syrup, as the chemical synthesis does. Hence, low-cost hemicellulosic hydrolysates might be potential substrates *(12)*. Xylose-utilizing yeasts, such as *D. hansenii*, *Candida guillermondii*, and *C. parapsilopsis*, produce high amounts of xylitol as the major product of xylose metabolism *(13)*.

Yeasts are relatively easy to isolate from natural habitats, being necessary to weigh out a sample of the substrate, suspend it in sterile water, and spread a suitable dilution on either a nonselective medium (usually acidified malt extract agar to suppress bacterial growth) to obtain organisms of the total population or on a selective medium to obtain cultures of particular species. After the plates have been incubated and the colonies have developed, they are ready to be identified **(Table 1)**.

Protoplast fusion has made a significant contribution to our understanding of the genetics and biochemistry of yeasts, and it has facilitated the creation of novel strains of yeasts that display enhanced biotechnological potential.

A protoplast may be defined as an osmotically fragile cell completely devoid of cell-wall material. Because of their inherent osmotic fragility, the protoplasts had to be maintained in an environment rendered isotonic by the addition of an osmotic stabilizer. In the absence of the restraining effect of an intact cell wall, a protoplast assumes a spherical shape in an isotonic buffer to minimize the

Table 1
Yeasts in Some Natural Habitats

Natural habitats	Osmotolerant yeasts
Corn silage	*Candida tropicalis*
	Candida shehatae
Viticulture residues	*Hansenula polimorfa*
Juice fruits	*Candida tropicalis*
Pickles	*Debaryomyces hansenii*
Fermenting honey	*Pichia farinosa*
	Torulaspora delbrueckii
	Candida magnoliae
Dry fruits	*Hansenula polymorpha*
	Zygosaccharomyces rouxii
Grape musts	*Torulaspora delbrueckii*
Soy sauce	*Zygosaccharomyces rouxii*
Salty environments	*Pichia miso*
	Pichia farinosa
	Debaryomyces hansenii
Bakery	*Torulaspora pretoriensis*
	Torulaspora delbrueckii

surface-to-volume ratio. The terms "protoplast" and "spheroplast" are often used interchangeably, but it should be recognized that the latter term is reserved for osmotically fragile cells enveloped in cell-wall material. As a result of the presence of such wall material, a spheroplast may retain its original cellular shape in isotonic buffer. Yeast cell-wall degradation and subsequent protoplast liberation are most frequently achieved enzymatically *(14)*.

The fusion technique requires a suitable method for the easy preparation of protoplasts and for the regeneration to the yeast cell. The usefulness of isolated protoplast physiological and biochemical research is based on the assumption that they are physiologically normal, retaining all the properties of the intact cells from which they are derived *(15)*.

In the fusion process, the yeast cell walls are enzymatically removed, and the resulting protoplasts are fused together using polyethylene glycol (PEG) and calcium ions (Ca^{2+}) before being embedded in an osmotically stabilizer agar under appropriate selected conditions. The mechanism by which PEG and Ca^{2+} bring about protoplast fusion is unknown, but it is thought that the PEG has the dual role of diminishing the electrostatic field between the lipid membranes and removing water for the protoplasts *(16)*.

For a better understanding of this work, we are dividing it into three parts:
1. Isolation and screening of xylitol-producing yeasts from agricultural residues.
2. Osmotolerant hybrids producing glycerol and arabitol obtained by protoplast fusion.
3. Production of polyols in batch culture.

2. Isolation and Screening of Xylitol-Producing Yeasts from Agricultural Residues

2.1. Materials

1. Agricultural residues, as corn silage, viticulture residues, and so forth.
2. Microorganisms: It is necessary to use a yeast strain as the positive control (e.g., *D. hansenii*) and to use another as the negative control (e.g., *C. utilis*).
3. Isolation medium: YM broth (Difco Laboratories, Detroit, MI), pH 5.0, and the same YM broth acidified to pH 3.5, also try a medium with 10% NaCl plus 5% glucose.

    Isolation medium for xylitol production: 6.7 g/L yeast nitrogen base (YNB), 5 g/L yeast extract, 20 g/L D-xylose, 0.03 g/L Rose Bengal, and 20 g/L agar, pH 3.5.

    Fermentation assays medium: 30 g/L D-xylose, 2 g/L yeast extract, and 6.7 g/L YNB *(15)*; pH 5.0.
4. Fermentation assays: Use 250-mL Erlenmeyer flasks containing 100 mL of medium.
5. Thin-layer chromatography (TLC): Use Merck Silicagel F_{254} plates.

2.2. Methods

1. Isolation and identification: Samples of 5–10 g have to be obtained and transported in sterile plastic bags. The samples are suspended in 50 mL of YM medium diluted 1:10 with sterile water containing 1 g/L Tween-80 *(18)*. After shaking 60 min at 25°C, the suspension is poured into sterile culture tubes and stored overnight at 4°C. Most of the supernatant is discarded, and the remaining (about 2 mL) is shaken in order to resuspend the settled cells. This suspension is streaked on acidified YM agar. The inoculated plates are incubated at 25°C, and after 3, 5 and 12 d, well-isolated colonies are transferred on YM agar until growth is observed. Colonies have to be restreaked and picked on selective medium with xylose as the sole carbon source.

    The yeast strains have to be identified according to their carbohydrate and nitrogen assimilation patterns, using the keys and description in **ref. *19*** and the computerized yeast identification program devised by Barnett, Payne, and Yarrow.
2. For the fermentation assays, the inocula are prepared by growing a loopful of cells from a stock culture in 50-mL Erlenmeyer flasks containing 15 mL of fermentation medium, incubated for 48 h at 30°C in a rotary shaker at 150 rpm. Fermentation flasks have to be inoculated to a final concentration of 10^7 cells/mL and incubated at 30°C in a rotary shaker at 150 rpm (*see* **Note 1**).

Production of Polyols by Osmotolerant Yeasts

3. After being spotted with 4 µL of samples and standards, the plates of TLC are developed using the double-ascending method in a solvent system consisting of ethyl acetate–isopropanol–water (130 : 57 : 23), at 30°C, for 35 min. After drying with hot air, the plates are sprayed with bromocresol green–boric acid (*see* **Note 2**).

3. Osmotolerant Hybrids Producing Glycerol and Arabitol Obtained by Protoplast Fusion

3.1. Materials

1. Yeast strains: Osmotolerant yeasts (salt tolerant or sugar tolerant) must be used as one of the parental strains (*see* **Note 3**).
2. Maintenance medium: 10 g/L yeast extract, 20 g/L peptone, 20 g/L glucose, and 10 g/L agar (YEPD).
    Regeneration media: (a) 10 g/L yeast extract, 20 g/L peptone, 700 g/L glucose, plus 0.6M KCl and 30 g/L agar (for sugar-tolerant hybrids); (b) 10 g/L yeast extract, 20 g/L peptone, 120 g/L glucose, 120 g/L NaCl, plus 1.5M KCl and 30 g/L agar (for salt-tolerant hybrids).
    Selective media: (a) 10 g/L yeast extract, 20 g/L peptone, 700 g/L glucose, plus 15 g/L agar (for sugar-tolerant hybrids; (b) 10 g/L yeast extract, 20 g/L peptone, 20 g/L glucose, 120 g/L NaCl, plus 15 g/L agar (for salt-tolerant hybrids).
3. Pretreatment solution: 0.01 M EDTA, 0.1 M Tris-buffer, 0.6M KCl (1.2 M –1.4 M for osmotolerant yeasts), and 1% β-mercaptoethanol (0.1 M for osmotolerant yeasts).
4. Osmotically stabilized solution: 0.6 M KCl (1.2 M–1.4 M for osmotolerant yeasts).
5. Enzyme solution: 0.08 M phosphate buffer (pH 5.8), 0.06 M KCl, and 5 mg/mL Novozyme 234.
6. Fusion mixture solution: polyethyleneglycol (PEG) MW 6000, 30% w/v; 10 mM CaCl$_2$ in 0.05 M Tris–HCl buffer, pH 8.1.
7. Polyol extraction: see **Subheading 2.3.**
8. Polyol determination: HPLC, Gilson, equipped with a pump 305, a differential refractometer 2142 LKB, and a recorder/integrator chromatopac CR601 (Shimadzu). Concentration of sugars were determined with a refractive index (RI) detector under the following conditions: Rezex ROA (Phenomenex) column (300 × 7.8 mm); temperature: 55°C; eluent, 0.02N sulfuric acid; flow rate, 0.6 mL/min; sample volume, 20 µL.

3.2. Methods

1. Protoplast formation
   a. Inoculate each of the strains into 10 mL YEPD and grow overnight at 30°C at 200 rpm.
   b. Harvest the cells in mid-exponential growth phase by centrifuging in Eppendorf tubes.
   c. Add pretreatment solution (1 mL), resuspend cells, and incubate at 37°C for 15 min.

d. Spin down and wash once with osmotically stabilized solution.
e. Add 1 mL of enzyme solution until protoplast formation is complete. Determine the progress of protoplasting by microscopic observation.
f. Spin down protoplasts of both parental strains at a relatively low speed and resuspend each pellet in 1 mL of osmotically stabilized solution.
g. If it is necessary to use nonviable protoplasts of one of the parental strains (*see* **Note 4**), resuspended protoplasts in the osmotically stabilized solution must be killed by heating at 70°C for 5–15 min (or use a petite mutant is *S. cerevisiae* is one of the parents).

2. Fusion and regeneration of the hybrids
    a. Mix both protoplast suspensions, centrifuge again at relatively low speed, discard the supernatant, and resuspend the pellet in 2 mL of the fusion mixture solution. Incubate for 30 min at 30°C.
    b. Add the fusion mixture to 100 mL of the osmotically stabilized, melted regeneration medium at 42°C, mix rapidly, pour as an overlay on Petri dishes with a base of the same osmotically stabilized regeneration medium, and allow to harden (*see* **Note 4**).
    c. Incubate the plates at 30°C for 4–5 d, until colonies appear.
    d. Pick colonies to selective media for isolation and characterization of the desired fusion products (*see* **Note 4**).

3. Polyol extraction and determination: Osmotolerant hybrids are cultured in Erlenmeyer flasks with a high-osmotic-pressure medium to determine polyol production. The intracellular and extracellular fractions of glycerol and arabitol is determined according to **ref. 20**. An appropriate amount of cell culture (usually 1.5 mL) has to be rapidly sampled and centrifuged in an Eppendorf centrifuge. The clear supernatant containing the extracellular polyols is transferred to a water bath and boiled for 10 min together with a parallel sample of the uncentrifuged cell culture. This latter sample contains total polyols (intracellular and extracellular fractions) and has to be cleared by centrifugation. Both samples are frozen until analyzing. Intracellular polyols are determined by subtracting the extracellular fractions from that of the total.

   Polyols are determined by usual techniques (e.g., HPLC).

   Total yields of polyols in the cells and in the supernatants should be compared.

4. Production of Polyols in Batch Cultures
4.1. Materials

1. Strains: Osmotolerant yeasts.
2. Fermentation medium: 10 g/L yeast extract, 20 g/L peptone, and 500 g/L glucose (*see* **Note 5**).
3. Fermentor: The fermentor should have automatic control of dissolved oxygen tension, pH, foam, and temperature.
4. Analytical determination: Glucose, ethanol, and polyols can be determined by HPLC.

4.2. Methods

1. Fermentation assays: The fermentation medium is inoculated with 12-h-old culture of the osmotolerant yeast to a final concentration of 0.25-g/L (dry weight) cells. The pH is maintained at 3.0 by the addition of either 0.5 N HCl or 0.5 N NaOH. The agitation speed is kept at 500 rpm and the temperature at 30°C. Dissolved oxygen tension is controlled at 60% saturation by supplying air automatically via a proportional integrative and derivative (PID) controller, if the fluctuations are lower than 5% (*see* **Note 6**).
2. Analytical determination: Samples are withdrawn periodically in order to control the time-course of fermentation until glucose is exhausted. Supernatants obtained from each sample are analyzed by HPLC. Under the conditions described using osmotolerant yeasts, high yields of polyols are achieved.

5. Notes

1. Fermentation assays have to be done in order to detect xylitol production in the isolated yeast strains. Keep in mind that some yeasts that utilize xylose do not produce xylitol; however, other yeasts such *P. farinosa* produces xylose from xylose, xylitol, and heptitols.
2. Xylose and xylitol are visualized as yellow spots on blue background, with the same retardation factor (R_f) values as the standard ones. Solutions of xylitol (ranging from 5 to 30 g/L) and xylose (20 g/L) have to be used as standards. TLC, a qualitative technique, is used as a rapid preselection step. Supernatant samples from the fermentation assays showing spots of at least the same size of that corresponding to the 5-g/L xylitol standard should be selected to perform further analysis by high-performance liquid chromatography (HPLC). It is worthwhile to point out that it is possible to separate xylose from xylitol with this TLC technique.
3. In order to obtain polyol-producing fusion products, it is necessary to use, as one of the parental strains for the fusion experiment, salt-tolerant yeasts (e.g., *D. hansenii*) or sugar-tolerant yeasts (e.g., *T. delbrueckii* or *Z. rouxii*) *(21,22)*. The other parental yeast strain could be, e.g., *S. cerevisiae*, taking into account the performance of this species in fermentation processes *(5,6)*.
4. When hybrids are isolated by complementation of physiological markers, the regeneration and selective media, and the culture conditions must have selection pressure in order to avoid parental strains development. For osmotolerant yeasts, the salt or sugar tolerance can be used as selective markers. For example, when fusing a salt-tolerant yeast (such as *D. hansenii*) with *S. cerevisiae*, the medium to be used for regeneration of the cell wall and isolation of the fusion products can be YEPD + 12% NaCl, and the incubation temperature at 37°C. *S. cerevisiae* does not grow on media containing a high concentration of salt and *D. hansenii* does not grow at 37°C, so only the hybrids survive *(5)*.

    When both parental strains do not complement their markers for isolating the fusion products, one important and interesting solution is to use protoplasts from one of them killed by heating. For example, when fusing a sugar-tolerant yeast (such as *T. delbrueckii*) with *S. cerevisiae*, the medium to be used for regeneration of the cell wall and isolation of the hybrids can be YEPD with 700 g/L glucose. *S. cerevisiae* is not able to grow in this medium, and protoplasts of *T. delbrueckii* are not viable, thus only fusion products can grow. As a control and to verify that the protoplasts of *T. delbrueckii* are not viable, an aliquot of the suspension should be spread on Petri dishes containing the regeneration medium and no growth must be observed *(6)*.

5. It is necessary to optimize the fermentation culture medium according to the strain used in order to improve polyol yields *(4)*.
6. According to the critical demand of oxygen of each yeast strain, it is necessary to assure enough availability of it in the culture medium. Another important parameter to consider in the production of polyols is the pH value during the process, usually in the range 3–4.

References

1. Brown, A. D. (1978) Compatible solutes and extreme water stress in eukaryotic microorganisms. *Adv. Microb. Physiol.* **17**, 181–242.
2. Lars, A., Nilsson, A., and Adler, A. (1988) The role of glycerol in osmotolerance of the yeast *Debaryomyces hansenii*. *J. Gen. Microbiol.* **134**, 669–677.
2a. Spencer, J. F. T. and Sallans, H. R. (1956) Production of polyhydric alcohols by osmophilic yeasts. *Can. J. Microbiol.* **2**, 72–29.
2b. Spencer, J. F. T., Roxburgh, J. M., and Sallans, H. R. (1957) Factors influencing the production of polyhydric alcohols y osmophilic yeasts. *Agricultural Food Chem.* **5**, 64–67.
5. Loray, M. A., Spencer, J. F. T., Spencer, D. M., and de Figueroa, L. I. C. (1995) Hybrids obtained by protoplast fusion with a salt-tolerant yeast. *J. Ind. Microbiol.* **14**, 508–513.
6. Lucca, M. E., Loray, M. A., de Figueroa, L. I. C., and Callieri, D. A. (1999) Characterization of osmotolerant hybrids obtained by fusion between protoplasts of *Saccharomyces cerevisiae* and heat treated protoplasts of *Torulaspora delbrueckii*. *Biotechnol. Lett.* **21**, 343–348.
7. Yagi, T. (1991) Effects of increases and decreases in the external salinity on the intracellular glycerol and inorganic ion content in the salt-tolerant yeast *Zyogsaccharomyces rouxii*. *Microbios* **68**, 109–117.
8. Prior, B. A. and Hohmann, S. (1997) Glycerol production and osmoregulation, in *Yeast Sugar Metabolism* (Zimmermann, F.K. and Entian, K.D., eds.), Technomic, Lancaster, PA, pp. 313–337.
9. Spencer, J. F. T. and Spencer, D. M. (1978) Production of polyhydroxy alcohols by osmotolerant yeasts, in *Primary Products of Metabolism* (Rose, A.H., ed.), Academic, London, pp. 393–425.
10. Agarwal, G. P. (1990) Glycerol, in *Advances in Biochemical Engineering/Biotechnology* (Fiechter, A., ed.), Sringer-Verlag, Berlin, pp. 95–128.
11. da Silva, S. S. and Afschar, A. S. (1994) Microbial production of xylitol from D-xylose using *Candida tropicalis*. *Bioprocess Eng.* **11**, 129–134.

12. Roseiro, J. C., Peito, M. A., Gírio, F. M., and Amaral-Collaço, M. T. (1991) The effects of the oxygen transfer coefficient and substrate concentration on the xylose fermentation by *Debaryomyces hansenii*. *Arch. Microbiol.* **156**, 484–490.
13. Gírio, F. M., Pelica, F., and Amaral-Collaço, M. T. (1996) Characterization of xylitol dehydrogenase from *Debaryomyces hansenii*. *Appl. Biochem. Biotech.* **56**, 79–87.
14. Cavanagh, K. and Whittaker, P.A. (1996) Application of protoplast fusion to the non-conventional yeast. *Enzyme Microb. Technol.* **18**, 45–51.
15. Spencer, J. F. T., Bizeau, C., Reynolds, N., and Spencer, D. M. (1985) The use of mitochondrial mutants in hybridization of industrial yeast strains. VI. Characterization of the hybrids, *Saccharomyces diastaticus* × *Saccharomyces rouxii*, obtained by protoplast fusion, and its behavior in simulated dough-raising tests. *Curr. Genet.* **9**, 649–652.
16. Curran, B. P. G. and Bugeja, V. C. (1996) Protoplast fusion in *Saccharomyces cerevisiae* in *Methods in Molecular Biology, Yeast Protocols* (Evans, I., ed.), Humana, Totowa, NJ, pp. 45–49.
17. Barbosa, M. F. S., Medeiros, M. B., Mancilha, M., Schneider, H., and Lee, H. (1988) Screening of yeasts for production of xylitol from D-xylose and some factors which affect xylitol yield in *Candida guilliermondii*. *J. Ind. Microbiol.* **3**, 241–251.
18. Middelhoven, W. J. (1997) Identity and biodegradative abilities of yeasts isolated from plants growing in an arid climate. *Antoine van Leeuwenhoeck* **72**, 81–89.
19. Kurtzman, C. P. and Fell, J. W. (1998) *The Yeasts, a Taxonomic Study.* 4th ed., Elsevier Science, Amsterdam.
20. Adler, L., Blomerg, A., and Nilsson, A. (1985) Glycerol metabolism and osmoregulation in the salt-tolerant yeast *Debaryomyces hansenii*. *J. Bacteriol.* **162**, 300–306.
21. Spencer, J. F. T., Spencer, D. M., Bizeau, C., Vaughan-Martini, A., and Martini, A. (1985) The use of mitochondrial mutants in hybridization of industrial yeast strains. V. Relative parental contributions to the genomes of interspecific and intergeneric yeast hybrids obtained by protoplast fusion, as the determined by DNA reassociation. *Curr. Genet.* **9**, 623–625.
22. Legmann, R. and Margalith, D. (1983) Ethanol formation by hybrid yeasts. *Appl. Microbiol. Biotechnol.* **1**, 320–322.

29

Identification of Yeasts from the Grape/Must/Wine System

Peter Raspor, Sonja Smole Mozina, and Neza Cadez

1. Introduction

Yeasts are the most significant microorganisms in all conversion steps of the grape/must/wine system. Because of their metabolic activity, yeasts play a central role in the must fermentation process and also in contamination and spoilage of the final products—wines. Consequently, only reliable and rapid identification of yeast species during process and quality control enables enologists to assess the role of yeasts as a main protagonist of alcoholic fermentation or as a contaminant.

Traditionally, yeasts have been identified on the basis of morphological, physiological and biochemical criteria. However, such identification has many limitations: more than 90 tests may be required *(1–3)* and the procedure is time- and material-consuming and requires an experienced person to exclude subjective judgment. Some commercially available sets of selected physiological/biochemical tests are used for identification of yeasts from different substrata, but they do not adequately identify wine yeasts *(4)*. In addition, distinction capacity is low; the differentiation of strains belonging to the same species is inaccurate or impossible at all. This is not acceptable in studying mixed-yeast-population dynamics during the must/wine fermentation process (differentiation of *Saccharomyces cerevisiae* strains originating from grapes, starter culture, etc.).

Recently, many simple or more sophisticated molecular techniques have been introduced for wine yeast characterization. These include restriction and/or hybridization analyses or direct sequencing of yeast chromosomal or mitochondrial DNA or amplification of yeast DNA with specific or nonspecific primers in combination with further analyses of amplified fragments (restriction fragment length polymorphism with polymerase chain reaction amplified ribosomal DNA [PCR-RFLP of rDNA], random amplified polymorphic DNA [RAPD], etc.). Among molecular typing methods we recognized the restriction analysis of amplified fragments of yeast ribosomal DNA as a suitable and rapid method with most species-specific results (although some exceptions exist). In addition, the electrophoretic karyotyping using pulsed field gel electrophoresis (PFGE) enables strain differentiation, especially of *Saccharomyces* sensu stricto yeasts with a large number of shorter chromosomes.

Considering the knowledge about the limited number of yeast species occurring in grape/must/wine systems [approx 100 species out of 470, according to Barnett et al.*(5)* or even less (15–20)] relevant for winemaking *(6)*, we developed a procedure for the identification of yeasts, isolated from grape/must/wine, on a species and strain level. The protocol is based on rapid molecular typing of yeasts, but also includes selected classical confirmation tests resulting from possible misidentification in the case of species nonspecific restriction patterns of amplified rDNA fragments (PCR-ribotypes) of yeasts. It remains technically rational, economically feasible, and applicable for the identification of a larger number of yeasts isolated in studies, including aspects of all conversion steps from grapes (or other sources of yeasts in the fermentation process) and mixed yeast population dynamics during fermentation to the assessment of contamination risk of final products.

The identification involves a few steps: isolation, enumeration, and morphological characterization of yeast strains, isolation of DNA for PCR amplification and PCR-RFLP of rDNA (PCR ribotyping), analyses of species-specific PCR-ribotypes, confirmation of species identification with selected physiological tests, and electrophoretic karyotyping for differentiation of isolates on a strain level (see flow diagram in **Fig. 1**).

2. Materials

2.1. Isolation and Enumeration of Yeasts

1. Acidified yeast–malt agar (Difco, USA). After autoclaving, cool to 60°C and add 1 M HCl to a final pH 3.7–3.8; mix and pour plates *(1)*. The medium can also be prepared as follows: 3.0 g/L yeast extract, 3.0 g/L malt extract, 5.0 g/L peptone, 10.0 g/L glucose, and 20.0 g/L agar. Follow the other instructions for preparation of Difco YM agar.
2. Membrane filter apparatus (Sartorius, Millipore), sterile filters, 0.45 µm.

2.2. Morphological Characterization of Yeasts

1. YM agar (Difco).
2. Wine diluted to 8-9% (v/v) of ethanol with YM broth (yeast extract [3.0 g], malt extract [3.0 g], peptone [5.0 g], glucose [10.0 g]) and filtered through membraneous filter aseptically.

Identification of Yeasts from Grape/Must/Wine System

3. Sporulation media: McClary's acetate agar: 1.8 g/L potassium chloride, 8.2 g/L sodium acetate tryhidrate, 2.5 g/L yeast extract, 1.0 g/L glucose, 15 g/L agar *(7)*. 5% Malt extract agar: 50 g/L malt extract, 20 g/L agar. Vegetable juice (V8, Campbell's) agar: Suspend 5 g of compressed baker's yeast in 10 mL of water and add to 350 mL of V8 vegetable juice, heat in a boiling water bath for 10 min, filtrate, and adjust pH to 6.8. Melt 14 g agar in 340 mL of demineralized water and add to V8 juice *(7,8)*.

2.3. PCR-RFLP of rDNA

2.3.1. Isolation of DNA

1. YEPD agar: 20 g/L glucose, 10 g/L yeast extract, 10 g/L peptone, 15 g/L agar.
2. Solution 1: 0.9 M sorbitol and 0.1 M EDTA. Preparation: Dilute 1 M EDTA 1:10 and dissolve 163.9 g/L of sorbitol.
3. Lysing enzyme (Sigma): stock solution 1 mg/mL.
4. Solution 2: 50 mM Tris–HCl and 20 mM EDTA.
5. 10% (w/v) sodium dodecyl sulfate (SDS).
6. 5 M Potassium acetate.
7. Isopropanol.
8. 70% (v/v) Ethanol.
9. TE buffer: 10 mM Tris–HCl (pH 7.5) and 1 mM EDTA.
10. RNase A solution (stock conc. 10 mg/mL TE buffer).
11. Ice-cooled 95% (v/v) ethanol.

2.3.2. PCR Amplification of rDNA

1. PCR buffer (100 mM Tris–HCl [pH 9.0], 15 mM MgCl$_2$, and 500 mM KCl).
2. 200 µM dNTP.
3. 1 µM solution of primer 1 (NS 1:5' GTAGTCATATGCTTGTCTC 3' or ITS 1:5' TCCGTAGGTGAACCTGCGG 3') and primer 2 (ITS 4: 5'TCCTCCGCT TATTGATATGC 3') *(9)*.
4. *Taq* polymerase (conc. 5 U/µL).
5. Bidistilled sterilized water.
6. PCR tubes and PCR cycler.

2.3.3. Restriction Analysis of Amplified rDNA

1. Selected restriction enzymes with corresponding restriction buffers (*see* **Note 1**).
2. Classical gel electrophoresis equipment, loading buffer, agarose, and DNA marker with the bands ranging from 100 bp and longer.
3. 1X TAE buffer: 40 mM Tris–acetate (pH 8.0) and 1 mM EDTA. Prepare stock solution 50X TAE: 242 g/L Tris base, 57.1 mL glacial acetic acid, 20 mL of 0.5 M EDTA (pH 7.5). Dilute before use.

4. Ethidium bromide: stock solution 1 mg/mL.
5. Ultraviolet transilluminator; Polaroid camera.

2.4. Confirmative Physiological Testing

The latest detailed and extensive protocols for preparation of materials for determining physiological and biochemical characteristics of yeasts as well as useful identification keys for selection of appropriate tests are given in Chapter 11 of **ref. 7**. Also, some older editions of yeast identification keys are useful *(1,2,5)*. The instructions for preparation of materials, selection of tests, and testing conditions should be followed exactly to reduce subjective factors as much as possible.

2.5. Electrophoretic Karyotyping

2.5.1. Isolation of Chromosomal DNA

1. YEPD agar: 20 g/L glucose, 10 g/L yeast extract, 10 g/L peptone, and 15 g/L agar.
2. 1 M EDTA, pH 7.5; 0.5M EDTA, pH 9.0; 50 mM EDTA, pH 7.5.
3. CPES buffer: 40 mM citric acid, 120 mM Na$_2$HPO$_4$ (pH 6.0), 1.2 M sorbitol, 0.5M EDTA (pH 7.5). Preparation: First prepare CPESa: 24.5 g/L citric acid, 42.7 g/L Na$_2$HPO$_4$ (pH 6.0), and 437 g/L sorbitol. Mix CPESa and 1 M EDTA, pH 7.5, in the ratio 1:1 and add dithiothreitol (DTT) (final concentration 5 mM) just before use.
4. Low-melting-point (LMP) agarose (Pharmacia LKB, Sweden, or others).
5. Lysing enzyme (Sigma): stock solution 1 mg/mL.
6. CPE buffer (the same as CPES, but without sorbitol and DDT).
7. Lysis solution: 0.45 M EDTA (pH 9.0), 10 mM Tris–HCl (pH 8.0), 1% Na-lauryl-sarcoine, 1 mg/mL Proteinase K (Boehringer Mannheim, Mannheim, Germany). Preparation: Mix 90 mL of 0.5 M EDTA (pH 9.0), 1 mL of 1 M Tris–HCl (pH 8.0), 5 mL of 20% Na-lauryl–sarcosine solution, and 4 mL of sterile demineralized water and dissolve 100 mg of Proteinase K. A 20% Na-lauryl–sarcosine solution should be sterilized before by filtration through a 0.2-µm membranous filter.

2.5.2. Pulsed Field Gel Electrophoresis

1. 0.5X TBE buffer: Stock solution: 5X TBE buffer (0.5 M Tris–HCl [pH 8.0], 325 mM boric acid, and 10 mM EDTA. Dilute in the ratio 1:10 before use.
2. Agarose for PFGE (Bio-Rad chromosomal grade or others).
3. Chromosomal DNA size markers (*S. cerevisiae*, *P. canadensis* [*H. wingei*], *S. pombe*, Bio-Rad).
4. 0.5 µg/mL ethidium bromide solution (stock solution 1 mg/mL).

Identification of Yeasts from Grape/Must/Wine System

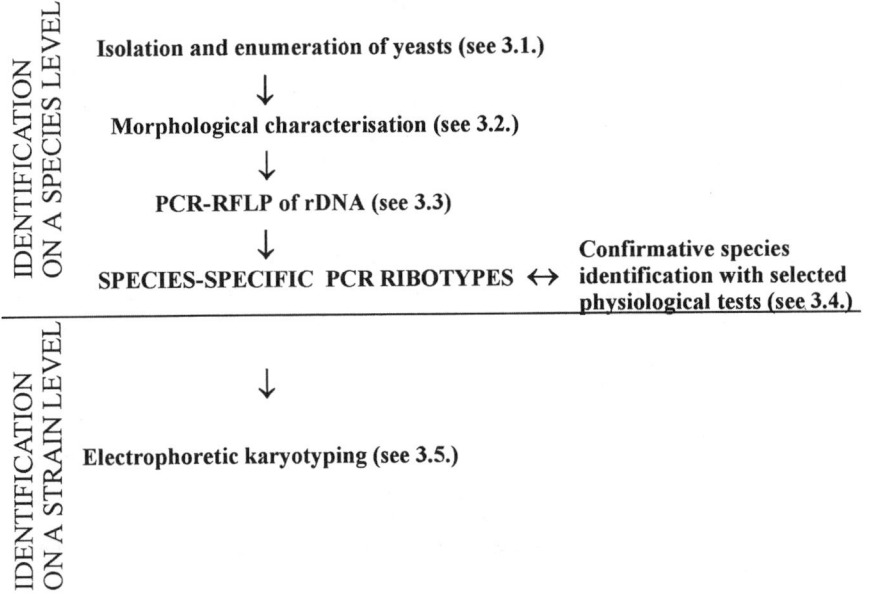

Fig. 1. Major steps in the isolation and identification of yeasts isolated from the grape/must/wine system.

3. Methods

The major steps in the isolation and identification of yeasts isolated from the grape/must/wine system is presented in **Fig. 1**.

3.1. Isolation and Enumeration of Yeasts

1. Pipet or weigh out 1 mL/1g of substrate (collected in sterile container), suspend it in sterile water, dilute accordingly to expected number (up to 10^{-7} in the case of fermenting musts), and spread it on acidified YM agar plates. Incubate 3–6 d at 25°C and check the growth in intervals, since filamentous fungi may overgrow the plates (*see* **Note 2**). If you isolate yeasts from final products (wines), filter 100–500 mL of wine through the filter membrane (0.45 µm) and incubate the membrane on the malt agar plate as described.
2. After incubation count distinctive colonies and determine total colony count. Restreak colonies of visually distinctive types on YM agar for isolation of pure cultures and determination of morphological characteristics of the colonies and vegetative cells (*7*). If you are interested in yeast community structure, select plates with the proper number of colonies and restreak all colonies from the plates for further identification, as many yeasts cannot be distinguished by visual examination (form and/or color of the colony).

3.2. Morphological Characterization

1. Observe the growth of yeast colonies on YM agar after 5–7 d of incubation: form, texture, color, surface, elevation, and margin of the colonies. Additional morphological characteristics could be observed when giant colonies are formed on YM agar after 3–4 wk and on the liquid substrate, where so-called "film-forming yeasts" form film already after 24–48 h or with a delay of 3–4 wk (*see* **Note 3**).
2. Suspend the overnight grown culture of yeast in water, place 10 µL on a microscope slide and cover with the cover slip. Observe the form (spherical, ovoidal, ellipsoidal, cylindrical, ogival, apiculate, triangular), position (single, paired, aggregated in clumps), and mode of asexual reproduction (budding [multipolar, bipolar, unipolar, fission], formation of conidia on stalks, etc.) of yeast cells under ×630 – ×1000 magnification. The shape may reflect the mode of reproduction and, in some cases, it is a characteristic of particular genera or species. Among wine yeasts, an example is the apiculate yeasts of *Hanseniaspora/Kloeckera*, which reproduce asexually by bipolar budding in basipetal succession on a narrow base (7).
3. Ascospore formation: Grow the cultures on McClary's acetate agar, 5% malt extract agar, and V8 agar for 14 d. Suspend the culture in water, place 10 µL on a microscope slide, and cover with the cover slip. Observe the form of asci and the form and number of ascospores in ascus (*see* **Note 4**).

3.3. PCR–RFLP of rDNA

3.3.1. DNA Isolation

1. Grow the isolated colonies on YEPD agar plates for 24–48 h at 28°C.
2. Harvest the cells with the loop and suspend them in 1 mL of demineralized water in Eppendorf tube to wash them and centrifuge (1500g, 5 min).
3. Resuspend the cell pellet in 500 µL of solution 1 and add 60 µL of lysing enzyme solution (conc. 1 mg/mL) (*see* **Note 5**).
4. Incubate at 30°C for 30 min. Collect the spheroplasts by centrifugation (10,000g, 5 min).
5. Resuspend the sediment in 500 µL of solution 2 and add 13 µL of 10% (w/v) SDS. Mix well on vortex and incubate at 65°C for 5 min.
6. Add 200 µL of 5 M potassium acetate and cool on ice for 5 min.
7. Centrifuge (19,000g, 15 min) to remove the precipitated protein and transfer 700 µL of clear supernatant to an Eppendorf tube. Add the equal volume of isopropanol, gently mix, and hold at room temperature for 10 min and collect the precipitated DNA by centrifuging (14,000g, 10 min).
8. Discard the supernatant and rinse the pellet with 500 µL of 70% (v/v) ethanol to remove the residual isopropanol. Centrifuge (14,000g, 5 min), discard the ethanol, and dry in the vacuum centrifuge until all traces of ethanol have been removed.
9. Dissolve the DNA in 50 µL of TE buffer with 1 µL of RNase A (stock conc. 10 mg/mL) and incubate at 37°C for 30 min.

10. Extract with 70 µL of ice-cooled 95% (v/v) ethanol, centrifuge (14,000g, 15 min), discard the supernatant, and dry the pellet in the vacuum centrifuge. Dissolve in 50 µL of TE buffer and store in the freezer until use.

3.3.2. PCR Amplification of rDNA

1. Preparation of reaction mixture for one sample (total volume is 20 µL):
   2 µL of isolated DNA in TE buffer
   2 µL of 10X PCR buffer (100 mM Tris–HCl, pH 9.0; 15 mM MgCl$_2$, 500 mM KCl)
   1.6 µL of 200 µM solution of each dNTP
   1 µL of 1 µM primer 1 (*see* **Note 6**)
   1 µL of 1 µM primer 2
   0.2 µL of *Taq* polymerase (conc. 5 U/mL)
   12.2 µL bidistilled sterilized water
   Prepare the reaction mixture for all samples tested, dispense 18 µL into PCR tubes and add 2 µL of DNA finally.
2. Place the PCR tubes for amplification of 18S + ITS rDNA (primer NS1/ITS4) or ITS region (ITS1/ITS4) in a thermal cycler under the following conditions:
   a. Primer NS1/ITS4: 5 min at 95°C followed by 35 cycles of 30 s at 95°C, 30 s at 60°C, and 3 min at 72°C. Finally, heat the mixture at 72°C for 7 min and cool to 4°C.
   b. Primer ITS1/ITS4: 5 min at 95°C followed by 35 cycles of 1 min at 95°C, 1 min at 56°C, and 2 min at 72°C. Finally, heat the mixture at 72°C for 7 min and cool to 4°C.
3. Check the concentration and specificity of the PCR product: Prepare 1% (w/v) agarose in 1X TAE buffer. Mix 1 µL of PCR product, 9 µL of distilled water, and 1 µL of loading buffer and load the gel. Add the molecular marker of appropriate size. Run the electrophoresis in 1X TAE buffer at 250 V for 20 min, stain the gel in 0.5 µg/mL of ethidium bromide solution, and observe the product under ultraviolet (UV) light (*see* **Note 6**).

3.3.3. Restriction Analysis of Amplified rDNA

1. Dispense 4 µL of the PCR product in each of four Eppendorf tubes and add 8 µL of the restriction mixture (for each tube, 2 U of frequently cutting restriction enzyme [*see* **Note 1**], 1.2 µL of 10X concentrated corresponding restriction buffer, and 6.8 µL of water) and incubate 2–3 h at optimal temperature for specific endonuclease (primarily at 37°C, but at 65°C for *Taq*I).
2. After incubation, add 2.0 µL of loading buffer to the restriction mixture and pipet the whole volume into agarose wells. Use 1.5% agarose gel when the 18S + ITS region is amplified or 3.0% agarose gel when only the ITS region is amplified.
3. Before running the electrophoresis, cool the 1X TAE buffer to 4°C to avoid the smiling effect.
4. Adjust the voltage to approx 10 V/cm of the gel and run the electrophoresis for 1–1.5 h.

5. Stain the gel in ethidium bromide solution (0.5 µg/mL) for 20–30 min. Rinse in distilled water or electrophoresis buffer for 30 min and document with Polaroid camera.

3.3.4. Analysis of Species-Specific Restriction Patterns of Ribosomal DNA (PCR Ribotypes)

With the PCR primers and restriction enzymes mentioned, species-specific restriction patterns are generated for most of wine yeast species (*see* **Notes 7** and **8**). Here, we recommend a comparison with already published data in the literature and comparative analyses of species-type strains.

3.4. Confirmative Species Identification with Selected Physiological and Biochemical Tests

For differentiation of yeast species with identical restriction patterns of amplified rDNA (and for confirmative identification of all other species), we recommend the observation of selected physiological and biochemical characteristics. The tests most often used are fermentation and/or assimilation of carbon sources, assimilation of nitrogen sources, requirements for vitamins, growth at various temperatures, resistance to antibiotics, and so forth (*7*). The advantage of our protocol is the possibility of a significant reduction in the number of tests if species-specific restriction patterns of amplified rDNA fragments are collected first. We recommend combining the phenotypic observation with the molecular identification to distinguish among species where differentiation on the basis of PCR ribotypes is doubtful or impossible (*see* **Notes 9** and **10**).

3.5. Electrophoretic Karyotyping with PFGE

3.5.1. Isolation of Chromosomal DNA

1. Grow the isolated colonies on YEPD agar plates for 24–48 h at 28°C.
2. Harvest the cells with the loop and suspend them in 1 mL of distilled water in an Eppendorf tube to wash them and centrifuge ($1500g$, 5 min).
3. Wash the cell pellet in 1 mL of 50 mM EDTA, pH 7.5, centrifuge ($1500g$, 5 min,) and discard the supernatant.
4. Resuspend the cells in 40 µL of CPES buffer and warm the prepared biomass at 42°C to prevent the solidification when 80 µL of melted 1% LMP agarose with 1 mg/mL of lysing enzyme is added (*see* **Note 11**).
5. Mix by pipetting, pour 100 µL of biomass into dry block formers with taped bottom and refrigerate for rapid setting.
6. Remove the agarose blocks from the formers by tapping them in appropriate containers (i.e., 2 mL cryovials) and incubate them in 1 mL of CPE buffer at 30°C for 1 h without shaking.

Identification of Yeasts from Grape/Must/Wine System

7. Replace CPE buffer with 50 mM EDTA (pH 9.0) and incubate for 15 min at room temperature. Repeat the rinsing step three times.
8. At final rinsing, replace EDTA with 1 mL of lysis solution with Proteinase K and incubate overnight at 50°C or at 37°C for 2–3 h with gentle shaking on the platform shaker.
9. Dilute N-lauryl–sarcosine by rinsing the plugs with 50 mM EDTA, pH 9.0, for 1 h.
10. Store the blocks in 0.5 M EDTA (pH 9.0) at 4°C. Blocks are stable in EDTA for more than a year.

3.5.2. Pulsed Field Gel Electrophoresis

1. Melt 1% (w/v) agarose in 0.5X TBE buffer, cool it to approx 50°C, and pour onto a level glass plate; place the comb and allow to set.
2. Load the sliced pieces of agarose blocks (approx. 1–2 mm wide, according to the concentration of DNA in the blocks) (*see* **Note 12**) against the bottom and front surfaces of the well and overlay with melted 1% (w/v) LMP agarose. Add appropriate size markers to each gel (*see* **Note 13**).
3. Cool the 0.5X TBE buffer in PFGE apparatus to 10–12°C and place the plate with the gel into the PFGE apparatus (*see* **Note 14**).
4. When the running time expires (*see* **Note 15**), switch off the power supply and slide the gel from the plate to a staining solution (0.5 µg/mL of ethidium bromide in distilled water or electrophoresis buffer) for 30–40 min. Rinse it in distilled water for 1 h or longer for improved contrast and photograph. The time of exposure to UV light is restricted to a maximum of 1 min because of degradation of DNA under UV light.

4. Notes

1. Recommended restriction enzymes are *Hae*III, *Cfo*I, *Hinf*I, *Msp*I, *Rsa*I, *Scr*FI, and *Taq*I.
2. Instead of acidified YM agar, special media could be used for isolation and maintaining of certain yeast species (addition of 0.5% $CaCO_3$ for *Brettanomyces* yeasts, higher pH for *Schizosaccharomyces*, or higher sugar concentration for osmotolerant *Zygosacharomyces* yeasts). If the problem of mold contamination is to be expected (i.e., botrytizied grapes), 12 mg/L of diphenyl should be incorporated into the growth media. When you expect bacteria to be present you may add oxytetracycline (0.01%) to reduce bacterial contamination. Another limitation of usually used rather nonselective isolation medium is doubtful isolation of certain species from samples containing yeast species in significantly different concentrations: for example, isolation of *S. cerevisiae* from grape surface or grape juice, or vice versa, and isolation of non-*Saccharomyces* species in later stages of fermentation. Selective lysin agar (Oxoid, England) could be used in the latter case (*6*).
3. Some macromorphological features of wine yeasts may be useful for species identification. An example is the production of certain pigments. Yeasts from

genera *Rhodotorula, Rhodosporidium, Cryptococcus, Sporidiobolus*, and *Sporobolomyces* produce nondiffusible red, orange, and pink carotenoid pigments. *Metschnikowia pulcherrima* strains produce diffusible pulcherrimin pigment. Yeasts belonging to these genera are usually isolated in vineyards, from grapes, wine-cellar equipment and also in early stages of must fermentation.

Another interesting morphological feature is the formation of giant colonies, which are prepared by applying a strong inoculum into the center of YM agar plate and judged after 3–4 wk. Their shape, size, and consistency, topography of surface, edges, and perpendicular section, and luster are observed and found characteristic for particular species.

The formation of film could be observed on the surface of wine with low alcohol content (8–10% v/v) after 3 and 30 d of incubation at 25°C. Typical examples of film-forming yeasts are *Pichia anomala, P. membranifaciens, Candida vini*, and *C. krusei*.

4. Ascospore morphology is often used for genus delimitation. Among wine yeasts needle-shaped, spheropedunculate, and ellipsoidopedunculate asci (spores) are characteristic for *M. pulcherrima* and *M. reukaufii*, hat-shaped ascospores for genera *Pichia* and *Dekkera*, and round/oval ascospores for *Kluyveromyces, Saccharomyces, Torulaspora, Zygosaccharomyces, Debaryomyces*, and so forth.

5. DNA isolation (**Subheading 3.1.1.**) should be adapted for yeast genera that produce large amounts of extracellular mucus (e.g., *Rhodotorula, Cryptococcus, Sporidiobolus*) or for genera with a particularly resistant cell wall (e.g., *Zygosaccharomyces, Schizosaccharomyces*). In **step 3**, increased concentration of lysing enzyme (12 mg/mL) and the use of Lyticase (12 mg/mL) in addition to lysing enzyme is recommended (and, sometimes, further purification with Proteinase K and phenol extraction is also required).

6. Comparing the amplification with NS1/ITS4 and ITS1/ITS4 primers, there is no difference in distinction capacity among all species tested (data not published) because in both cases, the same variable ITS region is amplified. The difference is in the length of the PCR product (in the case of NS1/ITS4, approx 2100 – 2500 bp and for ITS1/ITS4, approx 400–800 bp) and in the complexity of the restriction patterns.

7. With PCR-RFLP of rDNA, it is possible to distinguish among the majority of wine yeasts with some exeptions:
    a. In the genus *Debaryomyces*, it is not possible to distinguish among the species *D. hansenii, D. vanrijiae, D. udenii, D. castellii*, and *D. polymorphus*.
    b. It is not possible to differentiate between *Saccharomyces bayanus* and *S. pastorianus* (**10**).
    c. Species-specific restriction patterns for the species *Hanseniaspora uvarum* and *H. guilliermondii* is possible to obtain only with the restriction enzyme *Dde*I (**11**).
    d. Closely related species of the genera *Torulaspora* and *Zygosaccharomyces* have the same restriction patterns obtained with some restriction enzymes (i.e., *Msp*I: there are no differences among type strains *T. pretoriensis, Z. rouxii, Z. baillii, T. delbrueckii, Z. bisporus*, and *Z. microellipsoides*).

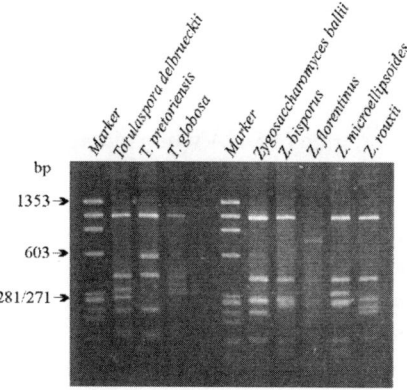

Fig. 2. Selected species-specific PCR-ribotypes of *Torulaspora* and *Zygosaccharomyces* wine yeasts generated with NS1 and ITS4 primers and restriction endonuclease *Cfo*I.

However, with the restriction enzymes *Hae*III, *Cfo*I, and *Sau*3A, it is possible to obtain species- and genus-specific patterns (**Fig. 2**) *(12)*.

8. There are also found some intraspecies variations in PCR-ribotypes, which are more frequent when dealing with strains isolated from geographically distinct environments and express possible varieties of the same species. Intraspecies differences so far found are in the species *K. lactis, D. polymorphus, P. membranifaciens, S. pombe, Z. microellipsoides, H. osmophila,* and *C. laurentii (11,12)*.

9. In the cases where final determination of yeast species is not possible just on the basis of restriction fragments of amplified rDNA (because of limited databases, lack of variation within different taxa and/or limited number of restriction enzymes used), a combined study of genotypic and phenotypic features is necessary. Some examples are as follows: Differentiation between teleomorphic and anamorhic yeast forms is possible only with the observation of spore formation because the restriction patterns of rDNA of teleomorph/anamorph pairs are the same [i.e., *Hanseniaspora, Kloeckera (13)*]. In addition, although differentiation among very closely related species such as *H. uvarum/K. apiculata* and *H. guilliermondii/K. apis* by molecular tools is very difficult (*see* **Note 7c** and **ref. 13**), we recommend the testing of growth at 37°C, which is characteristic for *H. guilliermondii/K. apis* but not for *H. uvarum/K. apiculata (14)*.

10. In attempts to identify closely related species in the genera *Torulaspora* in *Zygosaccharomyces*, the following tests are recommended: growth at 37°C, assimilation of galactose, melibiose, α-methyl-D-glucoside and trehalose, growth on agar with 1% acetic acid, growth with 0.1% cycloheximide and 16% NaCl/5% glucose *(14)*.

11. For yeasts belonging to the genera *Rhodotorula, Cryptococcus*, and *Sporidiobolus* and others that produce large amounts of extracellular polysaccharides, an increased concentration of lysing enzyme (12 mg/mL) and also the use of Lyticase (12 mg/mL) is recommended.
12. An overloaded concentration of DNA will yield a smear instead of clear electrophoretic fragments. It may also change the migration rate and result in the wrong estimation of chromosome size.
13. Chromosomal fragments of *S. cerevisiae, P. canadensis (H. wingei)* and *S. pombe* are suitable commercially available size markers for estimation the length of short and long yeast chromosomes, respectively.
14. Several apparatuses have been developed for PFGE that provide a homogenous electric field allowing DNA molecules to migrate in straight lines in the gel. The most widely used is the contour-clamped homogeneous electric field (CHEF) in which a hexagonal electrode array allows a switch through an angle of 120° between pulses. There are at least three commercially available machines of this type: Pulsaphor, Pharmacia; CHEF DR-11, Bio-Rad and Hexafield, BRL. Other PFGE devices use rotating gel electrophoresis (RGE), where the field is stationary but the gel is rotated between fixed points controlled by microswitches, rotating field electrophoresis (RFE), in which electrodes rotate around the gel boundary, and, finally, alternating field electrophoresis (TAFE), in which the gel stands vertically and the electric field passes through the thickness of the gel *(15)*.

    The separation process of PFGE is a subject of many parameters:
    - Concentration of agarose: The recommended concentration of agarose varies between 0.9% and 1.2%. At a lower agarose concentration, the resolution of fragments is lower although the run time can be shorter *(15)*.
    - The pulse time: if the pulse time is much shorter than the molecular reorientation time, the molecules do not change the direction and any separation resulting from molecular reorientation is lost. Typical pulse times vary from 0.1 s for molecules smaller than 10 kbp to 1000 s for those approaching 10 °Mbp *(16)*.
    - The field strength: For separation of very large DNA molecules, it is important to lower the voltage *(17)*.
    - The temperature: with increasing temperature, the mobility of DNA molecules is increased because of a decrease of viscosity of the buffer *(18)*.
15. The electrophoresis conditions are different for the *Saccharomyces* sensu stricto group and other wine yeast species because of the high number of rather small chromosomes (smaller than 2200 kbp) within the *Saccharomyces* sensu stricto group. Other wine yeasts have evidently larger chromosomes (even over 3 Mbp) and, consequently, different protocols are used for karyotyping *Saccharomyces* sensu stricto and other yeasts. As an example, two electrophoresis conditions of pulse intervals for contour-clamped homogenous field system (CHEF, Pulsaphor[TM]) are presented:

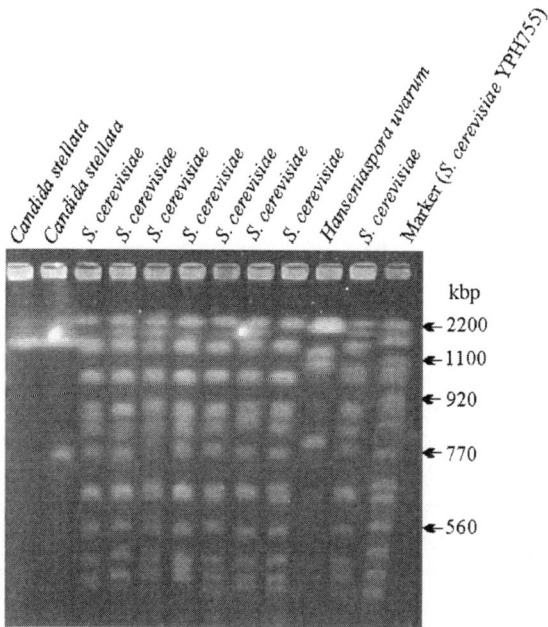

Fig. 3. Selected electrophoretic karyotypes of *S. cerevisiae*, *C. stellata*, and *H. uvarum* generated under electrophoresis conditions as described in protocol 1. (From K. Povhe-Jemec et al., unpublished results.)

Protocol 1
For *Saccharomyces sensu stricto* yeasts
Phase 1: 60 s, 15 h
Phase 2: 90 s, 8 h
Phase 3: 100 s, 1 h
Total run time: 24 h
Voltage: 170 V

Protocol 2
For non-*Saccharomyces* yeasts
Phase 1: 150 s, 24 h
Phase 2: 300 s, 24 h
Phase 3: 600 s, 20 h
Total run time: 68 h
Voltage: 100 V

Just to differentiate between *Saccharomyces* sensu stricto and non-*Saccharomyces* yeasts, protocol 1 is recommended (**Fig. 3**). Large chromosomal fragments of non-*Saccharomyces* yeasts remain on the top of the gel, whereas chromosomes of *Saccharomyces* are separated throughout the gel.

For differentiation among non-*Saccharomyces* yeasts, protocol 2 should be used. The method is not optimized for all genera of wine-associated non-*Saccharomyces* yeasts, but the chromosomal separation is good for the following genera: *Rhodotorula*, *Sporidiobolus*, *Zygosaccharomyces*, *Saccharomycodes*, *Debaryomyces*, *Hanseniaspora*, *Pichia*, and *Kluyveromyces*. For the genera

Cryptococcus, Dekkera, Metschnikowia, Schizosaccharomyces, and *Torulaspora,* further optimization is necessary because of the length of their longest chromosomal fragments, which exceeds 3.13 Mbp.

References

1. Barnett, J. A., Payne R. W., and Yarrow, D. (1990) *Yeasts: Characteristics and Identification,* Cambridge University Press, Cambridge.
2. Barnett, J. A., Payne, R. W., and Yarrow, D. (1996) *Yeast Identification PC Program,* version 4, Cambridge University Press, Cambridge.
3. Lachance, M. A. and Starmer W. T. (1998) Ecology and yeasts, in *The Yeasts, a Taxonomic Study* (Kurtzman, C. P. and Fell, J. W., eds.), Elsevier, Amsterdam, pp. 21–30.
4. Praphailong, W., Van Gestel, M., Fleet, G. H., and Heard, G. M. (1997) Evaluation of the Biolog system for the identification of food and beverage yeasts. *Lett. Appl. Microbiol.* **24,** 455–459.
5. Barnett, J. A., Payne, R. W., and Yarrow, D. (1983) *Yeasts: Characteristics and Identification,* Cambridge University Press, Cambridge.
6. Fleet, G. H. (1993) The microorganisms of winemaking — isolation, enumeration and identification, in *Wine Microbiology and Biotechnology* (Fleet, G. H., ed.), Harwood AP, Sydney, Australia, pp. 1–26.
7. Yarrow, D. (1998) Methods for the isolation, maintenance and identification of yeasts, in *The Yeasts, a Taxonomic Study* (Kurtzman, C. P. and Fell, J. W., eds.), Elsevier, Amsterdam, pp. 77–102.
8. Spencer, J. F. T., and Spencer, D. M. (1996) Maintenance and culture of yeasts, in *Yeast Protocols, Methods in Cell and Molecular Biology,* vol. 53 (Evans, I. H. ed.), Humana, Totowa, NJ, p. 7.
9. White, T. J., Bruns, T., Lee, S., and Taylor, J. (1990) Amplification and direct sequencing of fungal ribosomal RNA genes for phylogenetics, in *PCR Protocols: A Guide to Methods and Applications* (Innis, M. A., Gelfand, D. H., Sninsky, J. J., and White, T. J., eds.), Academic, London, pp. 315–322.
10. Smole Mozina, S., Dlauchy, D., Deak, T., and Raspor, P. (1997) Identification of *Saccharomyces* sensu stricto and *Torulaspora* yeasts by PCR ribotyping. *Lett. Appl. Microbiol.* **24,** 311–315.
11. Esteve-Zarzoso, B., Belloch, C., Uruburu, F., and Querol, A. (1999) Identification of yeasts by RFLP analysis of the 5,8S rDNA gene and the two ribosomal internal transcribed spaces. *Int. J. Syst. Bacteriol.* **49,** 329–337.
12. Cadez, N., Princic, M., Smole – Mozina, S., and Raspor, P. Unpublished results.
13. Smole Mozina, S., Cadez, N., and Raspor, P. (1998) rDNA RFLPs and AP-PCR fingerprinting of type strains and grape-must isolates of *Hanseniaspora (Kloeckera)* yeasts. *Food Technol. Biotechnol.* **36,** 37–43.
14. Kurtzman, C. P. and Fell, J. W. (1998) *The Yeasts, a Taxonomic Study,* Elsevier, Amsterdam.
15. Maule, J. (1998) Pulsed field electrophoresis. *Mol. Biotechnol.* **9,** 107–126.

16. Bustamante, C., Gurrieri., S., and Smith, S. B. (1993) Towards a molecular description of pulsed-field gel electrophoresis. *TibTech* **11,** 23–30.
17. Pharmacia LKB Technology (1990) *Instruction Manual PulsaphorTM System*, Pharmacia LKB Biotechnology, Uppsala, Sweden.
18. Mathew, M. K., Smith, C. L., and Cantor, C. R. (1988) High-resolution separation and accurate size determination in pulsed-field gel electrophoresis of DNA. 1. DNA size standards and the effect of agarose and temperature. *Biochemistry* **27,** 9204–9210.

30

Carotenogenic Microorganisms

A Product-Based Biochemical Characterization

José Domingos Fontana

1. Introduction

Carotenoid production or occurrence—including derivatives biosynthesized from precursors—is widespread in nature in both the Prokaryota and Eucaryota superkingdoms and more than 600 different chemical structures are reported (*1*), most of them as tetraterpenoids (C_{40}). Mammalian species lack the biochemical ability for carotenoid biosynthesis, but they convert some of them to vitamin A or perform other chemical modifications on the diet carotenoid input. At least four particular carotenoids (**Fig. 1**) are fully exploited for commercial applications because their production was consolidated through chemical synthesis: β-carotene (C_{40}; double cyclic ends), canthaxanthin (diketo-β-carotene), astaxanthin (dihydroxy-diketo-β-carotene), and apocarotenoic acid as its ethyl ester (C_{32}; single cyclic end). As diluted organosolvent solutions (e.g., 2–4 µg/mL), these pigments display yellow to orange deep colors. The former nonoxygenated product, also obtained from plant sources like carrots and from genetically improved strains of the molds *Blakeslea trispora* (*2*) and *Phycomyces blakesleeanus* (*3*), is mainly employed in pharmaceutical multivitamin formulations or as a food additive in margarines. The three other xanthophyls (oxygenated carotenoids) are mainly used in aquaculture (salmonoid fish farming) and poultry purposes as an enhancer of meat and egg-yolk color. Cantaxanthin is the pink–orange natural pigment in the edible mushroom *Cantharellus cinnabarinus* (Agaricaceae) and in flamingo feathers. Astaxanthin is naturally found in the orange–red basidiomicetous yeast *Phaffia rhodozyma* (now *Xan-*

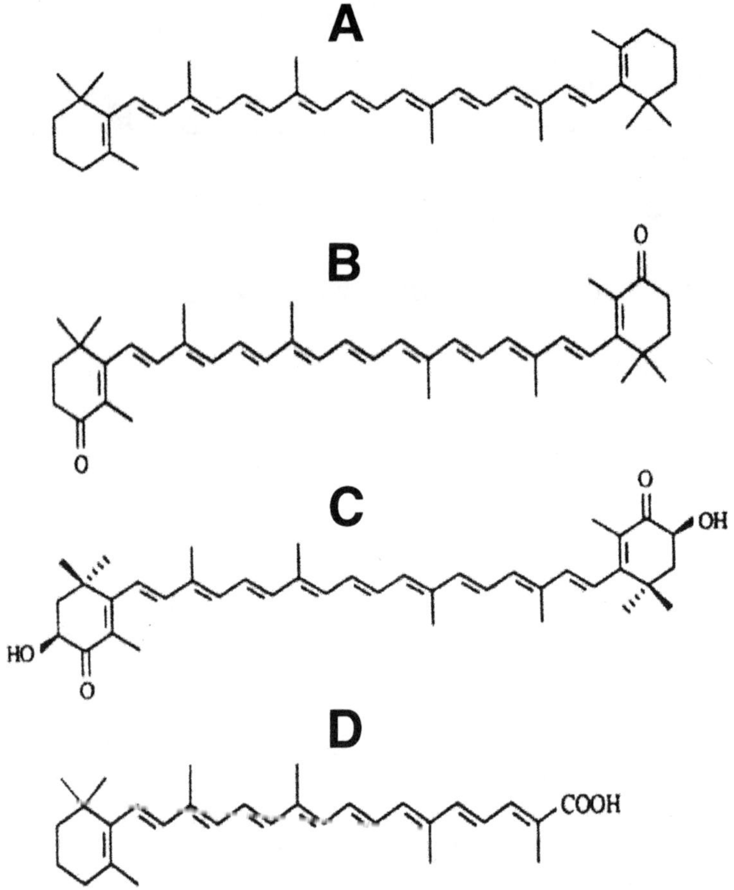

Fig. 1. The basic chemical structures for a hydrocarbon carotene (**A** = β-carotene) and for xantophylls (**B** = canthaxanthin, **C** = astaxanthin, and **D** = apocarotenoic acid, free form).

thophyllomyces dendrorhous) *(4)*, in the chlorophycean unicellular alga *Haematococcus pluvialis (5)*, and in the marine bacterium *Agrobacterium aurantiacum (6)*. The involved market appeal for carotenoids is strongly supported by scientific knowledge of their well-known biological activity for quenching and scavenging of free radicals (e.g., singlet oxygen and other active oxygen species), which are responsible for the undesirable effect of aging *(7)*. This is the main reason for a consolidated market estimated about US$ 455 million for 2000 only for astaxanthin and cantaxanthin. Both contributions for this market—chemical synthesis and microbial source—are experiencing an increase, but in the second parcel, a doubling is seen every 4 yr *(8)*.

Extraction, purification and, particularly, characterization of carotenoids from microorganisms may be carried out by methods exploring their polarity (i.e., differential solubility or partition between water and water-imiscible organosolvents but for those more intensively oxygenated, the tight interaction with protein carriers must be taken in account). A typical case is the strong association between astaxanthin and actomyosin in salmon flesh; a similar situation applies in the case of the yeast *P. rhodozyma*. For the sake of sample enrichment or purification, a mild saponification may be used *(9)*, although alternative natural occurrences like esters of xanthophylls (e.g., astaxanthin palmitate found in fish; bixin, a natural half-methyl ester of a C_{25} carotenedioic acid found in the seed tegument of the plant *Bixa orellana*) are, obviously, lost as such. Conversely, saponification allows the interference arising from chlorophylls (e.g., in the case of algal and plant samples) also to be eliminated. Concerning the characterization of carotenoids, thin-layer chromatography (TLC) allied to high-pressure liquid chromatography (HPLC) and/or spectroscopy is a very valuable strategy for carotenoid characterization, although more sophisticated techniques like ^{13}C- and ^{1}H NMR (nuclear magnetic resonance) are mandatory for the elucidation of the detailed fine structure determined by the complex geometrical, optical, and conformational isomerisms.

2. Materials

1. The yeast *Phaffia rhodozyma (Xantophyllomyces dendrorhous)* as the single species and *Rhodotorula rubra* may be obtained from ATCC (American Type Culture Collection) and the alga *Haematococcus pluvialis* from the collection from the University of Texas at Austin. An hypercarotenogenic and genetically improved strain of *P. blakesleeanus* was provided by Enrique Cerdá-Olmedo from the Department of Genetics at Seville University, Spain. Any bacterial isolate (e.g., "RB" and "A," presently used in TLC analyses) may be considered from the deep colored aspect of its colonies in a solid media culture (yellow → red pigmentation) provided the preliminary spectroscopy of an organosolvent extract (hexane, acetone or chloroform:metanol 2:1 v/v) appoints dominant maximal absorbance peaks in the range from 440 to 480 nm and that no marked spectral shift occurs upon mild acidification or alkalinization. Liquid culture media for yeast and bacteria growth and carotenoid-based pigmentation may be formulated with 2–3% sucrose or glucose as the carbon source with 0.1–0.2% urea or peptone as simple and complex nitrogen sources and 0.1% yeast extract providing all other micronutrients. *Phaffia,* particularly, affords satisfactory pigmentation in any sugar cane (beet, too) derivative like crude juice, brown sugar, and saccharified bagasse *(10)*. The final pH is adjusted to around 4.5 (yeast) and 6.0 (bacteria). Microalgal cultures may be carried out in salt-based improved media containing $NaNO_3$ and under strong illumination *(11)*.

2. Plant sources like corn kernels (*Zea mays*; dihydroxy-β-carotene or zeaxanthin), "urucum" seeds (*Bixa orellana*; C_{25} carotenedioic acid monomethyl ester bixin and its demethylated form norbixin), saffron (*Crocus sativus*; C_{20} carotenedioic acid or crocetin as the nucleus for the digentiobioside crocin), and paprika (*Capsicum annuum*; capsorubin and capsaxanthin, C_{40} hydroxy/keto derivatives of β-carotene) are useful for comparative standard preparation of the respective oxygenated carotenoids (xanthophylls).
3. Premade chromatoplates of finely divided silica gel on aluminum foils (gypsum as the binder) are the ideal stationary phase for TLC and may be obtained from suppliers like Merck (art. 1.05553), Schleicher& Schuell, and similar companies. Sample application of carotenoid solutions (1–2 µL from solutions containing >100 µg/µL) is preferable as 0.3- to 0.5-cm-long thin bands for resolution improvement through the help of Bunsen-stretched 5-µL glass capillaries. Low-polarity chromatographic mobile phases are based in the mixture of different proportions of toluene:ethyl acetate:acetone (e.g., 85:7:8; solvent A) or more complex organosolvent combinations (e.g., hexane:acetone:ethyl acetate: nitroethane:methanol:water; 79:19:5:3.5:3.5:0.25; solvent B) for superior resolution of carotenoid components. The glass chromatographic tank, internally aligned with a piece of filter paper in one of the faces to ensure vapor-phase saturation, should be bubbled with a nitrogen stream just after solvent pouring and then protected from light by an aluminum foil wrapper. A piece of plastic film (MagiPak) can be inserted between the tank and its lid to avoid loss of the more volatile components of the mobile phase and, hence, its alteration, which leads to less reproducible results. If individual carotenoid-component quantitation is intended, the freshly ran chromatoplate (natural colors) or after a chromogenic reagent spraying (iodine vapors or warmed acidified *p*-anisaldehyde) may be scanned in a densitometer (the flying spot scanning densitometer CS-9301PC apparatus from Shimadzu is a particularly fast and skill apparatus for such a purpose) following the selection of the appropriate scanning wavelength.
4. Spectroscopy and spectrophotometry may be carried out in any apparatus having a double-beam arrangement for the solvent blank and the colored sample for the purpose of light-absorbance measurement at a fixed and known maximum of absorbance (λ_{max} = 470–475 nm for canthaxanthin or astaxanthin) or for the sample scanning in the visible range from 360 to 600 nm. A spectrum recorder is the ideal accessory.
5. As a refinement of carotenoid analyses, a HPLC multimodular apparatus (e.g., 712 WISP and 600 E System Controller from Millipore-Waters, USA or LC-10-AD from Shimadzu, Japan) having a tunable absorbance detector and recorder is indicated. The ideal monitoring/recording may be obtained with a diode array because the simultaneous full spectral analysis may be done simultaneously for each resolved peak. A reversed-phase packed column (e.g., Supelcosil LC-18; 25 cm long; 5-µm silica particles mean diameter; from Supelco) isocratically irrigated with 1 mL/min (about 600 psi as operating pressure) of acetonitrile: chloroform:methanol:water = 60:25:10:5) allows satisfactory reso-

lution for many carotenoids of microbial and commercial interest, including lycopene (from tomatoes) in addition to the examples depicted in **Fig. 1**.

3. Methods
3.1. Sampling, Extraction, and Spectrophotometry

1. Microbial biomass, usually from the stationary phase of growth and other comparative samples (e.g., carotenoid-containing fruits and seeds) are used as dry starting materials through lyophilization. Heat application or hot-air ovens should be avoided for moisture removal because of the ease of oxidation of carotenoids. Although the dry state is not *a sine qua non* condition for the efficient carotenoid extraction, water is better in the next partition step of the moisture-free organosolvent extract regardless of the coapplication of an intermediate saponification step. In any instance, the conjunction of light, heat, and oxygen is very harmful to carotenoid native structure. Hence, whenever possible, a flux of nitrogen (e.g., solvent deaereation) is strongly recommended. For those carotenoids not readily available as purified standards, the above procedure may be also applied (e.g., for the preparation of canthaxanthin, astaxanthin, and apocarotenoic acid ethyl ester starting from the commercial bead-shaped formulations provided by F. Hoffmann-La Roche Ltd., and respectively designated as Carophyll-Red, -Pink, and -Yellow where each carotenoid is protected by a mix of starch-encapsulated gelatin/free sugar and ethoxyquin and ascorbyl palmitate as antioxidants).
2. To each portion of the moisture-free sample (e.g., 100 mg), 1 mL of dimethylsulfoxide (DMSO) is added and the mixture left for 1 h for complete swelling followed by the addition of 2 mL of acetone. Centrifugation at $\geq 3000\ g$ for 5 min renders a supernatant containing most of the apolar and polar (e.g., oxygenated) carotenoids. Extraction is brought to completion by repeating the swelling step in DMSO but now with the addition of 2 mL of chloroform:methanol 1:1. Combined supernatants are then adjusted to exactly 10 mL with acetone and an aliquot is read against a solvent blank (DMSO:acetone:chloroform methanol = 2:6:1:1) at the appropriated wavelength (e.g., at 444 *or* 472 nm for β,β-carotene-enriched samples considering the respective absortion coefficients [$E^{1\%}_{1\ cm}$] of 2500 *or* 2180 in hexane:acetone 9:1 *(11)* or 2550 at 451 nm in pure hexane *(12)*; for astaxanthin and $E^{1\%}_{1\ cm} = 2220$ at 492 nm when using acidified DMSO *(13)* or $E^{1\%}_{1\ cm} = 2100$ at 470 nm also for astaxanthin if the solvent is hexane:chloroform 95.5:4.5 v/v *(14)*. The concentration of carotenoid in the DMSO:acetone:chloroform:methanol mix (as in the case of an astaxanthin-enriched sample obtained from *Phaffia* or *Haematococcus*), computing the unit change from 1 g% = 10 g/L to mg/mL ($10^3 \cdot 10^1$) and a correction of $E^{1\%}_{1\ cm*}$ for solvent effects (*see* below) would then be from the absorbance reading at $\lambda_{max} - 472$ nm (e.g., $A = 0.6115$ for calculation ease):

[Astaxanthin] (in mg/L) = $A \times 10^4)/E^{1\%}_{1\ cm*}$ = $(0.615 \times 10^4)/2050*$ = 3 mg/mL = 3 µ/mL

which in turn is multiplied by 100 (since the total sample organosolvent extract is $Vt = 10$ mL but from only 1/10 of g of dry yeast sample) is equal to the amount of mg of astaxanthin/g of dry yeast (= 0.3 mg/g; present case). Accordingly, wild-type strains of the aforementioned astaxanthinogenic yeast most often accumulate (stationary phase of growth) about 0.3–0.6 µg astaxanthin/g dry cells *(15)*. For the sake of precision, an extinction coefficient $E^{1\%}_{1\,cm}$ between 2120 and 1910 (hexane with 2% ethanol or dichlorometane as cosolvent and a maximum absorption peak centered at λ_{max} = 472 nm) may be considered when working with other natural sources containing more than one isomeric form of astaxanthin (e.g., the shrimp *Penaeus*) *(16)*. In any case, one seldom deals with the occurrence of a pure carotenoid when processing natural sources. Hence, the statement that an absorbance of 0.25 corresponds to a carotenoid concentration of approx 1 µg/mL *(11)*. The isomer 3R,3R-dihydroxy-β,β-carotene-4,4-dione is the particular astaxanthin isomer in the yeast *Phaffia (Xanthopyllomyces)*. The nature of a selected solvent (or organosolvent mix) may result in deep modification of the absorbance readings. For instance, the comparative use of hexane *or* methanol in normalized solutions of lutein or zeaxanthin (xanthophylls that differ solely by the position of one double bond) of lycopene (a non-cycle-ended hydrocarbon carotene) leads to dramatic changes in the absorbance values (with neither ipsochromic nor bathochromic significant displacements of λ_{max}) but deeply dependent on the carotenoid concentration in the range from 1 to 10 µM. Lutein absorbance is minimally affected by solvent; zeaxanthin results in similar values until 3 µM but in almost doubled ones when above 6 µM (methanol leading to an apparent hyperchromic effect), and lycopene in methanol results, in any concentration range, in fourfold lower values (methanol now resulting in an apparent hypochromic effect). Solubility and microcrystallization must be considered to explain these differences besides the much lower contributions from the respective molar absorbities (ε) or extinction coefficients *(17)*. Concerning the auxochromic effects arising from polar substituents like R—OH, and R—COOH spectrophotometric scanning allows clear distinction between hydrocarbon carotenes and xanthophylls (*see* **Fig. 2**).

3.2. Thin-Layer Chromatographic Analysis

1. Thin-layer chromatographic (TLC) analysis (**Fig. 3A**), following spontaneous solvent evaporation allows the calculation of *R_f values for each carotenoid of interest (e.g., in solvent B, the following R_f are obtained: 0.88 for β-carotene; 0.82 for apocarotenoic acid ethyl ester; 0.51 for canthaxanthin; 0.36 for astaxanthin). If a known amount of each standard is applied from a calibrated capillary or microsyringe, then the use of a densitometer will allow quantitation of the components in the microbial sample as well as their respective percentage contribution to the whole carotenoid extract composition (**Fig. 4**, lower line; e.g., peak "A" of astaxanthin contributing to 50% of the carotenoids isolated from *Phaffia rhodozyma*). Because carotenoid bands will experience a progressive fad-

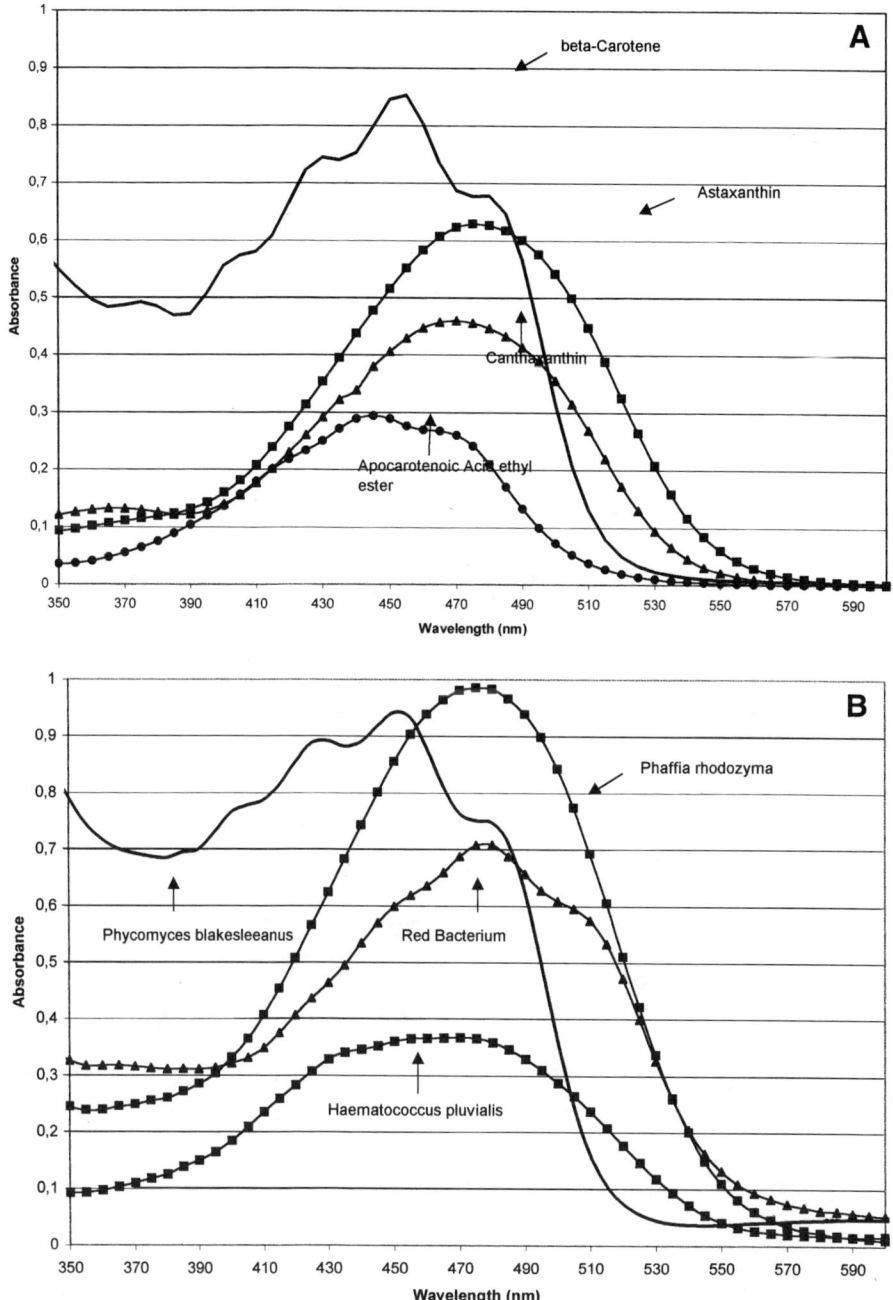

Fig. 2. Visible spectra for standard (**A**) and microbial-derived carotenoids (**B**).

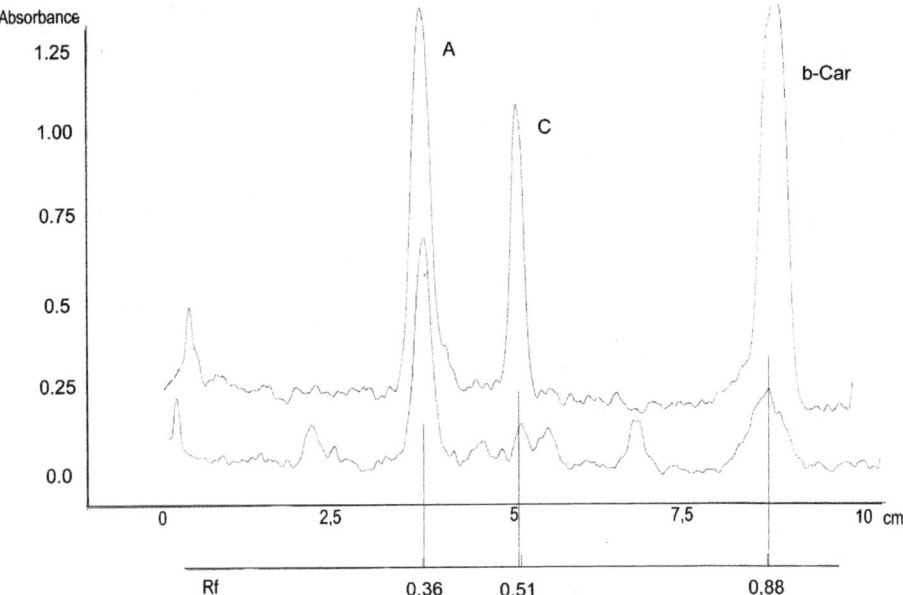

Fig. 3. Densitometric monitoring of TLC plates at 475 nm. Upper line: mix of astaxanthin (**A**), canthaxanthin (**C**), and β-carotene (**B–C**) standards from the TLC lane "m" and in order of increasing R_f values. Lower line: *Phaffia rhodozyma* crude extract from the TLC lane Pr.

ing on exposure to air, a permanent record of the freshly run plate with a color film may be obtained, which may be also used for densitometric purposes.

2. The contribution of liposoluble noncarotenoid components may be progressively evaluated by exposing the plate to a closed chamber saturated with iodine vapor (**Fig. 3B**) and/or spraying with a fine mist of 1% *p*-anisaldehyde in methanol:sulfuric acid 95:5 (**Fig. 3C**). In the case of the latter chromogenic reagent, careful heating on a warm plate (100°C) will change the variable natural yellow to orange–red colors of carotenoids to gray, whereas other noncolored lipid materials such as triacylglycerols (TAG), glycolipids (GL), phospholipids (PL), and sterols will stain deeply lilac to violet.

3.3. High-Pressure Liquid Chromatography

1. Isocratic elution allows satisfactory resolution for many carotenoid components (**Fig. 5**), but peak overlapping is unavoidable for those derivatives displaying similar polarities (for instance, the smaller peak preceding R_t [retention time] = 3.27 for astaxanthin is attributable to lutein, a dihydroxy-β-carotene). Better peak resolution for multicomponent carotenoid samples may be the attained using then gradient elution.

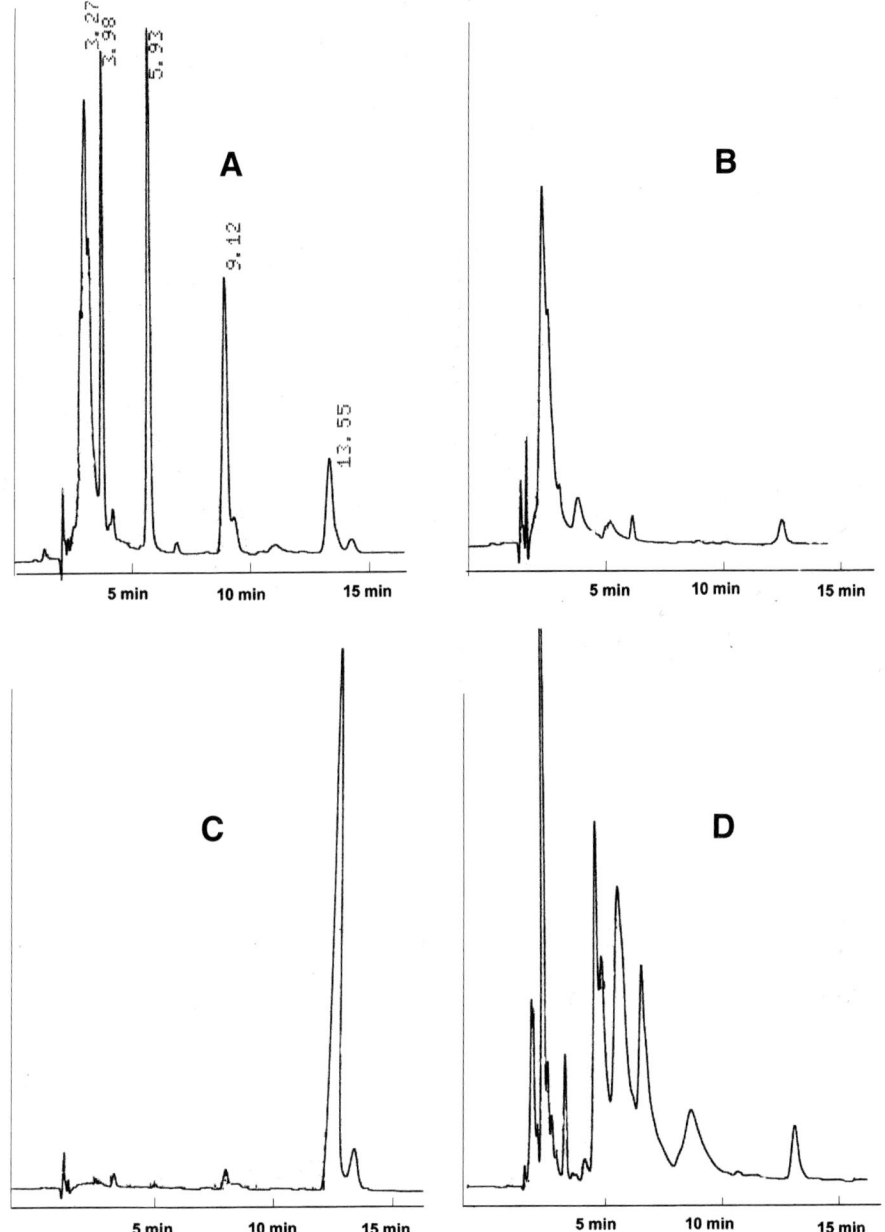

Fig. 4. High-pressure liquid chromatography with detector wavelength at 475 nm. Carotenoid standard mixture: astaxanthin, canthaxanthin, apocarotenoic acid ethyl ester, lycopene, and β-carotene (**A**), in increasing order of elution times. Samples from the organosolvent extracts of *Phycomyces blakesleeanus* (**B**), *Phaffia rhodozyma* (*Xanthopyllomyces dendrorhou*s) (**C**), and aged cultures of *Haematococcus pluvialis* (**D**).

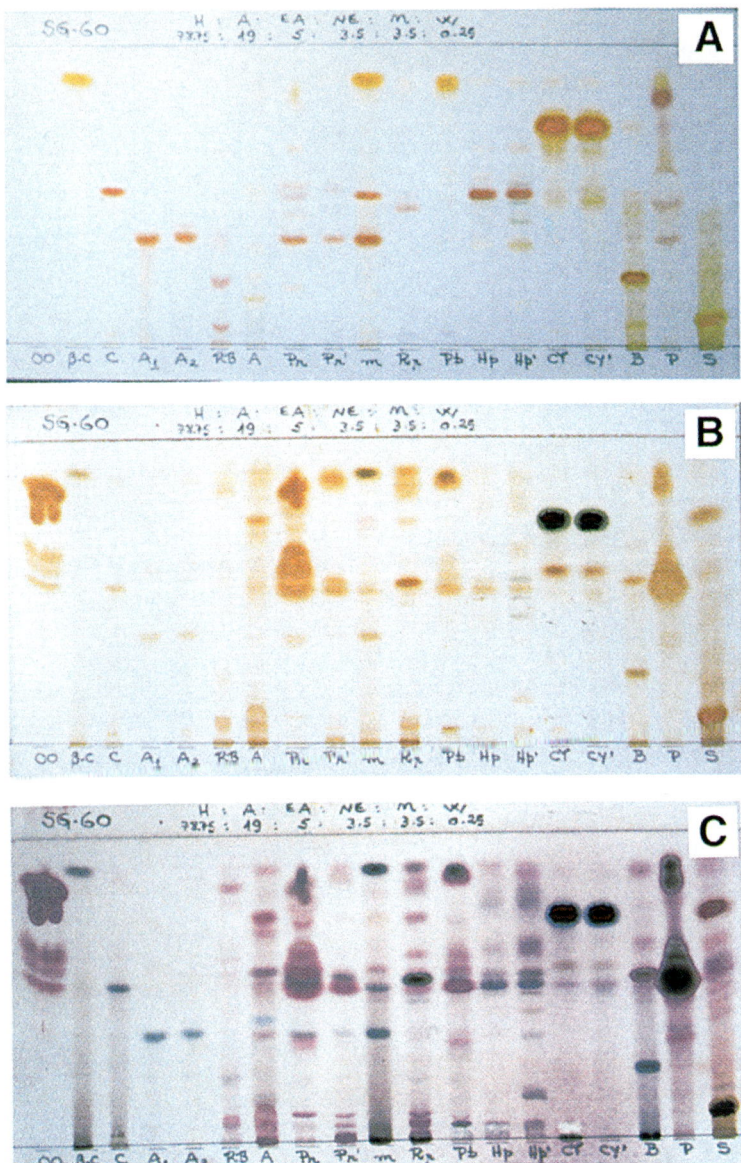

Fig. 5. Thin-layer chromatography as such (**A**) and after iodine vapors (**B**) or hot *p*- anisaldehyde staining (**C**). Carotenoid standards: β-C = all-*trans* isomer of β-carotene, from Sigma-Aldrich Co; C = canthaxanthin, isolated from Carophyll Red; A_1 and A_2 = astaxanthin, isolated or purified from Carophyll Pink; m = mix of these three standards, and CY and CY' = apocarotenoic acid ethyl ester, isolated from Carophyll

4. Notes

1. For hydrocarbon-carotenoid-enriched samples (e.g., β- and/or α-carotenes in *Phycomyces* and carrots; lycopene in tomatoes) previous swelling in DMSO is not absolutely necessary since there is no noticeable interaction between these carotenes and protein carriers and hence hexane, acetone, and/or chloroform:methanol have enough solubilizing power to yield quantitative yields.
2. Any turbidity interference observed in the organosolvent crude extract may be eliminated by two different ways: (1) increasing or decreasing the polarity by the addition of acetone *or* chloroform; (2) carrying out a cleaning partition with the addition of an large excess of water (> 4 volumes) and recovering the carotenoid component(s) in the lower chloroform phase, whereas the upper phase to be discarded contains the less soluble sample components or contaminants as well as most of the polar solvents (DMSO, acetone, methanol).
3. Taking 2500 as an average value of $E^{1\%}_{1\,cm}$ for carotenoid materials, an absorbance reading of 0.25 is equivalent to a concentration of 1 µg/mL (see the equation in **Subheading 2.1.**).
4. R_f is defined by the relation (distance traveled by a particular component)/(distance traveled by the solvent front), thus giving values <1.0 (only exceptionally equal to 1.0). Low R_f values seen in **Fig. 2A** for *Bixa orellana* (lane B) and *Crocus sativus* (lane S) extracts are in accordance with their less polar main carotenedioic acid components. Bixin's (C_{25}) relatively higher migration compared to crocin (C_{20}) is explained by the occurrence of a half methyl ester in the first diacid and/or the double contribution of two gentibiose substituents at both carboxy ends of the second diacid.
5. Thin-layer chromatography exposure to iodine vapors allows a fast, reversible, and reinforced staining for carotenoids as well as for most of the other noncolored liposoluble compounds. Apocarotenoic acid ethyl ester (**Fig. 3B**; lanes CY and CY') takes up iodine very strongly. The dominant contribution of fat material (TAG; PL) in *Phaffia* crude extract (**Fig. 3B,C**; lane Pr) can be easily realized comparing with the adjacent silica-gel-purified sample (**Fig. 3B,C**; lane Pr') following exposure to iodine vapors or to the anisaldehyde chromogenic spray (comparable also with lane OO, a triolein-rich olive oil sample).

Yellow. Microbial and plant sources: RB = a red bacterium isolated from laboratory surroundings by Tania Bonfim and Miriam Chociai; A = an inulinolytic yellow bacterium (**ref. 4**, p. 420); Pr and Pr' = crude and silica-gel-column chromatography-purified orange pigment prepared from *Phaffia rhodozyma*; Rr = *Rhodothorula rubra*; Pb = *Phycomyces blaskeleeanus;* Hp and Hp' = pigment mixture of aged cultures of *Haematococcus pluvialis*; B = crude extract of *Bixa orellana* seeds (main carotenoid component: bixin); P = crude extract from paprika (main carotenoids: capsorubin and/or capsaxanthin); S = saffron extract (main carotenoid: crocin). The first lane, OO = olive oil (triolein), is for the control of noncarotenoid liposoluble components or contaminants (e.g., TAG) when developing the TLC plate with iodine vapor or with the chromogenic spray of *p*-anisaldehyde.

6. It is advisable to include 0.05–0.1% of butylated hydroxytoluene (BHT) in the solvents as a protectant for carotenoids against oxidation from the extraction procedure through any purification step.
7. For those natural sources rich in fat materials other than carotenoids (e.g., TAG, PL) where previous saponification is advisable, the final concentration of NaOH or KOH in the water/organosolvent mix should not exceed 5% for a short period of treatment (>30 min) at temperatures not exceeding 60°C. TAG and PL may also be degraded with commercial lipases (e.g., Novo Lipozyme) which is a safer alternative.
8. ^{13}C-NMR spectroscopy indicates 19 signals (δ, in ppm) for β-carotene from 12.7 ppm until 137.8 ppm, including δ = 33.2 ppm for C-4 and C-4'. In canthaxanthin, a di-ketone derivative at C-4 and C-4' from β-carotene, this particular signal is, accordingly, displaced downfield at 198.7 ppm *(18)*.
10. Details about carotenoids occurrence, structure variability, and metabolism may be found in a fully illustrated brochure prepared at F. Hoffmann-La Roche Ltd., Basel, Switzerland *(19)*.

Acknowledgments

The author thanks A. R. C. Lima from L. Hoffmann-La Roche (Brazil) for the kind provision of the Carophyll series, Professor M. Baron for the partnership in the project CNPq-PADCT-SBIO/World Bank, Professor G. Yates for draft reading, S. V. Mendes and L. G. V. Krawiec for the untiring assistance in the preparation of illustrations, M. Passos for the help with TLC analyses, and Professor T. M. B. Bonfim and Professor M. B. Chociai for the cultivation of the "RB" bacterium isolate.

References

1. Pfander, H. (1993) Carotenoids: an overview. *Methods Enzymol.* **213-B,** 1–13.
2. Metha, B. J. and Cerdá-Olmedo, E. (1995) Mutants of carotene production in *Blakeslea trispora*. *Appl. Microbiol. Biotechnol.* **42,** 836–838.
3. Cerda-Olmedo, E. (1989) Production of carotenoids with fungi, in *Biotechnology of Vitamin, Growth Factor, and Pigment Production* (Vandamme, E., ed.) Elsevier Applied Science, Barking, UK, pp. 27–42.
4. Fontana, J. D., Guimarães, M. F., Martins, N. T., Fontana, C. A., and Baron, M. (1996) Culture of the astaxanthinogenic yeast *Phaffia rhodozyma* in low-cost media. *Appl. Biochem. Biotechnol.* **57/58,** 413–422.
5. Johson, E. A. and An, G. H. (1991) Astaxanthin from microbial sources. *Crit. Rev. Biotechnol.* **11(4),** 297–326.
6. Yokoyama, A., Izumida, H., and Miki, W. (1994) Production of astaxanthin and 4-ketozeaxanthin by the marine bacterium, *Agrobacterium aurantiacum*. *Biosci. Biotechnol. Biochem.* **58(10),** 1842–1844.
7. Palozza, P. and Krisnky, N. I. (1992) Antioxidant effects of carotenoids *in vivo* and *in vitro*: An overview. *Methods Enzymol.* **214-B,** 403–420.

8. Dean, K. L. (1992) IB market forecast: biopigments, in *Industrial Bioprocessing*, vol. 14, Technical Insights, Englewood, NJ) **14,** p. 4.
9. Schmitz, H. H., Poor, C. L., Gugger, E. T., and Erdman, J. W., Jr. (1993) Analysis of carotenoids in human and animal tissues. *Methods Enzymol.* **214-B,** 102–116.
10. Fontana, J. D., Czeczuga, B., Bonfim, T. M. B., Chociai, M. B., Oliveira, B. H., Guimarães and Baron, M. (1996) Bioproduction of carotenoids: the comparative use of raw sugar cane juice and depolymerized bagasse by *Phaffia rhodozyma. Bioresource Technol.* **58,** 121–125.
11. Schiedt, K. and Liaeen-Jensen, S. (1995) Isolation and analysis in *Carotenoids,* vol. 1-A: *Isolation and Analysis* (Britton, G., Liaeen-Jensen, S., and Pfander, H., eds.), Birkhauser, Basel, pp. 104–107.
12. Dawson, M. C., Elliott, D. C., Elliott, W. H., and Jones, K. M. (1991) *Data for Biochemical Research,* 3rd ed., Oxford Science Publications, Clarendon Press, Oxford, UK, p. 240.
13. Boussiba, S. and Vonshak, A. (1992) Enhacement and determination of astaxanthin accumulation in the green alga Haematococcus pluvialis. *Methods Enzymol.* **214-A,** 386–391.
14. Schuep, W. and Schierle, J. (1995) *Carotenoids,* vol 1-A, *Isolation and Analysis* (Britton, G., Liaeen-Jensen, S., and Pfander, H., eds.), Birkhauser, Basel, Switzerland, pp. 275–276.
15. Johnson, E. A. and Lewis, M. J. (1979) Astaxanthin formation by the yeast *Phaffia rhodozyma. J. Gen. Microbiol.* **115,** 1733–183.
16. Schiedt, K., Bischof, S., and Glinz, E. (1993) Metabolism of carotenoids and *in vivo* racemization of (3S, 3S')-astaxanthin in the crustacean *Penaeus. Methods Enzymol.* **214-B,** 148–168.
17. Zang, L.-Y., Sommerburg, O., van Kuijk, F. J. (1997) Absorbance changes of carotenoids in different solvents. *Free Radic. Biol. Med.* **23,** 1086–1089.
18. Breitmeier, E. and Voeter, W. (1978) *Carbon-13 NMR Spectroscopy,* VCH, Weinheim, Germany, p. 334–337.
19. Latscha, T. (1988) *Carotenoids – Their Nature and Significance in Animal Feeds,* Department of Animal Nutrition, F. Hoffmann-La Roche Ltd., Basel, Switzerland.

31

Genetic and Chromosomal Stability of Wine Yeasts

Matthias Sipiczki, Ida Miklos, Leonora Leveleki, and Zsuzsa Antunovics

1. Introduction

The conversion of grape juice into wine is a complex fermentation process in which yeasts play a central role. The composition of the yeast flora in the fermenting must vary according to geographic location, climatic conditions, and grape variety (e.g., **ref. 1**). During the early phase of fermentation, apiculate yeasts belonging to the species *Kloeckera apiculata* (*Hanseniaspora uvarum*) are dominant, but *Candida, Pichia, Rhodotorula, Kluyveromyces, Hansenula, Metschnikowia*, and *Saccharomyces* strains can also be detected *(1)*. Most of them die off when the ethanol concentration rises to around 4% *(2)* and leave *Saccharomyces* to complete the fermentation. The major *Saccharomyces* species found among wine yeasts is *S. cerevisiae*, but *S. bayanus* can also be detected. The *S. cerevisiae* flora itself is also variable, it can change in the course of fermentation and various strains can grow simultaneously as subpopulations (e.g., **refs. 3–6**).

Where the yeast flora comes from is a matter of scientific discussion (for a review, *see* **ref. 7**). The starting population composed mainly of non-*Saccharomyces* strains is probably determined by the yeasts that live on the grapes or the leaves of the vine. Many authors believe that the *Saccharomyces* strains also come from the grapes. Others, however, think that they originate from the surfaces of winery equipment and are thus winery-specific residential strains.

Considerable genetic instability has been observed in many *S. cerevisiae* strains isolated from fermenting wines. The wine strains are usually diploid but frequently undergo aneuploidization and/or chromosomal rearrangements leading to chromosomal length polymorphism *(8–11)*. These changes severely affect the pairing and segregation of homologous chromosomes in meiosis, which may result in low sporulation efficiency (0–75%) and poor spore viability (e.g., **refs. *10*** and ***12***). Because wine yeasts are usually heterozygous for several traits, the modification of chromosome number or the alteration of chromosome size (e.g., deletion) may entail changes in the proportion of alleles or can even eliminate favorable alleles. This variability may have significant impact on wine quality, because it allows the formation of yeast subpopulations with impaired fermentation parameters (e.g., fermentation power, the production of ethanol, secondary compounds, organoleptic and aroma components, etc.). To promote high velocity and reproducibility of fermentations, many wine producers inoculate their musts with starter cultures. These cultures can be prepared from yeasts isolated from fermenting wines or from commercially available dry wine yeasts. However, they ensure controlled fermentation and reproducible wine quality only if they are genetically stable and do not segregate.

Genetic stability can be tested by the examination of sporulation morphology and the viability of spores. The number and the size of the chromosomes can be determined by pulsed-field gel electrophoresis (PFGE) *(13)*, a technique suitable for separation of chromosome-size DNA molecules (electrophoretic karyotyping).

1.1. Genetic Stability Test by Sporulation and Spore Viability

Most wine yeasts are homothallic. The spores of a homothallic strain produce clones of vegetative cells that can mate with each other to form cells with a double amount of DNA. These cells also propagate vegetatively, but they can also convert to asci, harboring four spores, each of which contains half the DNA content of the sporulating cell. If the cell was diploid, its spores would be haploid. Prior to sporulation, the diploid cell replicates its DNA and undergoes a meiosis, in which the homologous chromosomes pair and then segregate into four haploid sets. Each set will be incorporated into a separate spore during sporulation. In natural fermentations, clones of various origin and genotypes (with differences in the number and size of chromosomes or in alleles of homologous genes) can grow simultaneously in the must. Matings between different clones produce hybrids heterozygous for all traits, in which the clones differ from one another. These heterozygotes may have superb propagation vigor and fermentation abilities as long as they propagate vegetatively but segregate into less efficient aneuploids and haploids when sporulating. Hybrids with chromosomes that cannot properly pair in meiosis (homologous chromosomes of different length or with translocations or deletions) may be poor in sporulation, and form asci with aberrant number and/or low viability of spores. Aneuploids also show sporulation deficiencies.

1.2. Karyotype Stability

As a result of recent technological developments, numerous pulsed-field gel electrophoresis techniques are now available, all with modifications of the basic principle of PFGE (for a review, see **ref. 14**). Contour-clamped homogenous electric field (CHEF) electrophoresis is a modified PFGE method *(15)* that provides high resolution of yeast and fungal chromosomal patterns. The number and position of the bands (electrophoretic karyotype) in the gel allow the determination of the number of the chromosomes and the assessment of their size. Even small size differences can cause detectable changes in the mobility of the chromosomal bands, which allows the detection of small chromosomal rearrangements. The intensity of bands is proportional to their DNA content; the bands corresponding to larger chromosomes show more intensive staining. Occasionally, faint bands can be seen between brighter bands. These may correspond to chromosomes that are present in single copies (aneuploidy), but mixed cultures containing a minor subpopulation of a different karyotype also give chromosomal patterns showing heterogeneous staining (e.g., **ref. 12**). Electrophoretic karyotyping can also be used to distinguish *Saccharomyces* strains from other yeast genera, because the species occurring in fermenting grape musts have characteristic karyotypes (e.g., **refs. 16–19**).

2. Materials and Equipment
2.1. Materials

1. YPG (yeast–peptone–glucose) medium: 1% yeast extract, 0.5% peptone, 2% glucose, and 2.5% agar (percentages are w/v).
2. YPGL: YPG without agar.
3. Presporulation medium: 1% yeast extract, 1% peptone, 2% glucose, and 1.5% agar (percentages are w/v).
4. Sporulation medium: 1% potassium acetate, 0.25% yeast extract, 0.1% glucose, 1.5% agar (percentages are w/v).
5. Helicase (snail gut enzyme) solution for the dissolution of asci: enzyme:sterile water = 1:20.
6. 50 mM EDTA, pH 8.0.
7. ECP buffer: 50 mM EDTA, 20 mM citrate–phosphate buffer, pH 7.5.
8. Agarose, purity suitable for pulse-field gel electrophoresis (e.g., Bio-Rad).
9. Low-melting agarose for embedding cells or spheroplasts in "plugs" (e.g., Sigma A-9414).
10. Lytic enzyme (lyticase, Sigma L-5263; zymolyase, ICN 320931, or Novozyme, Sigma L-1412) solution for spheroplasting cells. Recommended concentration: 1 mg/mL. Wine yeasts are rather heterogeneous in sensitivity to enzymatic cell-wall digestion.
11. ESP buffer: 0.5 M EDTA, 0.001 M Tris, and 1% N-lauroylsarcosinate, pH 8.0, with 0.5 mg/mL Proteinase K.

12. 0.5X TBE buffer: 45 mM Tris, 45 mM boric acid, and 0.5 mM EDTA, pH 8.3.
13. Ethidium bromide solution for staining gels: 0.5 µg/mL.

2.2. Equipment

1. Micromanipulator for dissecting asci and separating spores.
2. Hemocytometer or spectrophotometer for determining cell density.
3. Microfuge (refrigerated, if possible).
4. Equipment for pulse-field gel electrophoresis (e.g., Bio-Rad CHEF DRIII).
5. Molds for casting agarose gels and agarose plugs (should be parts of Bio-Rad CHEF DRIII set).
6. Ultraviolet (UV)-transilluminator for visualizing DNA in ethidium bromide-stained gel.

3. Methods

3.1. Sporulation Efficiency

1. Grow the cells of the desired strain on a YPG plate at 25°C for 48 h.
2. Inoculate a loopful amount of the culture onto presporulation medium and incubate at 25°C for 24 h.
3. Transfer cells onto sporulation medium.
4. Incubate at 25°C for 72 h, or until spores appear.
5. Take a sample with a small loop and suspend the cells in a drop of water placed on a microscope slide.
6. Evaluate sporulation efficiency and morphology by microscopic observation. A low percentage of spores (less than 20% asci) and/or high frequency of asci containing less than four spores (more than 10%) indicate that the strain has an "unbalanced" genome prone to segregate spores with variable genotypes.

3.2. Spore Viability

1. Suspend a small loopful amount of sporulating cells (*see* **Subheading 3.1.**) in 0.2 mL of helicase solution and incubate at 30°C until the ascus walls are dissolved (20–60 min should be sufficient).
2. Spread samples onto YPG plates and dissect asci by micromanipulation to liberate their spores.
3. Separate the spores on the surface of the medium so that each can form a separate colony.
4. Incubate at 30°C for 5–7 d and determine the percentage of colony formation. Values lower than 70% denote deficiencies in chromosome segregation and indicate genetic instability.

When a micromanipulator is not available, random spore analysis can yield information about spore viability. For this, the suspension of enzyme-treated asci is sonicated and the percentage of the free spores is determined in a hemocytometer. The appropriately diluted samples are then spread on YPG plates (100–150 potential colony-forming units per plate). After 7 d of incubation at 30°C, the colonies are counted and the efficiency of colony formation is determined. This test is based on the assumption that the vegetative cells and the asci that escaped disintegration will all form colonies, which, however, may not be the case.

3.3. Karyotype Stability During Vegetative Propagation

1. To test a yeast strain for chromosomal polymorphism, plate its cells onto YPG medium and isolate single-cell colonies after 7 d of incubation at 25–30°C.
2. Inoculate each colony separately into a flask containing YPGL and incubate the cultures to late logarithmic/early stationary phase.
3. Take a sample of appropriate size of each culture and subject it to electrophoretic karyotyping as described in **Subheading 3.5.**).
4. Compare the patterns of the chromosomal bands. If the patterns are not identical, the strain is polymorphic.
5. Select one clone showing the most common pattern and culture it by serial reinoculation into new medium (0.1 mL into 10 mL of YPGL) every 24 h over a period of 15 d. Assuming a generation time that allows 5 generations between every 2 inoculations, 80–85 generations will be produced by d 15.
6. Plate a sample from the last culture onto YPG plates and isolate 20 single-cell colonies. Subject each subclone to electrophoretic karyotyping (**Subheading 3.5.**) and compare the chromosomal patterns. If profiles different from that of the original clone are found, the strain is prone to undergo chromosomal rearrangements. Size modifications are usually more apparent among the smaller chromosomes.

3.4. Meiotic Karyotype Stability

1. To test a strain for the stability of its chromosomes during meiosis, plate cells onto YPG medium and select single-cell colonies after 7 d of incubation at 25–30°C.
2. Test each isolate for sporulation by streaking cells onto sporulation medium. Because wine yeasts are usually homothallic, most of the isolates will sporulate.
3. Select one of the sporulation-proficient clones and determine its chromosomal DNA profile as described in **Subheading 3.5.**
4. Streak the clone onto sporulation medium and after sporulation, dissect asci to liberate spores as described in **Subheading 3.2.**
5. Choose 5 complete tetrads or 20 clones from incomplete tetrads. Inoculate each clone into YPGL, grow it to a late logarithmic/early stationary phase, and subject it to electrophoretic karyotyping (**Subheading 3.5.**). Deviations from the chromosome pattern of the original clone and differences from the profiles of other spore clones indicate meiotic chromosome instability. Spores of incomplete tetrads usually show more variable karyotypes.

3.5. Electrophoretic Karyotyping

For the determination of the number and size of the chromosomes by electrophoresis, it is necessary to keep the chromosomal DNA molecules intact during the procedure of cell and chromosome dissolution and the subsequent electrophoretic separation. Therefore, the cells or spheroplasts are embedded in agarose, which protects the DNA against mechanical breakage while allowing the free diffusion of solutions necessary for lysis and digestion. To inhibit DNAses, all treatments are carried out at high concentrations of EDTA. To release the DNA from the chromosomes, the embedded cells or spheroplasts are treated with Proteinase K, which degrades most of the chromosomal proteins.

3.5.1. Growing Cultures for Karyotyping

1. Grow the cells in 50 mL of YPGL until a late logarithmic or early stationary phase at 30°C. Better resolution of chromosomal bands and lower background can be obtained with early stationary-phase cultures because they contain fewer cells being in the S-phase of the cell cycle. The S-phase cells replicate their chromosomes; thus, their DNA is more sensitive to physical breakage.
2. Determine cell density by either cell chamber (hemocytometer) counts or photometrically. Flocculation or poor separation of cells may cause a problem in cell counting. This can usually be overcome by one or two washes with 50 mM EDTA.

3.5.2. Preparation of Agarose-Embedded Cells

1. Centrifuge a sample of the culture that contains approx 5×10^9 cells. Wash the pellet twice with 50 mM EDTA.
2. Add 0.5 mL ECP buffer, resuspend the pellet, and transfer 200 µL of the suspension into a 1.5-mL microfuge tube.
3. Add 200 µL of molten 1.5% w/v low-melting agarose dissolved in ECP buffer and mix gently at 48°C.
4. Pipet the cell/agarose mixture into molds and let it harden at 4°C for 20 min.
5. Remove the cell/agarose blocks ("plugs") from the molds and transfer them into microfuge tubes for Proteinase K digestion (*see* **Subheading 3.5.4.**).

3.5.3. Preparation of Agarose-Embedded Spheroplasts

Saccharomyces strains vary in the yield of chromosomal DNA when intact cells are used for karyotyping. The differences can be attributed to the variable permeability of their cell walls. This difficulty can be overcome by spheroplasting (partial removal of the cell wall) before electrophoresis. The most convenient way of spheroplasting is the treatment of the cells with cell-wall lytic enzymes such as lyticase, zymolyase, or Novozym. Mechanical disruption has also been reported to be applicable and may be preferred when spheroplasting is not effective *(20)*.

Stability of Wine Yeasts

1. Centrifuge a sample of the culture that contains 10^9 cells and wash the cells twice in 50 mM EDTA.
2. Add 0.5 mL ECP buffer, resuspend the pellet, and transfer 200 µL of the suspension into a 1.5-mL microfuge tube.
3. Prepare spheroplasts from cells. Spheroplasts can be prepared in two ways: either in agarose block (plug) or in suspension.
   a. In agarose block
      - Add 200 µL of molten 1.5% w/v low-melting agarose dissolved in ECP buffer to 200 µL of cell suspension and mix gently at 48°C.
      - Pipet the cell/agarose mixture into molds and let it harden at 4°C for 20 min.
      - Remove the agarose blocks from the mold and transfer them into a microfuge tube containing 1 mL ECP buffer supplemented with lyticase.
      - Incubate at 37°C for 3–5 h or overnight.
      - Discard the buffer by a Pasteur pipet.
      - Rinse the plugs in 2–3 volumes of 0.5 mM EDTA and treat them with Proteinase K (*see* **Subheading 3.5.4.**).
   b. In suspension
      Many cells in the agarose will not be spheroplasted effectively and therefore will not release their chromosomal DNA during electrophoresis. Better efficiency can be obtained when spheroplasting is done in suspension and the spheroplasts are embedded into agarose afterward.
      - Add 1 mL ECP buffer containing 1.2 M sorbitol and lyticase to 200 µL cell suspension and mix gently.
      - Incubate at 25°C.
      - Check spheroplast formation microscopically at 15-min intervals: Mix 10 µL of suspension with 30 µL of water (osmotic shock) on a microscope slide and check if the cells lyse. When over 80% of cells lyse, most cells have spheroplasted.
      - Collect spheroplasts by gentle centrifugation and remove the supernatant with a Pasteur pipet.
      - Wash twice with ECP buffer containing 1.2 M sorbitol.
      - Resuspend pellet in 200 µL of the same buffer.
      - Add 200 µL of molten 1.5% w/v low-melting agarose to 200 µL of cell suspension and mix gently at 48°C.
      - Pipet the spheroplast/agarose mixture into molds and allow the gel to solidify.
      - Remove the cell/agarose blocks from the molds and transfer them into microfuge tubes for Proteinase K digestion (*see* **Subheading 3.5.4.**).

3.5.4. Release of DNA from Chromosomes

Chromosomes and other proteinous structures can be disintegrated by Proteinase K treatment in the presence of a detergent.

1. Transfer agarose blocks into 1 mL ESP buffer.
2. Incubate at 50°C for 24 h.
3. Remove buffer by pipetting and rinse the blocks twice with 0.5 M EDTA.
4. Use the blocks for electrophoretic karyotyping (*see* **Subheading 3.5.5.**) or store them at 4°C. They can be stored in this solution for several months at 4°C.

3.5.5. Electrophoresis

1. Prepare 1% agarose (w/v) in 0.5X TBE.
2. Assemble the gel casting mold and comb. Ensure that it is resting on a level surface. Pour the hot agarose into the mold and allow to set.
3. Remove the comb and load the blocks into the wells. Usually, half pieces are loaded.
4. Transfer the loaded agarose plate into the electrophoresis chamber (containing cooled 0.5X TBE) and fit it in correct orientation (movement of DNA is to the positive electrodes).
5. Adjust the running parameters according to the instruction manual of the apparatus. It is often difficult to choose electrophoretic conditions that give good resolution of both the smaller and larger chromosomal DNA molecules. A possible compromise for CHEF DRIII: voltage, 6 V/cm; angle 120°; pulse time, 60 s for 15 h and 90 s for 9 h.
6. Switch on the buffer pump and start electrophoresis. Check that the buffer is at the desired running temperature (13°C).
7. After electrophoresis, switch off the apparatus and transfer the gel into a tray containing distilled water with ethidium bromide. Stain for 30 min at room temperature.
8. Transfer the gel onto an UV transilluminator to check that DNA bands are stained sufficiently (*see* **Subheading 4.**).
9. Continue staining for another 15–30 min, if necessary.
10. Destain the gel for 30 min in distilled water to reduce background staining.
11. Transfer the gel onto the UV transilluminator and photograph it.

4. Notes

Ethidium bromide is highly toxic so use gloves when staining and viewing electrophoretic gels. Use goggles or other protective equipment while using the UV transilluminator.

References

1. Fleet, G. H. and Heard, G. M. (1993) Yeasts: growth during fermentation, in *Wine Microbiology and Biotechnology* (Fleet, G. H., ed.), Harwood, Chur, Switzerland, pp. 27–54.
2. Margalith, P. Z. (1981) *Flavor Microbiology*, Charles C. Thomas, Springfield, IL.
3. Schütz, M. and Gafner, J. (1993) Analysis of yeast diversity during spontaneous and induced alcoholic fermentations. *J. Appl. Bacteriol.* **75,** 551–558.
4. Querol, A., Barrio, E., and Ramon, D. (1994) Population dynamics of natural *Saccharomyces* strains during wine fermentation. *Int. J. Food Microbiol.* **21,** 315–323.
5. Versavaud, A., Courcoux, P., Roulland, C., Dulau, L., and Hallet, J. N.(1995) Genetic diversity and geographical distribution of wild *Saccharomyces cerevisiae* strains from the wine-producing area of Charentes, France. *Appl. Environ. Microbiol.* **61,** 3521–3529.

6. Nadal, D., Colomer, B., and Pina, B. (1996) Molecular polymorphism distribution in phenotypically distinct populations of wine yeast strains. *Appl. Environ. Microbiol.* **62,** 1944–1950.
7. Martini, A. and Martini, A. V. (1990) Grape must fermentation past and present, in *Yeast Technology* (Spencer, J. F. T. and Spencer, D. M., eds.), Springer-Verlag, Berlin, pp. 105–123.
8. Bakalinsky, A. T. and Snow, R. (1990) The chromosomal constitution of wine strains of *Saccharomyces cerevisiae*. *Yeast* **6,** 367–382.
9. Longo, E. and Vezinhet, F. (1993) Chromosomal rearrangements during vegetative growth of a wild strain of *Saccharomyces cerevisiae*. *Appl. Environ. Microbiol.* **59,** 322–326.
10. Guijo, S., Mauricio, J. C., Salmon, J. M., and Ortega, J. M. (1997) Determination of the relative ploidy in different *Saccharomyces cerevisiae* strains used for fermentation and "flor" film ageing of dry Sherry-type wines. *Yeast* **13,** 101–117.
11. Codon, A. C., Benitez, T., and Korhola, M. (1998) Chromosomal polymorphism and adaptation to specific industrial environments of *Saccharomyces* strains. *Appl. Microbiol. Biotechnol.* **49,** 154–163.
12. Miklos, I., Varga, T., Nagy, A., and Sipiczki, M. (1997) Genome instability and chromosomal rearrangements in a heterothallic wine yeast. *J. Basic Microbiol.* **37,** 345–354.
13. Schwartz, D. C. and Cantor, C. R. (1984) Separation of yeast chromosome-sized DNAs by pulsed field gradient gel electrophoresis. *Cell* **37,** 65–67.
14. den Dunnen, J. T. and van Ommen, G. J. B. (1993) Methods for pulsed-field gel electrophoresis. *Appl. Biochem. Biotechnol.* **38,** 161–168.
15. Chu, G., Vollrath, D., and Davis, R. W. (1986) Separation of large DNA molecules by contour-clamped homogeneous electric fields. *Science* **234,** 1582–1585.
16. Johnston, J. R., Contopoulou, C. R., and Mortimer, R. K. (1988) Karyotyping of yeast strains of several genera by field inversion gel electrophoresis. *Yeast* **4,** 191–198.
17. Vaughan-Martini, A., Martini, A., and Cardinali, G. (1993) Electrophoretic karyotyping as a taxonomic tool in the genus *Saccharomyces*. *Antonie van Leeuwenhoek* **63,** 145–156.
18. Versavaud, A. and Hallet, J.-N. (1995) Pulsed-field gel electrophoresis combined with rare-cutting endonucleases for strain differentiation of *Candida famata*, *Kloeckera apiculata* and *Schizosaccharomyces pombe* with chromosome number and size estimation of the two former. *Syst. Appl. Microbiol.* **18,** 303–309.
19. Frezier, V. and Dubourdieu, D. (1992) Ecology of yeast strains *Saccharomyces cerevisiae* during spontaneous fermentation in a Bordeaux winery. *Am. J. Enol. Vitic.* **43,** 375–380.
20. Kwan, H. S., Li, C. C., Chiu, S. W., and Cheng, S. C. (1991) A simple method to prepare intact yeast chromosomal DNA for pulsed field gel electrophoresis. *Nucleic Acids Res.* **19,** 1347.

32

Prediction of Prefermentation Nutritional Status of Grape Juice

The Formol Method

Barry H. Gump, Bruce W. Zoecklein, and Kenneth C. Fugelsang

1. Introduction

The Formol titration is a simple and rapid method for determination of the quantity of assimilable nitrogen in juice *(1)*. It provides an approximate, but useful, index of must nutritional status. The procedure consists of neutralizing a juice sample with base to a given pH, adding an excess of neutralized formaldehyde, and retitrating the resulting solution to an end point. The formaldehyde reacts with free amino groups of α-amino acids, causing the amino acid to lose a proton, which can then be titrated. Free ammonia is also titrated. Proline, one of the major amino acids in grapes that generally cannot be used by yeast under wine fermentation conditions, is partially titrated. Arginine, which contains four nitrogen atoms but only one carboxylic acid group, is titrated to the extent of the single acid functionality. Traditionally, barium chloride has been included to precipitate sulfur dioxide so that it does not interfere with the determination. If the juice is unsulfited or if the sulfur dioxide level is less than 150 mg/L, this part of the procedure may be ignored (*see* **Note 1**).

2. Analytical Methodology
2.1. Materials

1. Sodium hydroxide solution, 1 N.
2. Sodium hydroxide solution, 0.10 N, standardized against potassium hydrogen phthalate or equivalent.

3. Barium chloride solution, 1 N (0.05 formula weight/L) (*see* **Note 1**).
4. Formaldehyde, reagent grade, 37% (v/v or 40% w/v) neutralized to pH 8.0 with 1 N sodium hydroxide.
5. pH meter sensitive to ± 0.05 pH.
6. Calibration buffers for the pH meter.
7. Whatman No. 1 filter paper.

2.2. Method

1. Pour 100 mL of sample into a 200-mL beaker.
2. Neutralize the sample to pH 8.0 using 1 N sodium hydroxide and pH meter.
3. If sulfur dioxide is present, add 10 mL of the barium chloride solution and allow the sample to sit for 15 min (*see* **Note 1**).
4. Transfer the treated sample into a 200-mL volumetric flask. Bring to volume with deionized water and mix well.
5. Filter the solution through Whatman No. 1 filter paper with or without diatomaceous earth.
6. Transfer a 100-mL aliquot of the sample into a beaker, place calibrated pH/reference electrodes and a stirbar into the solution, mix, and readjust the pH to 8.0 with 1 N NaOh, if necessary.
7. Add 25 mL of the previously neutralized formaldehyde (pH 8.0) to the aliquot, mix, and titrate to pH 8.0 using 0.10 N sodium hydroxide (*see* **Notes 3** and **4**).
8. The concentration of assimilable nitrogen is calculated as follows:

(NH_4^{++} α–amino nitrogen) mg/LN = (mL of 0.1 N NaOH titrated) × 28 (*see* **Note 2**)

3. Practical Considerations and Recommendations

Fermentation problems may arise from numerous sources, including deficiencies in the fruit and processing (**Fig. 1**). Difficulties may arise from a combination of factors and a variety of sources. It is often the impact of two or more conditions that cause a problem of greater significance than would be predicted by a single parameter alone. Once yeast fermentative vigor and vitality have diminished, revitalization may be difficult, if not impossible. Thus, winemakers must approach each winemaking step with as complete an understanding as possible. The following is a review of practical issues influencing fermentation.

3.1. Vineyard

Fermentation problems are often vineyard-specific. Nitrogen deficiency in apparently healthy grapes can be severe. Drought, grapevine nutrient deficiencies, high incidences of fungal degradation, and level of fruit maturity all influence must nitrogen and vitamins. Cultivar, rootstock, crop load, and growing season may also influence juice or must nitrogen. Some varieties, such as Chardonnay, have a greater tendency toward deficiency. Higher total

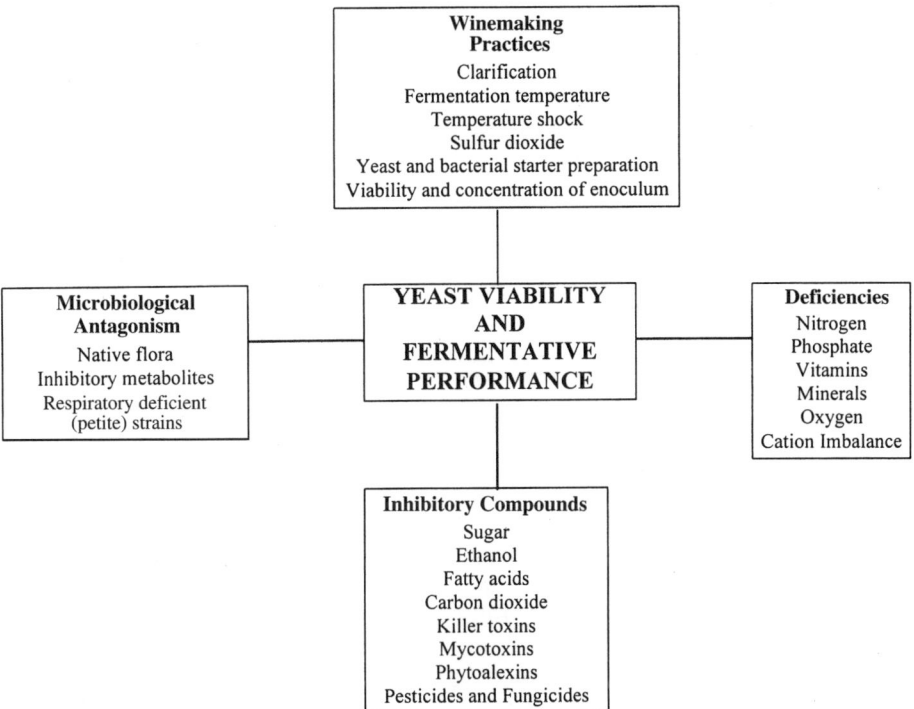

Fig. 1. Environmental and processing factors influencing viability and fermentative performance of wine yeasts.

nitrogen may also be associated with certain rootstocks. For example, grapes grown on St. George are higher in total nitrogen than those on AXR1.

As seen in **Fig. 2**, the concentration of α-amino nitrogen in Cabernet Sauvignon grapes changes as a function of maturity and crop load. Henick-Kling et al. *(2)* compared the concentrations of the two important sources of assimilable N (FAN and NH_4^+) among six cultivars at maturity over two seasons (**Table 1**). This study illustrated large variations from one season to the next in both free ammonia and α-amino nitrogen and significant differences in the concentration of both sources of nitrogen among cultivars.

Mold growth on fruit has been reported to cause fermentation problems as a result of the production of metabolites and the depletion of nitrogen *(3–5)*. *Botrytis cinerea* produces a group of mold-derived heteropolysaccharides collectively referred to as "Botryticine" *(6)*. The mycotoxins stimulate *Saccharomyces* sp. to produce high and inhibitory levels of acetic acid at the onset and during the latter stages of alcoholic fermentation *(5)*. *Botrytis cinerea* can consume 41% of the total amino acid concentration in the fruit, causing as

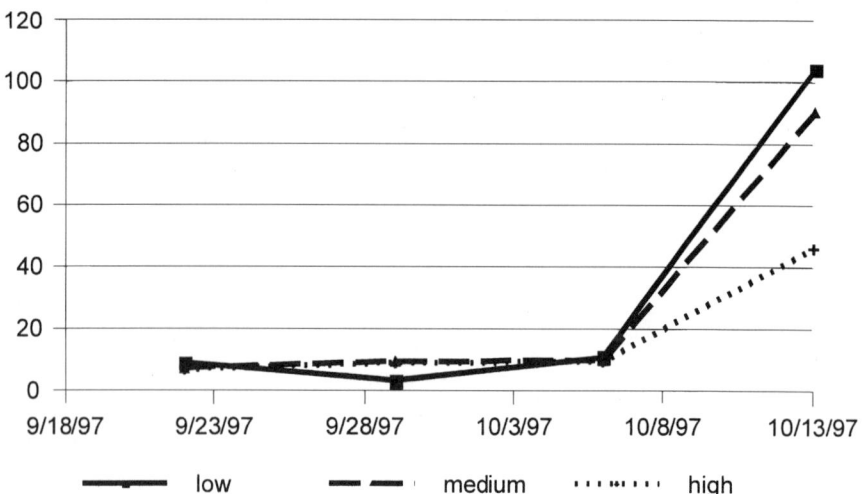

Fig. 2. α-Amino nitrogen (m*M*) of Cabernet Sauvignon grapes cluster thinned to low (2.6 kg/vine), medium (4.9 kg/vine), and high (5.3 kg/vine) crop level in 1997. (From B. W. Zoecklein and C. W. de Bordenave, unpublished).

much as a 51% reduction in proline *(7)*. Further, the presence of native yeasts, particularly *Kloeckera apiculata*, is known to deplete important vitamins such as thiamine. Fruit from diseased vines may also contain inhibitory levels of phytoalexins produced by the plant in response to the parasite *(8)*. These may be inhibitory toward *Saccharomyces* sp.

3.2. Yeast Strains

Strain differences among *Saccharomyces* sp. may be significant in terms of nitrogen requirements, the time frame for uptake and release of specific amino acids during fermentation, and the ability to ferment to dryness. Henschke and Jiranek *(9)* reported that the Montrachet strain had the highest nitrogen demand and exhibited the highest rate of amino acid and ammonium ion accumulation relative to sugar fermented among several strains studied. When considering utilizing unfamiliar strains, the winemaker is urged to consult the supplier's technical representatives.

3.3. Yeast Starter Population Density

Yeast populations should be large enough to overwhelm indigenous microflora and grow to $(2-5) \times 10^6$ yeast cells/mL juice (1–3% [v/v] of an active starter). These concentrations apply when the °Brix is below 24, the pH is above 3.1, and the temperature is above 13°C (55°F). Increases in the inoculum volume should be made when parameters are outside these values.

**Table 1
Survey Results from 1993 and 1994; Mean Content of Free Ammonia and Free Amino Nitrogen**

	Free ammonia (mg/L)		Free amino nitrogen (mg/L)	
Cultivar	1993	1994	1993	1994
Cayuga White	68	32	74	197
Chardonnay	46	55	151	177
Riesling	52	56	102	123
Seyval Blanc	19	14	82	156
Pinot Noir	52	88	135	116
Cabernet Sauvignon	49	69	74	142
Mean (all cultivars)	48	52	103	152

Source: **ref. 2**.

3.4. Yeast Preparation

Rehydration protocol should strictly adhere to the yeast supplier's recommendations to assure maximum viability. Viability and vigor decrease as rehydration temperatures vary above or below those recommended. After rehydration, the yeast should be added to the juice/must within 20–30 min or, alternatively, a carbohydrate source added. If this is not done, yeasts undergo a premature decline phase, resulting in an inoculum of low viable cell density. Significant yeast cell death occurs when temperature differentials between starter and juice/must are more than 5–7°C. Monk *(10)* reported that the addition of rehydrated yeast (40°C [104°F]) directly to a must at 15°C (60°F) kills approximately half of the cell population. In cases where the yeast is expected to ferment at low temperature, it is desirable to acclimate the starter to that temperature.

3.5. Nutrient Addition

Many juice/musts lack sufficient assimilable nitrogen and other components needed by yeast for fermentative growth. Although some have suggested that a minimum of 140 mg/L assimilable nitrogen is required by yeasts, others recommend 250 mg N/L or more. Morris et al. *(11)* suggested that concentration levels of 500–900 mg/L of assimilable nitrogen are required for healthy fermentations. Yeast with lower concentrations of N may perform well under optimum, but not adverse, conditions. Concentrations of 500–900 mg N/L give yeast the ability to produce cellular proteins needed to meet the worst environmental conditions (C. Cone, personal communications, 1996). Phosphate deficiency may also have a direct impact on yeast cell growth and fermentative performance *(9)*. Inorganic phosphate is required for synthesis of ATP/ ADP and nucleic acids. Supplementation should be carried out using a balanced source of diammonium phosphate (DAP) (25.8% ammonia, 74.2% phosphate), amino acids, minerals, and vitamins. Diammonium phosphate additions of 1 g/L (8.3 lbs/1000 gal) provide 258 mg/L fermentable nitrogen which exceeds the suppliers' recommended level. In the United States, the legal limit of DAP is 960 mg/L, which corresponds to 203 mg N/L.

3.6. Timing of Nutrient Additions

Amino acids are not incorporated equally by yeast and their incoporation may vary significantly among yeast strains *(12)*. Some are utilized at the beginning of the growth cycle, some later, and some not at all. Ammonia, on the other hand, is consumed preferentially to amino acids in growing populations. Stationary-phase yeast also vary significantly in terms of the order of amino acid incorporation, and do not always show preference for ammonia over amino acids *(12)*. Therefore, timing of DAP additions is important. A single large addition of DAP at the beginning may lead to an excessive fermentation rate and an imbalance in the uptake and usage of amino acids. To avoid this problem, multiple additions at 16° Brix and 10° Brix are preferred.

Although the addition of ammonium salts may not significantly benefit stationary-phase yeast *(13)*, the addition of specific amino acids may have a stimulatory effect and extend fermentative activity *(14)*. Single amino acids may be quickly utilized to resynthesize transporter proteins that are rapidly "turned over" during accelerated growth. Supplements added after about half the fermentation is completed may not be used by the yeast because alcohol prevents their uptake. For the same reason, adding nutrients to a stuck fermentation is seldom effective. With increasing ethanol concentrations, the permeability of the plasma membrane to hydrogen ions increases. This requires intracellular enzymes and ATPases to pump protons back out of the cell in order to balance the internal pH of the yeast cell against the external pH of the juice/ must. Because of the competing nature of these coupled transport systems, nitrogen is picked up by the cell only in the early stages of fermentation, stored in vacuoles and used on demand. Nitrogen added late in the fermentation may not be transported into the cell *(15)*. Once stopped because of nutrient stress, the fermentation may require significant effort to restart and finish.

3.7. Vitamin Addition

Juice/musts can be vitamin deficient as well as deficient in assimilable nitrogen when there is a high incidence of microorganisms (mold, yeast, and/ or bacteria). Growth of *Kloeckera apiculata* has been reported to rapidly reduce thiamine levels below those required by *Saccharomyces* sp. *(16)*. Further, the use of SO_2 may lead to additional reductions in levels of thiamine *(13)*. *Saccharomyces* sp. has been shown to synthesize all required vitamins, with the exception of biotin. However, vitamin supplementation has been demonstrated to be stimulatory *(17)*. Thus, it is usually desirable to add a mixed vitamin supplement with the nitrogen additions.

3.8. Yeast Hulls

Yeast hulls are by-products of the commercial manufacture of yeast extract. Consisting of cell walls and membranes, hulls are added to enhance fermentation rates and to restart stuck fermentations. Their mode of action has been described as lowering the concentration of inhibitory C_{8-10} fatty acids. Ingledew *(18)* reported that yeast hulls stimulate yeast populations by providing a source of C_{16} and C_{18} unsaturated fatty acids that act as oxygen substitutes under long-term fermentative conditions. Additionally, hulls may provide a source for some amino acids as well as surface area to facilitate release of potentially inhibitory levels of saturated CO_2.

3.9. Oxygen/SO_2

Although not directly stimulatory to fermentation, oxygen is required by yeasts for synthesis of cell-membrane precursors, including steroids (primarily ergosterol) and lipids (principally oleanoloic acid). Yeast propagated aerobically contains a higher proportion of unsaturated fatty acids and up to three times the steroid level of anaerobic yeast. Without initial oxygen, replication is usually restricted to four to five generations, as each yeast budding cycle reduces the sterol content of the membrane by approximately half. When the level reaches a critical point, replication stops and fermentation must continue with the population present at that point. Slight aeration of yeast starters may play an important role in subsequent fermentative performance. Wahlstrom and Fugelsang *(20)* reported increased cell density and more rapid fermentations when aerated starters were used compared with nonaerated starters.

The grape itself may supply at least a portion of the lipids needed by yeast during fermentative growth. Up to two-thirds of the cuticular waxes in some grape varieties are composed of oleanolic acid. This fatty acid has been found to replace the yeast's requirement for ergosterol supplementation under anaerobic conditions *(19)*. Thus, pomace contact, either prior to pressing in white wine production or extended during red wine fermentation, extracts this and other essential components from the grape cuticle.

In the absence of sulfur dioxide, grape-derived oxidative enzymes (tyrosinases) catalyze conversion of nonflavonoid phenols to their corresponding quinones. The reaction brings about rapid (but reversible) browning of the juice while consuming oxygen required by yeasts during the early stages of growth

(**Fig. 3**). Grape tyrosinase is readily and rapidly inactivated by addition of SO_2 to the juice/must. However, sulfur dioxide addition also inactivates thiamine. If additions of more than 50 mg/L SO_2 occur, thiamine (in the form of nutritional supplements) should be added to the fermenter.

3.10. Hydrogen Ion Concentration (pH)

Yeast growth occurs over the pH range from 2.8–8.0 *(17)*. However, cultures do not function equally well throughout this wide range. Biomass is produced best above pH 4.0 and slows as the pH decreases. Low pH reduces the tolerance of *Saccharomyces* sp. to ethanol. Kudo et al. *(21)* demonstrated a relationship between the concentrations of K^+ and H^+ and the completion of alcoholic fermentation. They suggested that a minimum K^+/H^+ of 25:1 is required. As pH drops below 3.2, the increase in H^+ raises the risk of premature arrest of fermentation. Added stress is placed on yeast at low pH values and is compounded by low nutrient concentrations, temperature extremes, high sugar, and/or high alcohol. Additionally, highly chaptalized juice has a limited buffering capacity. As a result, organic acid and CO_2 production during the initial stage of fermentation can drop the pH (C. Cone, personal communication, 1995). Juice/musts with pH <3.1 should receive an increased yeast inoculum.

3.11. Nonsoluble Solids

Nonsoluble grape solids serve as nutritionally important substrates and as oxygen reservoirs during the early stages of fermentation. Additionally, solids "hold" yeasts (native and inoculated strains) in suspension during the early stages of fermentation and before the evolution of large amounts of carbon dioxide. Conventional white juice processing calls for some level of suspended solids reduction prior to inoculation. However, reduction below 0.5% can result in nutrient deficiencies and promote premature sedimentation of yeast. The addition of bentonite may help to keep yeast in suspension during the initial stages of fermentation while helping to achieve protein stability. However, bentonite additions can also reduce must nitrogen and should be done in conjunction with supplemental nutrient additions. If processing protocol does not include preferentation bentonite additions, it may be necessary to mix tanks to achieve resuspension and dissipation of carbon dioxide.

3.12. Fermentation Temperature

Yeast growth at either end of the recommended temperature range affects the integrity and operation of the cell membrane. Growth at upper temperature limits brings about inactivation/denaturation of cell-membrane-associated transporter proteins and other enzymes whereas at low temperatures, fluidity/pliability is compromised *(22)*. Cell-membrane function is also affected by the

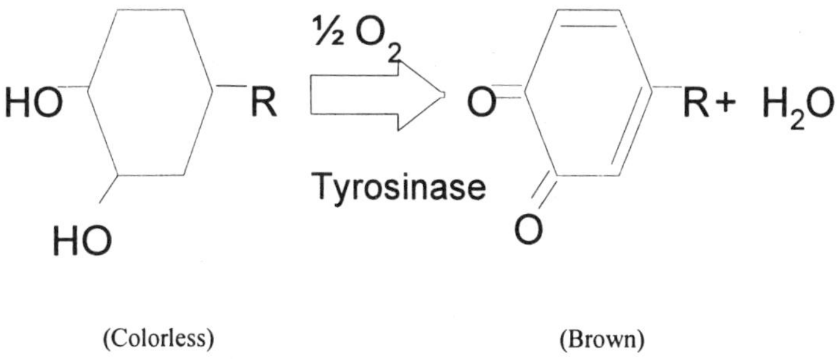

Fig. 3. Enzymatic oxidation of nonflavonoid phenols. (From **ref. 8**.)

presence of increasing concentrations of ethanol. The two antagonists act in synergy, narrowing the *Saccharomyces* sp. temperature tolerance range and, potentially, bringing about premature interruption of fermentation. For low (<10°C [50°F]) temperature fermentations, increased inoculum levels and nutrient additions are recommended.

3.13. CO₂ Toxicity

Carbon dioxide in concentrations of up to 0.2 atm stimulates yeast growth. Above this level, carbon dioxide becomes inhibitory. Pekur et al. *(23)* reported that, at increased pressures, carbon dioxide reduces the yeast's uptake of amino acids. Agitation can be used to help prevent growth-limiting accumulations of of CO_2.

3.14. Sugar Toxicity

Increased osmotic pressure associated with high sugar concentrations can inhibit yeast growth. Although *Saccharomyces* sp. are among the most tolerant species to high sugar concentrations, such environments are often nitrogen deficient. Fermentation under these conditions begins slowly and may stick prior to completion. In cases where sugar levels range from 25°Brix to 30°Brix, yeast starters should be prepared at greater than 5×10^6 yeast cells/mL. For >30°Brix musts, an additional 1×10^6 yeast/mL should be used. Ice wines and some late harvest wines require substantially more yeast inoculum, up to 20×10^6 yeast/mL (C. Cone, personal communication, 1996).

3.15. Glucose/Fructose

Grape juice usually contains approximately equivalent concentrations of glucose and fructose sugars. However, glucose is fermented preferentially to fructose. Stress can affect the yeast's ability to metabolize the last residual fructose. This problem appears to occur more frequently with the *S. bayanus* strains, which are glucophilic *(24)*. Fructose syrup should be used only as the last choice for chaptalization.

3.16. Alcohol Toxicity

Alcohol and its metabolic precursor acetaldehyde are toxic to all yeasts, including *Sacccharomyces* sp. Alcohol has a profound effect on all aspects of yeast metabolism, ranging from membrane integrity to nitrogen uptake and sugar transport. There are many environmental factors that act in synergy with alcohol to inhibit yeast growth, including low pH, high temperature, acetic acid, sugar, short-chain fatty acids, nitrogen depletion, and deficiency of sterols and vitamins. Acetaldehyde has also been reported to play a significant inhibitory role in the survival of *Saccharomyces* sp. during fermentation *(25)* and may increase the yeast's sensitivity to increasing concentrations of ethanol *(26)*. Light aeration during the growth phase stimulates synthesis of cell-membrane precursors, which helps maintain cell integrity. During fermentation, nitrogen supplementation of 250–500 mg /L is likewise helpful in mitigating the antagonistic affects of alcohol.

3.17. Microbially Compromised Fruit, Native Yeast/Bacterial Fermentations, and Late Starter Addition

Usually non-*Saccharomyces* species from the vineyard and winery-associated *Saccharomyces* sp. dominate the initial and early stages of fermentation of uninoculated musts. Their growth may result in significant depletion of nitrogen and vitamins such as thiamine. Among vineyard-related native species, *Kloeckera/Hanseniaspora* are typically found at highest population densities. *Kloeckera* sp. are tolerant of both low temperature and the presence of sulfur dioxide. The yeast can produce high levels of ethyl acetate while significantly depleting nutrient levels.

Inhibitory metabolites produced by mold and native yeast/bacteria growing on fruit or in the early stages of fermentation may have a significant effect on the fermentative performance of *Saccharomyces* species. Acetic and lactic acid bacteria and native yeast can produce potent inhibitors and deplete must nitrogen and vitamins levels. Acetic acid is a strong inhibitor of *Saccharomyces* sp., especially when combined with other antagonistic factors such as high alcohol. Acetic acid levels of >0.8 g/L in stuck wine may need to be reduced before attempting refermentation *(27)*. The technology to accomplish this goal is commercially available *(28)*.

Some *Saccharomyces* sp. and strains and some non-*Sacccharomyces* yeasts can produce killer toxins that inhibit other sensitive strains and may play a role in stuck fermentations. It is suggested that vigorous strains be used for high-risk fermentations. Increasing the level of yeast inoculum along with nitrogen supplementation of 250–500 mg/L may also help overcome these effects.

3.18. Pesticides and Fungicides

Pesticides and fungicides can influence fermentation by producing stress metabolites, such as reductive compounds, and by inhibiting and/or preventing fermentation. Not all yeasts and bacteria are affected in the same way. For example, there is a significant difference between systemic and contact fungicides with regard to residues. Vinification style influences residue concentrations. Prefermentation clarification and utilization of bentonite can affect the final concentration of contact fungicides in white wine fermentation. Close adherence to spray schedules, use of minimal applications, and avoidance of late-season applications are recommended.

3.19. Conclusions

Overall, the Formol method can provide a very useful index of the nutritional status of a juice or must. The simplicity of this procedure and its general ability to correctly describe the amount of assimilable nitrogen make it ideal for use in the winery production laboratory.

4. Notes

1. A series of samples of a juice with 0, 25, 50, and 150 mg/L SO_2 added were titrated using the Formol procedure but ignoring the addition of barium sulfate. The average titration values determined for the various levels of SO_2 addition differed by 5.3% or less. This would indicate that it is not necessary to add barium sulfate to a sulfited juice sample. This would also simplify the procedure, permitting one to take a 50-mL juice sample directly for titration.
2. The full equation for calculating assimilable nitrogen is

    $$\text{mg N/L} = (\text{mL of NaOH}) \times (0.10 \text{ meq OH}^-/\text{mL}) \times (1 \text{ meq N}/1 \text{ meq OH}^-)$$
    $$\times (200 \text{ mL}/100 \text{ mL-dilution factor}) \times 10 \text{ (to convert to liters)} \times 14 \text{ mg/meqN}$$
    $$= \text{mL} \times 28$$

    If a different concentration of base is used, the equation requires an additional term: Normality of NaOH used/0.1. The dilution factor in the equation (200 mL/100 mL) is changed if one uses a sample volume other than 100 mL and dilution to 200 mL. As there are ten 100-mL samples in a liter, the factor "× 10" is required to convert results to a mg/L basis.
3. A new bottle of formaldehyde may have a pH as low as approx 3.5. This will require about 0.5 mL of 1 N sodium hydroxide to neutralize. If the formaldehyde is not neutralized, significant overtitrations may result yielding high values for fermentable nitrogen. The pH of the formaldehyde will begin to drop with time and should be readjusted periodically to pH 8.0.

4. Taylor *(29)* noted that various authors have recommended other pH values, ranging between 6 and 9, for the initial neutralization pH and final titration pH. It has been our experience with standard solutions of various amino acids that working with pH 8.0 for both the initial neutralization of the sample and for the final titration end point minimizes errors. The use of a pH meter for end-point detection is an important feature of this method; the use of phenolphthalein as an indicator does not provide the same precision and accuracy.
5. Formol titrations of known concentrations of seven α-amino acids (alanine, arginine, serine, threonine, α-amino butyric, aspartic, and glutamic acid) and proline showed recoveries from 90% to 120% for the former and an approximate 17–33% recovery for proline. The percentage recovery appears to increase with the absolute amount of proline present. Similarly, Formol titration of known concentrations of ammonium chloride solutions (at approx 50 and 100 mg/L nitrogen) also exhibited quantitative recoveries.
6. The arginine recovery is quantitative for the single carboxylic group. However, because arginine contains four nitrogen atoms, its recovery understates the amount of nitrogen present. Ingledew *(18)* reported that arginine converts to ornithine and urea and that under the anaerobic conditions during fermentation, ornithine is almost quantitatively converted to proline. Thus, it would appear that arginine readily provides two assimilable nitrogens (in the urea formed) to the yeast, with a third nitrogen tied up as proline, and the fourth even more tightly bound possibly to an enzyme system (C. J. Muller, personal communication).
7. Formol titrations of known mixtures of the eight amino acids mentioned in **Note 5** also showed nearly quantitative recoveries when the titration factors (percentages recovered) for proline and arginine were considered. The Formol method, therefore, overstates the available nitrogen from proline and understates the available nitrogen from arginine. The positive and negative errors introduced with the titration of these two juice components are partially compensating. If the amount of proline in the must is much larger (approx 10 times) than the amount of arginine, the method will overstate the amount of available nitrogen. If the amount of proline is only double the amount of arginine, then the positive and negative errors in the titration essentially balance out.
8. Under oxidative conditions proline is oxidized to glutamic acid and becomes available to the yeast, and the Formol method will generally understate the ultimate amount of nitrogen available.
9. Direct titrations of 25-mL samples of juice with correspondingly smaller amounts of formaldehyde added were made. Other than the greater difficulty in accurately reading small volume increments from the burette, there was no significant impact on the procedure.

References

1. Giannessi, P., and Matta, M. (1978) Azoto amminico, in *Trattato di Scienza e Technica Enologica, Vol. I.* (Brescia, A. E. B., ed.) Analisi e Controllo dei mosti e dei Vini, Italy, pp. 87–88.

2. Henick-Kling, T., Edinger, W. D., and Larsson-Kovach, I.-M. (1996) Survey of available nitrogen for yeast growth in New York grape musts. *Wein-Wissenschaft* **51(3)**, 169–174.
3. Rapp, A., and Reuther, K. H. (1971) Der Gehalt von Aminosauren in gesunden und edelfaulen Beeren verschiedener Rebsorten. *Vitis* **10**, 51–58.
4. Dittrich, H. H. (1987) Die Garbeeinflussung 5.5 Stickstoff in *Mikrogiologie des Weines* (2nd ed.) (Ditrick, H. H., ed.). Ulmer, Stuttgart.
5. Doneche, B. J. (1993) Botrytized wines, in *Wine Microbiology and Biotechnology* (Fleet, G. H., ed.), Harwood Academic Publishers, Grey, Switzerland.
6. Dubourdieu, D., Pucheu-Plante, B., Mercier, M., and Ribereau-Gayon, P. (1978) Structure, role et localisation d'un glucane secrete par Botrytis cinerea dans la baie de raisin. *C. R. Acad. Sci. Paris,* **287D**, 571–573.
7. Sponholz, W. R. (1991) Nitrogen compounds in grapes, must, and wine in *International Symposium on Nitrogen in Grapes and Wine* (Rantz, J., ed.) American Society for Enology and Viticulture, Davis, CA, pp.67–77.
8. Smith, D. A., and Banks, S. W. (1986) Biosynthesis, elicitation and biological activity of isoflavonoid phtoalexins. *Phytochemistry* **25**, 979–995.
9. Henschke, P. A., and Jiranek, V. (1993) Metabolism of nitrogen compounds, in *Wine Microbiology and Biotechnology* (Fleet, G. H., ed.) Harwood Academic, Australia, pp. 27–54.
10. Monk, P. R. (1986) Rehydration and propagation of active dry wine yeasts. *Austral. Wine Ind. J.* **1(1)**, 3–5.
11. Morris, J. R., Main, G. and Threfall, R. (1996) Fermentations: problems, solutions and preventions. *Vitic. Enol. Sci.* **51(3)**, 210–213.
12. Manginot, C., Roustan, J. L., and Sablayrolles, J. M. (1998) Nitrogen demand of different yeast strains during alcoholic fermentation. Importance of stationary phase. *Enzyme Microb. Tech.* **23**, 511–517.
13. Lafon-Lafourcade, S. and Ribereau-Gayon, P. (1984) Developments in the microbiology of wine production. *Prog. Ind. Microbiol.* **19**, 1–45.
14. Manginot, C. and Sablayrolles, J. M. (1997) Use of constant rate alcoholic fermentations to compare the effectiveness of different nitrogen sources added during the stationary phase. *Enzyme Microb. Tech.* **20**, 373–380.
15. Bisson, L. F. (1996) Yeast and biochemistry of ethanol formation in *Principles and Practices of Winemaking* (Boulton, R. B., Singleton, V. L., Bisson, L. F., and Kunkee, R. E., eds.), Chapman & Hall, New York, p. 140.
16. Bataillon, M. and Rico, A. (1996) Early thiamine assimilation by yeasts under enological conditions: impact on alcoholic fermentation kinetics. *J. Ferm. Bioeng.* **82**, 145–150.
17. Fleet, G. H. and Heard, G. M. (1993) Yeasts—growth during fermentation, in *Wine Microbiology and Biotechnology* (Fleet, G. H., ed.), Harwood Academic, Australia, pp. 27–54.
18. Ingledew, W. M. (1996) Nutrients, yeast hulls and proline in wine fermentation. *Wein-Wissenschaft* **51(3)**, 141–146.
19. Brechot, P., Chauvet, J., Dupuy, P., Croson, M., and Rabatu, A. (1971) Acide oleanoique facteur de aroissance anaerobic de la levure du vin. *C.R. Acad. Sci.* **272**, 890–893.

20. Wahlstrom, V. L. and Fugelsang, K. C. (1988) *Utilization of Yeast Hulls in Winemaking.* Calif. Agric. Tech. Inst. Bull. 880103. California State University, Fresno.
21. Kudo, M., Vagnoli, P., and Bisson, L. F. (1998) Imbalance of potassium and hydrogen ion concentrations as a cause for stuck enological fermentations. *Am. J. Enol. Vitic.* **49,** 296–301.
22. Stanier, R. Y., Adelberg, E. A., and Ingraham, J.L. (1976) *Introduction to the Microbial World,* Prentice- Hall, Englewood Cliffs, NJ.
23. Pekur, G. N., Bur'yan, N. I., and Pavlenko, N. M. (1981) Characteristics of nitrogen metabolism in wine yeasts under different fermentation conditions. *Appl. Biochem. Microbiol.* **17,** 248–252.
24. Schultz, M. and Gafner, J. (1993) Sluggish alcoholic fermentation in relation to alterations of the glucose-fructose ratio. *Chem. Mikrobiol. Technol. Lebenom.* **15,** 73–78.
25. Stanley, G. A. and Douglas, N. G. (1993) Inhibition and stimulation of yeast growth by acetaldehyde. *Biotechnol. Lett.* **15,** 1199–1204.
26. Jones, R. P. (1989) Biological principles for the effects of ethanol. *Enzyme Microbiol. Biotechnol.* **11,** 130–153.
27. Rasmussen, J. E., Schultz, E., Snyder, R. E., Jones, R. S., and Smith, C. R. (1995) Acetic acid as a causative agent in producing stuck fermentations. *Am. J. Enol. Vitic.* **46,** 278–280.
28. Smith, C. R. (1999) *Vinovation, Inc. New Tools for an Ancient Craft,* Vinovation, Inc., Sebastopol, CA.
29. Taylor, W. H. (1957) Formol titration: an evaluation of its various modifications. *Analyst* **82,** 488–498.

33

Enological Characteristics of Yeasts

Fabio Vazquez, Lucía I. C. de Figueroa, and Maria Eugenia Toro

1. Introduction

The final product of spontaneous grape must fermentation (natural fermentation) is the result of the combined action of different yeast species and vine varieties. Both of them contribute, in different ways, to the organoleptic properties of wine *(1)*. At the first fermentation stages, lemon-shaped, low-ethanol-tolerant species (*Kloeckera, Hanseniaspora*) are predominant. During the fermentation process, these species are replaced by high-ethanol-tolerant *Saccharomyces cerevisiae* and other related species *(2,3)*. The variety and proportion of different yeasts in the must depend on factors such as geographic location, climatic conditions, and grape variety *(3,4)*.

It is possible to isolate pure cultures of yeast, which can be supplied to winemakers as pure culture slants and then be propagated in the winery to provide starters ("selected pure cultures").

On the basis of organoleptic testing of wines, different authors have claimed advantages for either natural fermentations or the use of selected cultures and active dry yeasts *(3)*. Production and use of dry wine yeasts started in the United States in the mid-1960s and expanded worldwide thereafter. Today, more than 100 different strains are commercially available. With a worldwide production of wine of 246,423,000 hL *(5)* and a yeast usage rate of 10–20 g/hL, the potential market for selected dry yeast can be estimated in the range of 5000 tons annually *(6)*. It soon became obvious that these strains are not the universal solution to the fermentation of grape musts. So, there is a need for the selection of strains better adapted to different regions of the world and their respective grape varieties. Moreover, Fleet *(7)* concluded that indigenous strains of *S. cerevisiae* are far better adapted to grow in grape must than any inoculated strain. It should be emphasized, however, that the assignment of all wine yeast strains to a single species does not imply that all strains of *Saccharomyces* are equally suitable for wine fermentation. Wine yeast strains differ mainly in their ability to contribute to the bouquet of wine and their fermentation performance *(8)*.

Selected culture has been therefore the objective of many investigations related to the desirable properties of wine yeasts and their influence on the composition of wine and its organoleptic characteristics. Numerous features of yeasts have been studied, some desirable are, as shown in **Table 1**, some always desirable, and others always undesirable *(9)*.

Desirable requirements described in **Table 1** are complex and difficult to define genetically without a better understanding of the biochemistry involved and need careful selection of appropriate yeasts of natural environments or culture collections. At present, no wine yeast in commercial use has all the characteristics listed in **Table 1**, and it is well established that wine yeasts vary according to their winemaking abilities.

When wine fermentation takes place with the addition of SO_2, the number of yeast species involved in the process is almost exclusively reduced to several *S. cerevisiae* resistant to this antiseptic. Strains belonging to the same species may present differences in their tolerance to SO_2 *(4,10,11)*.

Optimal conversion of carbohydrates to ethanol in wine making requires cells tolerant to high concentrations of both. The sugar conversion rate may be calculated as fermentation vigor (FV), which indicates the maximum yield of ethanol that a yeast strain can produce during fermentation with sugar excess *(12)*. Ethanol is clearly inhibitory for yeasts. Cell growth stops at relatively low ethanol concentrations and fermentation stops at relatively higher ones. Yeast tolerance to ethanol has been correlated with membrane fluidity and both phenomena have been associated with membrane lipid composition *(13)*. Some sugar-tolerant yeasts are also alcohol tolerant, but these two characteristics are not necessarily related *(14)*.

Some yeast strains liberate proteins or glycoproteins that kill sensitive yeasts. These proteins or glycoproteins are named killer factors. Yeasts are classified, according to the killer factor: killer (K), sensitive (S), and neutral (N). Different genera of yeast may present killer activity. Two species of double-stranded RNAs with different functions and distinct molecular sizes are encapsulated in viruslike particles, and they are responsible for producing the toxic proteins or glycoproteins in *S. cerevisiae*. Killer yeasts are immune to their own toxin but can be sensitive to toxins of other killer types. Killer character in *S. cerevisiae* is a genetically complex phenomenon because it depends on both cytoplasmic factors and chromosomal genes.

Killer action depends on the ratio of killer to sensitive cell amounts at the fermentation beginning, on the presence of protein-adsorbent substances,

Table 1
Characteristics of *Saccharomyces cerevisiae* Affecting Winemaking Process

Desirable	Undesirable
Ability to begin fermentation in high ethanol concentrations *(3,6,8–10,13,14,25,26,28,29)*	Production of sulfur dioxide and mercaptan *(6,9,10)*
Rapid initiation of fermentation immediately upon inoculation *(3,8,9,29)*	Formation of ethyl carbamate precursors (e.g., urea) *(6,9)*
Efficient conversion of grape–must sugar to ethanol *(3,6,8,9,12,14,29)*	Production of polyphenol oxidase *(6)*
Flocculating capacity at the end of fermentation to help clarification *(8,9,10,21,22,29)*	
Fermentation at low temperatures *(6,8–10,19,25,27,29)*	
Retention of viability during storage *(8,9)*	
Sulfur dioxide resistance *(3,4,6,8–10,25,29)*	
Killer factor or resistance to killer toxins *(6,8–10,15)*	
Film-forming capacity (sherry wines) *(9,26)*	
Malic acid degradation *(6)*	
Production of glycerol to contribute to the sensory qualities of wine *(1,6,8)*	
Uniform rate of fermentation *(8,9,29)*	
Low foam ability *(8,29)*	
Low volatile acid, acetaldehyde, sulfite, and a higher alcohol production *(8,25,29,30)*	
Low hydrogen sulfide or mercaptan production *(8,20,25,29)*	
Ability to begin fermentation in high sugar concentration *(3,25)*	
Production of desirable fermentation bouquet and reproducible production of the correct levels of flavor and aroma compounds *(3,8,9,29)*	

on the environmental conditions, and on the growth phase of the sensitive cells *(15)*.

Killer phenomenon in yeasts appears to be an important characteristic in industrial fermentations. In winemaking, there is a consensus about the consequences that these toxins have on other yeasts during fermentation. Killer strains have been found in wines from different regions of the world *(15–17)*. Generally, fermentation process in making white wine is controlled at low temperatures in order to obtain high-quality products. Many floral and fruity esters have low boiling points and are readily lost through evaporation at higher fermentation temperatures, and the wine assumes a vinous, heady bouquet. As the fermentation temperature is lowered, the risk of premature stoppage increases *(18)*. Therefore, a yeast strain that has good fermentation ability at low temperatures is desired. The study of these characteristics is important to establish a definition of cryophilic wine yeasts and to select and improve useful cryophilic wine yeasts more efficiently *(19)*.

Sulfidric problems remain as some of the most common in wine fermentation. Hydrogen sulfide results in an objectionable aroma when it is present in wine. Typically, yeast cells produce only enough H_2S to meet biosynthetic requirements. An excess of H_2S in the final fermentation product can presumably arise through either its increased formation or its reduced consumption by metabolism *(20)*.

Flocculation ability, whereby yeast cells aggregate in clumps that rapidly sediment in the culture medium, has received considerable attention owing to its industrial application *(21)*. This phenomenon is of great interest for industrial fermentations such as brewing, winemaking, and biological production of ethanol because it leads to an efficient separation of yeast cells from the fermenting medium *(22)*. The mechanism of flocculation in *S. cerevisiae* has been proposed to involve surface proteins (lectins) binding to carbohydrates receptors on neighboring cell walls *(23)*.

The aim of this work is the development of primary screening techniques through the modification of classical ones, together with the advances in the basic understanding of wine yeast physiology and growth.

The screening techniques proposed in this work are grouped in five sections: (1) efficient conversion of grape-must sugar to ethanol, (2) flocculating capacity at the end of fermentation to help clarification,(3) hydrogen sulfide production, (4) Killer factor or resistance to killer toxins, and (5) ability to start fermentation in high ethanol concentrations, low temperatures, high sugar concentration, and sulfur dioxide resistance

2. Materials
2.1. Strains

1. Yeasts to be enologically characterized in these screening tests may be isolated from natural environment (grape, grape must, or wine) or may belong to specific culture collections. Comparison with commercial wine yeasts is recommended.
2. Reference killer yeast: ATCC 36900 (NCYC 738).
3. Reference sensitive yeast: NCYC 1006.

2.2. Media

1. General propagation medium: 10 g/L yeast extract, 20 g/L peptone, and 20 g/L glucose (YEPD), agarized if it is necessary.
2. General inoculum medium: filter-sterilized commercial concentrated must, dilute to 13°Brix with distilled water, plus yeast extract (0.1 g/L).
3. Inoculum medium for SO_2 resistance test: 6.7 g/L yeast nitrogen base broth (YNB) and 10 g/L glucose.
4. Fermentation medium for the test of efficient conversion of grape-must sugar to ethanol: Commercial concentrated must should be diluted to 27°Brix by the addition of distilled water, plus yeast extract (1 g/L) and steam-sterilized at 90°C for 15 min (*see* **Note 1**).
5. Fermentation medium for the test of hydrogen sulfide production: Filter-sterilized commercial concentrated must should be diluted to 22°Brix with distilled water. Sodium metabisulfite is added to this medium to give a total SO_2 concentration of 25 mg/L.
6. Medium for SO_2 resistance: 6.7 g/LYNB broth and 10 g/L glucose. It is divided into nine lots and sodium metabisulfite is added to give total SO_2 concentrations of 0, 25, 50, 75, 100, 150, 200, 250, and 300 mg/L. This medium is filter-sterilized and the pH is adjusted to 4.0.
7. Medium for testing ability to begin fermentation at 30°Brix: Filter-sterilized commercial concentrated must should be diluted to 30°Brix with distilled water; the pH is adjusted to 4.0.
8. Medium for screening ability to start fermentation at high ethanol concentration and low temperature: Filter-sterilized commercial concentrated must should be diluted at 22°Brix. For the ethanol test, it must be adjusted to 8%, 9%, 10%, 11%, and 12% v/v with absolute ethanol (*see* **Note 2**).
9. Buffered-agarized YEPD medium: 10 g/L yeast extract, 20 g/L peptone, 20 g/L glucose, 0.03 g/L methylene blue, and 20 g/L agar, buffered with 0.1 M citrate/phosphate.
10. Must-agarized medium (MAM): grape must (commercial concentrated or fresh) at 22°Brix; 10 g/L yeast extract, 0.03 g/L methylene blue, and 20 g/L agar *(15)*.

2.3. Solutions and Devices

1. Washing solution for flocculating capacity assay: 0.25 M EDTA *(22)*.
2. Flocculation buffer: 0.51 g/L $CaSO_4$, 6.80 g/L CH_3COONa, and 4.05 g/L CH_3COOH. glacial. The solution should have a pH of 4.5 (adjust if necessary).
3. Trapping solution for hydrogen sulfide production test: 4.3 g/L $3CdSO_4 \cdot 8H_2O$, and 0.6 g/L NaOH.
4. Vaseline–paraffin overlay: 50% solid vaseline and 50% pure paraffin mixed and molten.
5. Müller valve: glass device that contains sulfuric acid (50%) that allows only CO_2 to escape from the system.

6. Cadmium hydroxide traps: Test tubes (200 mm; 25 mm inside diameter) containing 60 mL of the trapping solution are used. The glass tubing (250 mm; 3 mm inside diameter) is attached to the fermentation flask neck by a compression cap and O-ring. The outlet end is curved 90° and connected to Teflon tubing, to which a disposable 22.5-cm transfer pipet is attached. This micropipet is inserted through a plug and immersed into trapping solution to create a fermentation lock and trap of H_2S that emerged with gases, exiting the fermentation flask *(20,24)*.

3. Methods
3.1. Efficient Conversion of Grape-Must Sugar to Ethanol

1. Microfermentation conditions: Fermentation is carried out in 125-mL Erlenmeyer flasks with 87.5 mL of the fermentation medium (**Subheading 2.2., item 4**), seeded with 10^6 cells/ml (**Subheading 2.1., item 1**) from a 24-h culture of the inoculum medium (**Subheading 2.2., item 2**). After aseptic closing of the flask by the Müller valve, the weight loss is followed for several days until the end of fermentation (constant weight). Fermentation systems are incubated statically at 25°C.
2. Ethanol determination: It is expressed as volume percent and is indirectly estimated by multiplying the CO_2 weight loss in grams by a stoichiometric factor of 1.3642 *(12)*; *see* **Note 3**.

3.2. Flocculating Capacity at the End of Fermentation to Help Clarification

1. Sample conditions: When microfermentation is concluded (*see* **Subheading 3.1.**), cells must be recovered by centrifugation. In order to disperse cells, they must be washed twice with washing solution (**Subheading 2.3., item 1**), then twice with bidistilled water; centrifuged and the supernatant discharged.
2. Determination of sedimentation volume: Ten milliliters of flocculation buffer (**Subheading 2.3., item 2**) is measured into a 15-mL graduated (0.01-mL) centrifuge tube and 0.25 g (fresh weight) of the washed yeast is added. The yeast suspension is placed in a water bath at 20°C for 20 min. The suspension is then dispersed by vortexing, and 10 min later the volume of deposited yeast is measured.
3. Supernatant cell concentration measurement: When flocculation is completed, samples of the supernatant are taken and dispersed in washing solution (**Subheading 2.3., item 1**). Cell concentration is determined spectrophotometrically at 620 nm and expressed as percentage of total cell concentration *(22)*.

3.3. Hydrogen Sulfide Production

1. Fermentation conditions: A single yeast colony (**Subheading 2.1., item 1**) is seeded into 25 mL of propagation medium (**Subheading 2.2., item 1**) in a 125-mL Erlenmeyer flask, incubated overnight with shaking at 250 rpm and at 30°C.

Yeasts were subcultured to 2×10^6 cells/mL into 50 mL of inoculum medium **(Subheading 2.2., item 2)** in 250 mL baffled Erlenmeyer flasks and incubated with shaking (250 rpm) to the early stationary phase. This culture serves as inoculum for fermentation trials (5×10^6 cells/mL). The fermentation progress is monitored by CO_2 weight loss at 24-h intervals.

2. Trapping of hydrogen sulfide: An Erlenmeyer flask (500 mL) with 300 mL of inoculated fermentation medium **(Subheading 2.2., item 5)** is fitted with H_2S traps. Trapping is conducted in the dark to avoid photo-oxidation of CdS in the trap. A separate trap is used to each flask. This trapping assembly allows measurement of total H_2S produced during a given time period. Traps are replaced at 24-h intervals *(20,24)*; *see* **Note 4**.
3. Hydrogen sulfide determination: The H_2S content of the samples is immediately analyzed colorimetrically (672 nm). H_2S formation is calculated from a standard curve (*see* **Note 5**).

3.4. Killer Factor or Resistance to Killer Toxins

Assay for killer and sensitive phenotype: Strains to be tested for sensitivity **(Subheading 2.1., item 1)** are grown for 24-h on YEPD-agarized medium **(Subheading 2.2., item 1)**. Then, 10^5 cells (determined spectrophotometrically) are suspended in 1 mL of sterile water and mixed with 19 mL of molten media **(Subheading 2.2., items 9 and 10)**. The suspension is poured into a Petri dish and the agar is allowed to solidify. The plate is streaked with the reference killer yeast strain **(Subheading 2.1., item 2)** and incubated at 20°C for 72 h. A positive killer reaction (that means that the lawn yeast is sensitive) is recorded in those cases where an evident and clear zone of inhibition surrounded the streak on the plate. In a similar way but using a sensitive yeast strain **(Subheading 2.1., item 3)** lawn as a reference, a potential killer yeast may be tested.

3.5. Ability to Start Fermentation in High Ethanol Concentrations, Low Temperatures, High Sugar Concentration, and Sulfur Dioxide Resistance

These four desirable characteristics are screened with a similar methodology. For this reason, they are grouped in this section.

1. SO_2 resistance: Ten milliliters of the medium **(Subheading 2.2., item 6)** with 25 mg/L of SO_2 is inoculated with 10^6 viable cells **(Subheading 2.1., item 1)** from an overnight culture in inoculum medium **(Subheading 2.2., item 3)**. Tubes are overlaid with vaseline-paraffin that solidify immediately, incubated at 25–30°C and was checked every 24 h for CO_2 production. Assays are recorded as positive when the vaseline–paraffin overlay is displaced (*see* **Note 6**). Yeasts from positive tubes are used to inoculate higher SO_2 concentrations *(25)*.

2. Ability to start fermentation at 30°Brix: Ten-milliliter aliquots of grape-must medium **(Subheading 2.2., item 7)** are inoculated (10^6 cells/mL) with cells grown overnight in 13°Brix medium. Tubes are overlaid with the vaseline–paraffin mixture. Fermentation is followed in the same way, for 5 d, as was described in **Subheading 3.3., step 1**).
3. Ability to start fermentation in the presence of 8–12% ethanol: Aliquots of medium **(Subheading 2.2., item 8)**, adjusted with ethanol to the desired concentrations, are seeded with cells cultured as was described. The CO_2 production is evaluated by displacement of vaseline–paraffin overlay.
4. Fermentation at low temperature: This is determined in a similar way. If fermentation at 4°C is assayed, it must be checked at 24-h intervals for 10 d (*see* **Note 7**).

4. Notes

1. In all experiments, the proposed commercial concentrated musts may be replaced by filter-sterilized fresh must of the available grape variety. In the specific case of this experiment, if fresh must is used, it may be improved with glucose to the required concentration.
2. If this test is used for the selection of sherry wine yeasts, concentration of the grape juice has to be adjusted to 14.5% v/v of ethanol *(26)*.
3. With the results obtained in this experiment it is possible to calculate the following: fermentation vigor (FV), which indicates the maximum ethanol yield (as % v/v) that a yeast strain can produce by fermentation in the presence of a sugar excess; and fermentation rate (FR), calculated as the amount of CO_2 produced after 3 d of fermentation (CO_2/d). If volatile acidity (g acetic acid/L) is analyzed, fermentation purity (FP) can be evaluated. It indicates the amount of volatile acidity formed in relationship to ethanol produced (g volatile acidity/L percent of ethanol [v/v]), *(12)*.
4. New micropipets must be fitted at the time trap tubes are changed, because the used micropipets frequently contain precipitated CdS.
5. Production of H_2S may be assayed qualitatively by the blackening of lead acetate paper *(1)*.
6. Despite the fact that tests grouped in this section are qualitative, the moment when the displacement of the vaseline–paraffin overlay occurs should be recorded (first day, second day, etc.).
7. There is a general agreement that the production of floral or fruity wines demands, among other things, a fermentation temperature of 15°C or lower. Nevertheless, the temperature considered low by different authors varies from 7°C to 15°C, depending on the grape variety and the different kind of winemaking process *(18,19,27)*.

References

1. Romano, P., Suzzi, G., Comi, G., Zironi, R., and Maifreni, M. (1997) Glycerol and other fermentation products of apiculate wine yeasts. *J. Appl. Microbiol.* **82,** 615–618.

2. Ciani, M. and Picciotti, G. (1995) The growth kinetics and fermentation behaviour of some non-*Saccharomyces* yeast associated with wine making. *Biotechnol. Lett.* **17**, 1247–1250.
3. Martini, A. and Vaughan-Martini, A. (1989) Grape must fermentation: past and present, in *Yeast Technology* (Spencer, J.F.T. and Spencer, D.M., eds.), Springer-Verlag, New York, pp. 105–123.
4. Constantí, M., Reguant, C., Poblet, M., Zamora F., Mas, A., and Guillamón, J. M. (1998) Molecular analysis of yeast population dynamics: effect of sulphur dioxide and inoculum on must fermentation. *Int. J. Food Microbiol.* **41**, 169-175.
5. *Bull. OIV* (1995) **69**, 789–790.
6. Degree, R. (1993) Selection and commercial cultivation of wine yeast and bacteria, in *Wine Microbiology and Biotechnology* (Fleet, G. H., ed.) Harwood Academic Publishers, Grey, Switzerland, pp. 421–447.
7. Fleet, G. H. (1990) Growth of yeast during wine fermentation. *J. Wine Res.* **1**, 211–223.
8. Pretorius, I. S. and van der Westhuizen, T. J. (1991) The impact of yeast genetics and recombinant DNA technology on the wine industry. *S. Afr. J. Enol. Vitic.* **12**, 3–31.
9. Colagrande, O., Silva, A., and Fumi, M. D. (1994) Recent applications of biotechnology in wine production. *Biotechnol. Prog.* **10**, 2–18.
10. Ubeda, J., Briones, A. I., Izquierdo, P., and Palop, L. I. (1995) Predominant *Saccharomyces cerevisiae* strains in the fermentation of Airén grape musts with SO_2. *Food Sci. Technol.* **28**, 584–588.
11. Romano, P. and Suzzi, G. (1993) Sulphur dioxide and wine microorganisms in *Wine Microbiology and Biotechnology* (Fleet, G. H., ed.) Harwood Academic Publishers, Grey, Switzerland, pp 373–393.
12. Ciani, M. and Maccarrelli, F. (1998) Oenological properties of non-*Saccharomyces* yeasts associated with wine-making. *World J. Microbiol. Biotechnol.* **14**, 199–203.
13. Alexandre, H., Rousseaux, I., and Charpentier, C. (1994) Relantionship between ethanol tolerance, lipid composition and plasma membrane fluidity in *Saccharomyces cerevisiae* and *Kloekera apiculata*. *FEMS Microbiol. Lett.* **124**, 17–22.
14. Benítez, T., del Castillo, L., Aguilera, A., Conde, J., and Cerdá-Olmedo, E. (1983) Selection of wine yeasts for the growth and fermentation in the presence of ethanol and sucrose. *Appl. Environ. Microbiol.* **45**, 1429–1436.
15. da Silva, G. A. (1996) The occurrence of killer, sensitive, and neutral yeasts in Brazilian Riesling Italico grape must and the effect of neutral strains on killing behaviour. *Appl. Microbiol. Biotechnol.* **46**, 112–121.
16. Hidalgo, P. and Flores, M. (1994) Occurrence of the killer character in yeasts associated with Spanish wine production. *Food Microbiol.* **11**, 161–167.
17. Vazquez, F. and Toro, M. E. (1994) Occurrence of killer yeasts in Argentine wineries. *World J. Microbiol. Biotechnol.* **10**, 358–359.
18. Subden, R. E. (1987) Current developments in wine yeasts. *CRC Biotechnol.* **5**, 49–65.
19. Kishimoto, M., Oshida, A., Shinohara, T., Soma, E., and Goto, S. (1994) Effect of temperature on ethanol productivity and resistance of cryophilic wine yeasts. *J. Gen. Appl. Microbiol.* **40**, 135–142.

20. Jiranek, V., Langridge, P., and Henschke, P. A. (1995) Regulation of hydrogen sulphite liberation in wine-producing *Saccharomyces cerevisiae* strains by assimilable nitrogen. *Appl. Environ. Microbiol.* **61,** 461–467.
21. Mota, M. and Soares, E. V. (1994) Population dynamics of flocculating yeasts. *FEMS Microbiol. Rev.* **14,** 45–52.
22. Sieiro, C., Reboredo, N. M., and Villa, T. (1995) Flocculation of industrial and laboratory strains of *Saccharomyces cerevisiae*. *J. Ind. Microbiol.* **14,** 461–466.
23. Cubells Martínez, X., Narbad, A., Carter, A. T. and Stratford, M. (1996) Flocculation of the yeast *Candida famata* (*Debaryomyces hansenii*): an essential role for peptone. *Yeast* **12,** 415–423.
24. Thomas, C. S., Boulton, R. B., Silacci, M. W., and Gubler, W. D. (1993) The effect of elemental sulphur, yeast strain and fermentation medium on hydrogen sulphide production during fermentation. *Am. J. Enol. Vitic.* **44,** 211–216.
25. Parish, M. E. and Carroll, D. E. (1987) Fermentation characteristics of *Saccharomyces cerevisiae* isolates from *Vitis rotundifolia* grapes and musts. *Am. J. Enol. Vitic.* **38,** 45–48.
26. Martínez, P., Pérez Rodríguez, L., and Benítez, T. (1997) Velum formation by flor yeasts isolated from sherry wine. *Am. J. Enol. Vitic.* **48,** 55–62.
27. Kishimoto, M., Soma, E., and Goto, S. (1994) Classification of cryophilic wine yeasts based on electrophoretic karyotype, G+C and DNA similarity. *J. Gen. Appl. Microbiol.* **40,** 83–93.
28. Reed, G. and Nagodawithana, T. W. (1988) Technology of yeast usage in winemaking. *Am. J. Enol. Vitic.* **39,** 83–89.
29. Thornton, R. J. (1991) Wine yeast research in New Zealand and Australia. *CRC Biotechnol.* **11,** 327–345.
30. Romano, P., Suzzi, G., Turbanti, L., and Polsinelli, M. (1994) Acetaldehyde production in wine yeast. *FEMS Microbiol. Lett.* **118,** 213–218.

34

Utilization of Native Cassava Starch by Yeasts

Lucía I. C. de Figueroa, Laura Rubenstein, and Claudio González

1. Introduction

Cassava (*Manihot esculenta*) is a root crop of tropical American origin and is the fourth most important staple crop in the tropics. In the developing world, it is surpassed only by maize, rice, and sugarcane as a source of calories; cassava's starchy roots produce more food energy per unit of land than any other staple crop. Cassava is grown almost exclusively in the arid and semiarid tropics, where it accounts for approximately 10% of the total caloric value of staple crops. The cassava plant is extremely robust, is resistant to disease and drought, and can grow in relatively low-quality soils *(1)*.

This staple crop has many favorable characteristics, some of them are as follows: Cassava is well adapted to marginal soils (low fertility, high acidity) on which most other crops fail; cassava has the ability to tolerate environmental stress (drought), pest and disease attacks and to recover readily; compared to other staple crops, it gives relatively high yields and is an excellent source of carbohydrate; harvesting of cassava roots can take place from 6 to 36 mo after planting, thus providing the farmer with a permanent source of food *(1)*.

Different types of cassava satisfy a range of needs. There are early-maturing varieties and late-maturing varieties. There are giant-size cassava and pygmy types. There are also sweet cassavas and bitter ones to taste. Cassava roots are used as pig meal, as fuel, and also for brewing beer.

Part of the total world cassava production (about 22%) is incorporated into animal feed. A similar amount is converted into starch for industrial use and another portion to human food in some developing countries. In addition, new high-yielding varieties of cassava, with outputs of 100 ton/ha could provide the fermentation industry with an abundance of raw material *(2)*.

Cassava starch is composed of unbranched amylose (20 ± 5%) and branched amylopectin (80 ± 5%), both of which can be hydrolyzed acidically or enzymatically (either with pure enzymes or amylase-producing microorganisms) to release their constituent glucose and maltooligosaccharides. Both products are easily transported across the cell membrane and metabolized by yeasts *(3)*.

The ability to hydrolyze starch is reasonably widely distributed among the different genera and species of yeasts. Several laboratories around the world have been involved in recent years in the construction, isolation and/or evaluation of amylolytic yeast strains (**Table 1**). A major thrust of this work has been toward generating some amylolytic capability *in situ* during the fermentation stages of the processes. The advantage are those relating to: a reduced need to purchase or produce in separate process amylolytic enzymes, simplification of the processing stages that precede fermentation, improved efficiencies of starch conversion, and the possibility of producing amylolytic enzymes or the glucose released from the starch molecule for use in syrup manufacture, as by-products of alcohol fermentation *(4,5)*, or for obtaining a medium suitable for production of oenological yeasts.

The degradation products of starch are mainly used as basic carbohydrate sweeteners in the food industry. The extensive fermentation pattern revealed that during the fermentation of cassava starch by the fungus *Rhizopus oligosporus* or by the yeast *Endomycopsis fibuligera,* a higher concentration of glucose accumulation can be obtained *(6,7)*.

Enzymatic hydrolysis of raw starch at ambient temperatures without cooking has a tremendous economic advantages over high-temperature methods. Uncooked raw cassava starch can be hydrolyzed to glucose by amylolytic yeast enzymes. The amylolytic fermentative yeasts (Table 1) produce extracellular amylases (α-amylase, amyloglucosidase) and can utilize soluble or raw starch as the sole carbon source.

Cassava tubers, contain about 3% protein (dry basis), and their leaves yield 5–10 tons/ha dry matter, of which about 25% is good quality protein. This protein could either be extracted and purified for human consumption or fed directly to animals.

The use of cassava as a feedstock for single-cell protein (SCP) in developing countries would have social and macroeconomic as well as technical advantages, as its cultivation on a large scale would provide much-needed rural employment. Also, export of the finished SCP to developed countries could be a large earner of foreign currency *(8)*. The sugars formed by the amylolytic organisms in the cassava starch can be used as substrates for the generation of microbial biomass rich in protein. Mixed cultures of yeasts for the production of SCP have already been used elsewhere, with the Symba process being used commercially in Sweden. It is also possible to treat the cassava starch for SCP production by submerged fermentation of two organisms *Endomycopsis fibuligera* and *Candida utilis (9)*.

Table 1
Amylolytic Yeasts

Genera and species	Amylolytic enzymes			Ref.
	α-Amylases	Glucoamylases	Debranching	
Candida tsukubaënsis; now *Pseudozyma tsukubaënsis* (12)	1[a]	1[a]	—	21,22
Lipomyces kononenkoae	1[a]	1[a]	Transferase	23
Lipomyces starkeyi	1[a]	1[a]	–	24
Endomycopsis capsularis; now Saccharomycopsis capsularis (12)	1[a]	1[a]	—	25
Endomycopsis fibuligera; now *Saccharomycopsis fibuligera* (12)	1	2	—	26,27
Schwanniomyces castelli., S. alluvius, or *S. occidentalis;* now *Debaryomyces occidentalis* (12)	1	2	Glucoamylases	28,29
Pichia burtonii	1[a]	1[a]	—	26
Saccharomyces diastaticus; now *Saccharomyces cerevisiae var diastaticus* (12)	—	4[b]	—	30–33
Candida edax; now *Stephanoascus smithiae* (12)	1[a]		1[a]	34

[a] They produce at least one type of each enzyme.
[b] Three unlinked genes STA (1–3) and SGA 1 (specific glucoamylase for sporulation).

In the present chapter, we will describe five aspects of the utilization of native cassava starch by yeasts: (1) Isolation and screening of amylolytic yeasts from agricultural and industry starch residues, (2) a new method of screening and differentiation of amylolytic enzymes from yeast strains, (3) ethanol production from native cassava starch by mixed culture using amylolytic yeasts and bacteria, (4) hydrolysis of native cassava starch using amylolytic yeast enzymes for application to enological yeast production and (5) fusion of yeast protoplasts and isolated nuclei of filamentous fungi.

2. Materials

Agricultural and industry starch residues, as cassava roots, starch factory effluents, potato-processing wastewater, and so forth and as substrates for the isolation and screening of amylolytic yeasts.

2.1. Strains

1. Yeast strains for screening and differentiation of amylolytic enzymes (*see* **Note 1**).
2. Use of an amylolytic yeast and a bacteria strain for ethanol production by mixed culture (*see* **Note 2**).
3. Use of enological yeast when using hydrolyzed cassava starch as culture medium, and amylolytic yeasts for hydrolyzing native cassava starch (*see* **Note 3**).
4. When fusing yeast protoplasts and isolated filamentous fungi: yeast strains and filamentous fungi have to be used (e.g., strains of *Aspergillus*, *Rhizopus*, *Aureobasidium*, etc.) (*see* **Note 4**).

2.2. Media

1. General isolation medium: YM broth (Difco), pH 5.0, and the same YM broth acidified to pH 3.5.
2. Medium for isolation of amylolytic yeasts: 5 g/L yeast extract, 20 g/L cassava flour starch, 0.03 g/L rose bengal, 10 µg/mL erythromycin, 150 µg/mL ampicillin, and 20 g/L agar; pH 3.5.
3. Preparation of α-amylase-resistant starch (α-RS) medium for isolation of amylolytic yeasts: Two hundred grams of wheat starch is simultaneously gelatinized and liquefied at 80°C in 1 L of 2 mM $CaCl_2$ with 1000 U of commercial α-amylase. Then, the solution is cooled and incubated with more 1000 U of α-amylase preparation for 3 h at 55°C. The resulting insoluble material is collected, washed several times with water by centrifugation (8000g, 15 min), and then lyophilized; yield is approx 5% (*see* **Note 5**).
   Basal medium for isolation and selection with α-RS: 5 g/L α-RS, 1.0 g/L NH_4NO_3, 1.4 g/L KH_2PO_4, 0.5 g/L $CaCl_2$, 0.2 g/L $MgSO_4·7H_2O$, 0.05 g/L chloramphenicol, and 0.1 g/L yeast extract; pH 5 *(10)*.
4. Culture medium for screening and differentiation of amylolytic enzymes: 20 g/L yeast extract, 20 g/L starch, and 20 g/L agar in 10 mM acetate buffer; pH 4.5.
5. Yeast inoculum medium for mixed cultures: 10 g/L yeast extract, 20 g/L peptone, 50 g/L Lintner starch (YEP–starch medium); pH 4.5.
6. Bacteria inoculum medium for mixed cultures: 10 g/L yeast extract, 1 g/L $NH_4(SO_4)_2$, 1 g/L KH_2PO_4, 1 g/L $MgSO_4·7H_2O$, 100 g/L glucose, RMG medium; pH 5.6 (see **Note 6**).
7. Fermentation media for mixed cultures: They have to be prepared by replacing the glucose in RMG with different concentrations of native cassava flour (e.g., 50, 100, or 150 g/L, w/v); pH 6.2. Media are sterilized for 15 min at 121°C.
8. Medium for production of amylolytic enzymes: 2 g/L $NH_4(SO_4)_2$, 2 g/L KH_2PO_4, 0.5 g/L $MgSO_4·7H_2O$, 4 g/L yeast extract, 20 g/L starch; pH 5.0.

9. Basal cassava starch-hydrolyzed medium: It is obtained by adding 150 g/L of cassava flour into the amylolytic yeast culture.
10. Cultivation medium for filamentous fungi, Czapek's: 3.0 g/L $NaNO_3$, 1.0 g/L K_2HPO_4, 0.5 g/L $MgSO_4$, and 0.01 g/L $FeSO_4$, with 50 g/L glucose.
11. Regeneration medium for recovery of fusion products: 10 g/L Czapek's medium plus starch, 2 g/L yeast extract, 0.6 M KCl, and 30 g/L agar.
12. Selective medium for fusion products: 10 g/L Czapek's medium plus cassava raw starch, 2 g/L yeast extract, 0.6 M KCl, and 20 g/L agar.

2.3. Solutions

1. Solution for enzymatic activity assay: 10g/L starch, in 10 mM acetate/acetic acid buffer (pH 4.5) and 20 g/L 3,5-dinitrosalycilic acid (DNS), in 1 N NaOH solution (DNS reagent).
2. Pretreatment solution for protoplast formation from fungal mycelium: 0.01 M dithiothreitol (DTT) in citrate–phosphate buffer pH 7.3.
3. Osmotic stabilizer solution for protoplast formation from fungal mycelium: citrate–phosphate buffer pH 5.8 with 0.7 M KCl, as an osmotic stabilizer.
4. Enzyme solution for protoplast formation from fungal mycelium: citrate–phosphate buffer (pH 5.8) containing 0.7 M KCl and 7 mg/mL of Novozyme 234 (Novo Biolab, Denmark). KCl is added as an osmotic stabilizer.
5. Solution for lysing filamentous fungi protoplasts: 180 g/L Ficoll, 20 mM KH_2PO_4, 0.5 mM Ca^{2+}, 1mM phenylmethylsulfonyl fluoride (PMSF); pH 6.5.
6. Solution for purifying filamentous fungi nuclei: 70 g/L Ficoll, 20 mM KH_2PO_4, 0.5 mM Ca^{2+}, 1mM PMSF, 1 M sorbitol, and 200 g/L glycerol.
7. Fusion mixture solution for fusing yeast protoplasts with nuclei of filamentous fungi: 300 g/L polyethylene glycol 4000 (average molecular weight 3300–4000) in 10 mM $CaCl_2$ solution.

2.4. Zymograms

Polyacrylamide gels for electrophoresis (PAGE). Solutions: 10 g/L starch (pH 4.5) and Lugol solution.

3. Methods

3.1. Isolation and Screening of Amylolytic Yeasts from Agricultural and Industry Starch Residues

1. Isolation and Identification using YM broth. Samples of 5–10 g are obtained and transported in sterile plastic bags. The samples are suspended in 50 mL of YM medium (**Subheading 2.2., item 1.**) diluted 1 : 10 with sterile water containing 1 g/L Tween-80 *(11)*. After shaking 60 min at 25°C, the suspension is poured into sterile culture tubes and stored overnight at 4°C. Most of the supernatant is discarded, and the remaining (about 2 mL) is shaken in order to resuspend the settled

cells. This suspension is streaked on acidified YM agar. The inoculated plates are incubated at 25°C, and after 3, 5, and 12 d, well-isolated colonies are transferred on YM agar until growth is observed. Colonies are restreaked and picked on selective medium with cassava flour starch as the sole carbon source (**Subheading 2.2., item 2.**) and incubated at 30°C until growth occurred (*see* **Note 7**).

The yeast strains are identified according to their carbohydrate and nitrogen assimilation patterns, using the keys and description in **ref.** *12* and the computerized yeast identification program devised by Barnett, Payne, and Yarrow.

2. Isolation and identification using basal medium with α-RS. Samples are suspended in sterilized water and then 1 drop of each supernatant obtained on centrifugation is added to 5 mL of the medium (**Subheading 2.2., item 3.**) in a test tube. After shaking the tubes at 30°C for 1–3 d, 1 drop of the culture in each tube is spread on the medium solidified with 2% agar in Petri dishes, and incubated for 1–3 d. The colonies producing large "halos" are picked up and purified by streaking on selective medium with cassava flour starch as sole carbon source.

3. Enzyme determinations. Enzymatic extracellular amylolytic activity is determined by the DNS method (*see* **Note 8**). The reaction mixture contains 100 µL of supernatant obtained from samples of one culture centrifuged 5 min at 5000*g*, and 400 µL of the solution for enzymatic activity (**Subheading 2.3., item 1**). The mixture is incubated 15 min at 45°C. The reaction is stopped by adding 770 µL of DNS reagent, and the reducing sugars released are determined at 590 nm *(13)*. One unit of enzyme is defined as the amount that liberates 1 µmol of reducing sugars per minute per milliliter.

3.2. New Method of Screening and Differentiation of Amylolytic Enzymes from Yeast Strains

1. Culture conditions and recovery of the crude extracts. Yeasts are cultured in Petri dishes (with the medium described in **Subheading 2.2., item 4**) 48 h at 30°C. Once the colonies appear, they are scraped off and samples of agar of 1 cm diameter of each colony are taken. The agar is cut in small pieces and put into Eppendorf tubes. Samples are frozen at –20°C for 4–6 h and the supernatant is recovered by centrifugation for 10–15 min.

2. Enzymatic activity assays. Extracellular activities are determined by recovering of the supernatants. Amylolytic activity of yeasts is determined using the method in **Subheading 3.1., step 3**.

    The extracellular amylolytic activity is determined by measuring the reducing sugar groups released from starch by a colorimetric method, based on the reduction of 3,5-dinitrosalicylic acid. One unit of enzyme activity is defined as the amount that liberates 1 µmol of reduced group per minute per milliliter of enzyme sample.

3. Zymograms. The polyacrylamide gel electrophoresis (PAGE) is done according to Davis *(14)*. Enzymatic activity is detected in the PAGE gels by zymographic techniques. To detect amylolytic activity, the gels are incubated in the starch solution (**Subheading 2.4.**) for 4 h. Gels are stained with Lugol solution and the amylolytic activity zones are detected by the presence of clear areas on the blue gel (*see* **Note 9**).

3.3. Ethanol Production from Native Cassava Starch by Mixed Culture Using Amylolytic Yeasts and Bacteria

1. Inocula. Yeast strain is incubated for 16 h at 30°C in YEP–starch medium (**Subheading 2.2., item 5**). The bacteria strain is grown in RMG medium (**Subheading 2.2., item 6.**) for the same period, *see* **Note 6**. The cultures have to be centrifuged 5 min at 5000*g*, and the pellets (about 0.14 g of dry biomass) are resuspended in water (2 mL final volume) and used as the inoculum.
2. Fermentation assays. They are performed in 500 mL Erlenmeyer flasks containing 200 mL of medium (**Subheading 2.2., item 7**), at 30°C on a rotary shaker at 200 rpm. The culture is shaken in order to ensure its homogeneity because of the high viscosity of the medium containing native cassava starch (*see* **Note 10**).
3. Analytical and enzymatic determinations. Growth can be determined by optical density measurements at 660 nm. Starch is determined as follows: To 1 mL of sample, 1 mL of 1 *M* HCl was added. The mixture is boiled for 45 min and then neutralized with 1 mL of 1 *M* NaOH solution *(15)*. The reducing sugars released are measured colorimetrically using the 3,5-dinitrosalycilic acid (DNS) method *(13)*.

    Glucose can be measured enzymatically using a glucose–oxidase–peroxidase method. Ethanol can be determined in a previously distilled sample using an immersion refractometer.

    The enzymatic extracellular amylolytic activity is determined by the DNS method. Amylolytic activity of yeasts is determined using the method described in **Subheading 3.1., step 3**. One unit of enzyme is defined as the amount that liberates 1 µmol of reducing sugars per minute per milliliter.

3.4. Hydrolysis of Native Cassava Starch Using Amylolytic Yeast Enzymes; Application to Enological Yeast Production

Fermentation assays. Batch cultures are done using the medium described in **Subheading 2.2., item 8.** for obtaining yeast biomass and hydrolyzed cassava flour. It is better to use a fermentor, and the optimum adequate conditions are 450 rpm, 30°C for 12–24 h, according to the yeast strain used (*see* **Note 11**). After that, this culture with the hydrolyzed cassava flour (**Subheading 2.2., item 9.**) must be sterilized in order to destroy the amylolytic yeast cells.

Enological yeasts propagation is done using the sterilized hydrolyzed cassava starch as culture medium. Enological properties are evaluated by microvinifications.

3.5. Fusion of Yeast Protoplasts and Isolated Nuclei of Filamentous Fungi

1. Protoplasts formation from fungal mycelium. Cultures of filamentous fungi are grown in Czapek's medium with 50 g/L glucose as the carbon source (**Subheading 2.2., item 10**). These flasks are inoculated with 1×10^6 fungal spores/mL and

incubated for 18 h at 30°C in a rotary shaker operated at 200 rpm. The mycelium is recovered by filtration and washed three times with sterile water.

Washed mycelium is resuspended in the pretreatment solution (**Subheading 2.3., item 2**) and incubated for 1 h at 30°C without agitation. The mycelium is then filtered through filter paper and has to be washed three times with the osmotic stabilizer solution (**Subheading 2.3., item 3**). The biomass is resuspended in the enzyme solution (**Subheading 2.3., item 4**). The suspension is incubated at 30°C with agitation at 100 rpm for 4–5 h. Protoplast formation should be monitored under a phase-contrast microscope. The protoplasts suspension is filtered through a filtration device to remove the remaining mycelium and other debris and washed three times with the osmotically stabilized solution *(16,17)*.

2. Isolation of fungal nuclei. Filamentous fungi nuclei are isolated by differential centrifugation of suspensions of lysed protoplasts in Ficoll solutions at 4°C as follows:
   a. Protoplasts are lysed in the solution for lysing (**Subheading 2.3., item 5**). This solution lyses the protoplasts but not the nuclei that remain intact.
   b. Nuclei suspension is treated with the solution for purifying the nuclei (**Subheading 2.3., item 6**).

   The final purified pellet of nuclei is resuspended in 0.5 mL of the solution for purifying, and the nuclei should be used immediately for fusion with yeast protoplasts. Isolated nuclei should be stained with 4',6-diamino-2-phenylindole (DAPI) and observed using fluorescence microscopy *(16,17)*.

3. Fusion of yeast protoplasts with nuclei of filamentous fungi. Yeast protoplasts are obtained by standard methods. Protoplasts are resuspended in 0.5 mL of a 0.4-M $CaCl_2$ solution, mixed with the suspension of nuclei and centrifuged at 12,000g for 10 min. The supernatant is discarded and the pellet is resuspended in the fusion mixture solution (**Subheading 2.3., item 7**). The suspension is incubated at 25°C for 10 min *(16,17)*.

4. Regeneration of the fusion products. The fusion mixture is added to the osmotically stabilized, melted regeneration medium (**Subheading 2.2., item 11**) at 42°C, mixed, poured as an overlay on a base of the same osmotically stabilized regeneration medium, and allowed to harden. Plates are incubated at 30°C for 4–5 d until yeastlike colonies appear. These colonies are transferred to selective medium (**Subheading 2.2., item 12**) for isolation and characterization of the desired fusion products. Yeastlike colonies growing on the selective medium have to be selected *(16,17)* (*see* **Note 12**).

4. Notes

1. Screening and differentiation of amylolytic enzymes from yeast strains should be done by using yeasts that grow in media with starch as sole carbon source (**Table 1**).
2. It is necessary to use amylolytic yeast strains, better those that produce α-amylase as well as glucoamylase (**Table 1**) in order to hydrolyze native cassava starch, and, for example, *Zymomonas mobilis*, an efficient ethanol producing aerotolerant Gram-negative bacterium *(18)*.

3. It is necessary to use amylolytic yeast strains that produce α-amylase as well as glucoamylase in order to hydrolyze native cassava starch (**Table 1**).
4. In order to use filamentous fungi as α-amylases or glucoamylases donor strains, the best ones are *Aspergillus oryzae* and *Aspergillus niger*. Taking into account that the starch molecules to hydrolyze are from cassava, it is important also use strains such as *Aureobasidium pullulans* as the donor strain. This fungus produce the enzyme pullulanase with a very good debranching activity.
5. α-amylase-resistant-starch (α-RS) is a helical inclusion complex between short-chain amylose (degree of polymerization approximately 90) and lipid, similar to the complex between amylose and *n*-butanol *(10)*.
6. RMG medium is specific for *Z. mobilis*, one of the best ethanol-producing bacteria, and for this reason, it is recommended to use this microorganism for ethanol production using a mixed culture from native cassava starch *(7)*.
7. When colonies appear on the starch plates, they have to be refrigerated at 0–4°C for 3–4 d, in order to observe "halos" visible of hydrolysis of the starch around the colonies.
8. Amylolytic enzymes should be measured colorimetrically using the 3,5-dinitrosalycilic acid (DNS) method *(13)*.
9. The zymographic profiles obtained from solid-media crude extracts are comparable with those obtained after precipitating liquid media with acetone. The described technique permits the differentiation of different amylolytic enzyme activities. With this technique it is possible to differentiate strains of microorganisms with the same enzymatic amylolytic activity from their own zymographic profiles at the initial screening.

    The methodology described with the aim of recovering crude extracts from solid media allows the evaluation of numerous samples and the identification of amylolytic enzymes in a very simple and rapid way. This procedure can be used as a routine screening method without any special equipment. It also allows for the preliminary characterization of the strains of microorganisms by their zymographic profile *(19)*.
10. If amylolytic strains of *E. fibuligera* and *Z. mobilis* are used, the inocula have to be prepared as follows: *E. fibuligera* has to be incubated for 16 h at 30°C in a 250-mL shaken flask (200 rpm) containing 100 mL of YEP-starch. *Z. mobilis* has to be grown in RMG without agitation, for the same time and the same temperature. The cultures have to be centrifuged for 5 min at 5000*g* and the pellets (about 0.14 g of dry biomass) are resuspended in water, 2 mL final volume, and used as inoculum.
11. *Endomycopsis fibuligera* is an adequate amylolytic yeast strains for this purpose. With this strain, it is possible to obtain a complete cassava flour hydrolyzed in 12 h *(20)*. The culture medium described is very suitable and cost-effective for enological yeasts production.
12. Fusion of yeast protoplasts either with nuclei from other yeast strains or with nuclei from unrelated organisms has been known to be feasible for some time. In the case we are discussing, viable nuclei are isolated from the donor strain and fused in the conventional manner with protoplasts from the recipient yeast strain.

The fusion of yeast protoplasts with fungal nuclei is a useful technique to improve the methods of transferring genetic information between strains, especially those that are dissimilar or distantly related.

References

1. Komen, J. (ed.) (1990) *Cassava and Biotechnology, Proceedings of a Workshop*, Directorate General for International Cooperation, The Hague, Amsterdam.
2. Ejiofor, A. O., Chisti, Y., and Moo-Young, M. (1996) Culture of *Saccharomyces cerevisiae* on hydrolyzed waste cassava starch for production of baking-quality yeast. *Enzyme Microb. Technol.* **18,** 519–525.
3. Kristiansen, A. (1994) Integrated design of a plant, in *The Production of Baker's Yeast*, VCH Verlagsgesellschaft, Weinheim, Germany, p. 21.
4. Tubb, R. S. (1986) Amylolytic yeasts for commercial applications. *Tibtech*, April, 98–104.
5. Spencer, J. F. T. and Spencer, D. M. (1997) Taxonomy: the name of the yeasts, in *Yeasts in Natural and Artificial Habitats* (Spencer, J. F. T. and Spencer, D. M., eds.), Springer-Verlag, Berlin, pp. 11–22.
6. Garg, S. K. and Doelle, H. W. (1989) Optimization of cassava starch conversion to glucose by *Rhizopus oligosporus*. *Ñircen J.* **5,** 297–305.
7. Gonzalez, C., Delgado, O., Baigorí, M., Abate, C., de Figueroa, L. I. C., and Callieri, D. A. (1998) Ethanol production from native cassava starch by a mixed culture of *Endomycopsis fibuligera* and *Zymomonas mobilis*. *Acta Biotechnol.* **2,** 149–155.
8. MacLennan, D. G. (1976) Single cell protein from starch: a new concept in protein production. *Search* **7,** 155–163.
9. Manilal, V. B., Narayanan, C. S., and Balagopalan, C. (1991) Cassava starch effluent treatment with concomitant SPC production. *World J. Microbiol. Biotechnol.* **7,** 185–190.
10. Bergmann, F. W., Abe, J., and Hizukuri, S. (1988) Selection of microorganisms which produce raw-starch degrading enzymes. *Appl. Microbiol. Biotechnol.* **27,** 443–446.
11. Middelhoven, W. J. (1997) Identity and biodegradative abilities of yeasts isolated from plants growing in an arid climate. *Antonie van Leeuwenhoek* **72,** 81–89.
12. Kurtzman, C. P. and Fell, J. W. (1998) *The Yeasts, a Taxonomic Study*, 4th ed. Elsevier Science, Amsterdam.
13. Miller, G. L. (1959) Use of dinitrosalicilic acid reagent for determination of reducing sugars. *Anal. Chem.* **31,** 426–428.
14. Davis, B. J. (1964) *Ann. NY Acad. Sci.* **121,** 404–427.
15. Laluce, C. and Mattoon, J. R. (1984) Developments of rapidly fermenting strains of *Saccharomyces diastaticus* for direct conversion of starch and dextrins to ethanol. *Appl. Environ. Microbiol.* **48,** 17–25.
16. Vázquez, F., Heluane, H., Spencer, J. F. T., Spencer, D. M., and de Figueroa, L. I. C. (1997) Fusion between protoplasts of *Pichia stipitis* and isolated filamentous fungi nuclei. *Enzyme Microb. Technol.* **21,** 32–38.

17. Heluane, H., Vázquez, F., and de Figueroa, L. I. C. (1998) Fusion of yeast protoplasts and isolated nuclei. *Acta Biotechnol.* **18,** 353–359.
18. Swings, J. and De Ley, J. (1977) The biology of *Zymomonas*. *Bacteriol. Rev.* **41,** 1–46.
19. González, C., Martínez, A., Vázquez, F., Baigorí, M., and de Figueroa, L. I. C. (1996) New method of screening and differentiation of exoenzymes from industrial strains. *Biotechnol. Tech.* **10,** 519–522.
20. González, C., Vázquez, F., Toro, M. E., Baigorí, M., and de Figueroa, L. I. C. (1997) Hidrólisis de harina de mandioca empleando *Endomycopsis fibuligera*. Su aplicación en la producción de levaduras de interés enológico. *VII Mexican Congress of Biotechnology and Bioengineering and II International Symposium on Bioprocess Engineering*, Mazatlán, México.
21. Onishi, H. (1972) *Candida tsukubaënsis* sp. *Antonie van Leeuwenhoek*, **38,** 365–367.
22. de Mot, R., van Oudendijck, E., and Verachtert, H. (1984) Purification and characterization of an extracellular glucoamylase for the yeast *Candida tsukubaensis* CBS 6389. *Antonie van Leeuwenhoek* **51,** 275–287.
23. Spencer-Martins, I. and van Uden, N. (1979) Extracellular amylolytic system of the yeast *Lypomyces kononenkoae*. *Eur. J. Appl. Microbiol. Biotechnol.* **6,** 24–250.
24. Kelly, C. T., Moriarty, M. E., and Fogarty, W. M. (1985) Thermostable extracellular α-amylase and α-glucosidase of *Lipomyces starkeyi*. *Appl. Microbiol. Biotechnol.* **22,** 352–358.
25. Soni, S. K., Sandhu, I. K., Bath, K. S., Banerjee, U. C., and Patnaik, P. R. (1996) Extracellular amylase production by *Saccharamycopsis capsularis* and its evaluation for starch saccharification. *Folia Microbiol.* **4,** 243–248.
26. Sills, M. A., Panchal, Ch.J., Rusell, I., and Stewart G. G. (1987) Production of amylolytic enzymes by yeasts and their utilization in brewing. *CRC Biotechnol.* **5,** 105–116.
27. Futatsugi, M., Ogawa, T., and Fukuda, H. (1993) Purification and properties of two forms of glucoamylase from *Saccharomycopsis fibuligera*. *J. Ferment. Bioeng.* **6,** 521–523.
28. Oteng-Gyang, K., Moulin, G., and Galzy, P. (1981) A study of the amylolytic system of *Schwanniomyces castellii*. *Z. Allg. Microbiol.* **21,** 537–544.
29. Ingledew, W. M. (1987) *Schwaniomyces*: a potential superyeast? *CRC Biotechnol.* **5,** 159–176.
30. Tamaki, H. (1978) Genetic studies of ability to ferment starch in *Saccharomyces*: gene polymorphism. *Mol. Gen. Genet.* **164,** 205–209.
31. Errat, J. A. and Stewart, G. G. (1978) genetic and biochemical studies on yeast strains capable of utilizing dextrins. *J. Am Soc. Chem.* **36,** 151–161.
32. Pretorius, I. S., Chow, T., Modena, D., and Marmur, J. (1986) Molecular cloning and characterization of the STA2 glucoamylase gene of Saccharomyces diastaticus. *Mol. Gen. Genet.* **203,** 29–35.
33. Pugh, T. A., Shah, J. C., Magee, P. T., and Clancy, M. (1989) Characterization and localization of the sporulation glucoamylase of Saccharomyces cerevisiae. *Biochem. Biophys. Acta* **994,** 200–209.
34. Moussa, E. and Baratti, J. (1988) Isolation and characterization of an amylolytic yeast *Candida edax*. *World J. Microbiol. Biotechnol.* **4,** 193–202.

V

SPECIAL METHODS AND EQUIPMENT

35

Reactor Configuration for Continuous Fermentation in Immobilized Systems

Application to Lactate Production

José Manuel Bruno-Bárcena, Alicia L. Ragout de Spencer, Pedro R. Córdoba, and Faustino Siñeriz

1. Introduction

There is great commercial interest in using immobilization technology for fermentation processes. Microbial immobilization is one of the novel methods in fermentation technology, especially important in the food and beverage industry, which allows the use of increased cell concentrations in the bioreactor, reducing process cycle times and increasing volumetric bioreactor productivity as compared with traditional batch and chemostat methods of fermentation. Results from bioreactor studies have demonstrated that immobilized cells have advantages over free cells, such as protection from toxic substances, increased plasmid stability and increased catalytic activity (*1*). Because of these advantages, methods employing immobilized microbial cells are used extensively in many industrial applications.

In immobilized cell reactors, whole cells are fixed on a cheap artificial or natural inert support while the substrate and products are continuously flowing with the mobile phase. The methods and modes used for immobilized cell systems include, among others, attachment by crosslinking agents, entrapping in a polymer gel matrix and adsorption to a preformed support. The last system allows better results than the others because of a continuous replacement of senescent cells by new growth of intact cells in the biofilms that are naturally produced and adsorbed to the support.

This chapter describes a method of immobilization by adsorption of whole cells with emphasis on applying this technique to the production of lactic acid, an important chemical for the food, agricultural, and pharmaceutical industries *(2)*, using adherent phenotypes of lactic acid bacteria.

2. Materials

1. Basal medium (MB) *(3)* medium: 10 g/L yeast extract, 0.05 g/L $MgSO_4 \cdot 7H_2O$, 2.5 g/L $(NH_4)_2HPO_4$, and 0.005 g/L $MnSO_4 \cdot H_2O$. Glucose concentration can be varied according to the experiment. Glucose and Mn^{2+} must be autoclaved separately (*see* **Note 1**).
2. Organism: Different strains of lactic acid bacteria may be used for lactic acid production with immobilized cells (*see* **Note 2**).
3. Reactor: The reactor consists of a water-jacketed glass column (3.5 cm internal diameter × 14 cm height) filled with the desired support material. The water-jacketed reactor is connected to a temperature-controlled water bath. The column must be provided with an external loop having a vessel equipped with a stirring device, a pH electrode, and a pH control unit, through which the medium can be continuously recycled in the reactor. A recycling ratio of 50:1 may be employed using the same ascentional velocity (1.5 cm/min) in the reactors in all the assays. Ammonium hydroxide (1.5 M or higher) is used to control pH. The system can be assembled according to the scheme in **Fig. 1**.
4. Pumps: The media is fed to the column using a variable-speed pump (Watson Marlow 101 U/R). This pump is capable of delivering the medium at rates as low as 0.001 mL/min using a 0.5-mm bore silicone tubing. Additional pumps are used for withdrawing the spent broth and for recycling.
5. Supports: Different inert supports can be used for the immobilization of cells (*see* **Note 3**):
   a. Poraver®: Porous foam glass particles (having a mean diameter of 2–4 mm, a porosity of 60%, and a density of 0.2 g/cm³ (a gift from Dennert Poraver GmbH, Postbawer Heng, Germany). Wash thoroughly with tap water, select the beads without air bubbles inside, and dry at 150°C.
   b. Polyurethane foam: This foam is used for domestic purposes and is widely available. The usual porosity is 91%. Cut blocks of 10 × 10 × 10 mm, boil in water, wash five times with distilled water, dry, and use to fill the reactors.
   c. *Luffa cylindrica*: This is another form of sponge also used as a bath sponge. Cut into blocks and clean the pieces by boiling in water. Dry and use as support.
6. Solutions for lactate determination
   - Buffer glycine–hydrazine (SIGMA).
   - NAD (Yeast : N-7004 98%).
   - L (+)-Lactate (L-1750 Sigma 95%).
   - D(–)-Lactate (N-7004 Sigma 98%).
   - L(+)-nLDH (Beef heart, 1000 units/mL).
   - (–)-nLDH (*L. leishmanii*: L-3888 Sigma 300 U: 0.33 mL; 5.3 mg protein/mL; 290 U/mg protein).

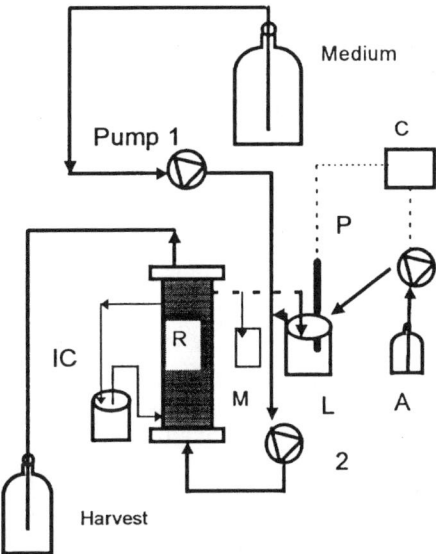

Fig. 1. Schematic diagram of the upflow packed-bed reactor. A, alkali; 1, feed pump for complete medium; 2, recycling pump; 3, alkali pump; C, pH controller; F, air filter; IC, heat exchanger; L, external pH controller device; M, sampling port; R, reactor; P, pH electrode.

3. Methods
3.1. Variables to Determine

The kinetic and yield parameters can be evaluated by the classical chemostat equations; for this reason, it is necessary to know the different volumes in the reactor.

1. Active volume of reactor. To determine the active volume, the support material is placed in the interior of the reactor without empty zones to obtain good packing bed. The active volume will be the volume of water that is fed through the bottom of the reactor from a graduated device such as a buret with appropriated connections that permits the reactor to fill up to the operating volume. The real active volume must include the volumes of all tubing, fittings components and the neutralization vessel if used.
2. Porosity. As described in the above paragraph, the porosity of the filling material that will support the microbial growth can be determined by filling a measured volume of a glass recipient with the inert support that will be placed in the final form in the operating reactor. The volume of liquid added to this device compared to the volume of column initially considered will represent the porosity of the material and can be calculated as

   (Volume of column occupied by the support/Volume of liquid added) × 100

3. Flow control. Install a flow meter in the portion of tubing between the feeding reservoir and the fed pump. Before sterilization interrupt the feeding line and connect a T-tube with an appropriate connection provided with a clamp to a graduated tube (10-mL pipet). This pipet can be used to calibrate the flow rate by measuring the time needed to deliver a given amount of liquid expressed in milliliters per minute or milliliters per hour.
4. Ascentional velocity. It is necessary to know the cross-sectional area through which the liquid will flow in the upflow mode through the reactor. The real cross-section area will be the geometrical surface multiplied by the porosity factor of the porosity of the support.

The ascending velocity expresses the volume of medium that is allowed to flow through the area section of the reactor per unit of time. This will be determined from the section of passage and the net flow of liquid ascending through the column. The units are centimeters per hour and will be calculated as

$$\text{Ascending velocity} = \frac{\text{Flow } (cm^3/h)}{\text{Area section } (cm^2)}$$

5. Recycle ratio. The operation of the reactor requires an external recycling loop, similar to the agitation of a chemostat, for liquid homogenization and to avoid pH gradients and dead zones. It is necessary for adequate control of the operation to know the ascending velocity, which depends on the recycle ratio (*see* **Note 4**). This ratio can be calculated, from the flow through the external loop (*see* **item 3**) and referring it to the actual feed flow rate. This ratio relates the times of the recycling flow with respect to the feed flow rate. It will be calculated as

Recycle ratio = Feeding flow (mL/h) / Flow through external loop (mL/h)

Example:

Reactor glass volume:	$(\pi/4) (3.5 \text{ cm})^2 (14 \text{ cm}) = 134.6 \text{ cm}^3$
Active volume (Poraver):	$(134.6 \text{ cm}^3) \times (60\%) / 100 = 80.7 \text{ cm}^3$
External loop and vessel volume, leave:	19.3 cm^3
Total active volume of:	100 cm^3
For a dilution rate of	$D = 0.13 \text{ h}^{-1}$
The feeding flow is	$(0.13 \text{ h}^{-1}) \times (100 \text{ mL}) = 13.2 \text{ mL/h}$
If recycle ratio is:	50:1
The external flow is:	$50 \times (13.2 \text{ mL/h}) = 660 \text{ mL/h} = 660 \text{ cm}^3/\text{h}$
Calculating the sectional area:	$(12.25 \text{ cm}^2) (0.6) = 7.35 \text{ cm}^2$
The ascentional velocity is:	$(660 \text{ cm}^3/\text{h}) / (7.35 \text{ cm}^2) = 98.8 \text{ cm/h} = 1.5 \text{ cm/min}$

3.2. Preparation of the Reactor

Once the reactor and silicone tubes and fittings are properly cleaned, the system is assembled according to the scheme in **Fig. 1** (*see* **Notes 5 and 6**).

First, fill the reactor with the desired support and connect and test the neutralizing device with the pH electrode, the agitator, and all the pumps. Fill the whole system from the bottom of the reactor with distilled water and sterilize all the system for 60 min at 121°C (1 atm).

Verify all the connections and replace the water with culture medium. Hold the system at the controlled temperature and pH for at least 24 h prior to inoculation with the selected strain previously grown at the temperature to be used for culturing.

3.3. Microorganism

The adhesive strain could be obtained in a chemostat. An example of a continuous culture yielding the adhesive phenotype of *Streptococcus salivarius* ssp. *thermophilus* involve 7 d at 0.057 h^{-1}, 6 d at 0.12 h^{-1}, 15 d at 0.19 h^{-1}, and 3 d at 0.26 h^{-1}. When the biofilm is visually detected, the dilution rate is increased stepwise up to 0.96 h^{-1}. For the final selection of the adhesive phenotype, use dilution rates between 0.96 and 6.0 h^{-1} *(4)*.

3.4. Cultivation of Microorganism

1. Aseptically inoculate the reactor in the bottom of the column, use a sterile syringe with 6% (v/v) of an overnight culture of adhesive strain and leave for 24 h in a batch mode.
2. Set pH to 5.5, temperature to 42°C, and stirring speed to 150 rpm in the external vessel.
3. Determine the feed flow and start the continuous culture by pumping the medium in upflow mode through the reactor at the lowest dilution rate.
4. Growth is monitored by the absorbance at 620 nm using a spectrophotometer. Obtain a sample from the sampling device and determine the density of free cells by measuring optical density. The dry weight of cells is determined from a definite volume of broth collected and filtered through a weighed 0.22-µm cellulose acetate membrane filter, washing with 0.9% saline solution and drying at 105°C.
5. Centrifuge the sample and determine residual sugar and total lactic acid in the supernatant by high-performance liquid chromatography. Determine glucose with an enzymatic kit (Sigma), and L(+) and D(−)-lactic acid according to the procedure described below.
6. Steady states must be maintained for a least 15 generations prior to any sampling. The samples must be withdrawn along five retention times at each dilution rate (*see* **Note 8**).
7. Change the dilution rate starting at the slowest retention time and repeat **steps 4** and **5**.
8. At the end of the process, determine the amount of cells immobilized on the support by the difference between the dry weight of the immobilized biocatalyst (a portion of support and cells) and that of the support prior to use. Dry the material at 105°C for 72 h.

Table 1
Determination of L (+)-Lactic Acid

Solutions	Sample	Blank	Standard curve			
			Standard 1	Standard 2	Standard 3	Standard 4
Buffer glycine–hydrazine	2.5 mL	2.5 mL	2.5 mL	2.5 mL	2.5 mL	2.5 mL
NAD+ (40 mM)	0.2 mL	0.2 mL	0.2 mL	0.2 mL	0.2 mL	0.2 mL
H$_2$O bidistilled	—	0.2 mL	0.150 mL	0.1 mL	0.050 m.	0
L(+)– Lactate (2 mM) (5)	[200 µL]	—	0.050 mL	0.1 mL	0.150 mL	0.2 mL
L(+) nLDH	20 µL	20 µL	20 µL	20 µL	20 µL	20 µL

Note: Allow to stand at 37°C for 30 min. Wavelength: 334–365 nm.

Table 2
Determination of D(–)-Lactic Acid

Solutions	Sample	Blank	Standard curve			
			Standard 1	Standard 2	Standard 3	Standard 4
Buffer glycine–hydrazine	3 mL	3 mL	3 mL	3 mL	3 mL	3 mL
NAD+ (40 mM)	0.2 mL	0.2 mL	0.2 mL	0.2 mL	0.2 mL	0.2 mL
H$_2$O bidistilled	—	0.2 mL	0.175 mL	0.150 mL	0.1 mL	0.050 mL
D(–) Lactate (10 mM) (5)	[200 µL]	—	0.025 mL	0.050 mL	0.1 mL	0.150 mL
D(–) nLDH	20 µL	20 µL	20 µL	20 µL	20 µL	20 µL

Note: Allow to stand at 37°C for 60 min. Wavelength: 334–365 nm.

Table 3
Dilution Factor of Samples to Determine

Lactate in each liter (g)	Dilution with water	Dilution factor
< 0.1	—	1
0.1–1	1+9	10
1–10	1+99	100
> 10	1+999	1000

3.5. Stereospecific Lactic Acid Determination

L(+)-and D(–)-lactate dehydrogenases are used to determine the concentrations of each isomer of lactic acid (*see* **Note 7**), in which the reaction products are pyruvate and NADH. The increase in NADH, as measured by the change in extinction at 340 (334, 365) nm, is directly proportional to the amount of lactate (**Tables 1** and **2**).

Samples of culture medium stored at –20°C are stable for long periods. Dilute a portion of the sample with distilled water using an appropriate dilution factor (**Table 3**).

4. Notes

1. In practice, it is often more convenient to sterilize the component parts separately and then reassemble them carefully before use. Sufficient medium for a whole experiment should be sterilized.
2. Adhesive variants can be selected in a chemostat according to the method described by Ragout et al. *(4)*. The attachment of the cells could be facilitated by growing in a mixed culture with a biofilm-forming organism such as *Streptomyces viridosporus (6)* and also the adsorption could be facilitated by use of a cationic polymer, polyethyleneimine *(7)*.
3. If the chosen support has a lower density than water, a good packed bed with a perforated stopper is needed.
4. To maintain the ascending velocity constant, it is necessary to change the flow through the external loop each time the feed flow rate is changed.
5. Depending on the configuration of the reactor design, samples may be taken from either the overflow system or from the recycling loop (**Fig. 1**). Care should be taken to avoid introducing contamination.
6. The pH electrode can be placed in the top of the column design, and adjustments to pH can be done automatically in the recycle loop.
7. Isomers of lactic acid can be determined using techniques described by Gutmann and Walhlefeld *(8)* for L(+)-lactate and Gawehn and Bergmeyer *(9)* for D(−)-lactate. These methods permit a stereospecific determination.
8. In a continuous-culture system, for every volume of fresh medium entering on the reactor, an equal volume of spent medium leaves. The volume of the reactor can be controlled by a simple overflow system in which the entering volume causes the discharge of the same volume of medium through the top by increasing the pressure in the system. In the same way, the volume can be controlled by a tube placed in the top of the reactor that draws off the excess liquid to the preset level, in this last operation mode, an air filter may be added at the top of the column.

References

1. Cassidy, M. B., Lee, H., and Trevors, J. T. (1996) Environmental applications of immobilized cells: a review. *J. Ind. Microbiol.* **16,** 79–101.
2. Kharas, G. A., Sanchez-Riera, F., and Severson, D. K. (1994) Polymers of lactic acid, in *Plastic from Microbes: Microbial Synthesis of Polymer Precursors* (Mobley D. P., ed.), Carl Hanser Verlag, Munich, pp 93–137.
3. Bruno-Bárcena, J. M., Ragout, A. L., Siñeriz, F. (1998) Microbial physiology applied to process optimization: lactic acid bacteria, in *Advances in Bioprocess Engineering II* (Galindo, E. and Ramírez, O. T., eds.), Kluwer Academic, Amsterdam, pp. 97–110.
4. Ragout, A., Siñeriz, F., Kaul, R., Guoqiang, D., and Mattiasson B. (1996) Selection of an adhesive phenotype of *Streptococcus salivarius* subsp. *thermophilus* for use in fixed-bed reactors. *Appl. Microbiol. Biotechnol.* **46,** 126–131.

5. Ragout, A. (1988) Estudio de la Fisiología de bacterias lácticas de interés industrial crecidas en cultivo continuo, Ph.D. Thesis. University of Tucumán, Tucumán, Argentina, pp. 66–68.
6. Dimerci, A., Pometto, A. L., III, and Johnson K. E. (1993) Lactic acid production in a mixed-culture biofilm reactor. *Appl. Environ. Microbiol.* **59,** 203–207.
7. Guoqiang, D., Kaul, R., and Mattiasson, B. (1992) Immobilization of *Lactobacillus casei* cells to ceramic material pretreated with polyethyleneimine. *Appl. Microbiol. Biotechnol.* **37,** 305–310.
8. Gutmann, I. and Walhlefeld, A. W. (1974) L(+) lactate. Determination with lactate dehydrogenase and NAD, in *Methods of Enzymatic Analysis, 2nd ed.* (Bergmeyer, H. U. and Gawehn, K. eds.), Academic, New York, pp. 1464–1468.
9. Gawehn, K. and Bergmeyer, H. U. (1974) D(–) lactate. Determination with lactate dehydrogenase and NAD, in *Methods of Enzymatic Analysis, 2nd ed.* (Bergmeyer, H. U. and Gawehn, K. eds.), Academic, New York, pp. 1469–1475.

36

Molecular Characterization of Yeast Strains by Mitochondrial DNA Restriction Analysis

Maria Teresa Fernández-Espinar, Amparo Querol, and Daniel Ramón

1. Introduction

The characterization of yeasts at the strain level is of relevance from an industrial point of view because numerous yeast strains belong to the natural flora of commercial fermented foods and beverages (bakery products, cheeses, cold meats, wines, and beers) and take part in fermentation processes. The addition of active dry yeasts is increasingly used to ensure the final quality of these products. In this sense, a fast and easy method is required for quality control of dry yeast production to ensure that the final product is identical to the original strain, and for control of the fermentation process to ensure that the latter really is conducted by the inoculated yeast. In addition, the importance of yeasts in the spoilage of foods is increasingly recognized (1). In consequence, accurate identification of these yeasts in the spoiled product as well as the detection of the origin of contamination are of great interest for food laboratories as an inevitable part of the control of the process and quality assurance. Between the techniques employed for yeast strain characterization, the mitochondrial DNA restriction analysis appears as one of the most sensitive method to differentiate among strains (2). This technique has been used successfully for strain differentiation (3–5). However, mitochondrial DNA purification is too complex and time-consuming to be used in industrial applications.

Recently, a new mitochondrial restriction analysis method that does not require previous isolation of mitochondria or purification of mitochondrial DNA has been developed in our laboratory for *Saccharomyces cerevisiae* strains (2,6). This technique has been used successfully by other authors to characterize strains of other yeast species *(7–9)*. This method relies on the different GC content of the nuclear and mitochondrial genomes, approximately 40% *(10)* and 20% *(11)*, respectively. In this way, G+C-rich restriction sites are much more frequent in the yeast nuclear DNA than in the mitochondrial genome because of its higher G+C-content. Then, digestion with G+C-rich restriction endonucleases yields an overdigestion of the nuclear DNA, which is cut in a few larger fragments. After electrophoresis separation, mtDNA restriction fragments are easily visualized in the upper part of the gel.

Here, we describe the application of this method to the characterization of a large number of wild and commercial yeast strains in a fast and reliable way. These considerations and the simplicity of the equipment required confer to the method a great industrial interest.

2. Materials

1. Culture medium: YEPD (yeast extract peptone dextrose): 1% yeast extract, 2% peptone, 2% glucose. The medium must be autoclaved at 121° C for 20 min.
2. Organisms: Different yeast strains may be used for mitochondrial DNA restriction analysis. The strains are maintained on Petri dishes with YEPD solid medium (*see* **Note 1**).
3. Restriction enzymes: Total DNA is digested with specific restriction endonucleases that recognize 4-bp or 5-bp GC-rich restriction sites; 4 bp: *Alu*I, *Cfo*I, *Hae*III, *Hpa*II, *Mae*I, *Mae*II, *Mbo*I, *Mvn*I, *Sau*3AI, *Rsa*I, *Taq*I; 5 bp: *Dde*I, *Hinf*I, *Ita*I, *Mae*III, *Sau*96I, *Scr*FI. (*See* **Note 2**.)
4. Microcentrifuge for 1.5-mL tubes.
5. Microwave oven. The microwave is used to prepare agarose gels by melting the agarose in the electrophoresis buffer until a transparent solution is obtained. A boiling-water bath or a hot plate can be used instead.
6. Electrophoresis apparatus. The equipment is composed of the following:
   * Plastic tray to pour the agarose slab gel.
   * Electrophoresis tank with a lid containing shielded electrical connections.
   * Combs to generate different number of wells of different sizes that allow to analyze many samples simultaneously.

   A power supply is necessary to plug the electrophoresis apparatus and allow the migration of the DNA toward the anode. The voltage applied is usually 100 V. If the gel is running overnight, the voltage should be 20–25 V.
7. Ultraviolet transilluminator for DNA visualization in the gel.
8. Molecular weight standard. Phage λ DNA cut with the restriction endonuclease *Pst*I (can be purchased from commercial sources). The set of DNA size markers (in bp) obtained is   11509, 5080, 4649, 4505, 2840, 2577, 2454, 2443, 2140, 1980, 1700, 1159, 1092, 805, 516, 467, 448, 339, 265, 247, and 210.
9. Solutions and buffers:
   a. Solutions for DNA extraction:
      * Solution I (1 M sorbitol, 0.1 M EDTA, pH 7.5).

Characterization of Yeast Strains

- Solution II (50 mM Tris–HCl, 20 mM EDTA, pH 7.4).
- 10% sodium dodecyl sulfate (SDS).
- 5 M potassium acetate: To 60 mL of 5 M potassium acetate, add 11.5 mL of glacial acetic acid and 28.5 mL of H$_2$O.
- TE (10 mM Tris–HCl, and 1 mM EDTA, pH 7.5).
- Zymolyase (2.5 mg/mL in solution I).
  b. 10X Gel loading buffer (50% glycerol in water, 0.25% bromophenol blue, and 0.25% xylene cyanol FF).
  c. Electrophoresis buffer: 50X TAE concentrated stock solution (242 g Tris base, 57.1 mL glacial acetic acid, and 100 mL 0.5 M EDTA, pH 8.0).
  d. Ethidium bromide stock solution (10 mg/mL in water). Use gloves and handle with care, ethidium bromide is mutagenic.

3. Methods
3.1. Microorganism Growth

Yeast cells must be grown for 12–16 h in 5 mL of YEPD with moderate agitation (200–250 rpm). Incubation temperature depends on the yeast species under study (usually between 25°C and 28°C) (*see* **Note 1**).

3.2. DNA Extraction

The total DNA is obtained by the method of Querol et al. *(2)* as follows:

1. Spin down the cells from the culture in a centrifuge at 3500 rpm for 5 min, wash with 5 mL of sterile distilled water, and resuspend in 0.5 mL of solution I.
2. Transfer the cells to a 1.5-mL microcentrifuge tube, to which 0.02 mL of a solution of Zymolyase 60 (2.5 mg/mL) is added. Incubate the tubes at 37°C for 30–60 min to obtain spheroplasts (*see* **Note 3**).
3. Pellet the spheroplasts in a microcentrifuge at maximum speed for 1 min and resuspended them in 0.5 mL of solution II. After resuspension, 0.05 mL of 10% SDS is added and the mixture incubated at 65°C for 30 min (*see* **Note 3**).
4. After incubation, add 0.2 mL of 5M potassium acetate and place the tubes on ice for 30 min (*see* **Note 3**).
5. Centrifuge the tubes at maximum speed in a microcentrifuge for 15 min (*see* **Note 4**).
6. Transfer the supernatant to a fresh microcentrifuge tube and precipitate the DNA by adding the same volume of isopropanol.
7. After incubation at room temperature for 5 min, centrifuge the tubes at room temperature for 10 min and wash the DNA with 70% ethanol, vacuum dry and dissolve in 50 µL of TE, pH 7.4.

3.3. Mitochondrial DNA Digestion

Yeast DNA (8–10 µL) is digested with restriction enzymes according to the instructions of the supplier in a final volume of 20 µL. Add to the reaction mixture 2 µL of a 500-µg/µL solution of pancreatic RNase DNase-free.

3.4. Mitochondrial Restriction Fragment Separation

Restriction fragments are separated by horizontal 0.8% agarose gel electrophoresis using 1X TAE electrophoresis buffer. Two microliters of 10X gel loading buffer must be added to each sample before loading. Marker DNA (λ PstI) should be loaded into slots on both sides of the gel.

3.5. Staining DNA in Agarose Gels and Mitochondrial DNA Patterns Visualization

After electrophoresis is completed (until the bromophenol blue and xylene cyanol FF contained in the gel loading buffer have migrated the appropriate distance through the gel), stain the gel by soaking it in a solution of the fluorescent dye ethidium bromide (0.5 µg/mL) at room temperature for 30–45 min. The background fluorescence caused by unbound ethidium bromide can be reduced by soaking briefly the stained gel in water. The DNA is then visualized on an ultraviolet transilluminator.

4. Notes

1. The method was initially developed to differentiate *S. cerevisiae* strains in order to solve enological problems that can substantially affect the quality and organoleptical characteristics of the wine. However, this method can be used also in the differentiation of industrial yeast strains involved in biotechnological process other than wine fermentation (beer brewing, bakery products, dairy products, cold meat products, active dry yeast production, etc.) as well as in the characterization of spoilage yeasts.
2. Because the GC content of mitochondrial DNA of the strains tested can be different depending on the species, it is recommended to test beforehand a great number of restriction enzymes in order to use those showing an extensive polymorphism for further analysis .
3. When the strains analyzed are *S. cerevisiae*, incubation at 37°C with solution I and zymolyase can be reduced to 20 min. In the same way, the amount of SDS used can be reduced from 0.05 to 0.013 mL. Incubation for 5 min at 65°C with the SDS and then 10 min on ice are enough.
4. If a microcentrifuge refrigerated is available, centrifugation at 4°C is recommended to avoid the resuspension of the precipitate.

References

1. Deak, T., Beuchat, L. R. (1996) *Handbook of Food Spoilage Yeasts* (Clydesdale, F. M., ed.) CRC, New York.
2. Querol, A., Barrio, E., and Ramón, D. (1992) A comparative study of different methods of yeast strain characterization. *System. Appl. Microbiol.* **15,** 439–446.
3. Lee, S.Y. and Knudsen, F. B. (1985) Differentiation of brewery yeast strains by restriction endonuclease analysis of their mitochondrial DNA. *J. Inst. Brew.* **91,** 169–173.

4. Dubordieu, D., Sokol, A., Zucca, P., Thalouarn, P., Datte, A., and Aigle, M. (1987) Identification des souches de levures isolées de vins par l'analyse de leur ADN mitochondrial. *Connaiss. Vigne Vin.* **21,** 267–278.
5. Vezinhet, F., Blondin, B., and Hallet, J. N. (1990) Chromosomal DNA patterns and mitochondrial DNA polymorphism as tools for identification of enológical strains of *Saccharomyces cerevisiae. Appl. Microbiol. Biotechnol.* **32,** 568–571.
6. Querol, A., Barrio, E., Huerta, T., and Ramón, D. (1992) Molecular monitoring of wine fermentations conducted by active dry yeast strains. *Appl. Environ. Microbiol.* **58,** 2948–2953.
7. Romano, A., Casaregola, S., Torre, P., and Gaillardin, C. (1996) Use of RAPD and mitochondrial DNA RFLP for typing of *Candida zeylanoides* and *Debaryomyces hansenii* yeast strains isolated from cheese. *System. Appl. Microbiol.* **19,** 255–264.
8. Guillamón, J. M., Sanchez, I., and Huerta, T. (1997) Rapid characterization of wild and collection strains of the genus *Zygosaccharomyces* according to mitochondrial DNA patterns. *FEMS Microbiol. Lett.* **147,** 267–272.
9. Ibeas, J. I., Lozano, I., Perdigones, F., and Jiménez, J. (1996) Detection of Dekkera–Brettanomyces strains in Sherry by a nested PCR method. *Appl. Environ. Microbiol.* **62,** 998–1003.
10. Barnett, J. A., Payne, R. W., and Yarrow, D. (1990). *Yeasts, Characteristics and Identification.* Cambridge University Press, Cambridge, UK.
11. Bernardi, G., Piperno, G., and Fonty, G. (1972) The mitochondrial genome of wild type yeast cells. I. Preparation and heterogeneity of mitochondrial DNA. *J. Mol. Biol.* **65,** 173.

37

Selection of Yeasts Hybrids Obtained by Protoplast Fusion and Mating, by Differential Staining, and by Flow Cytometry

Tohoru Katsuragi

1. Introduction

Chapter 38 deals with methods for hybridization of yeasts in different genera by protoplast fusion and for selection of hybrids by double-fluorescence staining and use of a micromanipulator. This chapter describes a similar but automated and rapid procedure for selection of hybrids; flow cytometry and cell sorting are used *(1)*. Mating, instead of fusion, followed by flow sorting as is done to obtain hybrids *(2)* also is described, together with some modifications.

2. Materials

2.1. Protoplast Fusion

Unless otherwise noted, materials used are as in Chapter 38.

1. Fluorescent dyes. Green and red fluorescent dyes are used in combination. (Approximate peak emission wavelengths, measured by fluorescence spectrophotometry with excitation at 488 nm, are given for the case of *Saccharomyces diastaticus* 251, stained as described later.) Use dyes at the concentration of 50 µg/mL, unless otherwise noted. Dyes are obtained from Molecular Probes, Inc. (Eugene, OR).
    a. Green fluorescence. Fluorescein isothiocyanate isomer I (FITC; 520 nm, strong fluorescence).
    b. Red fluorescence. Tetramethylrhodamine isothiocyanate (TRITC; 570 nm, moderate fluorescence). Also usable are rhodamine 6G (R6G; 550 nm, strong fluorescence), 4-chloro-7-nitrobenz-2-oxa-1,3-diazole (NBD; 5 µg/mL; 550 nm, moderate fluorescence); ethidium bromide (560 nm, weak fluorescence); rhodamine B (570 nm, moderate fluorescence); Nile Red (580 nm, weak fluorescence).

From: *Methods in Biotechnology, Vol. 14: Food Microbiology Protocols*
Edited by: J. F. T. Spencer and A. L. Ragout de Spencer © Humana Press Inc., Totowa, NJ

2. The sheath fluid used is osmotic buffer autoclaved and filtered through an in-line membrane filter (pore size, 0.2 µm).
3. Flow cytometer. Use a FACStar II flow cytometer (Nippon Becton-Dickinson K.K., Tokyo) with an argon-ion laser (488 nm; at 50 mW) and 530 (±30)- and 580 (±42)-nm band-pass filters for measurement of the green fluorescence signal (FL1) and the red fluorescence signal (FL2) respectively (*see* **Note 1**).
4. Collection tubes. Glass test tubes measuring 10 × 85 mm are wrapped in foil and sterilized in an autoclave (*see* **Note 2**).

2.2. Mating

Materials and models of combination are those of Bell et al. *(2)*.

1. Strains: Genetic markers are not needed for selection of hybrids. Parents must be haploid and of the opposite mating type. Strains of *Saccharomyces cerevisiae* used by Bell et al. *(2)* were from the Australian National Reference Laboratory in Medical Mycology (AMMRL), Sydney, and are used here as models.
   a. Strain PB1 AMMRL 57.9 (*MAT*a *trp1 his1 MAL6T::lacZ*).
   b. Strain SMC19-A AMMRL 57.11 (*MAT*α *MAL2-8^c MAL3 leu1 SUC3*).
2. Media
   a. Rich medium contains 2% (w/v) glucose, 1% (w/v) peptone, 0.5% (w/v) yeast extract, and 0.3% (w/v) KH_2PO_4. Solidify, if desired, with 2% (w/v) agar.
   b. YNB, as a minimum medium, contains 0.67% (w/v) yeast nitrogen base (Difco Laboratories, Detroit, MI).
3. Fluorescent dyes: Green and orange fluorescent dyes are used in combination. Cell Tracker (CT) probes and another dye can be obtained from Molecular Probes and Sigma Chemical Co. (St. Louis, MO), respectively. Typically used at the concentration of 10 µ*M* (*see* **Note 3**).
   a. Green fluorescence: CT-Green (CMFDA); CT-BODIPY; CT-Yellow-Green.
   b. Red fluorescence: CT-SNARF; PKH-26.
4. The sheath fluid used is Coulter IsoFlow (Beckman Coulter).
5. Flow cytometer. Bell et al. *(2)* used a FACSCalibur instrument (Becton Dickinson, Lane Cove, NSW, Australia) with an argon-ion laser (488 nm; set at 15 mW) (*see* **Note 4**). Linear gains are used for parameters of the forward scatter channel (FSC) and the side scatter channel (SSC). Fluorescence is monitored at the green fluorescence signal FL1 of 525 ± 5 nm and the red fluorescence signal FL2 at 575 ± 5 nm.
6. Collection tubes: Sterile tubes are used (*see* **Note 5**).

3. Methods

3.1. Protoplast Fusion

Protoplast fusion is done as in Chapter 38 except that fluorescence labeling is always done after protoplasting, unlike labeling with 4,6-diamino-2-phenylindole (DAPI), which was used to label cells before protoplasting. (DAPI is not used here because it does not fluoresce strongly with the 488-nm laser.)

3.2. Mating

3.2.1. Growth

1. Grow two yeast strains overnight on rich medium.
2. Collect, wash, and suspend cells in saline at the concentration of 4×10^8/mL.

3.2.2. Fluorescence Labeling

One yeast strain is stained with a green fluorescent dye; the other is stained with an orange fluorescent dye.

1. Prepare a stain mixture (1 mL) containing 975 µL of YNB medium, 25 µL of cell suspension, and 1 µL of a 10 mM stock solution of one of the CT dyes. As a rule, incubate the mixture at 30°C for 45 min in the dark. The working concentration will be 10 µM (*see* **Note 6**). For staining with PKH-2, *see* **Note 7**.
2. Centrifuge the mixture for 1 min at 12,000g and remove all supernatant that can be withdrawn with a pipet.
3. Wash the cells to remove unbound dye by three repetitions of the following procedure: suspend the pellet in 1 mL of YNB medium, incubate the suspension at 30°C for 30 min to allow unbound dye to mix with the medium, and obtain the cells by centrifugation as in **step 2**.

3.2.3. Mating Procedure

1. Suspend stained cells in 500 µL of 10X YNB medium.
2. Transfer two parents to a 1.5-mL Eppendorf tube and vortex the tube.
3. Pellet the cells by centrifugation for 1 min at 12,000g (*see* **Note 8**).
4. Place the tube in the dark at 20°C for 16 h.
5. Vortex the tube for 10 s to disrupt aggregates immediately before flow cytometry.

3.3. Flow Cytometry

1. Align and calibrate the instrument each day before use, following the instructions from the manufacturer and with Coulter Flow-Check fluorospheres (Beckman Coulter) as standard beads.
   a. For sorting with a jet-in-air cell sorter (FACStar II), fit the deflexion-plate cassette on the instrument.
   b. Adjust the pressure of the sheath fluid line to set the rate of flow.
   c. Adjust the power of the laser.
   d. Arrange the dichroic mirrors and optical filters.
   e. Focus the laser beam so that it precisely illuminates a small area.
   f. Monitor cytometric parameters (FSC, SSC, FL1, and FL2). (*See* **Note 9**.)
   g. Adjust the gain voltages of the detectors (voltages for the photomultiplier tubes) to the sensitivity needed for the standard beads.

h. For sorting with a jet-in-air cell sorter, set drop-drive frequency and adjust voltage applied to the deflexion plates so that the deflexion spots come in the opening of the tubes for right and left sorting.
    i. Estimate the sort delay time.
2. Place the sample in the cytometer and allow the sample to flow into the sensing area.
3. Adjust the pressure of the sample line so that the flow rate is no more than 1000 cells (events) per second during analysis.
4. Change the threshold to a level just below that of the lowest signals for yeast cells.
5. Inspect the cytometric parameters in dot plots on an FSC-versus-SSC cytogram and an FL1-versus-FL2 cytogram (*see* **Note 10**).
6. Accumulate data for at least 10,000 cells.
7. Define a sort window on the FL1-versus-FL2 cytogram so that the cells in the window have both FL1 and FL2 fluorescence. Select these as hybrids (*see* **Note 11**).
8. Change the flow rate to about 300 cells (events) per second or fewer for sorting.
9. Place collection tube(s) in the cytometer.
10. Allow more cells to flow through and sort possible hybrids into the collection tube.

3.4. Recovery of Hybrids

3.4.1. Hybrids from Fusion

Sorted cells are recovered from the collection tube by being spread on regeneration plates, which are incubated as usual.

1. Spread the contents of the collection tube on plates of regeneration agar.
2. Cover the agar plates by overlaying each with about 10 mL of molten soft agar.
3. Incubate for several days.
4. Transfer cells from a colony on a regeneration plate onto a plate or a slant of growth medium.
5. Incubate the cells for a few days more until cultures are thick enough for preparation of a master plate or of stock cultures.
6. Examine colonies of the sorted populations for confirmation of fusion by, in this case, incubating the cells on starch.

3.4.2. Hybrids from Mating

1. Sort about 20,000 cells.
2. Collect the cells by centrifugation.
3. Suspend the cells in 2 mL of YNB medium.
4. Place the cell suspension in the cytometer for a second round of sorting as in **step 1** (*see* **Note 12**).
5. Collect the cells by centrifugation in **step 2** (*see* **Note 13**).
6. Use sorted cells to inoculate a plate of rich medium and incubate the plate for 48 h for preparation of a master plate or of stock cultures.
7. Examine colonies of the sorted populations for confirmation of the mating by formation of colonies on an appropriate minimum medium and by the expression of *lacZ*, in this model case.

4. Notes

1. The Becton Dickinson FACStar II is a jet-in-air cytometer that uses the droplet-catcher sorting principle.
2. Glass tubes are used for droplet sorting with the jet-in-air cytometer to prevent static electricity from building up on the surface of tubes. They already contain a small amount of sheath fluid to protect cells from drying.
3. Dye stock solutions are prepared at the concentration of 10 mM in dimethyl sulfoxide, dispensed as single-use portions, and stored at $-50°C$ until use.
4. The Becton Dickinson FACSCalibur is a flow-in-capillary cytometer that uses the catcher-tube sorting principle; the Becton Dickinson FACSort instrument is similar. A flow-in-capillary cytometer always yields a high volume of sort fluid in the collector.
5. Centrifuge tubes are convenient because they can be used to pellet cells in the next step. Plastic tubes must be coated with bovine serum albumin before use, as described in the user's manual from Becton Dickinson to prevent adsorption of cells onto their surface.
6. The concentration to be used depends on the yeast strains and the physiological conditions of the cells and must be decided after preliminary staining of the cells with dyes in a range of concentrations.
7. PKH-26 did not stain either strain of yeast cells used here brightly under the various incubation conditions described in the manufacturer's instructions with the dye concentration in the range from 2–8 μM.
8. This procedure is to encourage mating by bringing the cells close together.
9. FSC is used as the trigger for acquisition of cytometric data. A threshold is tentatively set at the appropriate level.
10. For cancellation, if needed, of red and green components of the FL1 (green) fluorescence and FL2 (red) fluorescence, respectively, fluorescence compensation is done for each measurement with the operating software.
11. So that doublets or multicell clusters are not selected as hybrids, a sort window is defined also on the FSC-versus-SSC dot plot by the region with low FSC and low SSC values.
12. The model described here *(2)* is an example of the concentration of hybrids by repeated sortings: before sorting, 33% of the population were hybrids; after one round of sorting, the proportion of hybrids increased to 70%; after the second round, it increased to 96%.
13. If the number of sorted hybrids is expected to be a dozen or so as in rare-mating hybridization, concentrate sorted cells, using membrane filters with 0.2-μm pores.

Acknowledgments

The author thanks Professor Yoshiki Tani of the same school for collaboration in screening of microorganisms with cell sorters. He is indebted to Dr. John F. T. Spencer, coeditor of this book, for continuous encouragement throughout preparation of this chapter. Thanks also are due to Ms. Caroline Latta, Osaka City University Medical School, for critical reading of the manuscript.

References

1. Katsuragi, T., Kawabata, N., and Sakai, T. (1994) Selection of hybrids from protoplast fusion of yeasts by double fluorescence labelling and automatic cell sorting. *Lett. Appl. Microbiol.* **19,** 92–94.
2. Bell, P. J., Deere, D., Shen, J., Chapman, B., Bissinger, P. H., Attfield, P. V., et al. (1998) A flow cytometric method for rapid selection of novel industrial yeast hybrids. *Appl. Environ. Microbiol.* **64,** 1669–1672.

38

Selection of Hybrids by Differential Staining and Micromanipulation

Tohoru Katsuragi

1. Introduction

Microorganisms have been used intensively by the food industry as well as by pharmaceutical and other chemical industries. A great number of global companies take advantage of fermentation processes in the manufacture of a spectrum of useful products. Many of these processes involve yeasts, which have been used in the production of foods and alcoholic drinks long before the existence of microorganisms was known. Biochemical engineering of microbes has made possible the industrial production of a wide variety of metabolites on a commercial scale at low cost. These metabolites are a tiny proportion of the compounds produced in nature by living organisms. Progress in biological chemistry, genetics, and molecular biology has been put to practical use in the manufacture of foods. Such recent advances have led to the development of a number of commercial products. However, as with other agricultural products, consumers do not necessarily accept so-called engineered or recombinant foods, which involve genetically engineered organisms and enzymes in their production, and the processes themselves may not be generally acceptable because genetic engineering itself may be viewed with doubt. One example is the use of antibiotic resistance marker genes, considered undesirable when incorporated into industrial yeasts because the products are released into the environment. In addition, genetic markers introduced into parent strains for convenience in the selection of mutants may cause practical problems; a new gene once introduced may produce unnecessary metabolites and thereby disrupt the metabolism of the cell, use energy that could be used for growth, or consume a substance that would otherwise be a source of the desired product

(1). For these reasons, classical genetics and conventional techniques such as mating and protoplast fusion are preferable for use in strain improvement of yeasts. Mating followed by selection is a traditional technique still useful for strain improvement *(2,3)*. The use of protoplast fusion to produce hybrid yeasts is a way to solve the problem when a mating reaction does not occur. Protoplast fusion also makes possible the construction of new strains that are hybrids between different species or even different genera.

When two yeast strains with complementary genetic markers (e.g., nutritional complementation) are fused, their hybrids can be identified by growth on a selective medium. Many industrial yeast strains, however, lack selectable genetic markers, making identification of hybrids by genetic complementation difficult: when parent strains and hybrids can grow on the same medium, there is no easy way to distinguish between hybrids (those formed between the two parents) and nonhybrids.

In this chapter an effective technique for selection of hybrids obtained by protoplast fusion without use of genetic markers is described; instead, two fluorescent dyes are used for labeling *(4)*. Under a fluorescence microscope, hybrids are detected as cells that have been stained by both of the dyes, and these cells are selected by use of a micromanipulator *(5)*.

2. Materials

1. Strains: Genetic markers are not needed for selection of hybrids and there is no restriction as to what combinations of strains are chosen as parents. A model has been selected for description of the protocol here.
    a. *Saccharomyces cerevisiae* AKU 4111, the *sake* yeast Kyokai No. 7; a strain for ethanol fermentation.
    b. *Saccharomycopsis fibuligera* IFO 0106; a strain for saccharification.
2. Media
    a. Growth medium: 1% (w/v) glucose, 0.5% (w/v) peptone, and 0.3% (w/v) yeast extract at pH 5.5. Dispense as 5-mL portions in 16-mm test tubes or 100-mL portions in 500-mL Erlenmeyer flasks. Solidify, if desired, with 2% (w/v) agar.
    b. Regeneration medium: Growth medium containing 0.6 M KCl. For agar plates, dispense a 20-mL portion in a 90-mm Petri dish. Store regeneration agar in a flask until it is used to cover agar plates inoculated with protoplasts.
3. Buffers
    a. Buffer is 60 mM potassium phosphate buffer, pH 7.5, unless otherwise noted.
    b. Osmotic buffer: 0.6 M KCl as osmotic stabilizer in the buffer (*see* **Note 1**).
4. Fluorescent dyes obtained from Molecular Probes, Inc. (Eugene, OR) through Wako Pure Chemical Industries (Osaka). Keep prepared solutions in the dark or wrapped in aluminum foil to protect contents from light. They can be stocked for a few days in the refrigerator. Use a B-excitation cassette for microscopic observation.
    a. Fluorescein isothiocyanate isomer I (FITC), 50 µg/mL, dissolved in osmotic buffer; fluoresces green.

Selection of Hybrids 343

   b. 4,6-Diamino-2-phenylindole (DAPI), 150 μg/mL, dissolved in buffer; fluoresces yellow. For orange fluorescence (which, however, is not strong), use tetramethylrhodamine isothiocyanate (TRITC) or rhodamine 6G (R6G) (*see* **Note 2**).
5. Protoplasting solution. Dissolve Zymolyase-20T (Kirin Brewery Co., Tokyo) to the concentration of 5–100 μg/mL in osmotic buffer containing 10 mM 2-mercaptoethanol (*see* **Note 3**).
6. Fusion solution: 33% polyethylene glycol (PEG) 6000 (Wako) and 50 mM $CaCl_2$ in osmotic buffer.
7. Soft agar: 0.5% (w/v) agar in osmotic buffer. Pour over microscope slide glass to fix cells on it (*see* **Note 4**).
8. Microscope. Use a Diaphot 300 inverted microscope equipped with a Narishige single-handed micromanipulator (Nikon Corp., Tokyo) (*see* **Note 5**).
9. Micromanipulation tips: Commercially available or make tips from glass capillary tubes using a kit made by Narishige and obtainable from Nikon. Use a tube to make two tips with openings of about 5–15 μm with the use of the puller in the kit. Make opening flat, if desired, using a grinder.

3. Methods

Techniques used are sterile and all procedures are done at room temperature unless otherwise stated. Cells are collected by centrifugation at 2000g for 5 min when necessary. Starting with the staining procedure, containers are wrapped in foil when being handled in a lighted room.

Standard methods are presented here and modifications may be needed for particular microorganisms.

3.1. Cultivation

Culture at 30°C, with gentle shaking when broth is used. Collect cells by centrifugation with three steps of washing with buffer. Growth is expressed as the optical density at 610 or 660 nm. Dilute cultures 1:10 with saline and measure their absorbance. Multiply absorbance by the dilution rate to give the optical density.

The preliminary measurements described here often are helpful.

Optical density vs cell density curve. Dilute cell suspensions with saline, if necessary, for measurement of their optical density, and count under a microscope with use of a hemacytometer. Plot the optical density versus cell density to obtain a calibration curve.

Growth curve. Withdraw samples during cell growth and measure the optical density to plot the growth curve.

3.2. Fluorescence Labeling

1. Grow two yeast cultures in growth medium for 10 h to the early or mid-exponential growth phase (*see* **Note 6**).
2. Add 1 volume of DAPI solution to 9 volumes of one of the yeast cultures. Incubate the mixture for 4 h without shaking the culture.
3. Collect cells from the other yeast culture, wash them, suspend them at a density of 10^6–10^8/mL in osmotic buffer, and treat them as described in **Subheading 3.3.** Stain with FITC.

3.3. Protoplasting

1. Collect cells treated with DAPI, wash them with osmotic buffer, and resuspend them at a density of 10^6–10^8/mL.
2. Incubate the suspensions at 30°C for 10 min. Collect cells and suspend them at a density of 10^9/mL in protoplasting solution. Incubate the mixture for 30 min more. Collect the protoplasts, wash them with an osmotic buffer, and suspend them in the same buffer (*see* **Note 7**).
3. Collect (from **Subheading 3.2., step 3**) the protoplasts to be stained and suspend them in FITC solution. Incubate the mixture at 30°C for 3 min and then immediately collect the cells, wash with osmotic buffer, and suspend in the bufffer at a density of about 10^9/mL (*see* **Note 8**).
4. Estimate the rate of formation of protoplasts (*see* **Note 9**).

3.4. Protoplast Fusion

1. Mix portions of about 0.1 mL each of two suspensions of stained protoplasts prepared from different yeasts so that the protoplasts from the two parents are present in roughly equal numbers.
2. Collect protoplasts and suspend them in 0.2 mL of fusion solution.
3. Incubate the mixture at 30°C for 60 min. Immediately collect all cells, including the protoplasts that fuse together, by centrifugation at 3000*g* for 5 min and suspend them in osmotic buffer.

3.5. Microscopy and Micromanipulation

1. Melt soft agar in a flask, cool it to 40°C in a water bath, and use it to cover the surface of a regeneration plate that contains selected protoplasts, transferred as described next.
2. Transfer treated protoplasts from the suspension described in **Subheading 3.4., step 3** onto soft agar on a microscopic slide glass and spread them by gentle shaking of the slide from side to side (*see* **Note 10**).
3. Leave the slide undisturbed for the liquid portion to seep into the agar. The protoplasts will then be set firmly on the surface.

Selection of Hybrids

4. Fix a micromanipulator tip to the manipulator hand.
5. Set the slide on the stage and inspect the specimen.
6. Select a protoplast that has both green and red fluorescence as a probable fusant.
7. Aspirate a small volume of osmotic buffer inside the capillary and then introduce a bubble of air, which will separate the buffer inside from the droplet containing the protoplast to be picked up.
8. Move the micromanipulator tip over the slide glass into the field of view by coarse manipulation in the X and Y directions with the joystick. Bring the end of the tip up to the selected protoplast by fine manipulation. Touch the agar near the protoplast with the tip by manipulation in the Z direction.
9. Press the agar softly in the Z direction, causing fluid to ooze from the agar bed.
10. Detach the protoplast from the agar by aspirating it into the tip together with the fluid on the agar.
11. Detach the micromanipulator tip holder by hand and carefully move the tip to a position above a regeneration plate. Transfer its contents (the protoplast suspension together with osmotic buffer) onto the regeneration plate by pipetting.
12. Select more protoplasts as probable fusants by repetition of **steps 6–11**. If desired, use a new micromanipulator tip each time.

3.6. Regeneration

1. Cover the regeneration plate with about 10 mL of molten soft agar from a stock flask.
2. Incubate the plates at 30°C.
3. Within a week, check for regenerated yeast cells, seen as colonies on the plate.
4. Transfer individual cells separately from a colony onto growth agar when preparing master plates or slants for characterization (*see* **Note 11**).

4. Notes

1. Sorbitol may be used at the concentration of 1.0 M as an osmotic stabilizer instead of 0.6 M KCl.
2. Both TRITC and R6G can be used in the same way and at the same time as FITC. Staining must be done after protoplasting. *See* **Subheading 2** of Chapter 37.
3. The enzyme is used to remove yeast cell walls. Zymolyases (from *Arthrobacter luteus*) have β-1,3-glucan laminaripentaohydrolase and β-1,3-glucanase activities. The concentration is decided in preliminary experiments. Zymolyase 6000 (Kirin Brewery) is typically used at the concentration of 0.4 mg/mL. Other lytic enzymes of other origins or other commercial sources (e.g., snail gut juice or Novozyme SP234 from Novo-Nordisk Bioindustry, Chiba, Japan) and combinations of such enzymes can be used and may be more effective.
4. Soft agar can be prepared so as to have an even surface on the slide glass when die-cast with a plastic spacer plate 1.0 mm thick and measuring 20 × 40 mm on the edge, with a circular or square hole about 10 mm across. (The spacer can be made by hand from an acrylic plate with a cutter and reamer.) The spacer plate is

put on a slide glass, a slight excess of molten agar is poured into the hole, and another slide glass is placed on the spacer plate carefully so that air is not trapped under it. The agar gels quickly. Just before use of the plate, the slide glass on top is slid off sideways. Agar plates with a higher concentration of agar (e.g., 2%) can be useful also; sample protoplasts set rapidly and firmly on the surface because the agar bed readily absorbs small amounts of liquid.
5. A dark room may not be needed, although FITC and some other dyes are labile to light. Ordinary microscopes also work well with a micromanipulator if there is enough space between the specimen on the stage and the objective for insertion of a micromanipulator tip. The problem is greater with objectives giving greater magnifications. Objectives with a long focal length can be used to solve the problem.
6. The phase of exponential growth is suitable for the protoplast fusion of yeasts. Yeasts usually enter that phase after culture for 8–16 h. The time needed can be estimated from the growth curve, and preliminary experiments will show whether the early or mid-exponential phase is more suitable for fusion.
7. The protoplasts can be stored for a few days at 5°C in a solution of 1.2 M KCl containing 50 mM $CaCl_2$.
8. When on the protoplast membrane, FITC is unstable to light, as it is in solution, and careful handling is necessary. In addition, FITC sometimes may be eluted; in that case, the protoplasts are not washed thoroughly at this point but are washed only once.
9. The rate of protoplast formation (P) is estimated as follows. Equal volumes of suspensions of untreated and treated cells are spread separately on growth medium. Plates are incubated for 4–7 d. Colony counts will give the viable cell count for both cultures. The count for untreated cells on growth medium (m) minus that for treated cells on growth medium (n) gives the number of whole protoplasts, which is divided by m to give P:

$$P = (m - n)/m$$

Usually, the rate of regeneration of stained protoplasts is 75% or more. The rate of regeneration of fused protoplasts is not known, but it probably is almost the same as this percentage, changing in proportion.
10. The appropriate volume of the protoplast suspension to be transferred for isolation of the protoplasts from each other is decided on beforehand. Sometimes dilution with osmotic buffer is needed.
11. In the model used here for illustration, alcohol fermentation and saccharification are the desired activities for incorporation into the fusants. Such fusants can grow on starch (without sugars) and produce alcohol.

Acknowledgments

The author thanks Dr. Takuo Sakai, Department of Applied Biological Chemistry, Osaka Prefecture University, Sakai, Osaka (presently, Faculty of Agriculture, Kinki University, Nara) for providing information about

Selection of Hybrids

experimental details and for notes about these techniques. He is indebted to Dr. John F. T. Spencer, editor of this book, for continuous encouragement throughout preparation of this chapter. Thanks also are due to Ms. Caroline Latta, Osaka City University Medical School, for critical reading of the manuscript.

References

1. Spencer, J. F. T. and Spencer, D. M. (1983) Genetic improvement of industrial yeasts. *Annu. Rev. Microbiol.* **37,** 121–142.
2. Spencer, J. F. T. and Spencer, D. M. (1977) Hybridization of non-sporulating and weakly sporulating strains of brewer's and distiller's yeasts. *J. Inst. Brew.* **83,** 287–289.
3. Evans, I. H. (1990) Yeast strains for baking: recent developments, in *Yeast Technology* (Spencer, J. F. T. and Spencer, D. M., eds.), Springer-Verlag, Berlin, pp. 13–53.
4. Sakai, T., Koo, K.-I., Saitoh, K., and Katsuragi, T. (1986) Use of protoplast fusion for the development of rapid starch-fermenting strains of *Saccharomyces cerevisiae. Agric. Biol. Chem.* **50,** 297–306.
5. Sakai, T., Kanemoto, T., and Inoue, H. (1988) New method for selection of hybrid strains in protoplast fusion of yeast. *Chem. Express (Osaka)* **3,** 743–746.

39

Flotation Assay in Small Volumes of Yeast Cultures

Sandro Rogério de Sousa, Maristela Freitas Sanches Peres, and Cecilia Laluce

1. Introduction

Flocculation is a naturally occurring process of reversible aggregation (cell–cell aggregation) of yeast cells *(1)*, whereas flotation is defined as microbial enrichment in foams *(2)*, which can be carried out either by induced flotation (with addition of flotation agents) or spontaneous flotation (without addition of flotation agents). The spontaneous flotation of *Saccharomyces cerevisiae* was first described in 1991 by Gehle et al. *(3)*, who correlated flotation with the ability of the strain DSM 2155 to form cell aggregates. In 1996, Palmieri and collaborators *(4)* showed that flocculation and flotation were separate phenomena, by studying a new floating strain of *Saccharomyces cerevisiae*, which was highly hydrophobic but not flocculant. Batch flotation has been used by a number of authors for the separation of yeast cells *(4,5)*, bacteria *(6)*, algae *(7)*, bacterial spores, vegetative cells *(8)*, recovery of minerals, coal and crude petroleum *(9)*, deinking recycled pulp paper and de-oiling water *(10,11)*.

Measurement of flotation parameters has been described for yeast *(3,4)*: (1) cell recovery (percentage of the total cells from the medium present in the foam; **Subheading 3.3., step 1**), (2) enrichment factor (cell concentration in the liquid foam or "yeast cream" divided by the initial concentration in the medium before flotation; **Subheading 3.3., step 2**), expressing how much higher the concentration of the cells in the liquid foam is, compared with that in the initial medium, and (3) concentration factor (cell concentration liquid foam divided by the concentration in the residual medium after flotation; **Subheading 3.3., step 3**) expressing how much greater the concentration of cells in the liquid foam is, compared with that in the residual medium left in the column after flotation.

From: *Methods in Biotechnology, Vol. 14: Food Microbiology Protocols*
Edited by: J. F. T. Spencer and A. L. Ragout de Spencer © Humana Press Inc., Totowa, NJ

During the injection of air into the medium, a volume of liquid foam (Vs) proportional to the initial volume of medium (Vp) is formed, leading to a correlation factor $Vr. Vp^{-1}$ (factor of volumetric changes; **Subheading 3.3., step 4**), which varies from zero (Vr is zero, when all the medium spills out from the column) to 1 (residual medium $Vr = Vp$, when no foam is formed or a small volume of foam that does not spill out of the column is obtained). As part of the medium is drained off the culture as a liquid phase entrapped among the air bubbles, the cell concentration in the foam is dependent on the bubble size (lower or higher water content among the bubbles) generated by the culture *(12)* and the affinity of the cells for the air bubbles. Thus, the flotation efficiency (FLT_{EFF}; **Subheading 3.3., step 5**) is another parameter, which expresses the percentage of the total cells in a volume Vp of medium (added to the column initially) and transferred to the foam by flotation. The flotation efficiency varies with the changes in the affinity of the cells for the air bubbles and/or the volume of medium entrapped among the bubbles. The flotation ability the yeast cells changes with the total hydrophobicity of cell surface *(4)*, which can be measured by the affinity of cells (bacteria or yeasts) to the organic phase formed when *p*-xylene is added to the cell suspension *(4,13)*.

The flotation process is well known for being relatively economical and has been applied to separations in large volumes of liquid, such as in recovery of minerals and waste residue treatments. A great deal of research is lacking in this area for a better understanding of the flotation phenomenon and its application to industrial microbial processes in which complex media are used.

Simple and fast assays are helpful for physiological and genetic studies when great numbers of measurements are required. In this chapter, a simple system for measurement of batch flotation and the operation of the system for the determination of flotation parameters in cultures of yeast cells are described.

Comparing the flotation ability among yeast strains, effects of determination of medium composition on flotation, and choice of a suitable frother or collector can be obtained using this simple system.

2. Materials

1. Flotation system as described in **Subheading 3.1., step 1**.
2. Yeast culture grown for 12–48 h in a chemically synthetic medium, described by Brown et al. *(15)* and modified by Palmieri et al. *(5)*, prepared as follows (g/L of distilled water):
    a. Carbon source: glucose, 20 g.
    b. Salt mixture (150 ml added to 1 L medium): Ammonium sulfate (20.8 g), KH_2PO_4 (13.3 g), $MgSO_4 \cdot 7H_2O$ (3.6 g), NaCl (0.7 g), and $CaCl_2$ (0.6 g). Sterilize the mixture in autoclave for 20 min at 120°C (1 atm pressure).
    c. Trace element mixture (104 µL/L added to each liter of medium): H_3BO_3 (0.1 g), $MnSO_4$, $ZnSO_4 \cdot 7H_2O$ (0.7 g), $FeCl_3 \cdot 6H_2O$ (0.5 g), KI (0.1 g), and $CuSO_4 \cdot 5H_2O$

Flotation Assay in Small Volumes of Yeast Cultures

(0.1 g). Lower the pH of the mixture to values around 1.0, with the addition of HCl solution (to avoid salt precipitation during storage at 4°C) and sterilize in autoclave for 20 min at 120°C (1 atm pressure).

d. Growth factor mixture (2 mL mixture added to each liter of medium): myo-inositol (15 g), calcium pantothenate (2.4 g), pyridoxine–HCl (2.4 g), thiamine–HCl (8.4 g), and biotin (0.18 g). Sterilize the solution by filtration (filter membrane of 0.22 μm pore size from Millipore, GSWP 025 00).

3. Stationary culture of yeast cells grown at 30°C (10^7–10^8 cells/mL after 12–48 h) in a rotary shaker operated at 150 rpm.
4. Filtration membranes (filter membrane of 0.65 μm pore size from Millipore, DAWP 025 00).
5. Infrared lamp.

3. Methods

3.1. Setup of the Flotation System

1. Attach a glass compartment to a 50-mL graduated cylinder (30 cm high and 2.5 cm inner diameter) at a distance of 20 cm from the bottom of the column, so that the upper part of the glass cylinder is located inside the compartment for foam collection, as shown in **Fig. 1**.
2. Connect a flowmeter with a standard valve (Cole-Parmer, P-03216-55) at the air line between the compressor and the air outlet tube.
3. Attach a microporous sparger (Cole-Parmer, model H-01919-59) at the terminal end of the air outlet tube, located at the bottom of the glass cylinder (**Fig. 1**).

3.2. Operation of the Flotation System

1. Add 10–12 mL culture of known biomass (Vp) to the flotation column *(14)*.
2. Aerate for 2 min at 0.36–0.48 L/min airflow rate *(14)*.
3. Add 2–3 drops of amyl-alcohol to the foam before measurement of the biomass in the liquid foam (Cs).
4. For measurement of biomass in the initial culture (Cp), liquid foam (Cs), and residual volume of the medium (Cr) resulting from flotation, harvest and wash the cells with water by filtration, and dry up the Millipore membranes at constant weight under an infrared lamp for 3 h.

3.3. Calculation of the Flotation Parameters

1. Total cell recovery or flotation yield obtained as follows *(4)*:

$$FLT(\%) = [(Cp-Cr)/Cp^{-1}] \times 100$$

2. Enrichment factor calculated as follows *(3)*:

$$\text{Enrichment factor (undimensionless)} = CsCp^{-1}$$

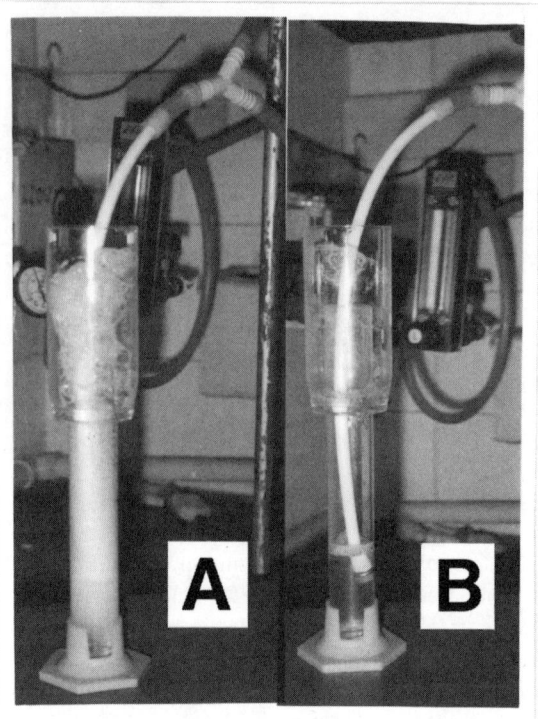

Fig. 1. Apparatus used for flotation: **A** = during flotation and **B** = after flotation. A "yeast cream" (0.1–0.2 mL), containing practically all the cells present in 11 mL of medium (0.5 mg/mL initial concentration), was collected in the foam compartment located at the top of the column. The flotation time was 2 min. (From ref. *14*.)

3. Concentration factor calculated as follows *(3)*:

    Concentration factor (undimensionless) = $CsCr^{-1}$

4. Factor of volumetric changes, calculated as follows:

    Factor of volumetric changes (undimensionless) = $VrVp^{-1}$

5. Flotation efficiency or FLT_{EFF} obtained as follows:

    $FLT_{EFF}(\%) = [(Cp-Cr)Vr(CpVp)^{-1}] \times 100$

where $FLT_{EFF}(\%)$ = [(biomass in the foam obtained from an initial volume of medium equivalent to Vr) × (total biomass in the medium before flotation or $CpVp)^{-1}$ × 100.

An efficiently cell recovery (≥90%) by flotation requires a highly hydrophobic cell wall (80%) *(4)*; and flotation parameters are dependent on both the initial cell concentration and flotation time *(14)*.

Acknowledgments

The authors thank Dr. J. T. F. Spencer for the opportunity of having a chapter included in this book and Ms. Doris Barnes for the careful reading of the manuscript.

References

1. Calleja, G. B. (1987) Cell aggregation, in *The Yeasts, Vol. 2,* 2nd ed., (Rose, A. H. and Harrison, J. S., eds.), Academic, London, pp. 165–238.
2. Dognon, A. and Dumonte, A. (1941) Concentration et séparation des microorganismes par moussage. *CR Soc. Biol.* **135,** 884–887.
3. Gehle, R., Sie, T. L., Kramer, T., and Schurgerl, K. (1991) Continuous cultivation and flotation of *Hansenula polymorpha* and *Saccharomyces cerevisiae* in an integrated pilot plant reactor–flotation column-system. *J. Biotechnol.* **17,** 147–154.
4. Palmieri, M. C., Greenhalf, W., and Laluce, C. (1996) Efficient flotation of yeast cells grown in batch culture. *Biotechnol. Bioeng.* **50,** 248–256.
5. Hashim, M. A., SenGupta, B., Kumar, S. V., Lim, R., Lim, S. E., and Tan, C. C. (1998) Effect of air to solid ratio in the clarification of yeast by colloidal gas aphrons. *J. Chem. Technol. Biotechnol.* **71,** 335–339.
6. Grieves, R. B. and Wang, S. L. (1966) Foam separation of *Escherichia coli* with a cationic surfactant. *Biotechnol. Bioeng.* **8,** 323–336.
7. Levin, G. V., Clendenning, J. R., Gibor, A., and Bogar, F. D. (1961) Harvesting of algae by froth flotation. *Appl. Microbiol.* **10,** 169–175.
8. Boyles, W. A. and Lincoln, R. E. (1958) Separation and concentration of bacterial spores and vegetative cells by foam flotation. *Appl. Microbiol.* **6,** 327–334.
9. Hornsby, D. and Leja, J. (1982) Selective flotation and its surface chemical characteristics, in *Surface and Colloid Science* (Matijevic, E., ed.), Plenum, New York, vol. 12, pp. 217–313.
10. Huls, B. J. (1994) When innovations occur: their effect on a large mining company, in *Innovations in Mineral Processing* (Yalcin, T. ed.), Acme, Sudbury, Canada, pp. 1–14.
11. Finch, J. A. (1995) Column flotation: a selected review–Part IV: novel flotation devices. *Min. Eng.* **8:** 587–602.
12. Gaden, E. L., Jr., and Kevorkian, V. 1956 Foams in chemical technology. *Chem. Eng.* **63,** 151–184.
13. Rosenberg, M., Gutnick, D., and Rosenberg, E. (1980). Adherence of bacteria to hydrocarbons: a simple method for measuring cell surface hydrophobicity. *FEMS Microbiol. Lett.* **9,** 29–33.
14. De Sousa, S. R. and Laluce, C. (2000) Flotation assays in small volumes of yeast cultures. *Biotechnol. Lett.*, in press.
15. Brown, S. W., Sugden, D. A., and Oliver, S. G. (1984) Ethanol production and tolerance in grand and petite yeast. *J. Chem. Tech. Biotechnol.* **34B,** 116–120.

40

Obtaining Strains of *Saccharomyces* Tolerant to High Temperatures and Ethanol

Maristela Freitas Sanches Peres, Sandro Rogério de Sousa, and Cecilia Laluce

1. Introduction

Strains tolerant to high temperatures, ethanol levels, and high concentrations of sugar are highly desirable for a fermentation process. The maintenance of a high degree of viability during the operation of a process, biomass storage, and pauses between batches is fundamental. Fermentations at 35–40°C or higher have the advantage of facilitating ethanol recovery, leading to significant savings on operational costs of refrigeration.

Mutants may arise either during cell-cycle progression (natural mutants) or adaptation to the process in response to environmental stresses *(1–5)*. Prolonged exposure to different types of stresses, such as anaerobiosis *(6)*, nutrient starvation at the stationary phase *(7)*, high levels of ethanol, low pH, and high temperature occur either during process operation or over long periods of biomass storage. Alterations in cell morphology induced by "fusel" alcohols have also been described *(8)*, and chromosomal rearrangements have been observed for *Saccharomyces cerevisiae* in a continuous process, when growth was limited by the concentration of organic phosphate *(9)*.

Mutation associated with positive selection can improve the process (enrichment with the improved mutant) and the original strain is supplanted by variants and/or contaminants from the environment in a matter of time. Thus, a continuous process or batch process with cell reuse could be a source of useful industrial strains. Industrial fermentations for ethanol and beverage production do not occur under aseptical conditions. A screening program for isolation and evaluation of yeasts for fuel ethanol production was carried out in our laboratory from 1985 to 1993, supported by the Fundação Banco do Brasil (funds from Brazil Bank, proc. FIPEC/1-1337-4 and proc. FBB/10-1078-2). Yeasts are usually submitted to nutritional and temperature stresses during the process operation at a number of Brazilian alcohol plants. Methods of strain screening and measurement of tolerance were compared and thermotolerant strains, which were also tolerant to high concentrations of sugar and external ethanol, were obtained *(10–15)*. At present, Brazilian alcohol plants are already using yeasts isolated during operation of the fermentation process.

Screening of yeasts for glucose fermentation at 40°C showed that strain 62 of *S. diastaticus* was the most thermotolerant yeast out of a total of 65 yeast strains from various genera *(16)*. Strain 62 was found capable of completely utilizing 15% glucose at 40°C, producing 6.38% ethanol (w/v) and showing good cell viability (80.5%) after 24 h fermentation. Similarly, an Indian distiller's yeast (*Saccharomyces*) produced 6.9–7% ethanol (v/v), fermenting 15% glucose *(17)*. Thermotolerant yeasts (genus *Saccharomyces*) isolated from Brazilian alcohol plants (10 thermotolerant strains out of a total of 460 isolates), showed higher ethanol yields at 40°C than the baker's yeast strains used as starters *(10,13)* for a long period. The screening procedure, used for samples collected from Brazilian alcohol plants, was based on measurements of the levels of ethanol and viability at the end of the fermentation of 25% sucrose in a rich medium at 40°C *(10)*. Under such stressful conditions, the screening was oriented toward the selection of thermotolerant yeasts, which were also tolerant to high ethanol and sugar concentrations.

The assay of the glycerol-3-phosphate dehydrogenase (not included among the procedures described in this chapter) seems to be helpful in the evaluation of strains tolerance to ethanol. Cells of strain MT-2, which accumulated less ethanol and biomass in the medium during growth, also showed high levels of intracellular glycerol-3-phosphate dehydrogenase *(15)*. Similarly, higher levels of glycerol were also obtained for thermotolerant strains of *S. cerevisiae*, during the fermentation of grape juice, as compared to levels attained using nonthermotolerant strains *(18)*. The activity of invertase attached to the cell wall is important when sucrose is the carbon source. It is also dependent on increases in temperature *(15)*.

1.1. Assessment of Thermotolerance

Ethanol and temperature have several effects on yeast and fermentation processes *(13,14,19)*. Levels of ethanol formation and viability are good indicators of thermotolerance in fermentation processes operated at high cell density; growth is another important indicator of tolerance at low cell density. Thermotolerance of the isolates (*S. cerevisiae*) obtained from Brazilian alcohol plants were assessed on the basis of maintenance of viability, biomass, and ethanol formation, during fermentation of 15% sugar-cane syrup (total reducing sugar, w/v) at 40°C *(13)*. Isolates that were thermotolerant under agitation conditions, became relatively thermosensitive (low viability) in fermentations carried out under nonagitated conditions *(13)*. The biorectors used in the alcohol plants for fuel ethanol production are not operated under agitated conditions. The ethanol yields depend on the interactive effects established between sugar concentration in the medium and temperature. As a result of increases in the total reducing sugar from 22.5% to 33.5% (nonagitated flasks), inhibition of ethanol formation and decreases in viability at 40°C were observed for the thermotolerant isolates, and the highest levels of ethanol were reached at 35°C (optimum temperature at high sugar concentration) *(13)*. Thus, ethanol formation at temperatures above 35°C are greatly limited by increases in sugar concentration, as well as by increases in the temperature.

A thermotolerant strain should first be evaluated in repeated batch fermentations in 20-mL test tubes before the final evaluation in a 5- to 30-L fermentor, as described in **Subheading 3.8.**, **step 1**, for the assessment of thermotolerance during cell recycling in repeated batch fermentations. **Figure 1** shows the changes in viability and ethanol formation, during the fermentation of 10% sugar-cane syrup at 40°C by the thermotolerant yeast 78 I (*S. cerevisiae*) in repeated batches (for 24-h cycles). The viability was high (ranging from 75–90%), the final ethanol was 5–6% (v/v), and the residual sugar was low (0.5–1.5 %, w/v). Growth was observed at the end of each 5-h cycle (early stationary phase), leading to rises in biomass from 11.0- to 14.5-mg/mL initial cells to 13.5- to 15.5-mg/mL final cells. The final biomass was constant after five fermentation cycles and the residual sugar was low ([0.5–1.0] ± 0.5%, w/v).

1.2. Assessment of Alcohol Tolerance

The inhibitory effect of ethanol on growth is greater than on fermentation and dependent on the strain and composition of the medium. It is also well known that the accumulation of ethanol in yeast cultures leads to decreases in viability, growth, and ethanol formation. The effects of ethanol on yeasts are still not extensively understood and some methods used to determine ethanol tolerance in yeasts are not universally accepted. A variety of definitions and methods used to determine ethanol tolerance under different conditions have been proposed for alcohol tolerance by a number of authors *(15,20–23)*. Methods for the measurement of internal ethanol are controversial and the information obtained provides little help for the understanding of the toxic effects *(24)*.

Production of high levels of ethanol requires a strain that should be tolerant to both high sugar concentration and levels of external ethanol, for yeast cells are continually being exposed to external ethanol during fermentation

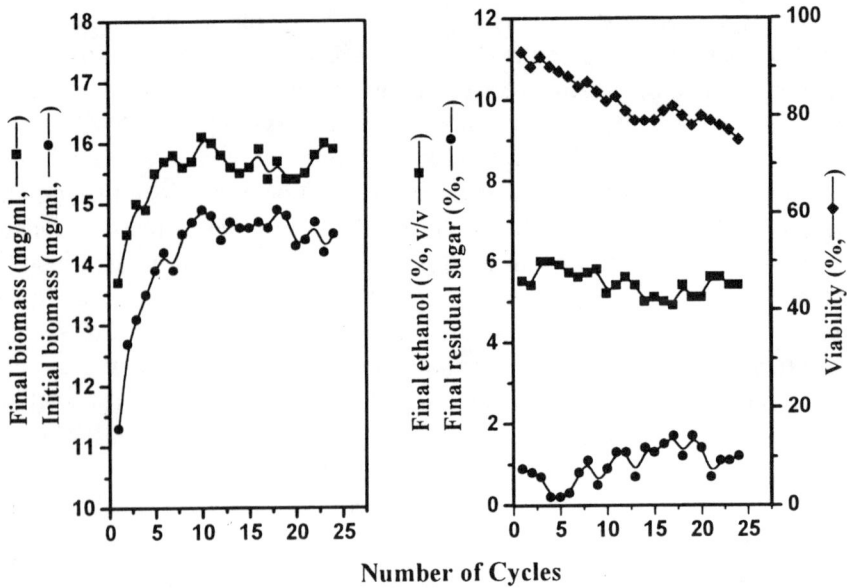

Fig 1. Effect of cell recycling (isolate 78I) in batch fermentations at 40°C using diluted sugar-cane syrup as the raw material containing 10% total reducing sugar, and 11- to 14-mg/mL initial cells at the beginning of the fermentation cycles.

processes. Thus, the resistance of *S. cerevisiae* strains to high levels of added ethanol (7% external ethanol, v/v) in a medium of high sugar concentration (19–20%, w/v) seems to be a reasonable criterion for the screening of alcohol-tolerant strains at 30°C and low initial cells in a rich medium *(15)*. As growth and fermentation take place simultaneously, biomass formation was another indicator of alcohol tolerance in the medium containing 19–20% initial sucrose and external ethanol added initially. Constant levels of final ethanol (16.5–20.3% ethanol produced plus the amount added) and total sugar consumption were observed, when ethanol was added to the medium over a range from 2–8% or 2–6% (v/v) initial concentration, depending on the strain. Inhibition of growth was observed for wall ethanol additions over a range from 2–9% initial ethanol *(15)*. In **Fig. 2**, strain MT-2 shows the accumulation of 17.3% final ethanol (produced and added to the medium, v/v), when the initial reducing sugar was 19% in the sugar-cane syrup and the ethanol added initially was 6% (v/v). Final ethanol increased, reaching the highest level (17.3%, v/v) at 6% added ethanol, whereas the produced ethanol was constant (11.3%, v/v), and a total sugar consumption observed over a range from 2–4% (v/v) ethanol added initially. The measurements of viability, ethanol, and residual sugar were made at the early stationary phase for each concentration of added ethanol, and the

Saccharomyces Tolerant to High Temperatures and Ethanol

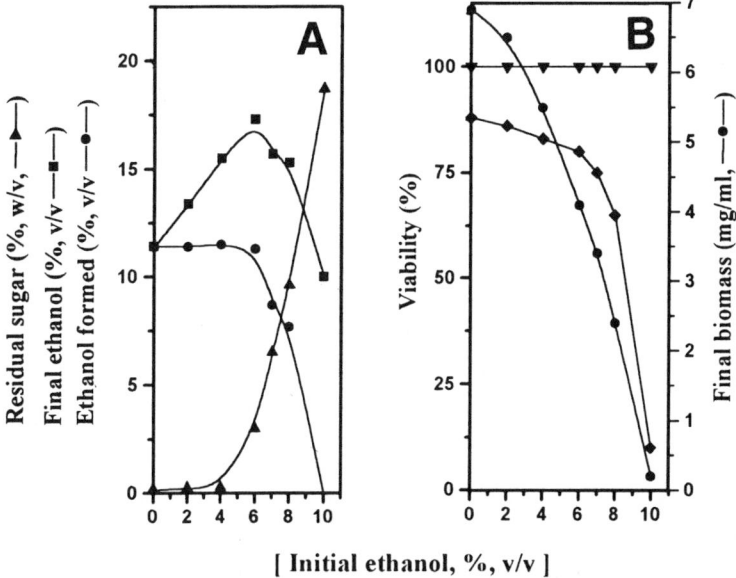

Fig 2. Effect of the concentration of ethanol (%, v/v), added initially to the medium, on fermentation (**A**) and biomass formed by strain MT-2 (**B**). Residual sugar and final ethanol (produced plus added initially) were measured at the late stationary phase (after 7 d). Symbols for viability: (▼), 2 d; (◆), 7 d.

biomass formation was greatly decreased above 6% ethanol added initially to the medium.

Losses in viability are minimized by the decreases in growth and fermentation rates, when the levels of external ethanol rises during fermentation or is added to the medium *(13,15,19)*. Thus, more ethanol can be obtained at the expense of less biomass formation in fermentation process, when both the level of external ethanol and the temperature of process are properly adjusted. Maintenance of high levels of viability, ethanol yields, as well as the control of the rate of the entire process can be obtained by adjustment of levels of the external ethanol and temperature.

The control of aeration is important in assessing strain tolerance to ethanol and temperature. Strains that produce high levels of ethanol, and maintain relatively high viability in agitated cultures may show great losses in viability in nonagitated cultures *(14)*. In addition, cell density utilized must be taken into consideration for the obtention of reproducible data in evaluation of tolerance of the isolates to ethanol and temperature. Less growth and increased product formation have been observed with the increases in cell density, as significant changes in the cell physiology occur when the cell density is raised *(25–28)*.

2. Materials
2.1. Media

1. YPD medium for plating of samples and maintenance of cultures prepared in 2 mL vials: 1% yeast extract, 2% peptone, 2% glucose, and 2% agar. Sterilize at 120°C and 1 atm pressure for 15 min.
2. Modified YPD medium containing rose bengal and propionic acid for obtention of isolates *(29)*: Dissolve the agar (2%) by boiling the medium and adding both the rose bengal (0.003%) and propionic acid (0.19%, v/v) before pouring into the plates (for use in **Subheading 3.5., step 2**). Sterilization is not required.
3. Raw material (10% total reducing sugar, w/v) for inoculum growth (to prepare 70 mL initial medium as described in **Subheading 3.4.**): 70 mL raw material (syrup or molasses) in a 250-mL Erlenmeyer flask. Sterilize at 120° C and 1 atm pressure.
4. Raw material (25% total reducing sugar, w/v) for screening of isolates at 40°C (to prepare 15 mL initial medium as decribed in **Subheading 3.5., step 5**): 9.87 mL diluted raw material (38% syrup or molasses, v/v) at pH 4.5 (adjusted with H_2SO_4) added to a loosely capped test tube (180 mm high and 15 mm inner diameter) before sterilization (120°C and 1 atm pressure for 15 min).
5. Raw material (15% total reducing sugar, w/v) for assessment thermotolerance to 40°C (to prepare 70 mL initial medium as described in **Subheading 3.6., step 1**, and **Subheading 3.7., step 1**): Add 52.5 mL raw material (20% syrup or molasses, w/v) at pH 4.5 (adjusted with H_2SO_4) in a 250-mL Erlenmeyer flask and close with rubber stopper perforated with a needle. Sterilize at 120°C and 1 atm pressure for 15 min.
6. Raw material (10% total reducing sugar [w/v] at pH 4.5) for the repeated batch fermentations at 40°C (20 mL initial volume as described in **Subheading 3.8., step 1**): Add 10 mL raw material (20% syrup or molasses, w/v) to a 50-mL centrifuge tube and close with a rubber stopper perforated with a needle. Sterilize at 120°C and 1 atm pressure for 15 min.
7. Raw material (20% total reducing sugar, w/v), for alcohol-tolerance studies at 30°C (to prepare 70 mL initial medium as described in **Subheading 3.9., step 1**): Place 50 mL raw material (28% total reducing sugar, w/v) in a 250-mL Erlenmeyer flask closed with rubber stopper perforated with a needle. Sterilize at 120°C and 1 atm pressure for 15 min and add 4.9 mL absolute ethanol after cooling at room temperature.

Use of the raw material gives more validation to the assessment of thermotolerant yeasts and/or alcohol tolerant yeasts for industrial applications. The fermentation process is greatly dependent on the medium and its nutritional supplementation *(30–32)*. The addition of nutritional supplementation will definately change the results. Improvements (data not shown) in growth and ethanol

Saccharomyces Tolerant to High Temperatures and Ethanol

formation by strain 781 grown on sugar-cane syrup (15% total reducing sugar) at 40°C were obtained with addition of 0.05% yeast extract and other extracts (additions equivalent to the presence of 3–4% solid supplement in the medium) obtained when concentrated suspensions of wheat bran, wheat germ, and castor oil bran were autoclaved for 10 min at 120°C and 1 atm. Saponified lipids (soy bean lecithin and peanut oil) led to remarkable increases in biomass (mainly below 40°C) and final viability when the 0.05–0.5% solid precipitate resulting from ethanol evaporation (after saponification) was present in the medium.

2.2. Other Materials and Solutions

1. Filtration membrane of 0.65 μm pore size from Millipore (DAWP 025 00).
2. Filtration membrane of 0.22 μm pore size from Millipore (GSWP 025 00).
3. Infrared lamp or an oven at 105°C.
4. Analytical balance.
5. Aluminum pans (for use with membrane filters in dry weight determinations).
6. Standard hemocytometer chamber.
7. Gas chromatography operating with flame ionization detector.
8. Sorvall centrifuge (model RC-5B) operating with the rotor SS-34.
9. Eppendorf microcentrifuge (model 5415C).
10. DNS solution *(33)*: 3.5-dinitrosalycilic acid (5 g) plus 100 mL NaOH (2 *N*), solid $KNaC_4H_4O_6$ (150 g), and distilled water to make up a 250-mL final solution.
11. Reducing sugar solution *(33)*: glucose (90 mg) plus fructose (90 mg) in 100 mL distilled water.
12. Methylene blue solution (g/100 mL distilled water) *(34)*: 0.025 g methylene blue, 0.9 g NaCl, 0.048 g KCl, 0.048 g $NaHCO_3$, 1 g glucose.
13. A shaker (New Brunswick Scientific Co. Inc., model no. G-25) for propagation of cultures.

3. Methods

3.1. Biomass Determination

1. Wash the cells obtained from 1- to 2-mL cell suspension with distilled water in a membrane filter (0.65 μm pore size) and dry up the filters containing the washed cells for 3–4 h under an infrared lamp or 16 h in an oven at 105°C. Weigh the filters (triplicates of preweighed aluminum pans) after the drying of the samples and express the dry weight in milligrams per milliliter.
2. Correlate dry weights (mg/mL) (abscissa) with the absorbance units (assay–blank) of the cell suspensions at 600 nm (ordinate) to obtain the standard curve for biomass determination.
3. Convert absorbance of the unknown samples (600 nm) into biomass by using the standard curve or multiplying the inverse of the tangent from the standard curve by the absorbance of the unknown samples.

3.2. Cell Viability (34)

1. Dilute the cell suspension ($[2–4] \times 10^8$ cells/mL) and mix 0.1 mL diluted suspension with 0.9 mL of methylene blue solution (**Subheading 2.2., item 12**).
2. Load the hemocytometer with the dye solution and count the cells, as recommended by the standard counting procedure *(35)*.
3. Count yeast cells in 10 min and express the viability as a percentage of the total cells that have lost the capacity of reducing the methylene blue. The vital cells are colorless and the dead (or low-vitality cells) are blue.

Measurements of viability must be obtained at the early stationary phase for evaluation of tolerance to the temperature, while it is still high *(15)*. Decreases in viability caused by both the fermentation time and temperature, have been observed at the stationary phase, whereas the biomass and ethanol levels remain constant. However, the losses in viability are greatly diminished in the presence of ethanol added initially to the medium, when the tolerance to external alcohol is assessed under conditions described in **Subheading 3.9.**

3.3. Total Reducing Sugar (33)

1. Prepare two tubes, one containing 1.5 mL water and 1.5 mL of diluted samples, and a second tube containing 3 mL water blank.
2. In order to obtain the standard curve, prepare five tubes containing 3 mL diluted standard solution (standard reducing solution prepared as described in **Subheading 2.2., item 11**, plus water), with different concentrations of reducing sugar.
3. Add 5 mL distilled water to all tubes.
4. Add 1.5 mL DNS solution (**Subheading 2.2., item 10**).
5. Boil the tubes in a water bath for 3 min, cool rapidly, add 10 mL distilled water, and read the optical density at 540 nm against the tube containing 3 mL water (blank).
6. The difference in absorbance between the tubes containing sugar and the blank gives the amount of reducing sugar, when read on the standard curve obtained (plot of reducing sugar [mg/mL] vs absorbance units).

3.4. Inoculum Growth

1. Transfer a loopful of cells from a fresh stock culture to a 250-mL Erlenmeyer flask containing 70 mL of diluted raw material (10% total reducing sugar, **Subheading 2.1., item 3**), and cultivate for 24 h at 30°C in a shaker operating at 170 rpm).
2. Decant the cell overnight in a refrigerator at 4°C, aspirate, and discard the supernatant. The concentrated cell suspension is used as the inoculum (48–60 mg/mL).
3. Determine the biomass of the cell suspension by filtering and washing the cells in a membrane filter from Millipore or by reading the absorbance, as described in **Subheading 3.1.**

3.5. Obtention and Screening for Tolerant Isolates (10)

1. Take samples (100 mL cell suspension) from more than one bioreactor in a fuel alcohol plant, operated daily for more than 1 mo in batch process with cell reuse (at the end of each batch) or at the fermented wort outlet in a continuous culture.

2. Plate each sample (three Petri dishes) to obtain colonies on YPD medium containing Rose Bengal and propionic acid (**Subheading 2.1., item 2**). Incubate the plates for 5 d at 30°C for single colony formation.
3. Select isolates (5–10 colonies) on the basis of their size, morphology (rough or smooth), and rose bengal assimilation by the cells (white to deep-pink colonies).
4. Transfer the isolates to vials containing YPD medium and maintain the stock cultures at 4°C after 2 d growth at 30°C.
5. Inoculate the tubes containing 25% total sugar as described in **Subheading 2.1., item 4**, and the cultures (4.5×10^8 cells/mL) are incubated for 12 h at 40°C. Levels of ethanol as high as 10.0–11.5% (v/v) are obtained for the tolerant isolates.

3.6. Thermotolerance (Biomass and Ethanol Yields and Viability as Determinant Factors) in Low-Cell Density Fermentation (14)

1. Inoculate 52.5 mL of diluted raw material (containing 20% sugar-cane juice or molasses, as described in **Subheading 2.1., item 5**) with the concentrated inoculum (**Subheading 3.1.**) plus water to reach a total volume of 70 mL culture containing 15% total reducing sugar and 0.025 mg/mL initial cells (dry weight). Transfer the cultures to a shaker at 40°C operating at 170 rpm.
2. Measurement of biomass after 4 d: values not exceeding 1.3–1.7 mg/mL (dry weights) in nonagitated cultures and 2.0–4.0 mg/mL in shaken cultures, when sugar-cane juice is used.
3. Measurements of ethanol: values not exceeding 7.5–9.9% ethanol (v/v) after 8 d in nonagitated cultures and 9.0–9.9% ethanol (v/v) after 5 d in agitated cultures, when sugar-cane juice is used.
4. Measurements of viability: values as high as 80–90 % after 4 d in nonagitated cultures and 88–100% after 3 d in agitated cultures, when sugar-cane juice is used.

3.7. Thermotolerance (Ethanol Yields and Viability as Determinant Factors) in High-Cell Density Fermentation (13)

1. Inoculate 52.5 mL raw material (20% sugar-cane juice or molasses, as described in **Subheading 2.1., item 5**) with concentrated inoculum (**Subheading 3.1.**) plus water to make up 70 mL culture containing 15% total reducing sugar and 20 mg/mL initial cells (dry weight). Incubate at 40°C without shaking.
2. Measurements of ethanol after 12 h: ethanol levels as high as 8–10% (v/v) can be obtained when sugar-cane syrup is used without nutritional supplementation.
3. Measurements of viability after 9 h: values as high as 50–90% can be obtained (depending on the strain) when sugar-cane syrup is used.

3.8. Thermotolerance (Ethanol Yield and Viability as Determinant Factors at 40°C) During Cell Recycling in Repeated Batch Process of High Cell Density

1. Add the inoculum (concentrated cell suspension prepared as described in **Subheading 3.1.**) and sterilized water (as described in **Subheading 2.1., item 6**)

to reach 20 mL initial culture containing 10% total sugar and 10 mg/mL initial cells when sugar-cane syrup is used.
2. Transfer the tubes to a water bath at 40°C and incubate for 6 h before sampling (0.1 mL cell suspension) for determination of viability.
3. Centrifuge the tubes at 8000g for 5 min and discard the supernatant after determination of the final ethanol (produced plus added initially to the medium) and residual sugar without nutritional supplementation.
4. Resuspend the cells by vortexing the pellet with part (10 mL) of the fresh medium (20% total reducing sugar), and add more fresh medium plus water to the suspension to obtain 20 mL of the final suspension containing 10% total sugar and 10.0 mg/mL initial cells. Incubate for another additional 5- to 6-h period.
5. Repeat **steps 2–4** of **Subheading 3.8.** (20 fermentation cycles) as illustrated in **Fig. 1**.

Repeated batch fermentations at 40°C (as described in **Subheading 2.1., item 6**, for the evaluation of maintenance of thermotolerance during cell reuse at high cell density) cannot be carried out when sugar concentration is high. Considerable growth is maintained during the repeated baths using 10% sugar-cane syrup as shown in **Fig. 1**. When the sugar concentration is raised above 22% (total reducing sugar) in single-batch experiments, remarkable inhibitory effects of the temperature on viability and ethanol formation were observed at above 35°C *(8,13)*.

Increases in temperature and ethanol induce mutation mainly in processes with cell reuse *(3,5)*. Strain 781 was submitted to a great number of fermentation cycles (82 cycles) and samples were plated on medium containing glycerol as sole carbon source for evaluation of respiratory competence (data not shown). Decreases in the size and increases in the number of colonies showing no growth on glycerol medium were observed during the repeated cycles. Above the 50th cycle, decrease in number of small colonies occurred followed by the appearance of colonies (normal size after growth on glycerol medium) showing improved tolerance (higher biomass yield, fermentation activity, and viability) to fermentation at 40°C.

3.9. Alcohol Tolerance (Biomass, Ethanol Formation, and Viability in the Presence of 7% Added Initially to the Medium, as Determinant Factors) in Low-Cell-Density Fermentation at 30°C (15)

1. Add concentrated inoculum (**Subheading 3.1.**) and sterilized water to Erlenmeyers flasks containing 50 mL raw material (28% total reducing sugar, w/v) and 4.9 mL absolute ethanol (as described in **Subheading 2.1., item 7**) to reach a 70-mL final volume of medium (20% initial reducing sugar, 7% initial ethanol [v/v], and 0.025-mg/mL initial cells). Incubate the cultures at 30°C in a shaker operated at 170 rpm.

2. Measurements of biomass, final ethanol, and viability at early stationary phase (4 d without to 5–7d with ethanol added initially): values of biomass as high as 3.0–3.5 mg/mL for biomass, 15–17% final ethanol (produced plus added initially), and 85–95% viability are obtained when sugar-cane syrup is used without nutritional supplementation.

As ethanol added initially to the medium (6–8% ethanol, v/v) has no effect on the final ethanol formation (depending on the strains, the medium, and other operational conditions) *(15,21)*, a limiting concentration of added ethanol above 6% (v/v) is recommended to make the effects of external ethanol on growth and fermentation, more evident as shown in **Fig. 2**.

Acknowledgments

The authors are grateful to Dr. J. F.T. Spencer for the opportunity of having this chapter included in this book and Ms. Doris Barnes for the careful reading of the manuscript.

References

1. Cairns, J., Overbaugh, J., and Miller, S. (1988) The origin of mutants. *Nature* **335,** 142–145.
2. Foster, P. L. (1993) Adaptive mutation: the uses of adversity. *Ann. Rev. Microbiol.* **47,** 467–504.
3. Watson, K. (1987) Temperature relations, in *The Yeast: Yeast and the Environment* (Rose, A. H. and Harrison, J. J. eds.), Academic, New York, pp. 41–71.
4. Jinks-Robertson, S., Greene, C., and Chen, W. (1998) Genetic instabilities in yeast, in *Genetic Instabilities and Hereditary Neurological Diseases* (Wells, R. D. and Warren, S. T., eds.), Academic, San Diego, pp. 485–507.
5. Laluce, C. (1997) Variability in yeast cultures and mutants generated by process conditions, in *Ciencia de Alimentos: Avanços e perspectivas na America Latina* (Rodriguez-Amaya, D. B. and Pastore, G. M., eds.), Faculdade de Engenharia de Alimentos, Campinas, Brasil, pp. 172–176.
6. Pahl, H. L. and Baeuerle, P. A. (1994) Oxygen and the control of gene expression. *BioEssays* **16,** 497–502.
7. Werner-Washburne, M., Braun, E., Johnston, G. C., and Singer, R. A. (1993) Stationary phase in the yeast *Saccharomyces cerevisiae. Microbiol. Rev.* **57,** 383–401.
8. Dickinson, J. R. (1996) "Fusel" alcohols induce hyphal-like extensions and pseudohyphal formation in yeasts. *Microbiology* **142,** 1391–1397.
9. Adams, J., Puskas-Rozsa, S., Simlar, J., and Wilke, C. M. (1992) Adaptation and major chromosomal changes in populations of *Saccharomyces cerevisiae. Curr. Genet.* **22,** 13–19.
10. Laluce, C., Bertolini, M. C., Hernandes, J., Martini, A., and Martini, A. V. (1987) Screening survey for yeasts that ferment sucrose at relatively high temperature. *Ann. Microbiol.* **37,** 151–159.

11. Ernandes, J. R., Matulionis, M., Cruz, S. H., Bertolini, M. C., and Laluce, C. (1990) Isolation of new ethanol-tolerant yeasts for fuel ethanol production from sucrose. *Biotechnol. Lett.* **12**, 463–468.
12. Bertolini, M. C., Ernandes, J. R., and Laluce, C. (1991) New yeast strains for alcoholic fermentation at higher sugar concentration. *Biotechnol. Lett.* **13**, 197–202.
13. Laluce, C., Palmieri, M. C., and Cruz, R. C. L. (1991) Growth and fermentation characteristics of new selected strains of *Saccharomyces* at high temperatures and high cell densities. *Biotechnol. Bioeng.* **37**, 528–536.
14. Laluce, C., Abud, C. L., Greenhalf, W., and Peres, M. F. S. (1993) Thermotolerance behavior in sugar cane syrup fermentations of wild type yeast strains selected under pressures of temperature, high sugar and added ethanol. *Biotechnol. Lett.* **15**, 609–614.
15. Peres, M. F. S. and Laluce, C. (1998) Ethanol tolerance of thermotolerant yeasts cultivated on mixtures of sucrose and ethanol. *J. Ferment. Bioeng.* **85**, 388–397.
16. D'Amore, T., Celotto, G., Russell, I., and Stewart, G. G. (1989) Selection and optimization of yeast suitable for ethanol production at 40°C. *Enzyme Microbiol. Technol.* **11**, 411–416.
17. Banat, I. M., Nigam, P., and Marchant, R. (1992) Isolation of thermotolerant, fermentative yeasts growing at 52°C and producing ethanol at 45°C and 50°C. *World J. Microbiol. Biotechnol.* **8**, 259–263.
18. Rainieri, S., Zambonelli, C., Tini, V., Castellari, L., and Giudici, P. (1998) The enological traits of thermotolerant *Saccharomyces* strains. *Am. J. Enol. Vitic.* **49**, 319–324.
19. van Uden, N. (1984) Effects of ethanol on temperature relations of viability and growth in yeasts. *Crit. Rev. Biotechnol.* **1**, 263–272.
20. Casey, G. P. and Ingledew, W. M. (1985) Reevaluation of alcohol synthesis and tolerance in brewer's yeast. *ASBC J.* **43**, 75–83.
21. Casey, G. P. and Ingledew, W. M. (1986) Ethanol tolerance in yeasts. *Crit. Rev. Microbiol.* **13**, 219–280.
22. D'Amore, T. and Stewart, G. G. (1987) Ethanol tolerance of yeast. *Enzyme Microb. Technol.* **9**, 322–330.
23. Kalmokoff, M. L. and Ingledew, W. M. (1985) Evaluation of ethanol tolerance in selected *Saccharomyce* strains. *ASBC J.* **43**, 189–196.
24. Pamment, N. B. and Dasari, G. (1989) Intracellular ethanol concentration and its estimation, in *Alcohol Toxicity in Yeasts and Bacteria* (Van Uden, N. F., ed.), CRC, Boca Raton, FL, pp. 147–192.
25. Quain, D. E. (1988) Studies on yeast physiology—impact on fermentation performance and product quality. *J. Inst. Brew.* **95**, 315–323.
26. Ratledge, C. (1987) Biochemistry of growth and metabolism, in *Basic Biotechnology* (Bu'Lock, J. D. and Kristiansen, B., eds.), Academic, London, pp 11–55.
27. Ratledge, C. (1991) Yeast physiology—a micro-synopsis. *Bioprocess. Eng.* **6**, 195–203.
28. Walker, G. M. (1998) *Yeast: Physiology and Biotechnology,* Wiley, Chichester, UK, pp.101–264.

29. Booth, C. (1971) Fungal culture medium, in *Methods in Microbiology, Volume 4* (Booth, C., ed.), Academic, London, pp. 49–94.
30. Dombek, K. M. and Ingram, L. O. (1986) Magnesium limitation and its role in apparent toxicity of ethanol during yeast fermentation. *Appl. Environ. Microbiol.* **52,** 975–981.
31. Thomas, K. C. and Ingledew, W. M. (1992) Production of 21% (v/v) ethanol by fermentation of very high gravity (VHG) wheat mashes. *J. Ind. Microbiol.* **10,** 61–68.
32. Jones, A. M. and Ingledew, W. M. (1994) Fuel alcohol production: appraisal of nitrogenous yeast foods for very high gravity wheat mash fermentation. *Process Biochem.* **29,** 483–488.
33. Naudin, O., Boudarel, M. J., and Ramirez, A. (1986) Measurement of yeast invertase during alcoholic fermentation. *Biotechnol. Lett.* **8,** 591–592.
34. Lee, S. S., Robinson, F. M., and Wang, H. Y. (1981) Rapid determination of yeast viability. *Biotechnol. Bioeng. Symp.* **11,** 641–649.
35. Kistler, A. and Michaelis, S. (1997) Counting yeast cells with a standard hemocytometer chamber, in *Methods in Yeast Genetics* (Adams, A., Gottschling, D. E., Kaiser, C. A., and Steams, T., eds.), Cold Spring Harbor Laboratory Press, Cold Spring Harbor, NY, Appendix G, p. 177.

41

Multilocus Enzyme Electrophoresis

Timothy Stanley and Ian G. Wilson

1. Introduction

Multilocus enzyme electrophoresis (MEE) is a method for characterizing organisms by the relative mobilities under electrophoresis of a large number of intracellular enzymes. These differences in mobility are directly related to mutations at the gene locus that cause amino acid substitutions in the enzyme coded by the gene. Differences in the electrostatic charge between the original and substituted amino acid will affect the net charge of the enzyme, and hence its electrophoretic mobility. Thus, it is possible to relate mobility differences to different alleles at the gene locus for the enzyme in question. These mobility variants are called electromorphs. The unique profile of electromorphs produced for each strain of organism is called an electromorph type (ET).

MEE was first used in the study of the population genetics of *Drosophila* (*1*) and humans (*2*). It soon became a standard technique in eukaryotic evolutionary biology. Later, MEE was applied to microorganisms, with extensive work being done on the genetic structure of natural populations of bacteria (*3*). More recently, it has been used in the epidemiological typing of bacteria and other microorganisms.

2. Materials

1. Breaking buffer (stock solution). Tris 10.0 mM, Disodium EDTA 1.0 mM. Adjust to pH 6.8 and store at 4°C. For working buffer add 0.5 mM NADP.
2. The sonicator used is a 150-W MSE model.
3. Good-quality starch is essential for successful electrophoresis. Sigma potato starch S4501 produces a gel of high mechanical strength.
4. See **Table 1** for electrophoresis buffers.

Table 1
Electrophoresis Buffers

Tris-citrate (pH 8.0)

Electrode buffer (pH 8.0)		Gel buffer	
Tris	83.2 g	Electrode buffer diluted	1:29
Citric acid monohydrate	33.0 g		
H_2O	1 L		

Tris-citrate (pH 6.3)

Electrode buffer (pH 6.3)		Gel buffer (pH 6.7)	
Tris	27.0 g	Tris	0.93 g
Citric acid monohydrate	10.07 g	Citric acid monohydrate	0.63 g
H_2O	1 L	H_2O	1 L

Tris malate (pH 7.4)

Electrode buffer (pH 7.4)		Gel buffer	
Tris	12.11 g	Dilute electrode buffer 1:10	
Maleic acid	11.61 g		
Disodium EDTA	3.72 g		
$MgCL_2\ 6H_2O$	2.03 g		
NaOH	6.0 g		
H_2O	1 L		

Potassium phosphate (pH 7.0)

Electrode buffer (pH 7.0)		Gel buffer	
K_2HPO_4 anhydrous	87.0 g	Dilute electrode buffer 1:9	
KH_2PO_4 anhydrous	68.0 g		
H_2O	1 L		

Lithium hydroxide (pH 8.1)

Electrode buffer (pH 8.1)		Gel buffer (pH 8.5)	
LiOH	2.52 g	Tris	3.63 g
Boric acid	18.55 g	Citric acid monohydrate	1.05 g
H_2O	1 L	Boric acid	185 mg
		LiOH	25 mg
		H_2O	1 L

Borate (pH 8.2)

Electrode buffer (pH 8.2)		Gel buffer (pH 8.7)	
Boric acid	18.55 g	Tris	9.21 g
NaOH	2.4 g	Citric acid monohydrate	1.05 g
H_2O	1 L	H_2O	1 L

Tris borate (pH 8.0)

Electrode buffer (pH 8.0)		Gel electrode	
Tris	60.6 g	Dilute electrode buffer	1:9
Boric acid	40.2 g		
Disodium EDTA	6.0 g		
H_2O	1 L		

5. The gel mold (**Fig. 1**), cutting tool (**Fig. 2**), and cutting table (**Fig. 3**) are not available commercially. They can be fabricated in any competent plastics workshop at little cost. The gel mold is made of 2 mm Perspex. Its dimensions depend on the make of electrophoresis tank used, in this case, the Multiphor II tank with a narrow cooling plate (Pharmacia LKB Biotechnology). The wire of the cutting tool is a guitar high E string. It is tensioned with two screws and locking nuts set into the sides of the tool. The cutting table has a central section recessed by 2.0 mm. The gel block is placed in this recess allowing thin sections of the gel to be sliced.
6. As a staining tray we use plastic sandwich boxes, but any plastic tray can be used. Because no lids are available, cover with plastic wrap to reduce evaporation during staining. For staining buffers and reagents see **Table 2**.

$0.2\ M$ Tris-HCL (pH 8.0)	
Tris	24.2 g
H_2O	1 L

Adjust pH with HCL

$0.2\ M$ Phosphate buffer (pH 7.0)	
a. $NaH_2PO_4 \cdot H_2O$	27.6 g
b. $Na_2HPO_4 \cdot 7H_2O$	53.6 g

Stock solution mix equal volumes of (a) and (b).
For working solution, dilute 1:25 with H_2O and adjust pH with NaOH.

Malic acid solution	
DL malic acid	288 g
NaOH	160 g
H_2O	1 L

Dissolve the malic acid in the H_2O, then cool with running cold water while slowly adding the NaOH. Mixing the acid with the alkali produces a very exothermic reaction, it is very important to keep the mixture cool, otherwise this reaction can become explosive.

Glycine–KOH buffer (pH 7.5)	
Glycine	11.3 g
H_2O	1 l

Adjust pH with 1 M KOH.

Methyl-thiazolyl blue (MTT) solution	
MTT	1.25 g
H_2O	100 mL

Table 2
Staining Methods for Enzymes

Enzyme	Buffer	Substrate	Coenzyme	Dye and catalyst	Inorganic ion	Coupling enzyme
Alcohol dehydrogenase (ADH)	0.2 M Tris-HCl pH 8.0 (50 mL) Isopropanol (2 mL)	Ethanol (3 mL)	NAD (20 mg)	MTT (1.0 mL) PMS (1.5 mL)		
Sorbitol dehydrogenase (SDH)	0.2 M Tris-HCl pH 8.0 (50 mL)	Sorbitol (125 mg) Sodium pyruvate (50 mg) Pyrazole (50 mg)	NAD (20 mg)	MTT (1.0 mL) PMS (1.5 mL)		
Glycerol 3-phosphate dehydrogenase (GPD)	0.2 M Tris-HCl pH 8.0 (25 mL) 0.5 g Agar in 25 mL hot buffer	DL-α-glyceraldehyde Disodium salt (650 mg) Sodium pyruvate (200 mg)	NAD (20 mg)	MTT (1.0 mL) PMS (1.5 mL)		
Mannitol 1-phosphate dehydrogenase (M1P)	0.2 M Tris-HCl pH 8.0 (50 mL)	Mannitol 1-PMS (1.5 mL) phosphate (5 mg)	NAD (20 mg)	MTT (1.0 mL)		
Lactate dehydrogenase (LDH)	0.1 M Glycine-KOH pH 7.5 (50 mL)	Fructose 1,6-diphosphate trisodium salt (10 mg)	NAD (20 mg)	MTT (1.0 mL)		

Enzyme	Buffer	Substrate	Cofactor	Dye	Salt
		DL Sodium lactate 60% (w/w) syrup (1.0 mL)		PMS (1.5 mL)	
Hydroxybutyrate dehydrogenase (HBD)	0.2 M Tris-HCl pH 8.0 (50 mL)	DL 3- (or 2- or 3-) hydroxybuterate (100 mg)	NAD (20 mg)	MTT (1.0 mL) PMS (1.5 mL)	NaCl (200 mg) 0.2 M MgCl$_2$ (1.0 mL)
Malate dehydrogenase (MDH)	0.2 M Tris-HCl pH 8.0 (44 mL)	2.0 M Malic acid (6.0 mL)	NAD (20 mg)	MTT (1.0 mL) PMS (1.5 mL)	
Malic acid (ME)	0.2 M Tris-HCl pH 8.0 (42 mL)	2.0 M Malic acid (6.0 mL)	NADP (10 mg)	MTT (1.0 mL) PMS (1.5 mL)	0.2 M MgCl$_2$ (2.0 mL)
Isocitric dehydrogenase (IDH)	0.2 M Tris-HCl pH 8.0 (46 mL)	0.2 M Isocitric acid (2.0 mL)	NADP (10 mg)	MTT (1.0 mL) PMS (1.5 mL)	0.2 M MgCl$_2$ (2.0 mL)
6 Phosphogluconate dehydrogenase (6PD)	0.2 M Tris-HCl pH 8.0 (40 mL)	6-phosphogluconic acid (20 mg)	NADP (10 mg)	MTT (1.0 mL) PMS (1.5 mL)	0.2 M MgCl$_2$ (10.0 mL)
Glucose 6-phosphate dehydrogenase (G6P)	0.2 M Tris-HCl pH 8.0 (50 mL)	Glucose 6-phosphate disodium salt (100 mg)	NADP (10 mg)	MTT (1.0 mL) PMS (1.5 mL)	0.2 M MgCl$_2$ (0.5 mL)
Threonine dehydrogenase (THD)	0.2 M Phosphate buffer pH 7.0 (50 mL)	L-Threonine (50 mg)	NAD (20 mg)	MTT (1.0 mL) PMS (1.5 mL)	
Glycolate oxidase (GOX)	0.2 M Tris-HCl pH 8.0 (25 mL) 0.5 g Agar in 25 mL hot buffer	Glycolic acid (50 mg)	Peroxidase (10 mg)	o-Dianisidine (10 mg)	

Table 2 (continued)

Enzyme	Buffer	Substrate	Coenzyme	Dye and catalyst	Inorganic ion	Coupling enzyme
Glyceraldehyde 3-phosphate (NADP) dehydrogenase (GP2)	0.2 M Tris-HCl pH 8.0 (50 mL)	Fructose 1,6 diphosphate (100 mg)	NADP (10 mg)	MTT (1.0 mL) PMS (1.5 mL)	Arsenic acid (75 mg)	Aldolase (50 U)
Xanthine dehydrogenase (XDH)	0.2 M Tris-HCl pH 8.0 (50 mL)	Hypoxanthine (100 mg)	NAD (20 mg)	MTT (1.0 mL) PMS (1.5 mL)		
Alanine dehydrogenase (ALD)	0.2 M Phosphate buffer pH 7.0 (50 mL)	DL-alanine (50 mg)	NAD (20 mg)	MTT (1.0 mL) PMS (1.5 mL)		
Glutamate (NAD) Dehydrogenase (GD1)	0.2 M Tris-HCl pH 8.0 (50 mL)	L-Glutamic acid (200 mg)	NAD (20 mg)	MTT (1.0 mL) PMS (1.5 mL)		
Glutamate (NADP) Dehydrogenase (GD2)	0.2 M Tris-HCl pH 8.0 (50 mL)	L-Glutamic acid (200 mg)	NADP (10 mg)	MTT (1.0 mL) PMS (1.5 mL)		
Leucine dehydrogenase (LED)	0.2 M Phosphate buffer pH 7.0 (50 mL)	L-Leucine (50 mg)	NAD (20 mg)	MTT (1.0 mL) PMS (1.5 mL)		
Aspartate oxidase (ASO)	0.2 M Tris-HCl pH 8.0 (50 mL)	D-Aspartic acid (200 mg) Flavin adenine dinucleotide disodium salt (10 mg)	Peroxidase (5 mg)	o-Dianisidine (10 mg)		
Amino acid oxidase (AAO)	0.2 M Tris-HCl pH 8.0 (50 mL)	D-Phenylalanine (200 mg)	Peroxidase (5 mg)	o-Dianisidine (10 mg)		

Enzyme	Buffer	Substrate	Coenzyme	Stain	Other
Aspartic acid dehydrogenase (ASD)	0.2 M Phosphate buffer pH 7.0 (50 mL)	L-Aspartic acid (50 mg)	NAD (20 mg)	MTT (1.0 mL) PMS (1.5 mL)	
Lysine dehydrogenase (LYD)	0.05 M Phosphate buffer pH 7.0 (50 mL)	L-Lysine (50 mg)	NAD (20 mg)	MTT (1.0 mL) PMS (1.5 mL)	
Glutathione reductase (GSR)	0.2 M Tris-HCl pH 8.0 (25 mL) 0.5 g Agar in 25 mL hot buffer	Oxidized glutathione (20 mg)	NADPH (5 mg)	MTT (1.0 mL) 2,6-Dichlorophenol-indophenol (2 mg)	
Catalase 1 (CAT)[a]	Na$_2$SO$_3$ (750 mg) H$_2$O (50 mL)	30% H$_2$O$_2$ (2.5 mL)			1.5% KI solution (50 mL)
Catalase 2 (CAT)[a]	H$_2$O (50 mL)	30% H$_2$O$_2$ (50 μL)			50:50 mixture 2.0% K$_3$Fe(CN)$_6$ 2.0% FeCl$_3$
Superoxide dismutase (SOD)[b]	0.2 M Tris-HCl pH 8.0 (50 mL)		NAD (20 mg)	MTT (1.0 mL) PMS (1.5 mL)	
Glutamic oxaloacetic transaminase (GOT)	0.2 M Tris-HCl pH 8.0 (50 mL)	Pyridoxal 5-phosphate (1 mg) L-aspartic acid (50 mg) α-Ketogluteraric (100 mg)		Fast Blue BB salt (100 mg)	

Table 2 (continued)

Enzyme	Buffer	Substrate	Coenzyme	Dye and catalyst	Inorganic ion	Coupling enzyme
Glutamic-pyruvic transaminase (GPT)	0.2 M Tris-HCl pH 8.0 (25 mL) 0.5 g Agar in 25 mL hot buffer	L-Alanine (50 mg) α-Ketogluteric acid (40 mg)	NAD (5 mg) NADP (2.5 mg)	MTT (1.0 mL) PMS (1.5 mL)		Glutamic dehydrogenase (40 mg)
Hexokinase (HEX)	0.1 M Glycine KOH pH 7.5 (46 mL)	D-Glucose (200 mg) ATP (50 mg)	NADP (10 mg)	MTT (1.0 mL) PMS (1.5 mL)		Glucose 6-phosphate dehydrogenase (10 U)
Creatine kinase (CTK)	0.1 M Phosphate buffer pH 7.0 (25 mL) 0.5 g Agar in 25 mL hot buffer	Phosphocreatine sodium salt (20 mg) ADP (30 mg) Glucose (40 mg)	NADP (10 mg)	MTT (1.0 mL) PMS (1.5 mL)	0.2 M MgCl$_2$ (2 mL)	Hexokinase (10 U) Glucose 6-phosphate dehydrogenase (10 U)
Nucleoside triphosphate adenylate kinase (AK3)	0.2 M Tris-HCl pH 8.0 (25 mL) 0.5 g Agar in 25 mL hot buffer	Glucose (40 mg) AMP trisodium salt (25 mg) GTP trisodium salt (15 mg) Phosphoenol-pyruvate (20 mg)	NADP (10 mg)	MTT (1.0 mL) PMS (1.5 mL)	0.2 M MgCl$_2$ (2.0 mL)	Hexokinase (15 U) Glucose 6-phosphate dehydrogenase (10 U) Pyruvate kinase (20 U)
Phosphoglucomutase (PGM)	0.2 M Tris-HCl pH 8.0	Glucose 6-phosphate	NADP (2 mg)	MTT (1.0 mL) PMS (1.5 mL)	0.2 M MgCl$_2$ (2.5 mL)	Glucose 6-phosphate dehydrogenase

Enzyme	Buffer	Substrate	Cofactor/Dye	Salts	Linking enzymes
		disodium salt (10 mg) 0.5 g Agar in 25 mL hot buffer (25 mL)			(50 U)
Glucose pyrophosphorylase (GPP)	0.2 M Tris-HCl pH 8.0 (50 mL)	UDP glucose sodium salt (50 mg) Pyrophosphate sodium salt (40 mg) Glucose 1,6-diphosphate tetraacyclohexylammonium salt (5 mg)	NADP (10 mg) MTT (1.0 mL) PMS (1.5 mL)	0.2 M MgCl$_2$ (2.0 mL) 0.054 M EDTA pH 7.5	Glucose 6-phosphate dehydrogenase (10 U) Phosphoglucomutase (30 U)
Esterase (EST)^c	0.2 M Phosphate buffer pH 7.0 (50 mL)	α-Naphthyl acetate or α-Naphthyl propionate or β-Naphthyl acetate or β-Naphthyl propionate (1.5 mL of 1% solution in acetone)	Fast Blue RR salt (20 mg)		
Alkaline phosphate (ALP)	0.1 M Tris-HCl pH 8.5 (50 mL) Polyvinylpyrrolidone (PVP-40) (100 mg)	β-Naphthyl acid phosphate (50 mg)	Fast Blue BB salt (50 mg)	NaCl (1.0 g) 0.25 M MnCl$_2$ (2.0 mL) 0.2 M MgCl$_2$ (1.0 mL)	

Table 2 (continued)

Enzyme	Buffer	Substrate	Coenzyme	Dye and catalyst	Inorganic ion	Coupling enzyme
Acid phosphatase (ACP)	0.05 M Sodium acetate-HCL pH 5.0 (50 mL)	α-Naphthyl acid phosphate (50 mg) β-Naphthyl acid phosphate monosodium salt (50 mg)		Fast Black K salt (20 mg)		
Peptidase (PEP)^d	0.2 M Tris-HCl pH 8.0 (25 mL) 0.5 g Agar in 25 mL hot buffer	Peptide (20 mg)		o-Dianisidine (20 mg)	0.25 $MnCl_2$ (0.5 mL)	Peroxidase (250 U) L-Amino acid oxidase (2.0 mg)
Guanine deaminase (GDA)	0.2 M Tris-HCl pH 8.0 (20 mL) 0.5 g Agar in 25 mL hot buffer	Guanine-HCl solution (3.0 mL)		MTT (1.0 mL) PMS (1.5 mL)		Xanthine oxidase (1 U)
Adenosine deaminase (ADA)	0.2 M Phosphate buffer (25 mL) 0.5 g Agar in 25 mL hot buffer	Adenosine (15 mg)		MTT (1.0 mL) PMS (1.5 mL)		Xanthine oxidase (1 U) Nucleoside phosphorylase (1 U)

Enzyme					
Cytidine deaminase (CDA)	H$_2$O (25 mL) 2% Agar (25 mL) containing cytidine (45 mg)	Cytidine (45 mg) (see Buffer)		MTT (6.0 mg) Dithiothreitol (10 mg)	
Aldolase lyase (ALD)	0.1 M Tris acetate pH 7.5 (50 mL)	Fructose 1,6-diphosphate (100 mg)	NAD (20 mg)	MTT (1.0 mL) PMS (1.5 mL)	Sodium arsenate (100 mg)	Glyceraldehyde 3-PO$_4$ dehydrogenase (50 U) Triphosphate isomerase (100 U)
Citrate synthase (CTS)	1.0 M Tris-HCl pH 8.0 (10 mL)	Oxalacetate acid (20 mg)	Acetyl-CoA (50 mg)	MTT (250 mg) in 0.2 M Tris-HCl pH 8.0 (50 mL) 2,6-Dichloro-phenol-indo-phenol (40 mL)		
Fumarase (FUM)	0.2 M Tris-HCl pH 8.0 (48 mL)	Fumaric acid (50 mg)		MTT (1.0 mL) PMS (1.5 mL)		Makic acid dehydrogenase (150 U)
Acontase (ACO)	0.2 M Tris-HCl pH 8.0 (15 mL)	cis-Acontic acid (30 mg)		NAD (50 mg)	0.2 M MgCl$_2$ (5.0 mL)	Isocitric dehydrogenase (5 U)
Glyoxalase (GLO)	0.2 M Phosphate buffer pH 7.0 (48 mL) Methylgloxal (2.5 mL)	Reduced gluta-thione (250 mg)		MTT (50 mg) 2,6-Dichloro-phenol indo-phenol (1.0 mL)		

Table 2 (continued)

Enzyme	Buffer	Substrate	Coenzyme	Dye and catalyst	Inorganic ion	Coupling enzyme
Triose phosphate isomerase (TPI)	0.2 M Tris-HCl pH 8.0 0.5 g Agar in 20 mL hot buffer	α-Glycerophosphate (650 mg) Sodium pyruvate (250 mg) Arsenic acid (50 mg)	NAD (40 mg)	MTT (1.0 mL) PMS (1.5 mL)		α-Glycerophosphate dehydrogenase (4 U) Lactate dehydrogenase (50 U) Glyceraldehyde 3-phosphate dehydrogenase (40 mL)
Mannose phosphate isomerase (MPI)	0.2 M Tris-HCl pH 8.0 (25 mL) 0.5 g Agar in 25 mL hot buffer	Mannose 6-phosphate (10 mg)	NAD (20 mg) NADP (10 mg)	MTT (1.0 mL) PMS (1.5 mL)	0.2 M MgCl$_2$ (1.0 mL)	Glucose 6-dehydrogenase (10 U) Phosphoglucose isomerase (50 U)
Phosphoglucose isomerase (PGI)	0.2 M Tris-HCl pH 8.0 (23 mL) 0.5 g Agar in 25 mL hot buffer	Fructose 6-phosphate disodium salt (18 mg)	NAPD (5.0 mg)	MTT (1.0 mL) PMS (1.5 mL)	0.2 M MgCl$_2$ (23 mL)	Glucose 6-phosphate dehydrogenase (10 U)
Glcosyltransferase (GFT)	0.1 M KH$_2$PO$_4$ pH 6.8 (50 mL)	Sucrose (5.0 g)			Sodium azide (12 mg)	
Shikimate dehydrogenase (SHK)	0.2 M Tris-HCl pH 8.0 (50 mL)	Shikimic acid (50 mg)	NADP (10 mg)	MTT (1.0 mL) PMS (1.5 mL)	0.1 M MgCl$_2$ (2.0 mL)	

Adenylate kinase (ADK)	0.2 M Tris-HCl pH 8.0 (25 mL) 0.5 g Agar in 25 mL hot buffer	ADP (25 mg) Glucose (100 mg)	NADP (10 mg)	MTT (1.0 mL) PMS (1.5 mL)	0.2 M MgCl$_2$ (0.5 mL)	Hexokinase (10 U) Glucose 6-phosphate dehydrogenase (15 U)
Carbamate kinase (CAK)[e]	0.2 M Tris-HCl pH 8.0 (25 mL) 0.5 g Agar in 25 mL hot buffer	ADP (25 mg) Glucose (100 mg) Carbamyl phosphate disodium salt (100 mg)	NADP (1.0 mg)	MTT (1.0 mL) PMS (1.5 mL)	0.1 M MgCl$_2$ (1.0 mL)	Hexokinase (20 U) Glucose 6-phosphate dehydrogenase (10 U)
β-Galactosidase (BGA)[f]	Phosphate citrate buffer pH 8.0 H$_2$O (30 mL)	6-Bromo-2-β-D-galacto-pyraniside (10 mg) dissolved in 5 mL methanol		o-Dianisidine tetrazotized zinc chloride complex		

[a]Incubate slice in H$_2$O$_2$ solution for 15 min at 37°C. Rinse in tap water several times then add inorganic ion mixture. Remove staining mix when negative staining bands appear.

[b]Incubate under fluorescent light or bright sunlight until negative staining bands appear.

[c]Pre soak slice in 50 mL phosphate buffer for 20 min then pour off. Add Fast Blue to 50 mL phosphate buffer, mix for 5 min, and add ester solution. Pour immediately onto slice. Many other esters can be used.

[d]Many different peptides can be used, e.g., L-alanyl-L-leucine, L-phenylalanyl-L-proline, L-leucyl-L-alanine, L-phenylalanyl-L-leucine (hard to dissolve), L-leucyl-L-leucine, L-alanyl-L-leucine, L-alanyl-L-threonine, L-alanyl-L-tyrosine.

[e]Stains for ADK as well as CAK. Run both gels at same time. Extra bands on CAK gel are carbinate kinase ETs.

[f]Add substrate solution to phosphate citrate buffer. Pour on slice incubate at 37°C for 15 min then pour off. Dissolve dye in 30 mL H$_2$O, adjust pH with sodium bicarbonate, pour on slice, incubate for 5 min if BGA was induced and 1 h if it was not.

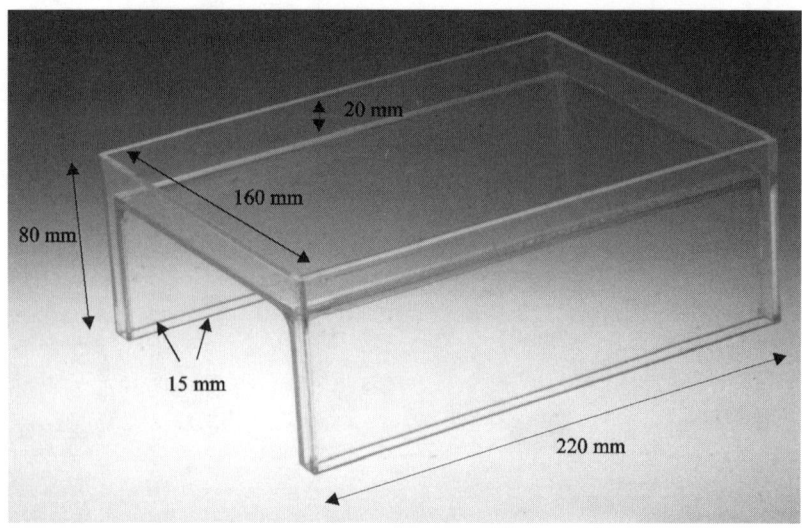

Fig. 1. Gel mold.

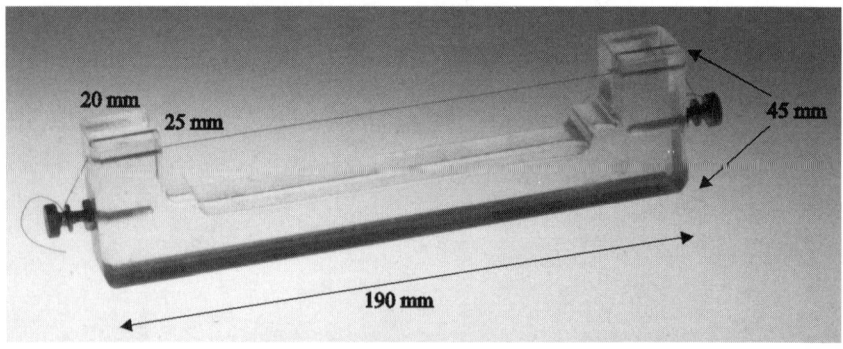

Fig. 2. Gel slicer.

Phenazine methosulfate (PMS) solution

PMS	1.0 g
H_2O	100 mL

Store both solution in the dark at 4°C until needed. Both MTT and PMS are very toxic and should be treated with care. If a fumehood is not available, gloves and face masks should be worn when working with these chemicals. Spillage onto bare skin should be washed off immediately.

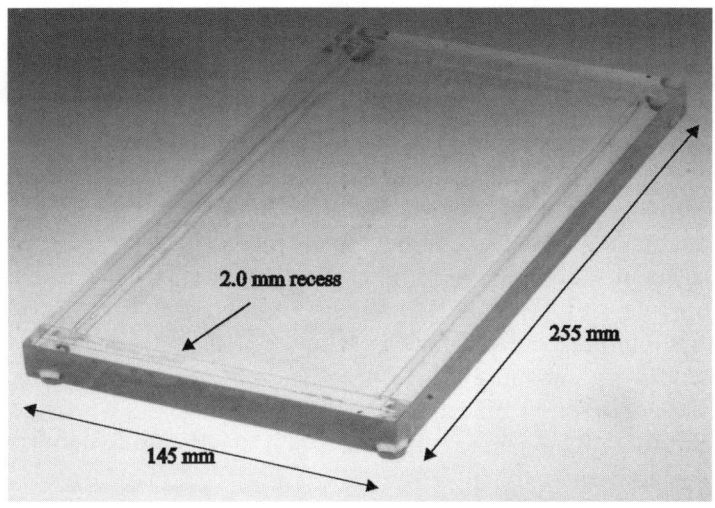

Fig. 3. Gel table.

0.05 M Sodium acetate buffer (pH 5.0)

Sodium acetate	6.8 g
H_2O	1 L

Adjust pH with HCl.

Phosphate citrate buffer (pH 5.0)

1.0 M phosphoric acid	10.2 mL
2.0 M NaOH	10.2 mL
Citric acid monohydrate,	1.03 g
H_2O	76.9 mL

0.1 M Tris acetate buffer (pH 7.5)

Tris	12.11 g
H_2O	1 L

Adjust pH with glacial acetic acid then add:

Potassium acetate	19.63 g
Cobalt chloride	333 mg
L-cysteine hydrochloride	35.2 mg

Guanine–HCl solution

Guanine–HCl	50 mg

Dissolve in 10 mL warm 0.1 M NaOH. Add to 40 mL H_2O.

3. Methods
3.1. Growth of Bacteria

The microorganisms to be studied must be in a pure culture and growing luxuriantly. Old cultures should not be used because although high weight of culture may be achieved, enzyme activity will be low, as a majority of cells will not be in a logarithmic growth phase. Solid or liquid media can be used, but solid media is the most convenient (*see* **Note 1**).

10^{11} Organisms are required for enzyme extraction. This is usually equivalent to 0.5 g (wet weight) of culture. When optimum growth of the organism is achieved the culture on each plate is scraped off using a plastic loop into a breaking buffer for enzyme extraction.

Take care not to scrape off too much growth media along with the culture, as this will add foreign enzymes to the electrophoresis gel, making interpretation of the stained gel difficult.

3.2. Extraction of Enzymes

Any method of extraction can be used as long as the extracted enzymes remain in their native state. Methods used include *(1)* freezing and grinding with fine sterile sand, *(2)* freeze/thawing with liquid nitrogen, *(3)* chemical or enzymic breakdown of the cell wall, and *(4)* sonication. The method chosen usually depends on the relative resistance of the organism under study to lysis (*see* **Note 2**). The most common method used is sonication.

1. The suspension of organism in breaking buffer is placed in a 2.0-mL Eppendorf tube and sonicated using a microtip probe with 15- to 30-s bursts of sonication interspersed with 30 s cooling in ice water. This cycle is repeated four to six times.
2. When sonication is complete the suspension is centrifuged at 30,000g for 30 min at 4°C.
3. The supernatant is then aliquoted in 100 µL amounts and stored at –70°C. Enzyme stability varies with different species, but there is usually insignificant loss of activity after storage at –70°C for several months.

3.3. Electrophoresis

Carry out MEE using any support medium that allows electrophoresis under native conditions. Starch, polyacrylamide and cellulose acetate have all been used. Of the three, starch is the most common. (*see* **Note 3**).

A large number of buffer systems have been described for MEE. **Table 1** shows the most common ones used *(3,4)*. Unfortunately, the choice of buffer depends on trial and error (*see* **Note 4**).

3.3.1. Starch Gel Preparation

1. Seal the legs of the mold (**Fig. 1**) with PVC insulating tape and level on a leveling table.

2. Suspend 82.0 g of starch with 820 mL of electrolyte buffer in a 2-L conical flask (*see* **Note 5**). Heat on a ring gas burner with wire gauze. While the mixture is heating, lift the flask off every 10 s and swirl the contents. At this stage the suspension should be white and fluid. After approx 5 min heating, the mix will suddenly become viscous and gray in appearance, it is important at this stage to continue to mix the contents of the flask vigorously. After approximately a further 2 min heating, the starch gel should become more translucent and less viscous. Small bubbles will start to form at the sides of the flask and its contents will start to steam gently. The gel is now ready to degas.
3. Quickly transfer the molten gel to a prewarmed thick-walled flask, with a side arm. Stopper the flask and degas the gel by applying a −80 kPa vacuum for 1 min to the side arm (*see* **Note 6**).
4. Pour the starch suspension into the mold until there is a convex meniscus. Remove any air bubbles trapped down the side legs of the mold with a Pasteur pipet.
5. While the gel is still molten, cover the starch surface with plastic wrap to reduce evaporation. Stick the wrap to the sides of the mold, and adjust the tension of the wrap to remove any wrinkles on the surface of the gel. Burst any large air bubbles trapped under the plastic wrap with a hypodermic needle. Place the mold at 4°C for 2–3 h until it has set. The gel will keep for 24 h, but it is best used as soon as it has set.

3.3.2. Loading the Gel

1. First remove the plastic wrap covering. Then make a vertical cut through the gel running parallel with one of the side arms 25 mm from the end wall. Carefully push a blunt spatula into the cut and make sure the gel is completely separated into two sections.
2. Add 30 µL of each extract to filter paper strips 20 mm × 5.0 mm. Allow the strip to dry until all surface moisture has gone, before carefully inserting the strip into the slit in the gel (*see* **Note 7**).
3. The first, middle, and last extract in the gel should be a control extract (*see* **Note 8**). The very last strip at one end of the gel should be loaded with 2.0% amaranth dye to act as a marker (*see* **Note 9**).

3.3.3. Running Conditions

1. Electrophoresis is carried out at a temperature of 4°C to prevent degradation of the catalytic activity of the enzymes under study (*see* **Note 10**).
2. Before putting the gel mold in the electrophoresis tank, remove the tape sealing the legs of the mold.
3. Place the mold in the tank with the extracts at the cathode side of the tank. Attach the leads from the tank to the power supply and apply a constant voltage across the gel. The voltage required depends on the buffer being used and the length of time available for electrophoresis (*see* **Note 11**).
4. Electrophoresis is complete when the marker dye has migrated approximately 5.0 cm.

3.3.4. Slicing the Gel

1. After electrophoresis switch off the power to the tank and remove the mold.
2. Trim the gel slab to the size of the recessed part of the cutting table by cutting right through the gel with a sharp thin blade. These cuts should be approximately 1.0 cm from the first and last extract and several centimeters above the distance traveled by the dye marker (*see* **Note 12**).
3. Remove the outer unwanted parts of the gel slab leaving the central portion containing the electrophoresed enzymes. Cut off the upper right-hand corner of the gel slab so that the correct orientation of the gel slices can be maintained during staining. Finally, remove the filter paper strips used to load the enzyme extracts into the gel.
4. Carefully lift the gel slab and place it, top side up, in the recessed section of the cutting tray. Make sure there are no air bubbles between the gel slab and the cutting table.
5. Rest the cutting tool at one end of cutting table with the cutting wire resting against the edge of the gel slab. Slowly draw the cutting wire through the gel slab using an even pressure. No downward pressure on the gel slab should be needed at this point.
6. When the cutting tool has completely cut through the gel slab, carefully lift the gel slab off the gel slice. This can be achieved by introducing a blunt instrument between the gel slab and the gel slice at one end, then slowly lifting up to break the surface tension between the two pieces of the gel. Place the gel slab on a glass plate until another slice is needed.
7. Remove the gel slice from the cutting table in a similar manner to that previously mentioned and place it in a staining tray at 4°C until required. Replace the gel slab on the cutting table and repeat the slicing operation reversing the direction of the cutting tool after each slice is cut. Continue until sufficient slices have been cut (*see* **Note 13**).

3.4. Staining and Recording Results

The location of enzymes in the gel after electrophoresis can be determined by the enzymes catalyzing specific reactions with a substrate that produces a colored product. For some enzymes, such as catalase, the substrate is colored instead of the product. The commonest methods of staining are electron transfer dyes and modified histochemical stains *(2,3,5)*. Specific staining methods are shown in **Table 2**. All stains are incubated at 37°C unless otherwise stated.

1. Add the reactants to the buffer and allow to dissolve. Add MTT and PMS just before the staining mixture is added to the gel slice.
2. Gently rock the staining tray until the gel slice floats free in the staining solution. Cover the tray and incubate, usually in the dark at 37°C.
3. If an agar overlay is needed, the reactants are mixed with half the volume of buffer. In the other half, melt 0.5 g of Oxoid No. 1 agar. Allow the agar to cool to

55°C. Mix both solutions together just before pouring over the gel. Leave the gel for 2 min on a level surface to set before incubation.
4. Regularly check the development of the staining reaction until optimum development has occurred. This varies greatly between enzymes (*see* **Note 14**).
5. When staining is complete, score the relative migration of the enzymes by eye. The enzymes are scored in order of increasing anodal migration, but this can only be done after all gels have been run. Until that is accomplished, the control extracts are scored as 6 and all other results are related to this result, i.e., electromorphs that migrate less than the control are scored 5, the next 4, and so on. Electromorphs that migrate further score 7, the next 8. Do not mistake a weak reaction for a negative result (*see* **Note 15**). **Figure 4** shows a stained gel of the enzyme malate dehydrogenase. In this example, the controls are in lanes 1, 8, and 18 and so are scored as 6. Extracts 5, 8, 11, 15, and 16 also score 6. Extract 7 scores 5. Extracts 2–4, 6, 9 and 12–14 score 7. Extract 17 scores 8. The extract in lane 10 does not produce a band and so is scored as 0.
6. After scoring the gel, a permanent record can be made by either photography or by fixation with an acid–alcohol wash (*see* **Note 16**).
7. After all extracts have been run once further, gels are run to check that null results are not due to weak reactions. Re-extraction of some extracts may be needed. Extracts with the same mobilities are also rerun together as a check on intergel comparability.
8. Once all gels have been run and all results collected, these results must be converted to a form that can be analyzed, i.e., they must be numbered in order of decreasing anodal migration. First tabulate the results. Then, taking the results of each enzyme in isolation, the electromorph with the greatest migration is scored 1, the next furthest 2, and so on until all the results have been converted.

3.5. Data Analysis

The type of analysis carried out on MEE data depends on the aims of the study in question.

Genetic diversity (h) can be calculated using the formula

$$h = 1 - \sum X^2_i [\, n/(n-1)]$$

X_i = frequency of the *i*th allele at the locus

n = number of isolates or ETs

$n/(n-1)$ = correction for small sample size

Genotypic diversity can be derived from the same formula. In that case X_i is the frequency of the *i*th ET and n is the number of Ets.

Genetic distance (*D*) between pairs of isolates or ETs can be calculated using several different coefficients (**6**). It is usually calculated as the proportion of loci at which dissimilar alleles occur, i.e., the number of mismatches. These

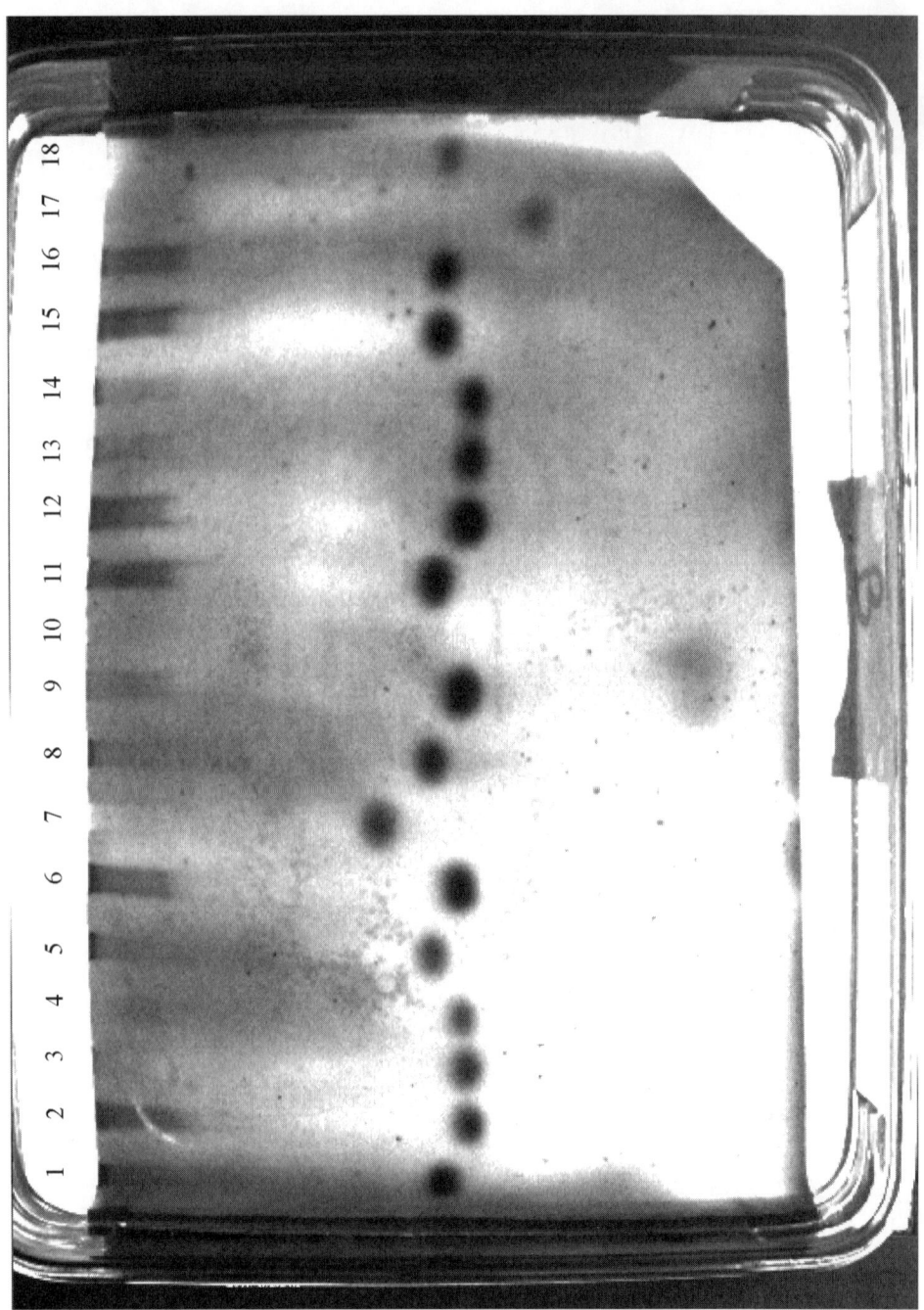

coefficients can be either unweighted or weighted. If they are weighted, the contribution of each locus to *(D)* is multiplied by the reciprocal of the mean genetic diversity at the locus in the total sample being analyzed. Weighting in this manner emphasises the significance of variation at loci that are have low genetic diversity. A variety of statistical methods can be used to produce a graphical representation of the relatedness between isolates or ETs *(6)*.

Figure 5 shows a dendrogram of genetic distance between pairs of ETs calculated using the proportion of mismatches method. **Figure 6** shows a three-dimensional graph of a principal coordinate analysis of the same data. Both were produced by a macro subroutine running on the SAS statistical package *(7)*. There are many other computer packages available that can be used for the analysis of MEE data *(8–10)*.

4. Notes

1. The use of liquid and solid culture is a personal choice. Higher yields can usually be achieved with liquid culture, but this is offset by the more complicated procedure needed in harvesting the organism and the greater problems of contamination.
2. Freezing/grinding with fine glass beads (Sigma, cat. no. G4694) is one of the simplest methods, but it is time consuming and can result in low yields of enzyme *(3)*. Freeze thawing with liquid nitrogen is quick and produces good yields of enzyme but, as there are few references to its use, its suitability for a wide range of organisms is in doubt *(11)*. Some organisms with thick cell walls resist physical disruption. In this case, enzymatic digestion of the cell wall can be used, such as lysozyme with staphylococci *(9)* and lyticase with yeasts *(10)*.
3. Each support medium has its own advantages. Cellulose acetate needs little preparation and has low sample volumes and short run times *(12)*. Polyacrylamide can produces high-definition bands and the pore size of the gel can be varied *(13)*. Starch is the most popular support media. This is due to the ability to cast thick gels, which can be sliced several times, allowing a different enzyme to be assayed with each slice. It is also the cheapest of the support media.
4. Gel buffers affect greatly the resolution of enzymes. One of the major factors is the pH of the buffer. The further the pH of the buffer is from the isoelectric point (pK) of the enzyme, the quicker the enzyme will move through the gel, leading to greater separation between electromorphs. Unfortunately, the pK of enzymes are only known for a few common organisms, such as *Escherichia coli*. As there is more variation between types of enzymes than between enzymes of different species, it is best to try buffers used for the same enzyme in other studies even if they are from different species.

Fig. 4. *(opposite)* Stained gel of malate dehydrogenase. Extracts of *Campylobacter* stained for the enzyme malate dehydrogenase. The control extracts are in lanes 1, 8, and 18. Lane 10 shows a null result.

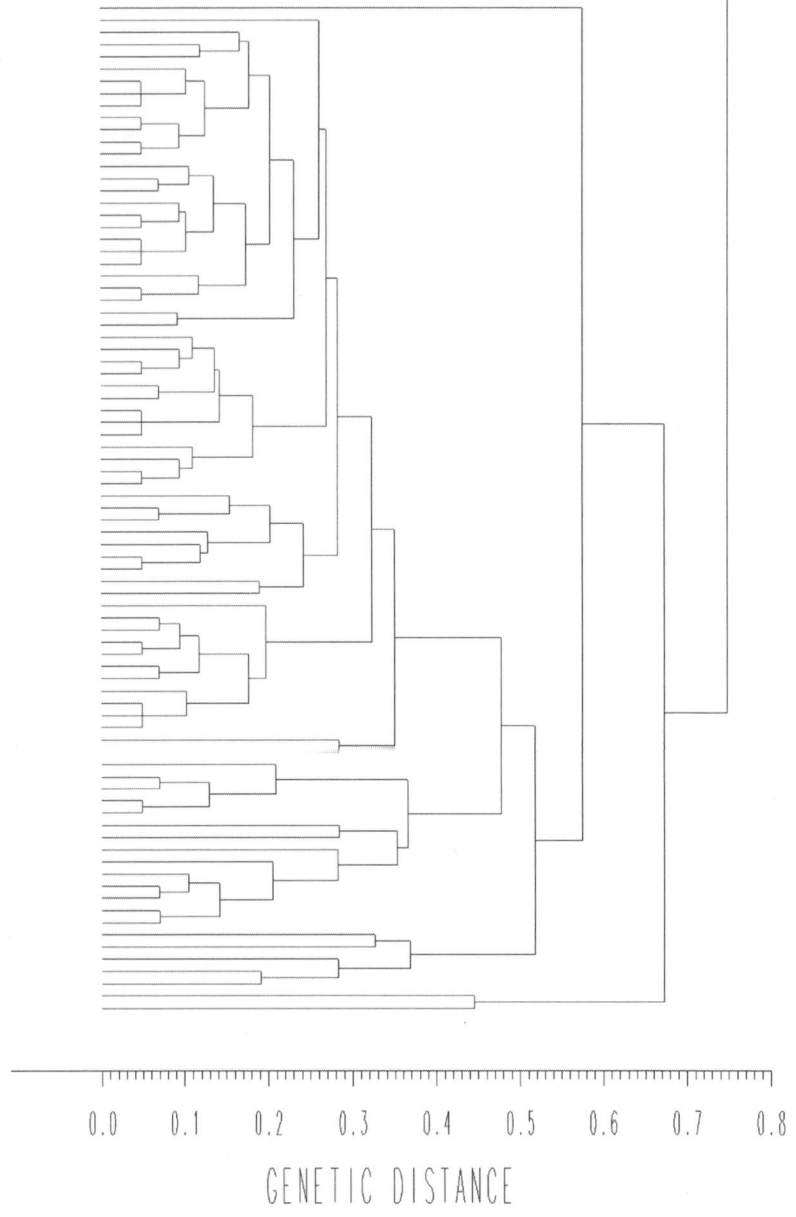

Fig. 5. Dendrogram of genetic distance. The dendrogram shows the genetic distance at which varius Ets are related. Genetic distance is estimated by moving from left to right across the dendrogram until two lines join. The relevant genetic distance is then read off the scale.

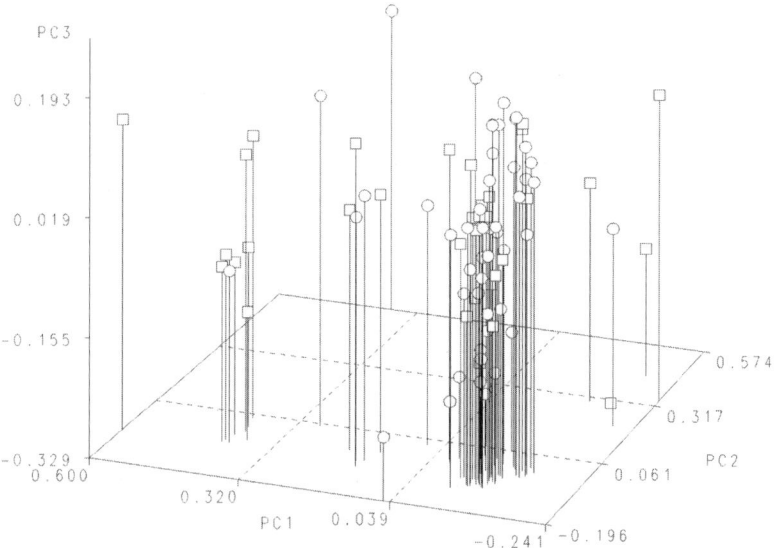

Fig. 6. Cluster diagram of *Campylobacter* Ets. The diagram shows the relationship between *Campylobacter* Ets in three dimentions. One large and one small cluster can be seen. ○, human isolates of *Campylobacter*. □, chicken isolates of *Campylobacter*.

5. The actual amounts of starch and buffer depend on the size of the gel mold. I usually use a 10% starch gel, but this depends on the particular gelling properties of the starch used. This can vary even between different lots from the same manufacturer. The actual concentration of starch used can vary between 12.0–9.5%.
6. The preparation of a starch gel requires skill. The point at which a gel is perfectly cooked is difficult to define and the times given in the method are only approximate. Undercooked gels will not set properly and will have low mechanical strength. Overcooked gels are dense, will distort during electrophoresis, and are difficult to slice. Undergassed gels are full of small bubbles. A perfect gel will be transparent, free from lumps, and pours smoothly. With care it is usually possible to produce satisfactory gels after only a few attempts.
7. To aid inserting the filter paper strips, use a blunt probe such as a spatula to pry open the gel at the cut. If the gel is loaded left to right, move the probe to the right of the strips as they are inserted.
8. Control extracts are needed for intergel comparison of enzyme mobility. Controls are chosen from extracts at the start of the trial. They need to give a strong reaction and have average mobility with all the enzymes tested.
9. The dye marker acts as a check that electrophoresis has taken place as until stained the enzymes are colorless. It is also used as a guide in trimming the gel.
10. The gel can be kept cool by several methods. A large basin of ice can be placed on the tank, but this is unsatisfactory for overnight runs unless the ice can be

replaced when it melts. A better method is to put the entire electrophoresis tank in a refrigerator. With some tanks, a cooling plate can be placed below the mold and connected to an external cooling unit.
11. The higher the voltage, the quicker the run time, but higher voltage increases the heat generated by electrical resistance in the gel. This can degrade the enzymes, but more importantly, high heat will break down the physical characteristics of the gel causing it to contract and split. For this reason, I have found it is very difficult to load run and stain a gel within an 8-h working day. An alternative is to run the gel overnight at a voltage of 90–100 V.
12. Most enzymes run slower than the dye marker, but occasionally some enzymes such as peptidase have very mobile electromorphs which may run faster, so it is always advisable to trim the gel leaving as much as possible above the dye marker.
13. As with cooking the gel, slicing takes practice. The most important factor is a good gel. The mechanical strength of a well-cooked gel is surprisingly high. Even the seemingly flimsy gel slices can be lifted by one end in the fingers. If the gel does break into pieces it is usually possible to piece them together in the staining tray like a jigsaw. The top slice should always be discarded due to its uneven surface. The bottom slice can be used, but false negative results can occur if the filter paper strips are not pushed right to the bottom of the gel. Because of this possibility, the bottom slice should only be used if no other slices are available.
14. Catalase is unusual as it only takes 30 s to develop. It must be read quickly as the bands fade rapidly. The other stains take 30–60 min to develop. Weak extracts may need to be left for several hours.
15. The reason for this complicated procedure is that until all samples are run it is impossible to know which electromorph has the greatest mobility. By arbitrarily scoring the control as 6, and relating all other results to this score, it is possible to construct a result table which, when all the results are produced, can be converted to a form ready for analysis (*see* **Subheading 3.5.**).
16. Fixation is not very satisfactory. Gel contract and whiten, weak bands may be lost, and it is difficult to store large numbers of fixed gels. Photographing the stained gel to produce a permanent record is the method of choice. Photographing gels with a digital camera and storing the resulting images on a computer is a very useful way of cataloguing large numbers of gels.
17. Intergel comparison is possible but errors can occur in scoring between gels. Running check gels with extracts that seem to have the same mobility is the best way to overcome this although it is possible to make some intergel comparisons with photographed gels.

References

1. Lewontin, R. C. and Hubby, J. L. (1966) A molecular approach to the study of the hetrozygosity in natural populations. II. Amount of variation and degree of heterozygosity in natural populations of *Drosophila pseudoobscura*. *Genetics* **54**, 595–609.

2. Harris, H. and Hopkinson, D. A. (1976) *Handbook of Enzyme Electrophoresis in Human Genetics*, North-Holland Publishing, Amsterdam.
3. Selander, R. K., Caugant, D. A., Ochman, H., et al. (1986) Methods of multilocus enzyme electrophoresis for bacterial population genetics and systematics. *Appl. Environ. Microbiol.* **51,** 873–884.
4. Morizot, D. C. and Schmidt, M. E. (1990) Starch gel electrophoresis and histochemical visualisation of proteins, in *Electrophoretic and Isoelectric Focusing Techniques in Fisheries Management* (Whitmore, D. H., ed.) CRC Press, Boston, pp. 23–80.
5. Rothe, G. (1994) *Electrophoresis of Enzymes: Laboratory Methods*, Springer-Verlag, Berlin, New York.
6. Sneath, P. H. A. and Sokal, R. R. (1073) *Numerical Taxonomy*, W. H. Freeman, San Francisco.
7. Graves, L. M. et al. (1994) Comparison of ribotyping and multilocus enzyme electrophoresis for subtyping of *Listeria monocytogenes* isolates *J. Clin. Microbiol.* **32,** 2936–2943.
8. Rodriguez, E., De Meeus, T., Mallie, M., et al. (1996) Multicentric epidemiological study of *Aspergillus fumigatus* isolates by multilocus enzyme electrophoresis *J. Clin. Microbiol.* **34,** 2559–2568.
9. Musser, J. M. and Kapur, V. (1992) Clonal analysis of methicillin-resistant *Staphylococcus aureus* strains from intercontinental sources: association of the *mec* gene with divergent phylogenetic lineages implies dissemination by horizontal transfer and recombination. *J. Clin. Microbiol.* **30,** 2058–2063.
10. Pujol, C., Joly S., Lockhard, S. R., et al. (1997) Parity among the randomly amplified polymorphic DNA method, multilocus enzyme electrophoresis, and southern blot hybridization with the moderately repetitive DNA probe Ca3 for fingerprinting *Candida albicans. J.Clin. Microbiol.* **35,** 2348–2358.
11. Hall, L. M. C., Whiley, R. A., Duke, B., et al. (1996) Genetic relatedness within and between serotypes of *Streptococcus pneumoniae* from the United Kingdom: analysis of multilocus enzyme electrophoresis, pulsed-field gel electrophoresis, and antimicrobial resistance patterns. *J. Clin. Microbiol.* **34,** 853–859.
12. Richardson, B. J., Baverstock, P. R., and Adams, M. (1986) *Allozyme Electrophoresis*, Academic Press, London, New York.
13. Hames, B. D. (1998) *Gel Electrophoresis of Proteins*, 3rd ed., Oxford University Press, London.

42

Bacteriocin Production Process by a Mixed Culture System

Suteaki Shioya and Hiroshi Shimizu

1. Introduction

Lactic acid bacteria (LAB) have been received much attention as bacteriocin producers. Antimicrobial proteins and peptides produced by bacteria, termed bacteriocin, are widely acknowledged to be important contributors to those organisms that survive dominate or die in microbial ecosystem such as our food supply or digestive tract. Also, there is tremendous interest in their use as a novel means to ensure the safety of minimally processed foods, because bacteriocins are proteins and natural *(1)*.

In order to enhance the productivity of those bacteriocins from LAB, one of the key issues is to remove lactate from the fermentor. Removal of lactate prevents the growth inhibition due to the increase in lactate concentration and the decrease in pH. In this chapter, one of methods to remove lactate is described by showing the data of nisin production from *Lactococcus lactis* as a typical example, but the method can be easily modified to other bacteriocin production process using LAB.

Nisin is an antimicrobial peptide produced by certain *Lactococcus* species. Nisin has been accepted as a safe and natural preservative in more than 50 countries *(2–4)*. Nisin inhibits the vegetative growth of a range of Gram-positive bacteria. Because nisin, in particular inhibits food-borne pathogens, e.g., *Listeria monocytogenes, Staphylococcus aureus*, and psychrotrophic enterotoxigenic *Bacillus cereus*, the effectiveness of nisin as a food preservative of these strains in various preservation conditions have been investigated in detail *(5–7)*. Not only the use of nisin-producing LAB as fermentation starters, but also the direct addition of nisin to various kinds of foods such as cheese, margarine, flavored milk, canned foods, and so on, are permitted *(3)*. The development of effective nisin production systems using LAB is a new field of interest. The most important problem in nisin production is the inhibition of growth due to the increase in lactate concentration and the decrease in pH. To avoid growth inhibition by the decrease in pH due to the accumulation of lactate in LAB fermentation processes, pH control methods by the addition of alkali or by the extraction of lactate using organic solvents have been reported *(8,9)*. However, the methods of extraction with organic solvents could not be used for food additive production. In this regard, removal of lactate in an appropriate way from the reactor will be preferable for pH control. Continuous culture with separation systems, such as membrane *(10–12)* or electrodialyzer *(4,13,14)*, have been reported. Instead of a separation system, a mixed culture system is utilized for removal of lactic acid *(15)* and explained in this chapter. The method can be easily modified to other bacteriocin production process using LAB.

2. Materials
2.1. Microorganisms and Media

L. lactis subsp. *lactis* ATCC 11454 (American Type Culture Collection, Rockville, MD) was used as a nisin-producing microorganism. *Kluyveromyces marxianus* MS1 was isolated from kefir grains in our laboratory and identified according to tests for morphological and biochemical properties *(16)*. *Staphylococcus aureus* IFO 12732, which was obtained from the collection at the Institute for Fermentation Osaka (IFO), Japan, was used as a indicator microorganism in the bioassay measurement of nisin concentration. The compositions of media for growth of microorganisms are summarized as follows. Medium A, used for seed culture and preculture of *L. lactis* (pH7.0), contained 5 g/L of maltose, 5g/L of polypeptone (Nihonseiyaku) and 5 g/L of yeast extract (Difco Laboratories, Detroit MI). Medium B, used for seed culture and preculture of *K. marxianus* (pH 7.0), contained 10 g/L of L-lactate, 10 g/L of polypeptone, and 10 g/L of yeast extract. Medium C, used for primary culture of *L. lactis*, contained maltose at 10 g/L in anaerobic culture without pH control and at 33–37 g/L in anaerobic culture with pH control and aerobic culture with pH control (pH 6.0), 10 g/L of polypeptone and 10 g/L of yeast extract. Medium D, used for primary culture of *K. marxianus* (pH 6.0), contained 40 g/L of L-lactate, 10 g/L of polypeptone and 10 g/L of yeast extract. Medium E, used for mixed culture of *L. lactis* and *K. marxianus* (pH 6.0), contained 42 g/L of maltose, 10 g/L of polypeptone and 10 g/L of yeast extract. Medium F, used for bioassay of nisin (pH 7.0), contained 10 g/L of glucose, 5 g/L of polypeptone, 5 g/L of yeast extract and 5 g/L NaCl. Selective medium G, used for determination of *L. lactis* colony-forming unit (CFU), (pH 7.0), contained 5 g/L of maltose, 5 g/L of polypeptone, 5 g/L of yeast extract, 5 mg/L of cycloheximide (Wako) and 15 g/L of agar 15 g/L. Selective medium H, used for determination of *K. marxianus* CFU (pH 7.0), contained 5 g/L of glucose, 5 g/L of polypeptone, 5 g/L of yeast extract, 5 mg/L of streptomycin (Nacalai tesque), and 15 g/L of agar. Initial maltose concentration in the pH-controlled anaerobic, pH-controlled aerobic, and mixed cultures were 33.2 g/L, 36.8 g/L, and 41.4 g/L, respectively.

2.2. Equipment and Facility

Precultures of *L. lactis* was performed in a 500-mL Erlenmeyer flask containing 200 mL of medium inoculated with 200 mL of the seed medium. The flask was statically incubated at 30°C for 10 h and harvested cells were inoculated to the primary culture medium. Preculture of *K. marxianus* was performed with 100 mL medium in 500 mL Erlenmeyer flask with 100 mL inoculation of the seed medium. The flask was incubated in the same way as the seed culture. After 16 h, the cells were harvested by centrifugation at 10,000g for 15 min and used to inoculate the primary culture medium.

Primary cultures were performed in a 5-L jar fermenter (EPC Control Box, Eyla, Japan) equipped with temperature, pH, dissolved oxygen concentration (DO), and gas flow control systems. The working volume was 2 L. The partial pressure of CO_2 in the exhaust gas was measured using a CO_2 gas analyzer (Horiba, cat. no. VBI-210). Air or nitrogen was supplied to the fermenter for aerobic or anaerobic cultivation conditions, respectively.

The cascade controller was developed for the systems that have more than one output with one manipulation *(17)*. In this study, the cascade control strategy was applied in order to control pH level via DO control by manipulating the agitation speed. Control strategy was coded by the N88BASIC (NEC) on a personal computer (NEC cat. no. PC-9801BX). The control strategy and inoculum conditions in the mixed culture are described later.

3. Method
3.1. Design of a Mixed CultureSystem

In order to remove lactate from the fermenter, a mixed culture system can be constructed and a pH cascade control system for the mixed culture system can be designed by following the procedure described as follows.

1. Select carefully a combination of carbon sources and microorganisms. For a targeted bacteriocin including a LAB, first select a carbon source, which should be commercially available and a microorganism which should or prefer not to assimilate the carbon source for the LAB but assimilate lactic acid quickly.

2. For this purpose, screen the microorganisms from some sources such as fermented food, because the fermented food is a good source for bacteriocin producing LAB as well as lactic acid-assimilating yeasts and bacteria. For nisin production, the carbon source is maltose and a yeast; *K. marxianus* was isolated from kefir grains. For another candidate, the carbon source is lactose and a yeast; *Candida vinaria* was isolated from cheese. This design is important step for a successful application of the method to other bacteriocin production process.

3.2. Assessment of the System Based on Kinetic Data

3.2.1. Analytical Methods of Experiments

Cell concentration of the pure cultures was measured as dry cell mass and optical density (OD). As for dry cell mass, the cell was filtered by membrane filter (pores size 0.45 mm, Advantec) and dried in the oven at 70°C. OD was measured at 660 nm by UV spectrophotometer (UV-2000, Hitachi). The viable cell concentrations of *L. lactis* and *K. marxianus* in mixed culture were determined as colony forming units (CFU) on selective media G and H, respectively. Relationship between dry cell concentration (DW) and CFU was approximated as a linear line by the least square method correlation coefficients of data and estimated values by the determined linear line were also calculated. Concentrations of L-lactate, acetate, and formate in the medium were analyzed enzymatically by F-kit lactate, F-kit acetate, F-kit formate (Boehringer Mannheim), respectively. Ethanol concentration was measured by gas chromatography (Hitachi G-3000). Glucose concentration was measured using a glucose analyzer (Model 2700, YSI Inc.). Maltose concentration was measured after hydrolysis to glucose as follows: 100 mL of 2N-HCl was added to the same volume of the sample and boiled at 100°C for 20 min. Two hundred milliliters of 1 N NaOH was added and glucose concentration was measured using the glucose analyzer. The calibration curve for ethanol concentration and maltose concentration were determined as linear lines by the least square method. Accuracy of measurement was evaluated by correlation coefficients.

Nisin concentration was measured by a bioassay method based on the method of Matsuzaki and co-workers *(18)* as follows: 5 mL of medium F was inoculated with *S. aureus* IFO12732 and incubated on a reciprocal shaker at 30°C, 100 strokes/min for 12 h. Fifty microliters of the cell suspension of *S. aureus* and 50 μL of the sample solution were added to 5 mL of fresh medium and incubated on the reciprocal shaker under the same conditions. After 12 h, the cell concentration was measured as optical density at 660 nm (OD660) by UV-spectrophotometer (Hitachi U-2000). A calibration curve was made for each new nisin concentration measurement, using commercially available nisin as a standard (Sigma, St. Louis, MO, 1000 international units (IU)/mg·solid; nisin content 2.5 wt %). The sample was diluted so that the value of OD_{660} was in the range of 0.1–1.5 absorbance units, because in this range nisin concentration was linearly related to OD_{660}. The calibration curve for bioassay of the nisin was made at each fermentation experiment by the least square method. Nisin concentration was represented by weight concentration (mg/L) and a nisin concentration of 1 mg/L was equivalent to 40 IU/mL *(12)*.

3.2.2. Collection of Kinetic Data of Pure Cultures

3.2.2.1. Anaerobic Pure Culture of *L. lactis*

In order to investigate kinetic parameters for both microorganisms, pure cultures should be done with and without pH control. The effect of DO concentration on kinetic parameters should be also analyzed from the experimental data. For anaerobic pure culture of *L. lactis* without pH control,

1. Inoculate *L. lactis* in the 5-L jar fermenter aseptically.
2. Monitor pH change every one minute without control. Store online data in the hard disk of a computer that is connected to fermentation equipment. Flow nitrogen gas in order to make anaerobic condition.
3. Take 10 mL sample every hour.
4. Measure offline data of concentrations of dry-cell (DW), nisin, L-lactate, and maltose.

For the case with pH control, the same methods as ones discussed above except for pH control are employed.

3.2.2.2. Aerobic Culture of *L. lactis*

Since the aerobic cultivation of LAB is not common, but the aerobic growth of, and production of lactate by, have been reported *(19)*. For lactate to be effectively assimilated by *K. marxianus,* aerobic conditions must be used. Hence growth of *L. lactis* under aerobic conditions was investigated.

1. Inoculate *L. lactis* in the 5-L jar fermenter aseptically.
2. Control pH at 6.0 by addition of NaOH.
3. Flow air and control DO at 6 mg/L by changing agitation speed.
4. Online and offline data should be measured and stored.

One of the results is shown in **Fig. 1**.

3.2.2.3. Aerobic Culture of *K. Marxianus*

In order to determine the specific lactate consumption rate of *K. marxianus*, aerobic fermentation should be performed, especially the data of the effect of the DO concentration on rate of lactate consumption are key ones to develop a cascade controller.

1. Inoculate *K. marxianus* in the 5-L jar fermenter aseptically.
2. Control pH at 6.0 by addition of NaOH.

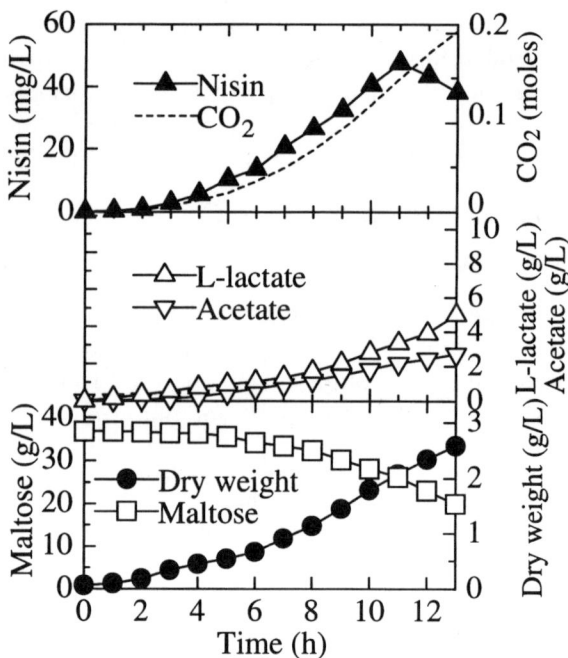

Fig. 1. Aerobic growth of *L. lactis* with pH control achieved by the addition of NaOH.

3. Flow air and control DO at several points (0–8) mg/L by changing agitation speed.
4. Online and offline data should be measured and stored.

3.2.2.4. Calculating Kinetic Parameters

The kinetic parameters, such as specific rates, were evaluated using the least square method. An example for the specific production rate of nisin is as follows: the material balance of nisin production is represented as

$$\frac{d(pV)}{dt} = \rho_N(VX_L) \qquad (1)$$

where p, V, X_L, and ρ_N are the nisin concentration, culture volume, cell concentration of *L. lactis*, and specific production rate of nisin, respectively. **Equation 1** was integrated as

$$\int_{p_0V_0}^{pV} d(pV) = \int_0^t \rho_N(VX_L)\,dt \qquad (2)$$

where the suffix 0 indicates the initial values of the variables. If ρ_N is constant, **Equation 2** can be rewritten

$$pV - p_0V_0 = \rho_N \int_0^t (VX_L)\,dt \qquad (3)$$

If the plot of pV vs the integral of VX_L is a linear relationship, it concludes that ρ_N is constant estimated from the slope of the linear line by the least square method. Algorithm of investigation of kinetic parameters is summarized as follows.

1. Give material balance equations for interested materials, e.g., cell, substrates, and products.
2. Integrate the equations with respect to time.
3. Plot the data so that the interested kinetic parameter becomes the slope of the curve.
4. Check the linearity of the curve.
5. If the plot shows the linear relation, estimate the slope by the least square method. Spreadsheet software like Excel (Microsoft Co.) is available.
6. Evaluate regression lines statistically. If the lag, log, and stationary phases existed, data for making one linear line should be determined. Check the consistency of the basis for determination by the correlation coefficients.

All of the kinetic parameters of *L. lactis* and *K. marxianus* are shown in **Tables 1 and 2**, respectively. The correlation coefficients for all the parameter estimations in the least square methods are also shown in these tables. The basis for determining the kinetic parameters over the first 5 h (0–5) and from 5 h to the completion of the experiment at 8, 12, or 11 h was mainly due to that the growth under anaerobic condition without pH control was ceased after 5 h. All the correlation coefficients of kinetic parameters were higher than 0.9, and it was concluded that the estimated values of kinetic parameters were consistent.

3.2.3. Assessment of the System

3.2.3.1. SIMULATION MODEL

In order to develop a cascade pH controller, a simulation model for dynamics of pH in the mixed culture of *L. lactis* and *K. marxianus* is useful. The dynamics of concentrations of lactate, L, and acetate, A, are represented by **Eqs. 4 and 5**, respectively.

$$\frac{dL}{dt} = \rho_L X_L - \nu_L X_K \quad (4)$$

$$\frac{dA}{dt} = \rho_A X_L \quad (5)$$

where X_L, ρ_L, and ρ_A are cell concentration, specific production rates of lactate and acetate of *L. lactis*, respectively, and X_K, and ν_L are cell concentration and specific lactate consumption rate of *K. marxianus*, respectively. From overall electroneutrality balance, total cation concentration in the medium have to be equal to anion concentration. H^+ ion concentration (10^{-PH}), OH^- ion concentration (10^{pH-14}), dissociated lactate ion concentration, and dissociated

Table 1
Kinetic Parameters of *L. lactis* Under Various Conditions

	Anaerobic conditions			Aerobic conditions			
	Without pH control	With pH control		With pH control		Mixed culture	
μ_L^a 1/h	(0–5 h) 0.30 ($r^2 = 0.979$)[b]	(0–5 h) 0.73 ($r^2 = 0.988$)	(5–8 h) 0.25 ($r^2 = 0.982$)	(0–5 h) 0.45 ($r^2 = 0.967$)	(5–12 h) 0.20 ($r^2 = 0.977$)	(0–5 h) 0.63 ($r^2 = 0.985$)	(5–11 h) 0.22 ($r^2 = 0.976$)
ρ_L^c g-lactate/g-cell/h	0.81 ($r^2 = 0.976$)	0.70 ($r^2 = 0.955$)	0.67 ($r^2 = 0.995$)	0.67 ($r^2 = 0.931$)	0.34 ($r^2 = 0.997$)	—[d]	—[d]
ρ_A^e g-acetate/g-cell/h	NM	0.29 ($r^2 = 0.994$)	0.16 ($r^2 = 0.994$)	0.32 ($r^2 = 0.986$)	0.23 ($r^2 = 0.989$)	0.28 ($r^2 = 0.997$)	0.24 ($r^2 = 0.984$)
ρ_N^f mg-nisin/g-cell/h	4.0 ($r^2 = 0.940$)	9.4 ($r^2 = 0.972$)	3.9 ($r^2 = 0.985$)	7.6 ($r^2 = 0.989$)	5.2 ($r^2 = 0.984$)	9.7 ($r^2 = 0.998$)	5.8 ($r^2 = 0.994$)
$Y_{L/M}^g$ mol-lactate/mol-maltose	3.7 ($r^2 = 0.95$)	2.0 ($r^2 = 0.94$) (0–8 h)		1.29 ($r^2 = 0.88$) (0–8 h)		—[d]	
$Y_{A/M}^h$ mol-acetate/mol-maltose	NM	0.96 ($r^2 = 0.91$) (0–8 h)		1.55 ($r^2 = 0.98$) (0–8 h)		0.94 ($r^2 = 0.99$) (0–11 h)	

[a] Specific growth rate; [b] correlation coefficient of the least square method; [c] specific lactate production rate; [d] cannot be evaluated; [e] specific acetate production rate of; [f] spec fic nisin production rate; [g] lactate production yield with respect to maltose; [h] acetate production yield with respect to maltose. NM, these values were not measured.

Table 2
Kinetic Parameters of *K. marxianus*

	Pure culture		Mixed culture	
	DO (mg/L)			
	0.5	2≥		
μ_K^a 1/h	0.12 ($r^2 = 0.987$)[b]	0.38 ($r^2 = 0.988$)	0.48 ($r^2 = 0.991$) (0–5 h)	0.31 ($r^2 = 0.987$) (5–11 h)
v_L^c g-lactate/g-cell/h	0.13 ($r^2 = 0.965$)	0.71 ($r^2 = 0.997$)	—[d]	—[d]

[a]Specific growth rate; [b]correlation coefficient of the least square method; [c]specific consumption rate of maltose of *K. marxianus*; [d]cannot be evaluated.

acetate ion concentration were balanced with ion concentration of acid and base, except lactate and acetate as

$$X_{acid} - X_{base} = 10^{-pH} - 10^{pH-14} - \frac{(L/MW_L)}{1 + 10^{-pH+pK_L}} - \frac{(A/MW_A)}{1 + 10^{-pH+pK_A}} \quad (6)$$

where X_{acid}, X_{base}, MW_A, MW_L, pK_L, and pK_A are total dissociated ion concentrations of acid and base, except lactate and acetate, molecular weights of acetate and lactate, and pK values of lactate and acetate, respectively. Third and fourth terms in the right-hand side of **Eq. 6** corresponds to dissociated lactate ion concentration, and dissociated acetate ion concentration, respectively. By rewriting the right-hand side of **Eq. 6** as a nonlinear function of pH, L, A as $f(pH, L, A)$ and differentiating **Eq. 6** with respect to time (t), **Eq. 7** is obtained.

$$[d(X_{acid} - X_{base})/dt] = [(\partial f/\partial pH)] \cdot [(dpH)/(dt)] + [(\partial f/\partial L)] \cdot [(dL)/(dt)] + [(\partial f/\partial A)] \cdot [(dA)/(dt)] \quad (7)$$

Because the changes in X_{acid} and X_{base} are negligible compared with the changes in concentrations of lactate and acetate, the dynamics of the pH change with time is described as

$$dpH/dt = -[(\partial f/\partial L) \cdot (dL/dt) + (\partial f/\partial A) \cdot (dA/dt)]/(\partial f/\partial pH) \quad (8)$$

By substituting **Eqs. 4** and **5** to **Eq. 8**,

$$dpH/dt = -[(\partial f/\partial L)(\rho_L X_L - v_L X_K)/MW_L + (\partial f/\partial A)(\rho_A X_L)/MW_A]/(\partial f/\partial pH) \quad (9)$$

The terms $(\partial f/\partial pH)$, $(\partial f/\partial L)$, and $(\partial f/\partial A)$ in **Eq. 9** are positive and the production rate of acetate at the beginning of the batch culture is negligible. Thus, it is

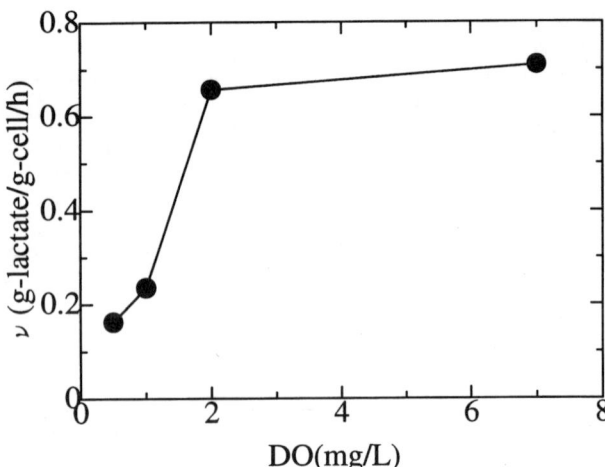

Fig. 2. Effect of DO level on the specific rate of lactate consumption by *K. marxianus*.

easily understood that if the rate of lactate production by *L. lactis* is higher than the rate of lactate consumption by *K. marxianus*, i.e., the term $(\rho_L X_L - v_L X_K)$ is positive, pH decreases. On the other hand, when the consumption of lactate is higher than production, i.e., the term $(\rho_L X_L - v_L X_K)$ is negative, pH increases. Then, the lactate consumption rate by *K. marxianus* must be increased or decreased, depending on whether the pH level is above or below the set point of 6.0 as long as lactate exists in the medium. The specific lactate consumption rate (v_L) by *K. marxianus* can be controlled within a limited range (0–0.7 g-lactate/g-cell/h) by changing the DO level (0–2 mg/L) as shown in **Fig. 2**. The DO concentration in the medium can be controlled by manipulating the agitation speed of the impeller. It was found that the pH level in the medium of the mixed culture system can be controlled by changing the set point of DO control. This type of controller is categorized as a cascade controller *(17)*.

Using **Eq. 9**, dynamics of pH can be predicted. Kinetic parameters can be available from the fermentation data analysis. Algorithm for development of simulation model of dynamics of pH in the mixed culture is summarized as follows.

1. Give all the material balance equations in terms of all the fatty acids that are produced and/or assimilated by one or more than one microorganism.
2. Establish an overall electroneutrality balance equation.
3. By differentiation of the electroneutrality balance equation, get an ordinary differential equation of the pH.
4. Simulate pH dynamics with these equations numerically by the Runge–Kutta method if necessary.
5. Examine the roles of both microorganisms for pH control from the dynamic equation and develop the controller.

3.2.3.2. Inoculum Size of Mixed Culture

So that the lactate concentration was constant at initial time of the operation, i.e., the start of the fermentation, the decision of the ratio of inoculum sizes of *L. lactis* and *K. marxianus* is very important. If lactate accumulates to a high level in the medium at the beginning of mixed culture, the growth rate of *L. lactis* is dramatically decreased, causing a fatal condition. For a reliable mixed culture operation, the ρ_L was overestimated as 2.0 g-lactate /g-cell/h, which was three times higher than the actual data (*see* **Table 1**). On the other hand, v_L was initially set to 0.5 g-lactate/g-cell/h. Then, the ratio of X_L and X_K was set to 0.25, which can be derived by setting the right-hand side of **Eq. 4** to zero, which is equivalent to that lactate concentration being constant. For the same reasons, the right-hand side of **Eq. 9** becomes zero because the third term is negligibly small at the beginning, which means that pH does not change. After all, before the primary culture experiment started, concentrations of both microorganisms in the precultures were determined by measuring the OD_{660} and inoculum size was set as $X_L/X_K = 0.25$ in the following experiments. The algorithm for determination of inoculum size is follows.

1. Examine the specific production rate of fatty acids (lactate) by LAB and specific consumption rate of the fatty acids, which is assimilated by another microorganism.
2. For a reliable mixed culture operation, overestimate the specific production rate of lactate.
3. From the dynamic equation of pH, decide the initial cell concentrations of both microorganisms so that the pH does not change.
4. Measure cell concentration in the preculture and adjust inoculum size for the primary culture according to **step 3**.

3.3. pH Cascade Controller Design

3.3.1. System Analysis Using the Simulation Model

A cascade controller of pH coupled with DO control developed here is shown in **Fig. 3**. Proportional and integral (PI), and proportional, integral, and differential (PID) controllers were used as precompensators of DO and pH in the cascade controller, respectively. The dynamic response of boxes A and B can be analyzed using the simulation model. The analysis is useful for design of the controllers, especially selection of controllers and determination of control parameters. The dynamic response of pH due to the DO change, shown in a box A in **Fig. 3**, is as follows: when the DO changes, v_L changes according to the relationship between DO and v_L shown in **Fig. 2**. Noted that **Fig. 2** shows a static relationship. However, when the DO level changes dynamically, there is a dynamic time delay from DO change to v_L, because cell activity for lactate assimilation occurs after changes in activities of many enzymes due to change

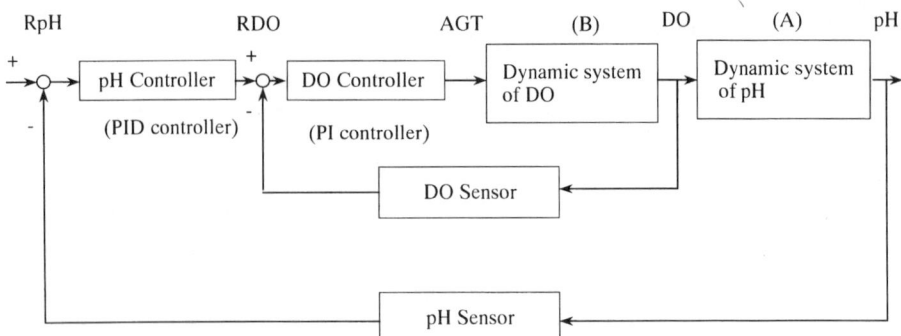

Fig. 3. Scheme of a cascade pH controller incorporating DO control. Two outputs of pH and DO were measured by pH and DO sensors, respectively. Manipulation was only via agitation speed (AGT), which changed DO, and DO changed pH via lactate consumption by *K. marxianus*. The pH was controlled by the DO set point, RDO.

in DO concentration. The change in pH is based on the dynamics described by **Eq. 9**. The dynamic response of the agitation speed of the impeller to the DO level, shown in box B of **Fig. 3**, is as follows: when the agitation speed changes, the mass transfer rate of oxygen from air bubble to liquid changes, with a very small time delay, and the dynamics of DO are described by the balance between oxygen supply from air bubble to liquid and consumption rate of oxygen by both microorganisms.

3.3.2. PI (DO) Control Strategy

For the DO controller shown in **Fig. 3**, the digital PI controller used is represented by

$$AGT(t) = K_p\{[RDO(t) - DO(t)] + \frac{\Delta_t}{T_i} \sum_{i=0}^{+} [PDO(i) - DO(i)]\} \quad (10)$$

where $AGT(t)$, $DO(t)$, $RDO(t)$, Δt, K_p, and T_i are the agitation speed of the impeller and DO at time (t), and the set point of DO control, sampling time, proportional gain and integral time in the controller, respectively. In the conventional PI controller, $RDO(t)$ was usually treated as a constant value, but it was given as the output of pH controller in the cascade controller. The PI controller was actually used as a velocity form by rewriting **Eq. 10** as

$$AGT(t) = AGT(t-1) + Kp\{[DO(t-1) - DO(t)] + [RDO(t) - RDO(t-1) + \frac{\Delta_t}{T_i}[RDO(t) - DO(t)]\} \quad (11)$$

where the initial value of AGT, AGT(0), was set to 100 rpm. In the velocity form of the controller, the $AGT(t)$ was realized by correcting $AGT(t-1)$. Although **Eqs. 10** and **11** are completely equivalent, the overshoot would be reduced in the velocity form, when manipulating variables of AGT have upper and lower limits.

3.3.3. PID (pH) Control Strategy

For the pH controller, the PID controller was used shown as

$$\text{RDO}(t) = K_p\{[\text{RpH} - \text{pH}(t)] + \frac{\Delta t}{T_i}\sum_{i=0}^{t}[\text{RpH} - \text{pH}(i)] + \frac{T_d}{6\Delta t}[\text{pH}(t-6) - \text{pH}(t)]\} \quad (12)$$

where $\text{pH}(t)$, RpH, and T_d are pH at time t, the pH set point, and derivative time in the controller, respectively. In this controller, the derivative of pH was estimated by subtracting the present pH from the pH data at time $(t-6)$. The PID controller was used as a velocity form in the same way as the PI (DO) controller as

$$\text{RDO}(t) = \text{RDO}(t-1) + K_p\{\text{pH}(t-1) - \text{pH}(t) + \frac{\Delta t}{T_i}[\text{RpH} - \text{pH}(t)] + \frac{T_d}{6\Delta t}[-\text{pH}(t) + \text{pH}(t-1) + \text{pH}(t-6) - \text{pH}(t-7)]\} \quad (13)$$

RDO at the initial time, RDO(0), was set as 1.0 mg/L. Because there was a large delay between DO change and specific lactate consumption rate of *K. marxianus* responses, the derivative correction term in the controller was very important for detecting the pH change, whereas there was a small delay from the agitation change to the DO response.

The control parameters of the PI (DO) controller—sampling time (Δt), proportional gain (K_p), and integral time (T_i) — are set to 0.5 min, 40 rpm·L/mg, and 2.0 min, respectively. Control parameters of the PID (pH) controller — set point of pH (RpH), sampling time, (Δt), proportional gain (K_p), integral time (T_i), and derivative time (T_d) — are set to 6.0, 0.5 min, 0.5 mg/L, 12.5 min, and 6.0 min, respectively. The control parameters of PI and PID controllers were tuned so that fluctuation of pH was less than 6.0 ± 0.5. The design algorithm of the cascade controller is as follows.

1. Design the whole structure of the cascade controller.
2. Select type of controller, e.g., PI, PID, or more advanced controller for precompensators.
3. Tune up controller parameters until you get the satisfactory control result.

3.3.4. Typical Results

The time course of a mixed culture of *L. lactis* and *K. marxianus* is shown in **Fig. 4**. The pH set point was 6.0. The DO set point was not obtained automatically but manually changed in this case for the first trial. The lactate concentration was kept at almost zero throughout the experiment. After 2 h, the DO level was increased and the v_L was enhanced because the pH decreased slightly from the set point of 6.0. The cascade control of pH succeeded. Both *L. lactis* and *K. marxianus* were growing exponentially and ended at 11 h. The nisin production reached a maximum of 98 mg/L.

Figure 5 shows the automatic cascade control results for the coupling of pH with DO control in the mixed culture. Because RDO(0) was set to 1.0 mg/L,

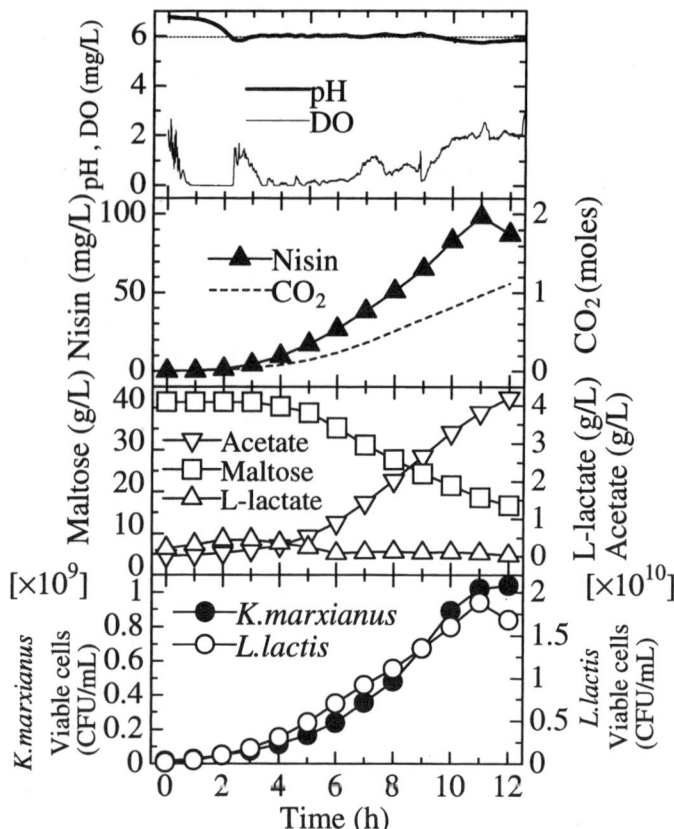

Fig. 4. Nisin production in a mixed culture of *L. lactis* and *K. marxianus*. The pH was controlled at 6.0 by the cascade controller.

the actual value of DO decreased rapidly at the beginning of the experiment and returned to the set point within 2 h. When the pH decreased below the set point at around 3 h, the RDO increased according to **Eq. 13**. As a result, the v_L was recovered and pH increased again to around 4 h due to the decreasing the lactate concentration. By changing the RDO, the pH was reliably and accurately controlled at the set point of 6.0.

4. Concluding Remarks

In order to control the pH in an antimicrobial peptide (nisin) production process by a LAB, *L. lactis* subsp. *lactis* (ATCC11454), a novel method was described, in which neither addition of alkali nor a separation system such as ceramic membrane filter and electrodialyzer was adopted. A mixed culture of *L. lactis* with *K. marxianus*, which was isolated from kefir grains, was utilized

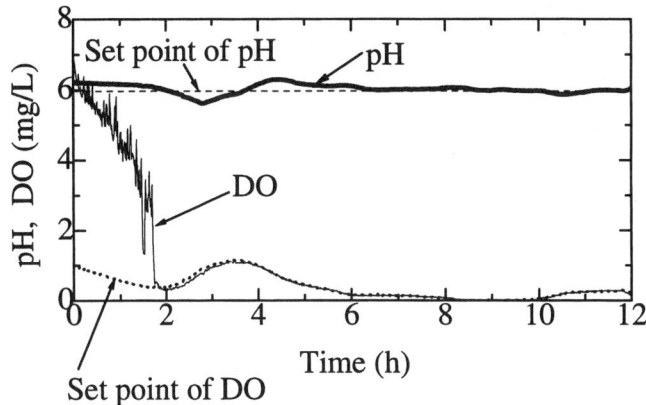

Fig. 5. Control of pH and DO by the cascade controller without addition of NaOH in the mixed culture system.

in the system. Because the pH of the medium could be controlled by the lactate consumption of K. marxianus and the specific lactate consumption rate of K. marxianus could be controlled by changing the dissolved oxygen (DO) concentration, a cascade pH controller coupled with DO control was described. As a result, pH and lactate were kept at low levels and nisin accumulated in the medium to a high level, compared with other pH control strategies such as processes without pH control and with pH control by addition of alkali. It should be stressed that by careful selection of the system, the principle presented here can be made available for use with another carbon source and other combination of microorganisms for other bacteriocin production.

5. Nomenclature

A	acetate concentration (g/L)
$AGT(t)$	agitation speed of impeller at time t (rpm)
$DO(t)$	dissolved oxygen concentration at time (mg/L)
K_p	proportional gain of PI or PID controller (rpm · L/mg) or (mg/L)
MW_A	molecular weight of acetate
MW_L	molecular weight of lactate
L	lactate concentration (g/L)
p	nisin concentration (mg/L)
pK_A	pK value of acetate
pK_L	pK value of lactate
$RDO(t)$	set point of DO at time t for DO controller (mg/L)
RpH	set point of pH for pH controller (−)
Δt	sampling time (min)
T	temperature (°C)

T_i integral time of PI or PID controller (min)
T_d derivative time of PID controller (min)
V culture volume (L)
X_{acid} total dissociated ion concentration of acid (mol/L)
X_{base} total dissociated ion concentration of base (mol/L)
X_K cell concentration of *K. marxianus* (g/L)
X_L cell concentration of *L. lactis* (g/L)
μ_K specific growth rate of *K. marxianus* (h^{-1})
μ_L specific growth rate of *L. lactis* (h^{-1})
ν_L specific lactate consumption rate of *K. marxianus* (g-lactate/g-cell/h)
ρ_A specific acetate production rate of *L. lactis* (g-acetate/g-cell/h)
ρ_L specific lactate production rate of *L. lactis* (g-lactate/g-cell/h)
ρ_N specific nisin production rate of *L. lactis* (mg-nisin/g-cell/h)

References

1. Hoover, D. G. and Steenson, L. R. (1993) *Bacteriocins of Lactic Acid Bacteria*, Academic Press, New York, pp. 1–2.
2. Hurst, A. (1981) Nisin. *Adv. Appl. Microbiol.* **27**, 85–123.
3. Broughton, J. B. (1990) Nisin and its uses as a food preservative. *Food Technol.* **44**, 100–117.
4. Ishizaki, A. and Vonktaveesuk, P. (1996) Optimization of substrate feed for continuous production of lactic acid by *Lactococcus lactis* IO-1. *Biotechnol. Lett.* **18**, 1113–1118.
5. Ryan, M. P., Rea, M. C., Hill, C., and Ross, R. P. (1996) An application in cheddar cheese manufacture for a strain of *Lactococcus lactis* producing a novel broad-spectrum bacteriocin, lacticin 3147. *Appl. Environ. Microbiol.* **62**, 612–619.
6. Thomas, L. V. and Wimpenny, J. W. T. (1996) Investigation of the effect of combined variations in temperature, pH, and NaCl concentration on nisin inhibition of *Listeria monocytogenes* and *Staphylococcus aureus*. *Appl. Environ. Microbiol.* **62**, 2006–2012.
7. Beuchat, L. R., Clacero, M. R. S. C., and Jaquette, C. B. (1997) Effects of nisin and temperature on survival, growth, and enterotoxin production characteristics of psychrotrophic *Bacillus cereus* in beef gravy. *Appl. Environ. Microbiol.*, **63**, 1953–1958.
8. Yabannavar, V. M. and Wang, D. I. C. (1991) Extractive fermentation for lactic acid production. *Biotechnol. Bioeng.* **37**, 1095–1100.
9. Honda, H., Toyama, Y., Takahashi, H., Nakazecko T., and Kobayashi, T. (1995). Effective lactic acid production by two-stage extractive fermentation. *J. Ferment. Bioeng.* **79**, 589–593.
10. Ohara, H., Hiyama, K., and Yoshida. T. (1993) Lactic acid production by filter bed-type reactor. *J. Ferment. Bioeng.* **76**, 73–75.
11. Shi, Z., Shimizu, K., Iijima, S., Morisue, T., and Kobayashi, T. (1990) Adaptive on-line optimizing control for lactic acid fermentation. *J. Ferment. Bioeng.* **70**, 415–419.

12. Taniguchi, M., Hoshino, K., Urasaki, H., and Fujii, M. (1994) Continuous production of an antibiotic polypeptide (nisin) by *Lactococcus lactis* using a bioreactor coupled to a microfilteration module. *J. Ferment. Bioeng.* **77,** 704–708.
13. Nomura, Y., Yamamoto, K., and Ishizaki. A. (1991) Factors affecting lactic acid production rate in built-in electrodialysis fermentation, and approach to high speed batch culture. *J. Ferment. Bioeng.* **71,** 450–452.
14. Vonktaveesuk, P., Tonokawa, M., and Ishizaki, A. (1994) Stimulation of the rate of L-lactate fermentation using *Lactococcus lactis* IO-1 by periodic electrodialysis. *J. Ferment. Bioeng.* **77,** 508–512.
15. Shimizu, H., Mizuguchi, T., Tanaka, E., and Shioya, S. (1999) Nisin production by a mixed-culture system consisting of *Lactococcus lactis* and *Kluyveromyces marxianus*. *Appl. Environ. Microbiol.* **65,** 3134–3141.
16. Kultzman, C. P. and Fell, J. W. (1998). *The Yeast, A Taxonomic Study*, Elsevier, Amsterdam, The Netherlands.
17. Stephanopoulos, G. (1984) *Chemical Process Control*, Prentice-Hall International Inc., Englewood Cliffis, NJ.
18. Matsusaki, H., Endo, N., Sonomoto, K., and Ishizaki. A. (1995) Purification and identification of a peptide antibiotics produced by *Lactococcus lactis* IO-1. *J. Fac. Agr., Kyushu Univ.* **40,** 73–85.
19. Borch, E. and Molin, G. (1989) The aerobic growth and product formation of *Lactobacillus*, *Leuconostoc*, *Brochothrix*, and *Carnobactrium* in batch cultures. *Appl. Microbiol. Biotechnol.* **30,** 81–88.

VI

REVIEWS

43

Nutritional Status of Grape Juice

Bruce W. Zoecklein, Barry H. Gump, and Kenneth C. Fugelsang

1. Introduction

The chemical and physical environment of grape juice during fermentation, coupled with competition from indigenous yeast and bacteria, can present significant challenges to the growth of *Saccharomyces cerevisiae*. Individually or collectively, these factors may impact both yeast growth and the conversion rate of sugar to alcohol, leading not only to the formation of objectionable odor- and flavor-active metabolites but, potentially, protracted, incomplete, or "stuck" fermentations as well. Sluggish and stuck fermentations can be described as those where the rate of sugar utilization is extremely slow, especially near the end, and/or where residual fermentable sugar is left in the wine. Such wines create significant management problems. Table wines containing biologically available levels of sugar (≥0.2% w/v glucose + fructose) may undergo spontaneous refermentation. Further, unexpectedly sweet wines may be an unacceptable departure from wine style.

Although the underlying causes of stuck fermentations may vary (*see* **Fig. 1** and Production Notes in Chapter 32, this volume), metabolic failure ultimately results from diminished and eventually blocked capacity of the yeast cell membrane to transport glucose and fructose into the cell *(1,2)*. Incorporation of both sugars is accomplished by the action of a group of membrane-associated permeases, the hexose transport proteins *(3,4)*. Continued operation requires unimpeded mobility of the carrier proteins and sugars across the cell membrane. Fermentative growth and a variety of environmental factors, including decreasing availability of critical nutrients and increasing concentrations of ethanol and other inhibitory metabolites, may significantly alter membrane fluidity, thereby lowering the incorporation of glucose and fructose *(5,6)*.

From: *Methods in Biotechnology, Vol. 14: Food Microbiology Protocols*
Edited by: J. F. T. Spencer and A. L. Ragout de Spencer © Humana Press Inc., Totowa, NJ

This review focuses primarily on yeast-assimilable nitrogen concentration, one of the underlying causes of problem fermentations. However, as suggested by **Fig. 1**, many factors may interactively contribute to protracted and/or interrupted fermentation. An abridged review of the impact of each factor on fermentation is presented in Chapter 32, this volume.

2. Total Nitrogen

Nitrogen compounds in grapes play important roles as nutrients for microorganisms involved in winemaking and wine spoilage and as aroma and aroma precursors *(7)*. Nitrogen is taken up by the vine roots as nitrate and reduced by the nitrate reductases system to ammonia, transported and stored subsequently as amino acids *(7)*. Compared with fermentable carbon generally present in grapes at >20% (w/v), total nitrogen levels range from 0.006% to 0.24%, of which only 0.0021–0.08% is biologically available to fermenting yeasts *(8)*. Thus, nitrogen may become an important growth-limiting constraint for microorganisms.

Fermentation rate (conversion of sugars to alcohol and carbon dioxide) is directly related to biomass. Yeasts follow a classic growth profile beginning with a lag phase followed by a period of rapid growth that culminates in the stationary phase, where the population density remains relatively high. Stationary-phase yeasts are responsible for the majority of alcoholic fermentation *(9)*. Eventually, depletion of carbon and nitrogen, coupled with accumulation of toxic metabolites, leads to cell death and the decline phase.

During the stationary phase nitrogen uptake and utilization are directed toward cell maintenance; for example, transporter proteins have a high turnover rate during this stage of growth and thus require continued resynthesis *(5)*.

The total nitrogen content of juice and wine consists of protein and nonprotein fractions. Protein nitrogen comprises from 1% to 13% of the total N *(10)*, whereas polypeptides may account for more than 21%. Because *Saccharomyces* sp. lacks both the extracellular proteases and transport enzymes necessary for protein incorporation *(11)*, neither fraction plays a significant nutritional role. However, some native, non-*Saccharomyces* yeasts and bacteria are capable of producing proteases *(12)*. Whether protease activity is sufficient to provide nitrogen to more than the species involved is unknown.

3. Assimilable Nitrogen

The nitrogenous components of grapes and juice that are metabolically available to yeasts are present as ammonium salts (NH_4^+) and primary or "free α-amino acids" (FAN). Combined, the two groups are referred to as Yeast-Assimilable Nitrogenous Compounds or YANC *(13)*. Thus, a complete evaluation of the nutritional status of juice or must requires measurements of both fractions.

Nutritional Status of Grape Juice

In grapes, NH_4^+ ranges from near 30 to more than 400 mg/L *(14,15)*, whereas in wine, levels of less than 50 mg/L have been reported *(16)*. Numerous studies have demonstrated the priority of NH_4^+ uptake by yeasts relative to amino acids. Jiranek et al. *(17)* and Monk et al. *(18)* reported that NH_4^+ was not only incorporated preferentially to α-amino acids but also altered the established pattern of amino acid uptake.

All of the 20 commonly occurring amino acids are found in grapes and wine. Their total concentration ranges from 0.4 to 6.5 g/L *(19)*. Of these, only the α-amino acid FAN fraction is directly assimilable by yeasts. This fraction includes arginine, serine, threonine, α-amino butyric, aspartic and glutamic acids. Collectively, this group comprises 35–40% of the total N and 75–85% of the total amino acids *(19)*. Arginine is typically present at levels ranging from 5 to 10 times that of the other amino acids and represents 30–50% of the total nitrogen utilization *(20)*.

Minimum levels of FAN required for successful completion of alcoholic fermentation range from 120 to 140 mg/L for musts with sugar concentrations of 160 to 240 g/L *(21,22)*. FAN levels of 400–500 mg/L or greater are required for maximum fermentation rate *(20)*. During growth, yeast utilizes 1–2 g/L amino acids *(23)*. Depending on the particular amino acid, the yeast's stage of growth and the presence and activity of necessary transport enzymes, amino acids may be (1) directly incorporated into proteins, (2) degraded for either their nitrogen or carbon components, or (3) stored in vacuoles for later utilization *(24)*.

Amino acid uptake by *Saccharomyces cerevisiae* requires two amino acid transport systems *(20)*. During the early stages of growth, permeases specific to individual L-amino acids (i.e., arginine permease) are active. Once the concentration of NH_4^+ has decreased and amino acids have been incorporated, a general amino acid permease (GAP) is induced. Unlike specific permeases, GAP is group specific, transporting both D and L isomers of basic (arginine, histidine, lysine) and neutral (glutamine, alanine, serine, asparagine, threonine) amino acids, with the exception of proline *(25)*. GAP's role has been described as that of a "nitrogen scavenger" that becomes operative in the latter stages of fermentation upon depletion of the more available forms of nitrogen *(26)*.

The patterns of amino acid incorporation from grape juice have been reported by Castor *(27)* and others *(18,28–32)* and are known to vary depending on the presence or absence of ammonium salts *(20)*. In the absence of NH_4^+, the first group of amino acids incorporated includes arginine, asparagine, aspartate, glutamate, glutamine, serine, and threonine (group A), followed by histidine, isoleucine, leucine, methionine, and valine (group B). Alanine, phenylalanine, glycine, tryptophan, and threonine (group C) represent the last amino acids to be incorporated. The presence of NH_4^+ delays both the timing and extent of amino acid incorporation.

Not all amino acids are directly utilizable by yeast during fermentation *(20)*. Metabolic block(s) may be transitory or permanent. Proline is an example of an amino acid whose utilization is linked to environmental conditions. Proline is present in relatively high concentrations (700–800 mg/L) but is not biologically available to *Saccharomyces* sp. before or during alcoholic fermentation. Proline utilization requires two enzymes, a permease and an oxidase. The permease, initially required for uptake, is inhibited by the levels of NH_4^+ present before and during the early stages of fermentation. By the time permease inhibition is released, redox conditions have dropped and the oxidase needed for ring cleavage is not functional *(33)*. Where oxidative conditions exist, proline becomes metabolically available, undergoing cleavage to glutamic acid and transamination to glutamate prior to entering biosynthetic pathways *(34)*. The importance of proline to the cell is not restricted directly to nutritional requirements. Along with glycine and the disaccharide trehalose, proline has also been reported to help "protect" the cell against osmotic stress *(35)*.

Other amino acids that cannot be directly utilized as sole nitrogen sources by *Saccharomyces* include lysine, cysteine, histidine, and glycine *(36–38)*. In the case of lysine, inhibition results from accumulation of the intermediate α-amino adipate semialdehyde *(37)*. Non-*Saccharomyces* yeasts are capable of utilizing lysine as a sole carbon source. This technique may be employed for identification in mixed cultures. As described by Zoecklein et al. *(8)*, the sulfur-containing amino acid cysteine can be degraded to yield sulfide (H_2S), ammonium, and pyruvate via reverse operation cysteine synthetase *(39)*. Cysteine accumulation, however, is inhibitory to *Saccharomyces (40)*.

4. Production Considerations Influencing FAN

Most grape growing and winemaking decisions can influence FAN concentration. These include grape variety *(14,41,42)* and rootstock selection *(43)*, climate, soil type *(41,44)*, fertilization, and irrigation practices *(45–47)*. Arginine and proline are the main amino acids in the fruit if the fertilization of the vine is low. With higher fertilization (>3 g N/plant), the amino acid amide glutamine increases dramatically *(7)*. Therefore, the nitrogen available for yeast fermentation can be different in distinct wine growing regions of the world *(7)*. Grape maturity is also an important issue influencing the concentration of FAN, in that underripe and overripe fruit may be low in FAN *(23)*. Butzke *(48)* evaluated the yeast-assimilable nitrogen status of *Vitis vinifera* musts from the western United States in 1996. The concentration ranged from 40 to 559 mg N/L, with an average of 213 mg N/L. Primary amino acid content ranged from 29 to 370 mg N/L (average, 135 mg N/L) whereas ammonium (NH_4^+) ranged from 5 to 325 mg N/L (average, 70 mg N/L).

Microbiological deterioration of fruit can also influence initial FAN levels *(23,49)*. The growth of *Botrytis cinerea* may lower the concentration of amino acids by up to 61% *(7)*. Increasing interest in prefermentation maceration and native fermentations (both yeasts and lactic acid bacteria) has lead to increased concern regarding depletion of FAN required by *Saccharomyces* sp. Native yeast and bacteria, present initially at relatively low population densities, require significant amounts of FAN and vitamins to build biomass. By the time *Saccharomyces* sp. populations become established, levels of available nitrogen may be less than those required to achieve complete fermentation.

Winemaking practices such as whole-cluster pressing rather than conventional crushing/stemming, coupled with juice clarification, are effective techniques for reducing nonsoluble solids levels in white juice. Reduction of nonsoluble solids concentrations may deprive yeasts of nutritionally important substrates *(50–52)* and oxygen reservoirs during the early stages of fermentation *(24)*. Fining agents may serve to further deplete vital nitrogen sources. Guitart et al. *(52)* evaluated processing and several commonly utilized prefermentation fining agents with regard to amino acid reduction. They reported that silica gel additions removed the highest concentration of amino acids, followed by enzyme treatment, cold-clarification, bentonite, and centrifugation. Koch *(53)* reported reductions of total N in white juice of 10–15%, following cold clarification. Where prefermentation bentonite additions were made, reductions in *total* nitrogen of over 50% were observed. In terms of amino acid reduction, Rapp *(54)* reported 15–30% reduction in amino acids following bentonite additions of 1 g/L.

5. Correcting Nitrogen Deficiency

Yeast-assimilable nitrogenous compound deficiency in fermenting juice/must is often corrected by the addition of assimilable nitrogen in the form of diammonium phosphate, or DAP (25.8% NH_3, 74.2% PO_4 w/w), and/or one of several commercially available nitrogen supplements. Commercial nitrogen supplements typically contain DAP (25–50% w/w) in addition to more complex forms of nitrogen such as yeast extract, vitamins, and yeast hulls. Because the concentration of nitrogen compounds may vary with the product, it is recommended that winemakers consult the Material Safety Data Sheet (MSDS) for formulation information prior to use.

Utilization of nitrogen supplements is regulated. In the United States and among OIV nations, the maximum addition of ammonium salts (DAP) is 960 and 300 mg/L, respectively. In Australia, supplementation is linked to phosphate levels in wine. The maximum level of inorganic phosphate (P_i) is 400 mg/L *(20)*.

Traditionally, winemakers employing nitrogen supplementation add the product along with yeasts at the start of fermentation. As noted, incorporation of ammonia and amino acids occurs primarily during the yeast's growth phase, with limited uptake thereafter. Further, the presence of NH_4^+ delays the uptake of amino acids. Given this, a better plan is to supplement at a stage after the yeast has incorporated available forms. This may take the form of incremental additions starting at 48 h (for reds) and 72 h (for whites) postinoculation or a single addition midcourse during the fermentation (*see* Chapter 32, this volume). Sablayroles *(15)* reported that a single addition made midway through fermentation was as effective as a single addition at the start and circumvented the issue of amino acid uptake. Nitrogen additions as low as 60 mg/L (as ammonium phosphate) stimulated yeast activity within 1 h after addition *(15)*.

6. Ethyl Carbamate Formation

Excessive amounts of nitrogen compounds in juice and wines can impact wine aroma *(55)* and the formation of ethyl carbamate. Ethyl carbamate, or urethane, is a carcinogen that occurs naturally in fermented foods, including wine, as a result of the fermentative and assimilative activities of microorganisms. Toxicity studies will be completed by the U.S. Food and Drug Administration by the year 2001 prior to recommending maximum acceptable levels (Gahagan, 2000, personal communication). At present, the U.S. wine industry has established a voluntary target level of <15 µg/L (ppb) for table wines and <60 µg/L for dessert wines.

Ethyl carbamate is produced from the reaction of urea and ethanol. Although several precursors of urea and, subsequently, ethyl carbamate have been identified *(56)*, quantitatively the most important source is the amino acid arginine. Upon incorporation, arginine is first converted to ornithine and urea, which is directly utilizable by fermenting yeasts *(34)*. In the presence of urea and ethanol, formation of ethyl carbamate increases exponentially as a function of temperature *(57,58)*.

Formation of urea occurs during the early to midphase fermentation. This corresponds to the point at which the fermentation of dessert-style wines (such as Port style) are typically arrested by additions of alcohol. Yeast strains exhibit variability in terms of urea uptake and excretion during fermentation *(59)*.

Grapes from high-vigor vines and/or heavily fertilized vineyards have high levels of arginine (>400 mg/L). Modifying vineyard fertilization practices, utilizing yeast strains that release less urea, and timing the fortification of dessert wines when urea concentrations are low may reduce ethyl carbamate formation *(58)*. Commercial ureases produced from *Lactobacillus fermentatum* are available for postfermentation treatment of wines *(60)*.

7. Conclusions

Several studies have highlighted the importance of nitrogen for successful fermentations *(61–65)*. There is a strong positive correlation between must N and wine aroma and flavor intensity. Specifically, wines produced from high-N musts have a higher concentration of esters and lower concentration of fusel alcohols. Both are positive factors contributing to wine quality. Successful management of nitrogen deficiency requires that the winemaker identify the potential for problems early in the winemaking process. Routine and easily performed estimates of assimilable nitrogen (FAN + NH_4^+) during juice and must processing would be a valuable tool for the winemaker. Historically, several analytical methods have been proposed to measure total nitrogen. These have included ninhydrin *(66)* and the trinitrobenzene sulfonic (TNBS) method described by Crowell et al. *(67)*. Utilization of TNBS has largely been discontinued because of difficulties in obtaining the chemical as well as waste-management issues. Further, these methods yield erroneously high results because of the inclusion of variable concentrations of protein and peptide nitrogen, proline, and other amino acids, which are not readily incorporated by *Saccharomyces*. A new spectrometric procedure utilizing ortho-phthalaldehyde derivatives of the α-amino acids is now available. Component analysis of individual amino acids provides insight into the potential for problem fermentations but is expensive and slow. A useful method is described in Chapter 32, this volume.

References

1. Salmon, J. M. (1989) Effect of sugar transportation inactivation in *Saccharomyces cerevisiae* on sluggish and stuck enological fermentations. *Appl. Environ. Microbiology* **55,** 9536–9538.
2. Lagunas, R. (1993) Sugar transport in *Saccharomyces cerevisiae. FEMS Microbiol. Rev.* **104,** 229–242.
3. Cirillo, V. P. (1962) Mechanism of glucose transport across the yeast cell membrane. *J. Bacteriol.* **84,** 485–491.
4. Salmon, J. M. (1997) Enological fermentation: kinetics of an isogenic ploidy series derived from an industrial *Saccharomyces cerevisiae* strain. *J. Ferm. Bioeng.* **83,** 253-260.
5. Lafon-Lafourcade, S. and Ribereau-Gayon, P. (1984) Developments in the microbiology of wine production. *Prog. Ind. Microbiol.* **19,** 1–45.
6. Fleet, G. H. and Heard, G. M. (1993) Yeasts—growth during fermentation in *Wine Microbiology and Biotechnology* (Fleet, G. H., ed.), Harwood Academic, Australia, pp. 27–54.
7. Sponholz, W. R. (1991) Nitrogen compounds in grapes, must, and wine in *International Symposium on Nitrogen in Grapes and Wine* (Rantz, J., ed.) American Society for Enology and Viticulture, Davis, CA, pp.67–77.
8. Zoecklein, B. W., Fugelsang, K. C., Gump, B. H., and Nury, F. S. (1995) *Wine Analysis and Production.* Chapman & Hall, New York.
9. Monk, P. R. (1986) Rehydration and propagation of active dry wine yeasts. *Austral. Wine Ind. J.* **1(1),** 3–5.

10. Correa, I., Polo, M. C., Amigo, L., and Ramos, M. (1988) Separation des proteines des mouts de raison au moyen de techniques electrophoretiques. *Bull. OIV.* **39,** 1475–1489.
11. Rosi, I., Costamagna, L., and Bertuccioli, M. (1987) Screening for extracellular acid protease(s) production by wine yeasts. *J. Inst. Brew.* **93,** 322–224.
12. Lagace, L. S. and Bisson, L. F. (1990) Survey of yeast acid proteases for effectiveness of wine haze reduction. *Am. J. Enol. Vitic.* **41,** 1246–1249.
13. Dukes, B. C. and Butzke, C. E. (1998) Rapid determination of primary amino acids in grape juice using an o-phthaldiadehyde/N-acetyl-L-cysteine spectrophotometric assay. *Am. J. Enol. Vitic.* **49,** 125–134.
14. Henick-Kling, T., Edinger, W. D., and Larsson-Kovach, I. -M. (1996) Survey of available nitrogen for yeast growth in New York grape musts. *Wein-Wissensch.* **51(3),** 169–174.
15. Sablayroles, J. M. (1996) Sluggish and stuck fermentations. Effectiveness of ammonium nitrogen and oxygen additions. *Wein-Wissensch* **51(3),** 147–151.
16. Ough, C. S. (1969) Substances extracted during skin contact with white must. I. General wine composition and quality changes with contact time. *Am. J. Enol. Vitic.* **20,** 93–100.
17. Jiranek, V., Langridge, P., and Henschke, P. A. (1990) Nitrogen requirement of yeast during wine fermentation in *Proceedings of the Seventh Australian Wine Industry Technical* Conference (Williams, P. J., Davidson, D. M., and Lee, T. H., eds.), Australian Industrial Publishers, Adelaide, pp. 166–171.
18. Monk, P.R., Hook, D., and Freeman, B. M. (1987) Amino acid metabolism by yeast in *Proceedings of the Sixth Australian Wine Industry Technical Conference* (Lee, T. H., ed.) Australian Industrial Publishers, Adelaide, pp. 129–133.
19. Wurdig, G. and Woller, R. (1989) *Chemie des Weines. Handbuch der Lebensmittel-technologie.* Ulmer, Stuttgart.
20. Henschke, P. A. and Jiranek, V. (1993) Metabolism of nitrogen compounds in *Wine Microbiology and Biotechnology* (Fleet, G. H., ed.) Harwood Academic, St. Leonards NSW, Australia, pp. 27–54.
21. Bely, M., Sablayrolles, J-M., and Barre, P. (1990) Automatic detection of assimilable nitrogen deficiencies during alcoholic fermentation in oenological conditions. *J. Ferment. Bioeng.* **70(4),** 246–252.
22. Bely, M., Sablayrolles, J. M., and Barre, P. (1991) Automatic detection and correction of assimilable nitrogen deficiencies during alcoholic fermentation in enological conditions in *Proceedings of the International Symposium on Nitrogen in Grapes and Wine* (Rantz, J., ed.) American Society for Enology and Viticulture, Davis, CA, pp. 211–214.
23. Dittrich, H. H. (1987) *Mikrobiologie des Weines. Handbuch der Lebensmitteltechnologie,* 2nd ed. Ulmer, Stuttgart.
24. Fugelsang, K. C. (1996) *Wine Microbiology.* Chapman & Hall, New York.
25. Grenson, M., Hou, C., and Crabeel, M. (1970) Multiplicity of amino acid permeases in *Saccharomyces cerevisiae.* IV. Evidence for a general amino acid permease. *J. Bacteriol.* **103,** 770–777.

26. Woodward, J. R. and Cirillo, P. V. (1997) Amino acid transport and metabolism in nitrogen-starved cells of *Saccharomyces cerevisiae. J. Bacteriol.* **130,** 714–723.
27. Castor, J. G. B. (1953) The free amino acids of musts and wines. II. The fate of amino acids of must during alcoholic fermentation. *J. Food Res.* **18,** 146–151.
28. Ingledew, W. M. and Kunkee, R. E. (1985) Factors influencing sluggish fermentations of grape juice. *Am. J. Enol. Vitic.* **36,** 65–76.
29. Bisson, L. F. (1991) Influence of nitrogen on yeast and fermentation of grapes, in *Proceedings of the International Symposium on Nitrogen in Grapes and Wine (Seattle)* (Rantz, J., ed.), American Society for Enology and Viticulture, Davis, CA, pp. 78–89.
30. Monteiro, F. F. and Bisson, L. F. (1991) Biological assay of nitrogen content of grape juice and prediction of sluggish fermentations. *Am. J. Enol. Vitic.* **42(1),** 47–57.
31. Monteiro, F. F. and Bisson, L. F. (1992a) Nitrogen supplementation of grape juice. I. Effect on amino acid utilization during fermentation. *Am. J. Enol. Vitic.* **43(1),** 1–10.
32. Monteiro, F. F. and Bisson, L. F. (1992b) Nitrogen supplementation of grape juice. II. Effect on amino acid and urea release following fermentation. *Am. J. Enol. Vitic.* **43,** 11–17.
33. Ough, C. S. (1968) Proline content of grapes and wine. *Vitis* **7,** 321–331.
34. Ingledew, W. M. (1996) Nutrients, yeast hulls and proline in wine fermentation. *Wein-Wissensch.* **51,** 141–146.
35. Manginot, C. and Sablayrolles, J. M. (1997) Use of constant rate alcoholic fermentations to compare the effectiveness of different nitrogen sources added during the stationary phase. *Enzyme Microbial. Tech.* **20,** 373–380.
36. Watson, T. G. (1976) Amino acid pool composition of *Saccharomyces cerevisiae* as a function of growth rate and amino acid nitrogen source. *J. Gen. Microbiol.* **96,** 263–268.
37. Cooper, T. G. (1982) Nitrogen metabolism in *Saccharomyces cerevisiae* in *The Molecular Biology of the Yeast Sacchromyces* (Strathern, J. N., Jones, E. W., and Broach, J. B., eds.), Cold Spring Harbor Laboratory, New York, pp. 39–99.
38. Large, P. J. (1986) Degradation of organic nitrogen compounds by yeasts. *Yeast* **2,** 1–34.
39. Tokuyama, T., Kuraishi, H., Aida, K., and Uemura, T. (1973) Hydrogen sulfide evolution due to pantothenic acid deficiency in the yeast requiring this vitamin, with special reference to the effect of adenosine triphosphate on yeast cysteine desulfhydrase. *J. Gen. Appl. Microbiol.* **19,** 439–466.
40. Maw, G. A. (1965) The role of sulfur in yeast growth and in brewing. *Wallerstein Lab. Commun.* **21,** 49–68.
41. Huang, Z. and Ough, C. S. (1989) Effect of vineyard location, varieties and woodstocks on the juice amino acid composition of several cultivars. *Am. J. Enol. Vitic.,* **40,** 135–139.
42. Huang, Z. and Ough, C. S. (1991) Amino acid profiles of commercial grape juices and wines. *Am. J. Enol. Vitic.* **42,** 261–267.
43. Ough, C. S. and Tabacman, H. (1979) Gas chromatographic determinations of amino acid differences in Cabernet Sauvignon grapes and wines as affected by rootstocks. *Am. J. Enol. Vitic.* **30,** 306–311.

44. Etievant, P., Schlich, P., Bouvier, J.-C., Symonds, P., and Bertrand, A. (1988) Varietal and geographical classification of French red wines in terms of elements, amino acids and aromatic alcohols. *J. Scie. Food Agric.* **45**, 25–41.
45. Bell, S. J. (1991) The effect of nitrogen fertilization on growth, yield, and juice composition of *Vitis vinifera* cv. Cabernet Suavignon grapevines in *Proceedings of the International Symposium on Nitrogen in Grapes and Wine (Seattle)* (Rantz, J., ed.), American Society for Enology and Viticulture, Davis, CA, pp. 206–210.
46. Bell, A. A., Ough, C. S., and Kliewer, W. M. (1979) Effects on must and wine composition, rates of fermentation and wine quality of nitrogen nitrogen fertilization of *Vitis vinifera* var. Thompson seedless grapevines. *Am. J. Enol. Vitic.* **30**, 124–129.
47. Ough, C. S. and Bell, A. A. (1980) Effects of nitrogen fertilization of grapevines on amino acid metabolism and higher alcohol formation during grape juice fermentation. *Am. J. Enol. Vitic.*, **31**, 122–123.
48. Butzke, C. E. (1998). Survey of yeast assimilable nitrogen status in musts from California, Oregon and Washington. *Am. J. Enol. Vitic.* **49**, 220–224.
49. Dittrich, H. H. and Sponholz, W. R. (1975) Die aminosaureabnahme in Botrytis-infizierten Traubenbeeren und die bildung hoherer alhohole in deisen Mosten bei ihrer vergarung. *Wein-Wissensch.* **30**, 188–210.
50. Ferenczy, S. (1966) Etude des proteines et des substances azotees. Leur evolution au cours des traitements oenologiques. Conditions de la stabilite proteique des vins. *Bull. OIV* **39**, 1311–1336.
51. Houtman, A. C. and duPleissis, C. S. (1981) The effect of juice clarity and several conditions promoting yeast growth on fermentation rate, the production of aroma components and wine quality. *S. Afr. J. Enol. Vitic.* **2**, 71–81.
52. Guitart, A., Orte, P. H., and Cacho, J. (1998) Effect of different clarification treatments on the amino acid content of Chardonnay musts and wines. *Am. J. Enol. Vitic.* **49(4)**, 389–396.
53. Koch, J. (1963) Proteines des vins blancs. Traitements des precipitations proteiques par chauffage et a l'aide de la bentonite. *Ann. Technol. Agric.* **12**, 297–313.
54. Rapp, A. (1977) Uber den Gehalt der Aminosauren in Weinbeeren, Traubenmost und Wein, Bundesausschuß fur Weinforschung, Germany, pp. 136–151.
55. Rapp, A. and Versini, G. (1991) Influence of nitrogen compounds in grapes on aroma compounds in wine in *Proceedings of the International Symposium on Nitrogen in Grapes and Wine* (Rantz, J., ed.), American Society for Enology and Viticulture, Davis, CA, pp. 156–164.
56. Monteiro, F. F., Trousdale, E. K., and Bisson, L. F. (1989) Ethyl carbamate formation in wine: Use of radioactively labeled precursors to demonstrate the involvement of urea. *Am. J. Enol. Vitic.* **40**, 1–8.
57. Stevens, D. F. and Ough, C. S. (1993). Ethyl carbamate formation: reaction of urea and citrulline with ethanol in wine under low to normal temperature conditions. *Am. J. Enol. Vitic.* **44**, 309–312.
58. Ough, C. S. (1993) Report on ethyl carbamate for the Wine Institute. Ethyl carbamate/urease enzyme preparation. A compendium from June, 1993 seminars.

59. An, D. and Ough, C. S. (1993) Urea excretion and uptake by various wine yeasts as affected by various factors. *Am. J. Enol. Vitic.* **44,** 35–40.
60. Yoshizawa, K. and Takahashi, K. (1988) Utilization of urease for decomposition of urea in sake. *J. Brew. Soc. Jpn.* **83,** 142–144.
61. Agenbach, W. A. (1977) A study of must nitrogen content in relation to incomplete fermentations, yeast production and fermentation activity, in *Proceedings of the South African Society for Enology and Viticulture (Cape Town)*, South African Society for Enology and Viticulture, Stellenbosch, pp. 66–88.
62. Sablayrolles, J.-M. (1992) Importance de l'azote assimilable de l'oxygene sur le deroulement de la fermentation alcoolique. *Biol. Oggi* **6,** 155–160.
63. Sablayrolles, J.-M., and Dubois, C. (1996) Effectiveness of combined ammoniacal nitrogen and oxygen additions for completion of sluggish and stuck fermentations. *J. Ferm. Bioeng.* **92,** 377–381.
64. Schulze, U., and Liden, G. (1996) Physiological effects of nitrogen starvation in an anaerobic batch culture of *Saccharomyces cerevisiae. Microbiology* **142,** 2299–2310.
65. Bission, L. F. (1996). Yeast and biochemistry of ethanol formation, in *Principles and Practices of Winemaking*, (Boulton, R. B., Singleton, V. L., Bisson, L. F., and Kunkee, R. E., eds.), Chapman & Hall, New York, p. 140.
66. Lie, S. (1973) The EBC-Ninhydrin method for determination of free *alpha*-amino nitrogen (FAN*). J. Inst. Brew.* **79,** 37–41.
67. Crowell, E. A., Ough, C. S., and Bakalinsky, A. (1985) Determination of alpha-amino nitrogen in musts and wines by TNBS methods. *Am. J. Enol. Vitic.* **36(2),** 175–177.
68. Dukes, B. C. and Butzke, C. E. (1998) Rapid determination of primary amino acids in grape juice using an o-phthalaldehyde/N-acetyl-L-cysteine spectrophotometric assay. *Am. J. Enol. Vitic.* **49,** 125–134.

44

Problems with the Polymerase Chain Reaction

Inhibition, Facilitation, and Potential Errors in Molecular Methods

Ian G. Wilson

1. Introduction

This chapter discusses the findings of many studies in food, clinical and environmental microbiology including approaches that have been used to overcome inhibition and facilitate amplification for detection and typing.

Over 70,000 published papers involve the polymerase chain reaction (PCR) *(1)*, and few areas of biological science remain untouched by the invention of this technique *(2–4)*. Other methods for amplifying nucleic acids have been described *(5–7)* such as Qβ replicase *(7)*, ligase chain reaction (LCR) *(8,9)*, single-stranded sequence replication (SSSR or 3SR) *(10,11)*, strand displacement amplification (SDA) *(12,13)* and nucleic acid sequence-based amplification (NASBA) *(14,15)*, but these have received less attention.

Problems may sometimes occur with PCR *(16)*. These can be categorized as follows:

1. False positives resulting from accidental contamination with nucleic acids
2. False negatives resulting from reaction inhibition or attenuation
3. Misleading identification or false categorization in typing procedures that involve amplification or other genetic typing methods

Factors that inhibit the amplification of nucleic acids by PCR are present with target DNA obtained from many sources. The inhibitors generally act at one or more of three essential points in the reaction: interference with the cell lysis necessary for extraction of DNA, nucleic acid degradation or capture, and inhibition of polymerase activity for amplification of target DNA. Although a wide range of inhibitors is reported, the identity and mode of action of many remain unclear. These effects may have important implications for clinical and public health investigations, especially if they involve food and environmental screening. Common inhibitors include various components of body fluids and reagents encountered in clinical and forensic science (e.g., hemoglobin, urea, heparin), food constituents (e.g., organic and phenolic compounds, glycogen, fats, Ca^{2+}) and environmental compounds (e.g., phenolic compounds, humicacids, heavy metals). Other, more widespread, inhibitors include constituents of bacterial cells, nontarget DNA, and contaminants and laboratory items such as pollen, glove powder, laboratory plasticware, and cellulose.

This chapter lists and discusses the problems associated with PCR and methods that can overcome them in food, clinical, and environmental microbiology. Aspects of clinical and environmental microbiology are important in this discussion because inhibitors from both these sources may be present in foods derived from animals and plants. The text examines the mechanical reasons for PCR problems, and the tables present information from the literature according to the substrate being investigated, in chronological order. It is beyond the scope of this chapter to discuss in detail the various physical, enzymic, and chemical methods used in the extraction, purification, and quantitation of nucleic acids. These are presented and discussed in commercial literature and elsewhere *(17–21)*. **Subheading 4.** in this chapter also considers the problems that may lead to errors when using a variety of methods for microbial identification and typing, not all of which involve PCR.

2. False Positives: Nucleic Acid Contamination

Despite early indications of great sensitivity, there are circumstances in which the ability to detect minute quantities of DNA may also be a negative aspect of the procedure. The most commonly reported problem of PCR is false-positive results because of DNA cross-contamination *(16,22)*. Contamination may arise from impure cultures, contaminated reaction components including commercial polymerases *(23)*, and laboratory contamination transferred by aerosol, gloves, pipets or other means. It can be overcome by ultraviolet (UV) light *(24)*, sodium hypochlorite *(25)*, and photochemical or enzymic methods *(26–29)*. Compact flow cabinets containing a UV light source are commercially available and provide a suitable environment for working with open tubes and eliminating unwanted DNA. Physical separation of preamplification and postamplification reactions in different areas of the laboratory is an important aspect of minimizing stray nucleic acid sequences. Negative control tubes should always be used to check for false-positive reactions. Some automated commercial systems utilize sealed reaction vessels to reduce the risk of unwanted DNA entering.

There are also problems of using amplification methods in clinical microbiology where the presence of an organism's DNA may not indicate its clinical significance in causing disease. Even when detection is a true positive (i.e., not a laboratory or clinical contaminant), this does not necessarily indicate viability or causality.

3. False Negatives

False negatives resulting from reaction inhibition have been widely reported and reviewed *(30–32)*. Inhibition may be total or partial and can become apparent as complete reaction failure or as reduced sensitivity of detection. In some cases, inhibition may be the cause of false-negative reactions because workers have not incorporated internal controls in each reaction tube. Early evidence of exquisite sensitivity using mammalian cells *(33)* involving detection of a single molecule of DNA from a hair was not followed by similar sensitivity when PCR was applied to many microbial (and some mammalian) situations where poor sensitivity, specificity, and reproducibility have been reported *(21,34–38)*. As discussed in detail in the following sections, there may also be potentially important effects in PCR typing reactions *(39)*, and difficulties can occur in post-PCR manipulation *(40)*. Although systematic study of inhibition has seldom been the focus of published investigations, many workers have reported these effects in the course of other studies *(16,37,38,41–44)*. Considering the prevalence of this problem, it is surprising that few systematic and mechanistic studies of PCR inhibition have been reported. Rossen et al. *(30)* contributed a comprehensive study of PCR inhibition, identifying inhibitory factors in foods, bacterial culture media, and various chemical compounds. These included organic and inorganic chemicals, detergents, antibiotics, buffers, enzymes, polysaccharides, fats, and proteins. Subsequent reviews have surveyed the literature extensively and provided much more information *(31,32)*.

3.1. Factors Affecting Detection Level

Cultural detection methods allow quantification of microorganisms with considerable precision. This may be of importance in clinical, food, and environmental investigations when distinguishing commensal organisms from those causing disease. Quantification of PCR product cannot necessarily be equated with number of organisms present in the original sample *(16)*.

The minimum number of cells containing a single-copy target gene theoretically detectable in a 10-µL sample aliquot added to a 90-µL reaction mix is one. This corresponds to 100 colony-forming units (CFU)/mL, a similar sensitivity to plate counts, and, in practice, may be reduced through sampling error. The random distribution of cells near the detection limit means that a single cell or gene copy may not actually be present in some aliquots that have been calculated to contain such numbers. Food, clinical, and environmental samples commonly reduce sensitivity through a wide range of inhibiting substances. Complete failure and false-negative amplification reactions are reported in many cases *(45,46)*, but often sensitivity is merely reduced *(47)*, sometimes considerably *(48)*. Typically, a variety of mechanisms reduce sensitivity by several orders of magnitude. Wernars et al. *(37)* found the sensitivity of detection was at least 10-fold poorer than the theoretical minimum, varying between 10^3 and >10^8 CFU/0.5 g in different brands of soft cheeses. Sensitivity was reduced 1000-fold in milk powder *(38)* where 10^5 CFU/mL were required for detection of *Staphylococcus aureus* despite fewer than 10 cells being detectable in diluent. Protein-breakdown products in aged cheeses caused sensitivity problems in the detection of lactic acid bacteria *(49)*. Ten plaque-forming units (PFU) per gram of enteroviruses in diluent were detectable directly in an reverse-transcriptase (RT)-PCR assay *(50)*, but only 10^3 PFU/g when extracted from oyster meat. Such sensitivity is typical of the lower detection limits that are achieved in practice for clinical and environmental samples and foods.

Dilution of samples, as with enrichment culture in the presence of potential growth inhibitors such as spices and disinfectants, provides a simple method that can facilitate amplification, albeit with reduced sensitivity. The sensitivity of PCR can be exploited to amplify target DNA that is still present when inhibitors have been diluted out *(51–54)*.

Conversely, PCR detection of intracellular bacteria may be more sensitive than expected because of multiple bacterial cells in each infected blood cell *(55)*. This may provide the opportunity to achieve greater sensitivity than that suggested earlier. Amplification of short-lived multiple-copy RNA species may provide increased sensitivity relative to DNA targets in some situations.

Authors sometimes are unclear about whether the number of cells detectable is per milliliter or per reaction volume (normally 100 µL). There are apparent contradictions in the published works that discuss inhibition. One substance may be reported as being both an inhibitor and a facilitator in different systems. At different concentrations, the same compound has been reported to act as both facilitator and attenuator in the same system *(56,57)*. Both bacterial cells and nontarget DNA have been shown to be both inhibitory and noninhibitory in different systems. Steffan and Atlas *(58)* were able to detect a 15–20 copy target sequence in *Pseudomonas cepacia* cells at 1 CFU/g against 10^{11} nontarget organisms in samples of sediment. This represents detection 10^3-fold more sensitive than probing nonamplified samples despite high numbers of potentially inhibitory nontarget organisms. Andersen and Omiecinski *(59)* found that >10^5 CFU of indigenous bacteria did not inhibit specific amplification of target DNA from enteroinvasive *Escherichia coli* (EIEC).

Dickinson et al. *(60)* found that high background levels of bacterial DNA were not inhibitory to specific amplification and that 10^2–10^4 CFU/mL could be detected in several food types. Sensitivity was lower for cheese than coleslaw or chicken. These authors speculate that nontarget DNA may have improved sensitivity by acting as a carrier for target DNA during precipitation. Nevertheless, bacterial cells can have an inhibitory effect on PCR *(44, 61–63)*. A study of the efficiency of DNA extraction from soil and PCR inhibition by humic compounds *(63)* showed that high levels of nontarget DNA could inhibit PCR. Sensitivity may vary with the microbial species detected, using identical extraction protocols, because recovery may be different between species *(63,64)*. Alvarez et al. *(51)* found that 10^3–10^4 CFU/m^3 of environmental organisms interfered with detection of *E. coli* during bioaerosol sampling. Freeze–thaw lysis and sample dilution allowed rapid and sensitive detection of a single cell. The multiplex amplification of VT1, VT2, and *eae*A genes was not diminished equally for all target sequences when nontarget cell populations were added in a study of *E. coli* O157:H7 *(65)*. The authors recommend caution when interpreting negative PCR reactions when high concentrations of nontarget cells are suspected.

Where nonspecific bacterial effects inhibited an antigen–antibody reaction, skim milk was found to overcome this *(66)*. It also acts as a blocking agent preventing the nonspecific adhesion of bacteria to immunomagnetic beads *(66)* and is used to similar effect for blocking nonspecific reactions during membrane hybridization. Skim and full cream milk have also been found to be inhibitory to PCR *(38,67,68)*, partly because of DNA loss during extraction procedures and partly because of amplification inhibition. In PCR, milk components may block DNA and shield it from access by polymerase.

Reaction conditions have long been recognized as important to sensitive and reproducible amplification. Contamination also must be considered as a possible cause of reaction failure.

3.2. Reaction Conditions

Suboptimal reaction conditions may arise for a number of reasons. The primary ones are inappropriate primers, improper time or temperature conditions, variable polymerase quality and incorrect Mg^{2+} concentration. These factors should be optimized by running several temperature profiles based on calculated primer melting points and Mg^{2+} dilution series reactions before relying on PCR for sensitive and specific detection from a given medium.

Amplification Assistant, an Internet troubleshooting resource, is available at www.promega.com/amplification/assistant. This site allows users to describe the type of reaction they are trying to perform, describe the main problem, and receive suggestions about the cause of the problem and its solution *(69)*.

Temperature inconsistencies across the block of a thermal cycler have been found to be responsible for poor amplification *(39,70)* and this has been overcome by the addition of formamide, which reduces the melting temperature of DNA *(70)*. This is not an ideal solution and further studies could reveal that the specificity of the reaction may suffer from this treatment. Excellent thermal consistency is important for sensitive and specific testing and should be considered when choosing between the many models of thermal cycler available. The reproducibility of cycle time and temperature should be checked, particularly when moving to a different machine or tubes with different thermal transfer properties, and the sample block should be cleaned to ensure consistent performance.

Physical modifications to the PCR reaction such as hot start may be helpful for improving the yield and specificity of reactions. Adding reactants when the reaction has reached denaturation temperature is particularly useful when detecting low-copy-number molecules amid a high background of nontarget DNA because it reduces the formation of primer oligomers and misprimed reactions that compete with the specific amplification of the targeted sequence.

Dimethylsulfoxide (DMSO) has been shown to improve reaction yield during RT-PCR *(71)*. Sensitivity was improved up to 25-fold for some viral target RNAs. This report involved the use of r*Tth* polymerase, and no inhibitors were mentioned. The optimal concentration of DMSO was dependent on the template and facilitated both the reverse transcriptase and PCR reactions. It was suggested that DMSO may enhance PCR by eliminating nonspecific amplification, alteration of the thermal activity of the polymerase, or improve the annealing efficiency of primers by destabilizing secondary structures within the template.

The Mg^{2+} ions are a vital cofactor for *Thermus aquaticus* (*Taq*) polymerase and their concentration will affect the success and specificity of amplification. The sequestration of Mg^{2+} ions by various compounds and interference by Ca^{2+} ions may inhibit amplification *(41,72)*. The Ca^{2+} ions in milk may be a cause of its inhibitory properties.

A number of DNA polymerases are now commercially available. These originate from several extreme thermophiles and exhibit differences in processivity and fidelity, making some more suitable than others for specific tasks. Batches of polymerase may be pooled to overcome variability in enzyme quality. Production of PCR product may be improved in some situations if mixtures of different polymerases are used. This may allow improved yield because some enzymes may be less susceptible than others to specific inhibitors and will permit amplification that would not otherwise occur.

Inhibitory effects may be minimized by optimizing the reaction conditions and ensuring that appropriate quantities of all reactants are freely available

Problems with the PCR

within the reaction mixture by minimizing the presence of any contaminants and known inhibitors.

3.3. Endogenous and Exogenous Contamination and Other Factors

Contamination and other factors intimate to the reaction, even apparently inert components, may cause inhibition at any of the points of molecular interaction to be discussed. Their modes of action are not yet understood, but there may be chemical or physical interference with the availability or activity of an essential reaction component. Contamination may be endogenous to the reaction components (e.g., sample, enzyme, tubes) or exogenous (e.g., bacteria, dust, pollen).

The most obvious origin of PCR inhibitors in endogenous contamination is compounds present in insufficiently purified target DNA. Inhibition can also arise from other endogenous sources, including reaction components. Commercial preparations of *Taq* polymerase, including a low-DNA product, have been shown to be contaminated with eubacterial DNA that originates in neither *E. coli* nor *Taf (23)*. In particular, this may reduce the effective application of PCR using broad-range eubacterial primers because of the production of false-positive reactions. For more specifically targeted reactions, such contamination generally is of little importance. This study demonstrated that enzyme-contaminating sequences can be destroyed by UV irradiation. It also showed that microcentrifuge tubes from different manufacturers gave greatly different results. Strong inhibition has also been identified in some brands and batches of PCR reaction tubes by other workers, but the cause could not be discovered *(73)*. Considerable variability in the performance of *Taq* polymerase within batch, by concentration, and between suppliers has been documented in a commentary that details many parameters affecting multiple arbitrary amplicon profiling *(39)*. In addition to reaction failure, enzyme contamination may be manifested as spurious background bands during amplification-based DNA fingerprinting using methods such as random amplified polymorphic DNA (RAPD). By nature of their low stringency, such typing techniques rely on both specific and nonspecific priming and combine the detection of artifactual variation with true polymorphism. Contamination, concentration-dependent effects, spurious bands, and inhibition may complicate the interpretation of banding patterns and give rise to misleading results.

Like reaction tubes, other apparently inert components may be inhibitory. Cellulose and nitrocellulose filters were found to inhibit PCR *(74)*. In this study, polycarbonate filters proved not to be inhibitory, perhaps because of differences in binding properties for DNA or its contaminants relative to those of cellulose-based filters. Mineral oil has been shown to have an inhibitory effect on PCR when irradiated with UV light *(75)*. The inhibition is dependent on the UV dose. Unirradiated mineral oil has been reported as a facilitator in "oil-free" reactions containing high concentrations of nonionic detergents *(76)*. This author suggested that components of detergent preparations (monomers, micelles, or impurities) could be responsible for adverse effects on the specificity of annealing. It is likely that detergents allow the greater solubilization of inhibitors that might otherwise aggregate and precipitate during preparation or in the reaction tube. Oil overlays may facilitate amplification by segregating inhibitors at the oil–water interface and remain an option in thermal cyclers designed for "oil-free" reactions.

Reaction failure may also be caused by exogenous contamination. This differs from the other sources of inhibition discussed earlier that result from compounds present in insufficiently purified DNA or from contaminants in reaction components. However it may involve the same mechanisms of inhibition. Inhibition may the result of even <10 grains of pollen that may enzymically digest an essential reaction component *(46)*, glove powder *(45,77)*, which may nonspecifically bind DNA, or other factors that enter reaction tubes when hygienic conditions are not sufficiently controlled.

Contamination can be prevented by Good Laboratory Practice (GLP) and scrupulous attention to aseptic technique that also serves to protect from cross-contamination of target sequences, although these can be dealt with using UV irradiation or uracil *n*-glycosylase (UNG) *(78)*. Restriction enzymes have been shown to be inhibited, or their specificity altered, by multiple uracil substitutions in restriction sites *(79)*. This may also have implications for the fidelity of PCR typing methods, as discussed later. Laminar-flow cabinets may be of use in preventing airborne contamination and cabinets equipped with a UV light source are available specifically to minimize contamination during PCR work. Ironically, the good practice of changing gloves may in some cases actually predispose reactions to failure. Powders from gloves were shown to have variable inhibitory effects on PCR, depending on manufacturer *(45)*. If it is suspected that this is the source of problems, washing gloves and the selection of nonpowdered brands may be helpful.

3.4. Mechanisms of Inhibition

The inhibition of amplification may be the result of a number of factors, none of which has been investigated thoroughly. Details of the inhibitory effects of some compounds that have been reported in food, clinical, and environmental systems may be found in **Tables 1–3**. Inhibitors in clinical and environmental are also relevant to food studies because of the animal and plant origins of foods. Inhibition may originate from poorly controlled reaction conditions, from the sample itself, from contaminants in reagents, containers,

Table 1
Inhibitors and Facilitators in Food Samples

Substrate	Target organism	Inhibitor	Facilitator	Ref.
Milk	*Listeria monocytogenes*	Unknown	Enzymatic digestion, membrane solubilization	81
Skim milk	*S. aureus*	Thermonuclease, proteins, bacterial debris	NaOH, NaI, physico-chemical extraction, nested PCR	38
Soft cheeses	*L. monocytogenes*	Brand-specific inhibitors, denatured protein	Phenol extraction, Qiagen column	37
Various foods	*E. coli*	Bean sprouts, oyster meat	Magic Minipreps	59
Foods and cultures	*S. aureus, E. coli, Salmonella* spp		Lectin affinity chromatography	173
Various foods	*L. monocytogenes*	Various	—	30
Skim milk	*S. aureus*	Thermonuclease, proteins, bacterial debris	NaOH, NaI, physico-chemical extraction, nested PCR	44
Solution	*S. aureus* proteins, bacterial debris	Thermonuclease, nested PCR	NaOH, NaI, physico-chemical extraction,	21
Oyster meat	Poliovirus 1, Norwalk virus, hepatitis A virus	Polysaccharides, glycogen	Organic flocculation, CTAB, PEG extraction	177
Meat	*Brochothrix thermosphacta*	Fetuin, meat components	Lectin binding	178
Milk	*L. monocytogenes*	Proteinase	BSA, proteinase inhibitors	116
Soft cheese	*L. monocytogenes*	Unknown	PEG/dextran extraction	179
Various foods	*L. monocytogenes*	Unknown	Enrichment, NASBA	15
Various foods	*L. monocytogenes*	Unknown	NaI/alcohol precipitation	180
Oysters	Enteroviruses	Unknown	Freon, PEG, chloroform, CTAB	181

Table 1 (cont.)
Inhibitors and Facilitators in Food Samples

Substrate	Target organism	Inhibitor	Facilitator	Ref.
Raw milk	*Clostridium tyrobutyricum*	Unknown	Chemical extraction, centrifugation	182
Shellfish	SRSVs	Unknown	PEG, Freon, ultracentrifugation	183
Various foods	Various bacteria	Unknown	Enzymic/physico-chemical extraction	60
Oyster meat	Poliovirus, hepatitis A virus, Norwalk virus	Acidic polysaccharides, glycogen	Freon, DMSO, glycerol, PEG, Pro-Cipitate	50
Drinking water	Enteroviruses, hepatitis A virus	Humic acid/organic compounds	ProCipitate, PEG, antibody capture	184
Raw milk	*Brucella* spp.	Milk proteins	Physico-chemical extraction, nested PCR	68
Cold-smoked salmon	*L. monocytogenes*	Food components, sucrose, ovalbumin, phenolic compounds	Ether and column extraction, Tween-20	86
Ground beef	*E. coli*	Food components	Enrichment culture, Taqman™	185
Milk	*L. monocytogenes*	Ca^{2+}	Chelation, $[Mg^{2+}]$	41
Raspberries	*Cyclospora* sp. and *Eimeria* spp.	Capture by fruit epidermal hairs in raspberry wash sediment?, humic or polyphenolic compounds?	Various treatments partially effective	85

Table 2
Inhibitors and Facilitators in Clinical Samples

Substrate	Target organism	Inhibitor	Facilitator	Ref.
Feces	E. coli	>10^3 bacterial cells	Ion-exchange column	120
CSF	Treponema pallidum	Cellular debris causing nonspecific amplification	Nested primers	109
Whole blood	Mammalian tissue	>4 µL blood/100 µL reaction (hemoglobin)	<1–2% blood per reaction	112
Feces	Rotavirus	Unknown	Dilution, cellulose fiber paper	186
Clinical specimens	Cytomegalovirus	Unidentified components	Glass bead extraction	97
Human blood and tissue	Human genes	DNA-binding proteins	Thermophilic protease from Thermus rt41A	99
—	Mammalian tissue genetics		Organic solvents, DMSO, PEG, glycerol	187
—	Mammalian tissue genetics	Thermal cycler inconsistencies	Formamide	70
Clinical specimens	T. pallidum	Various unknown factors	Various substrate-specific physico-chemical methods	108
	Many	Many	Many	30
Forensic semen samples	Interference of vaginal microflora with sperm genotyping	Genotyping errors and selective or total PCR inhibition by vaginal microorganisms	—	62
	Salmonella	Various body fluids	Immunomagnetic separation	111
Feces	Various enteric viruses	Unknown	Size-exclusion chromatography, physico-chemical extraction	188

Table 2 (cont.)
Inhibitors and Facilitators in Clinical Samples

Substrate	Target organism	Inhibitor	Facilitator	Ref.
Clinical specimens	Herpes simplex virus	Endogenous inhibitors, random effects	Repurification, coamplified positive control	105
Feces	E. coli	Nonspecifc inhibitors, urea, hemoglobin, heparin, phenol, SDS	Additional primers and reaction cycles, booster PCR	115
Tissue culture	Cytomegalovirus, HIV	Glove powder	—	45
Suspensions, skin biopsies	Mycobacterium leprae	Mercury-based fixatives, neutral buffered formalin	Reduced fixation times, ethanol fixation	106
Clincal specimens	M. tuberculosis	Unknown inhibitors in pus, tissue biopsies, sputum, pleural fluid	Physico-chemical extraction	110
—	Mammalian tissue genetics	Unknown contaminant of reverse transcriptase	Additional DNA	56
Formalin-fixed paraffin tissue extraction	Hepatitis C virus	Ribonucleotide vanadyl complexes	Phenol–chloroform	189
Nasopharyngeal aspirates and swabs	Bordetella pertussis	Unknown inhibitors	Phenol–chloroform extraction	190
Human mononuclear blood cells	HIV 1	Detergents	Mineral oil	76
Blood stain	Human mitochondrial DNA	Unidentified heme compound, hemin	Bovine serum albumin	104
Feces	Vibrio cholerae	Unknown contaminants	Dilution	191

Table 2 (cont.)
Inhibitors and Facilitators in Clinical Samples

Substrate	Target organism	Inhibitor	Facilitator	Ref.
Blood	Various	Heparin	Alternative polymerases and buffers, Chelex, spermine, $[Mg^{2+}]$, glycerol, BSA, heparinase	72
Sputa	*Mycoplasma pneumonia*	N-acetyl-L-cysteine, dithiothreitol, mucolytic agents	—	192
Human tissue	HLA-DRB1 genotyping	Pollen (<10 grains), glove impure DNA, heparin, powder, hemoglobin	—	46
Clinical specimens	*M. tuberculosis*	Unknown inhibitors	Competitive internal control	193
Bovine semen	Bovine herpesvirus-1	Unknown	Dilution, adjustment of ionic conditions	54
Feces	*Salmonella*	Hemoglobin degradation products, bilirubin, bile acids, faeces	Immunomagnetic separation	107
Dental plaque	Many	Unknown guanidium isothiocyanate, ethanol, acetone	Diatomaceous earth,	36
Ancient mammalian tissues	Cytochrome-*b* gene	Unknown	Ethidium bromide, ammonium acetate	95
Sputum	*M. tuberculosis*	Hemoglobin and others	Anion-binding resin	194

Table 2 (cont.)
Inhibitors and Facilitators in Clinical Samples

Substrate	Target organism	Inhibitor	Facilitator	Ref.
Sputum and bronchial samples	*M. tuberculosis*	Unknown	—	*195*
Endocervical specimens	*Chlamydia trachomatis*	Components of endocervical matrix	Qβ replicase	*196*
Aqueous and vitreous fluids	Viral nucleic acids	Unknown	Dilution, chloroform extraction, non-*Taq* polymerases	*118*
Bovine feces	*Cryptosporidium*	Unkown	Spin columns	*197*
Solution	*Bacteroides*	Heme; bilirubin; bile salts; humic substances (complex polyphenolics); EDTA; SDS; Triton X-100; hemin, tannic, fulvic and humic acids; bacterial extracts; proteases	BSA, gp32	*117*
Gastric mucosa, feces, dental plaque, oral rinses	*Helicobacter pylori*	Feces, plaque	Nested primers	*48*

Table 3
Inhibitors and Facilitators in Environmental Samples

Substrate	Target organism	Inhibitor	Facilitator	Ref.
River sediment	*Pseudomonas cepacia*	None	Physicochemical extraction	64
Suspension	*S. typhimurium*, *Listeria monocytogenes*	Unknown	Lectin affinity chromatography	172
Water	Various bacteria	Cellulose and nitrocellulose filters	PTFE filters	74
Sediments	*E. coli*	Humic substances	Gel filtration	53
Water	*E. coli*	Low sensitivity	Membrane filtration, 75 solid-phase PCR cycles and radiometric detection	198
Soils and sediments	*E. coli*	Humic substances, iron	Dilution	52
Plant tissue	Plant tissue	Acidic plant polysaccharides and buffer additives, SDS	Tween-20, DMSO, PEG 400	199
Soil	*P. cepacia*	Humic compounds interfere with lytic enzymes	Physicochemical extraction, cation-exchange resin	187
Soil	Various	Humic compounds, nontarget DNA	Ion-exchange chromatography, T4 gene 32 protein	63
	Plant material	Concentration-dependent inhibition and facilitation by polyamines (spermine and spermidine), glycerol, formamide	Variable with concentration	94

Table 3 (cont.)
Inhibitors and Facilitators in Environmental Samples

Substrate	Target organism	Inhibitor	Facilitator	Ref.
Soil	*P. fluorescens*	Humic substances	Physico-chemical extraction	200
	E. coli	Phenolic compounds, poly(vinyl pyrrolidone)	Retardation of phenolic poly(vinylpyrrolidone) agarose gel electrophoresis	57
Sediment	*P. putida*	Sediment, bacterial cells	Physico-chemical extraction, PVPP	47
	Hypoxylon truncatum	RNA interferes with amplification of RAPD markers	RNAse	201
Sediment, soil and water	RNA from various species	Humiclike compounds	Lysozyme/hot phenol extraction and gel filtration	202
Sludge and soil	Enteroviruses	Unknown	Beef extract, size-exclusion chromatography	203
Soil	Enteroviruses	Humic and fulvic acids, heavy metals	Physical separation, solvent extraction, size-exclusion/ion chromatography, double and seminested PCR	204
Bioaerosols	*E. coli*	Bacterial cells, nontarget DNA	Dilution, filtration	51
	E. coli	>10³ CFU m bacterial cells, nontarget DNA	Freeze–thaw lysis, dilution	61
Waters	*Cryptosporidium*	Formaldehyde, potassium dichromate, feces	Flow cytometry, magnetic antibody capture	205
Sewage/effluent	*Cryptosporidium, Giardia*	Humic/fulvic acids	Percoll-sucrose centrifugation, vortex flow filtration, nested PCR	206
Environmental samples	Various	Humic substances	Sepharose and Sephadex	207

or disposables, or from unintentional contamination during reaction preparation. Some sources of inhibition are well known to the biomedical community, but there is a dearth of systematic and biomechanistic studies that offer greater insight into the physical causes of the problem. Such studies are essential for the advancement of the science and the improved application of this powerful technology. Inhibition may involve multiple causes and complex interactions that are difficult to distinguish.

Some compounds have been reported as both inhibitors and facilitators in different studies and even in the same system *(56,57)*. It is essential that these problems are understood and resolved. An automated high-throughput filtration assay for the identification of RNA polymerase inhibitors has recently been described *(135)* and could be applied to RT-PCR inhibition. Other drug discovery assays could be employed for rapid and quantitative screening of many classes of inhibitor. The mechanisms of inhibition may be grouped into three broad categories by their point of action in the reaction. These are discussed in the following subsections. The categories are by no means absolute because an inhibitor may act in more than one way and the relationships among chemical, enzymic, and physical factors often cannot be distinguished given the poor current knowledge on the subject. It is likely that many inhibitors act through various physical and chemical means by interfering with the interaction between DNA and polymerase. This functional framework could serve as a focus for future systematic studies of the biomechanical origins of inhibition and lead to simple technological improvements to facilitate the use of rapid and sensitive DNA diagnostics.

3.4.1. Failure of Lysis

An elementary aspect of DNA amplification is that the failure to expose nucleic acids as targets for amplification will result in reaction failure. Loss of cell-wall integrity may not be enough to permit amplification of DNA, and enzymic degradation of cellular debris will often be necessary *(21)*. Protocols exist that rely on some or all of physical, chemical, and enzymic methods for cell lysis. Inadequate lysis may result from inadequate lysis reaction conditions, enzyme inactivation, or lytic enzymes of poor quality or consistency.

Early evidence was presented that PCR could successfully take place on unpurified DNA released from cells by boiling *(80)*. Such an approach gives a considerable time saving over more elaborate extraction protocols. Nevertheless, if whole cells are loaded into the reaction tube and released DNA fails to be sufficiently separated from structural and DNA-binding proteins by boiling, PCR inhibition may result. Extraction by boiling alone has been noted to reduce sensitivity, because of the above mechanism above or poor lysis efficiency, and may give rise to spurious bands in some cases *(136)* or prevent amplification altogether *(137)*. The high salt concentration in *Listeria* selective media was found to be the reason for unlysed cells and false-negative reactions using the Accuprobe DNA probe test *(138)*, and PCR could be similarly affected. Scanning electron microscope studies on coccidian parasites that contaminated raspberries suggested that a possible cause of PCR inhibition was the capture of oocysts by a mat of fruit epidermal hairs in raspberry wash sediment *(93)*.

In work with *S. aureus*, I have found that lysis using lysostaphin may be inconsistent. Subsequent observations in this laboratory showed that a genetically engineered version of this enzyme that had become available worked more reliably. Proteolytic enzymes and denaturants may degrade enzymes used for lysis. Phenolic compounds *(91)* from the sample or carried over from organic DNA purification procedures can inhibit the reaction by denaturing the lytic enzymes *(125)* and failing to expose the DNA. It may be that the stage of the growth cycle and nutrient conditions are important for the susceptibility to lysis of some cells. These factors have been little studied.

3.4.2. Nucleic Acid Degradation and Capture

Degradation or sequestration of target or primer DNA can also be a cause of failed reactions. Amplified DNA also may be degraded, and band smearing is sometimes evident on running amplified products after storage *(44)*. These effects may occur by physical, chemical, or enzymic processes. The primary structure of DNA is susceptible to instability and decay, mainly the result of hydrolysis, nonenzymic methylation, oxidative damage and enzymic degradation *(139)*. Amplification of long DNA sequences becomes increasingly difficult as DNA strands fragment after cell death. Nucleases may enter the reaction through careless handling, from the sample material, from various bacteria in the sample, and, in some cases, from the target organisms themselves *(44,140)*. The DNA of some Gram-negative bacteria is protected in nuclease-resistant vesicles that are involved in the export of genetic material *(141)*. Nonspecific autodegradation of DNA can occur in the presence of restriction endonucleases *(142)*. It was reported *(21,38,44)* that staphylococcal thermonuclease was not destroyed during thermal cycling and limited the sensitivity of PCR. Nucleases are produced by many other bacteria, but staphylococcal DNase exhibits uncommon heat stability and was able to hydrolyze genomic and primer DNA during amplification reactions. DNAse activity was reported to prevent pulsed-field gel electrophoresis (PFGE) typing of Lior biotype II (DNAse-positive) *Campylobacter jejuni* isolates. Treatment in formaldehyde solution for 1 h neutralized the DNAse activity and allowed typing of these strains to proceed *(140)*.

Unavailability of target or primer DNA by nonspecific blocking or sequestration may inhibit amplification or cause misleading band variations during typing based on PCR. Bacterial cells or debris, proteins, and polysaccharides that have caused inhibition in many studies may do so by physical effects such as making the target DNA unavailable to the polymerase. Milk proteins were reported as inhibitory *(68)* and may also act in this way by restraining DNA in high-molecular-weight complexes. An extraction procedure using benzyl chloride and the differential solubility and precipitation of DNA and polysaccharides was shown effectively to overcome inhibition by the high-molecular-weight polysaccharides from fungi that prevent enzymic DNA manipulation *(143)*. High concentrations of reverse transcriptase or an unidentified compound associated with some preparations of the enzyme have been shown to inhibit RT-PCR of RNA *(56)*. The inhibitory effect was demonstrated to vary between manufacturers and between batches. The addition of DNA overcame inhibition and these authors believe that a contaminant of some enzyme preparations binds to DNA, thus blocking PCR.

In a study of how forensic sperm genotyping was undermined by inhibition or allele dropout as a result of the presence of vaginal microorganisms, two possible mechanisms were suggested *(62)*. Efficient primer extension may be prevented by small, sheared single-stranded lengths of microbial DNA binding to the target sequence. Alternatively, the effective primer concentration could be reduced by nonspecific binding to nontarget microbial DNA. In other studies *(64)*, high numbers of nontarget organisms were not found to be detrimental to PCR. This may depend on the DNA sequences involved and on fragmentation size during extraction. Also, it has been shown that chemical insults such as the spermicide nonoxinol-9 do not prejudice the forensic examination of DNA *(144)*.

It was speculated *(126)* that polyamines binding to DNA prevented polymerases accessing the template. These workers found that the polyamines spermine and spermidine, along with formamide and glycerol, had a concentration-dependent effect on the yield and specificity of PCR. Spermidine had previously been reported to facilitate amplification by precipitating inhibitors *(115)*. Formamide has also been reported to improve amplification in tissue typing *(70)*.

Airborne allergens from latex gloves frequently contaminate the laboratory environment *(96)*. Glove powder has been reported to inhibit amplification and may do so by nonspecific DNA binding *(45)*. Because the binding of DNA by mineral compounds such as glass *(98)* and diatomaceous earth *(17,36,146)* is the basis of a number of nucleic acid extraction methods, the unavailability of DNA because of binding to potentially complex sample components such as sorption to sediments in environmental samples is a rational explanation of inhibition.

Nucleic acid sequestration and degradation may be overcome by physicochemical separation of target DNA from destructive compounds as soon as possible after cell lysis. Heat *(30)* and proteases *(99)* may be of use for destroying nucleases. If used, proteases must themselves be eliminated before polymerase is added to prevent enzymic inactivation of the polymerase.

Humic compounds are the most commonly reported group of inhibitors in environmental samples (**Table 3**) and appear to have deleterious effects on several reaction components and their interaction *(125)*. Public health and ecological investigations that involve environmental sampling may be hampered by inhibition from humic compounds. It was shown that as little as 1 µL of humic-acid-like extract was enough to inhibit a 100-µL reaction mix and that this was unlikely to be the result of chelation of Mg^{2+} by humic compounds *(52,53)*. Sephadex-spun columns helped facilitate PCR in this work. In the extraction of DNA from ancient human bone, it was found that 5 ng of ancient DNA was inhibitory to the amplification of 1 ng of recent DNA, because of coextraction with humic compounds or Maillard products of reducing sugars *(147)*. These workers found that solvent extraction, ethanol precipitation, the addition of bovine serum albumin (BSA), gelatin, and high concentrations of *Taq* polymerase all failed to facilitate amplification, although ion-exchange chromatography removed inhibitors. Young et al. *(57)* explained the commonly reported inhibition by soil humic compounds as follows. The phenolic groups of humic compounds denature biological molecules by bonding to N-substituted amides or oxidize to form a quinone that covalently bonds to DNA or proteins. The addition of poly(vinyl polypyrrolidone) (PVPP) or poly(vinyl polypyrrolidone) (PVP) overcame the inhibition and allowed separation of humic compounds from DNA during agarose gel electrophoresis. PCR yield was reduced by the addition of >0.5% PVP, however.

Recent work on facilitating amplification in the presence of humic acids, and perhaps other inhibitors, in soils involved introducing a component with higher affinity for the inhibitor than that of the essential reaction component that was inhibited *(148)*. These workers consider inhibition to be the result of interference in the interaction between polymerase and target DNA. Of nine proteins tested, BSA proved the most effective in overcoming inhibition, as had previously been indicated by other workers.

Minimum inhibitory concentrations (MICs) were calculated for effects of humic acids on *Taq* polymerases *(63)*. Up to an eightfold difference in MICs was found depending on the source of the humic acid and the commercial producer of the enzyme. The inhibitory effect was reduced by the addition of T4 gene 32 protein. Humic acids inhibited not only lytic enzymes *(125)* and polymerase activity during PCR but also DNA–DNA hybridization, restriction enzyme digestion of DNA, and transformation of competent *E. coli* cells. In this study *(63)*, high concentrations of nontarget DNA were also identified as inhibitory to PCR. This may prevent specific interaction between polymerase and target DNA.

Post-PCR restriction analysis may be inhibited by excess PCR primers *(149)*. Primer extension during PCR may sometimes be terminated by single-base mismatches between the primer and target DNA strands. This inhibitory effect on enzymic extension has been exploited to assess oligonucleotide and target sequence complementarity *(150)*.

In clinical samples many substances may be present that cause inhibition by one or more methods that are not understood *(95,98,101,102,104,106,107,110,114)*. These may come from the body or from sample preparation. Blood was reported as inhibitory if present at 4% or more of reaction volume. These workers recommended keeping blood below 1–2% of the 100-µL reaction volume to enable amplification of sequences in blood *(96)*. Serum proteins may act as blocking agents and prevent access to target DNA by polymerases. Various components of blood may cause inhibition, and the degree may possibly vary with their differential production during disease processes. Body fluid residues may be inhibitors in tests of animal products.

3.4.3. Polymerase Inhibition

As discussed earlier, humic compounds are widely reported as causing inhibition and some authors have identified this as being the result of interference with lytic enzymes *(125)*, binding to DNA and proteins *(54)*, interferring with the binding between target DNA and polymerase *(63)*, and binding of polymerase to clay particles with possible interference at the catalytic site *(151)*. Heat-mediated activation has been used to release *Taq* polymerase from affinity-immobilized inhibition to enable hot-start PCR *(152)*. Proteolytic enzymes and denaturants may also inactivate polymerase and must be promptly inactivated if used in cell lysis. Urea may cause inhibition by denaturing polymerases *(105)*.

Phenolic compounds *(91)* from the sample or carried over from organic DNA purification procedures can inhibit the reaction by binding to *(57)* or denaturing the polymerase. Proteinases and denaturants used for cell lysis may be carried over and inactivate polymerase if DNA purification is not adequate. Proteinase was prevented from inhibiting amplification in one study *(84)* by the addition of proteinase inhibitors and BSA. Kreader *(121)* also was able to restrict inhibition by factors including proteases through the addition of BSA that provides an alternative substrate for catalysis by these enzymes. Cheese proteases were found to inactivate *Taq* polymerase but could be removed by hot NaOH extraction *(30)*. Bile acids and salts may cause problems with both clinical samples and enteric bacteria enriched in culture media containing these compounds. Bacterial proteases and nucleases in feces, as well as cell debris, bile acids, and other factors may prevent amplification by physico-chemical and enzymic effects.

Post-PCR restriction analysis may be inhibited by excess PCR primers *(149)*. Chain extension by polymerase may be interrupted by primer-target noncomplemantarity. Primer extension during PCR may sometimes be terminated by single-base mismatches between the primer and target DNA strands. This inhibitory effect on enzymic extension has been exploited to assess oligonucleotide and target sequence complementarity *(150)*. Careful consideration must be given to primer design to avoid this form of premature termination.

Wiedbrauk et al. *(119)* reported that the detection of viral nucleic acids in intraocular fluids was inhibited by a mechanism that was not primer-specific and not the result of DNase activity or the chelation of Mg^{2+}. The unknown inhibitor was resistant to boiling for 15 min and affected *Taq* polymerase but not *Tth* or *Tfl* polymerases. Amplification was enabled by a single phenol–chloroform extraction.

Physico-chemical separation or inactivation may be used to overcome factors causing inhibition by these methods *(37,94,153)*. Specific inhibitors of inhibitors and competing substrates may also be of use *(121)*. For many situations, dilution of inhibited samples provides a rapid and straightforward way of permitting amplification. This dilution exploits the sensitivity of PCR by reducing the concentration of inhibitors relative to target DNA and is analogous to the dilution of substrates containing antimicrobial compounds prior to culture. In a systematic study of inhibitory substrates, careful selection of polymerase has been show to assist in overcoming inhibition caused by many classes of sample *(154)*.

4. Misidentification, Bias, and Error in Molecular Methods

Molecular methods are increasingly relied upon for food and environmental microbiology, for epidemiological investigations, and in medical and veterinary microbiology *(155–159)*. In some situations, these are used for rapid diagnostic purposes. In others, they are used for investigating genetic functionality or for typing bacteria that have been isolated previously, and less stringent quality criteria may sometimes be acceptable. Typing is seldom required for the control of outbreaks, but, on occasions, prompt availability of such data can be useful in outbreak management. Molecular typing methods are powerfully discriminatory and their success has been widely discussed.

Questions have been raised about errors in molecular studies that allegedly may affect 1–5% of these publications, scientific integrity, and the effectiveness of the peer-review process to eliminate poor science *(1)*. Very little attention has focused on the limitations of molecular methods because of their well-publicized successes and because of scientists' concerns about loss of funding and publication if they question the confidence of molecular results *(1,160)*. Ecological studies and taxonomic decisions also are heavily dependent on molecular characterization, and fundamental taxonomic errors are possible if poor quality data are relied upon. Errors may occur at any stage of the

Problems with the PCR   449

investigation, from selection bias in culture and amplification reactions, poor quality assurance, sequence infidelity, misleading electrophoresis and sequence data, genetic mutations and rearrangements, and interpretation of the laboratory, ecological, and clinical significance of results. A large proportion of the molecular tests conducted in clinical laboratories are for mycobacteria and HIV and use commercially prepared kits that have been carefully validated and require appropriate internal controls. Testing for food pathogens more often relies on in-house methods, and more thought may be needed to ensure that quality is adequately controlled and assured.

Sequence information has been widely published, and rapidly expanding online databases of probes and primers make information available to workers who may be less experienced in genetics than those who performed such work in previous years *(32,69,161,162)*. The expertise and expense necessary to obtain sequence information will continue to decrease with the commercial availability of new microfabricated DNA analytical instruments with costs similar to a single diagnostic test *(163,164)*. Although such developments are to be welcomed, the popularization of powerful methods may lead to poor results, misleading interpretations, and the entry of erroneous information into databases that are relied upon by other workers. Misleading results can arise by a number of mechanisms, and it is vital that scientists be aware of the pitfalls of molecular identification and typing. High degrees of standardization and reproducibility should be expected from methods used for critical diagnostic tests. For some typing methods, lower standards may be acceptable.

Some authors have suggested that Koch's postulates are to be superseded on the basis of genetic information *(165,166)*. Despite the apparent attractiveness of this approach, there are problems because the positivity of a sample can vary with the specimen or tissue examined *(167)*, and other workers have defended against conclusively accepting PCR results without evidence of reinfectivity, serology, and epidemiology *(168)*. Sequence data and other aspects of analysis require stringent quality assurance. Fundamental principles of quality management for molecular amplification methods and typing interpretation have been published, but are not as widely and consistently practiced as they should be *(39,169,170)*.

The consequences of failing to recognize the possible errors in molecular diagnosis and epidemiology are potentially large. Food companies, hospitals, and communities may be affected by misinformed outbreak management, individual patients may suffer, and the scientific literature and databases may be burdened with false information, leading to cumulative errors and the expense of wasted time. There has been little study of many of these areas and little information is available to enable assessment of these issues. This chapter is based on a review that considers the challenges these problems pose to the wider and more certain use of molecular methods *(160)*. Scientists and public health officials should be aware of potential problems when investigating and controlling foodborne outbreaks of disease. Laboratory scientists should be aware of these issues and implement appropriate measures to ensure confidence in their results. Only by confronting the challenge these issues present can the quality and confidence of laboratory results be improved. This review identifies potential sources of error so that workers can put in place controls that will assure the quality of data they generate.

Cultural methods introduce selection bias into the analysis of complex microbial communities. The impact may be small in qualitative investigations but is probably significant in many quantitative examinations. This problem in the analysis of complex microflora has been reported previously *(166)* and affects many areas, including the study of foods, the food processing environment, biofilms, and the microflora of the gastrointestinal tract. Other sources of error that can affect both pure and polymicrobial cultures are discussed in the following subsections. Examples of some reported sources of error are listed in order of publication in **Table 4**.

4.1. Sampling Errors, Differential Enrichment, and Recovery During Culture

The outcomes of molecular methods are not dependent solely on molecular aspects of the work. The quantity and quality of specific nucleic acids may be affected by conditions prior to their isolation. Sampling methods that fail to take account of nonuniform distribution of microorganisms, overgrowth by competitors, transport and storage conditions, selective enrichment, and even nonselective culture will not maintain microorganisms in the same proportions as they existed in their original environment.

Cultural methods may select groups of cells on the basis of species, subtype, or even metabolic state. This is particularly true of pre-enrichment and selective-enrichment procedures used to detect low numbers of organisms *(178)*. Selective-enrichment culture introduces toxic and thermal stresses to select the species of interest. Other species are generally prevented from growing to high numbers, but the species of interest may be stressed also, and the degree of stress may differ between strains.

Enrichment methods are used mainly in food and environmental microbiology, but they are of increasing relevance to clinical microbiology for organisms such as *E. coli* O157:H7. For many of the most studied and best understood microorganisms, these enrichment effects may be small, but for some species, the bias may be considerable. Campylobacters are often cultured using different media for food and fecal samples. This leads to potential bias in typing and epidemiological studies conducted on isolates. The quantities and

Table 4
Reported Errors in Molecular Methods

Method	Source of error	Effect	Ref.
Nested PCR for *Treponema pallidum*	DNA persistence in CSF after antibiotic therapy	Positive PCR reaction 3 yr after therapy	98
PCR for hepatitis C virus	Primers too specific to amplify mutant strains	False-negative test	208
Mixed human and microbial DNA	Microbial DNA interfered with human DNA typing	Total inhibition and/or dropout of the major human amplification fragment-length polymorphism allele	62
Restriction endonuclease digestion of PCR templates	Inhibition of cleavage by point mutation	Erroneous genotyping	209
RFLP analysis of mycobacterial isolates	Clinical mixed infection, reading error, laboratory cross-contamination	Different banding pattern for each patient	210
AP-PCR fingerprinting	Colony age	Number, clarity, and reproducibility of bands decreases with colony age	211
False positives	Contaminated *Taq* polymerase		212
Variations in yield	Differential efficacy of amplification between strains.		
Sporadic bands	Stochastic variations in bacterial numbers in sample		
Poor discrimination and difficulty interpreting clinical significance			
Cross-reactions	Poor probe specificity/crossreaction		
RAPD	Inaccuracies in sequence/hybridization predictions Melting–annealing	Poor reproducibility	213

Table 4 (cont.)
Reported Errors in Molecular Methods

Method	Source of error	Effect	Ref.
PFGE	temperature transition interval during thermal cycling	Altered banding pattern	137
	Point mutation, insertion and deletion		
	Gel interpretation	Misleading epidemiology	
PCR of 16S rRNA genes (SSU rDNAs)	Differential gene amplification in mixed cultures, dependent on cycle number	Quantitative errors and sequence bias	140
Computational specificity analysis of ribosomal small subunit-derived signature sequences	Published sequences nonspecific, ambiguous sites	Poor sensitivity, specificity, resolution, and reliability	128
RAPD	Lack of standardization	Poor reproducibility	124
Ribotyping	Inherent	Poor discrimination	
Review of AP-PCR, RAPD, DNA amplification fingerprinting	Primer concentration	Altered banding pattern	39
	Polymerase variation	Differential product amplification	
	Preferred synthesis of unrelated loci	Unrelated, comigrating RAPD products, biased phylogenetic analysis	
	Indentification of discriminatory and Poor specificity and reproducible primers	reproducibility	
rep-PCR (ERIC)	Various parameters	Poor interlaboratory reproducibility	
	Various	Minimal amplification of Gram-positive organisms	

Table 4 (cont.)
Reported Errors in Molecular Methods

Method	Source of error	Effect	Ref.
	Taq polymerase contamination/specific activity variation, pipetting volume variations	Background bands dependent on *Taq* concentration and supplier	
	PCR temperature steps	Number and intensity of bands, strain specificity of amplification	
		Poor specificity, spurious bands	
	Low annealing temperatures and low template concentration; ERIC primers generally unsuitable for Gram-positive organisms		
	Use of unstandardized whole-cell preparations	Spurious bands	
	Low primer concentration	Reduced number and intensity of bands	
	High primer concentration	Mispriming, nonspecific amplification	
	Primer:template DNA ratio and number of PCR cycles	Nonspecific amplification. Must be optimized for different species	
	Choice of primers	Reproducibility	
	RNA contamination	Mispriming	
	Thermal cycler variations	Poor reproducibility	
Soil bacteria: rep-PCR fingerprinting	Enrichment culture bias (cf. direct plating)	Distorted evaluation of genomic diversity	*214*
Review	Various biases and errors	Various	*133*
Review of PCR-based rRNA analysis	Sample collection, transportation and handling	Changes in microfloral composition	*157*
	Cell lysis and DNA extraction	Selective DNA recovery and misleading results	

Table 4 (cont.)
Reported Errors in Molecular Methods

Method	Source of error	Effect	Ref.
	PCR inhibition	False negatives	
	Differential amplification	Biased reflection of microbial diversity	
	Formation of artefactual PCR products—chimeras, deletions, and point mutations	Misleading results	
	Base substitution errors in early amplification cycles	Accumulating sequence errors and misleading results	
	Contamination	False positives	
	16S rRNA sequence variations as a result of *rrn* operon heterogeneity	Biased reflection of microbial diversity	
	Separation of amplified 16S rRNA genes	Coamplification of different sequences of identical size if not purified by cloning	
		DNA repair during cloning may lead to formation of artificial 16 S rRNA genes	
	Analysis of sequence data	Absence of data for other species	
		Poor quality and misleading sequences in databases	
		Natural and artificial chimeras	
Rotavirus PCR typing	Three-base mutation at 3' end of primer binding site	Typing failure	*162*
Chimeric alignment of 16S	Chimeric DNA from >1 species	Misclassification of sequences and species	*215*

Table 4 (cont.)
Reported Errors in Molecular Methods

Method	Source of error	Effect	Ref.
rRNA artifacts; computational methods			
Computer matching of AP-PCR fingerprints	Disagreement on presence–absence scoring by different operators	Incorrect matching of isolates	*159*
PCR using universal cold-shock protein primers	Spurious amplification of low-level contaminant organisms	Spurious bands	*216*
Amplification	Various	Inhibition and other effects	*31*
SSCP for BRCA1 mutations	Primer concentration	Poor reproducibility	*217*
PFGE	Spontaneous genomic rearrangements	Variation in PFGE profiles and misleading relationships in epidemiological studies	*145*
Ribotyping *E. coli*	Frequent faint bands	Uncertain types	*160*
RAPD-PCR of *Aeromonas hydrophila*	Comigration of different sequences	Misleading results and phylogenetic relationships	*155*

sequences of nucleic acids vary during the cell growth phases and may affect the sensitivity and specificity of molecular methods.

In viral load assays conducted to predict the prognosis of patients with HIV and hepatitis C virus, the expense of commercial test kits precludes monitoring with the frequency that might be useful. The sample is a biased one, taken perhaps every 3 months and may not be helpful for showing variations in viral load in the period between samples. As quantitation is the vital aspect of viral load monitoring, factors that alter the number of virions such as delays in transport and processing may lead to errors in assessing this prognostic indicator. Similar considerations apply when sampling shellfish beds for viruses. Representative samples must be taken from the bed to allow for spatial variations, quantity of virus will be affected by the species of shellfish and its filtration rate, and transport and processing should be prompt and temperature controlled.

There is no clear solution to the problems of sampling and culture bias other than to standardize methods as much as possible, to minimize delays, and to be aware of the potential for error.

4.2. Selective Lysis of Bacteria

Bacteria have different susceptibilities to lysis. Some species require specific treatments to lyse them effectively (e.g., *S. aureus* require lysostaphin). Variability in lytic enzyme activity may be a source of poor results *(31)*. Clumping *(188)* and changes in cell-wall structure during growth will affect susceptibility to lysis. In mixed populations, the predominant nucleic acid recovered may not represent the most numerous microbial species present in the sample. The species, metabolic state and position in the growth curve and cell cycle are probably of importance. The amount of nucleic acid recovered by different isolation methods varies substantially *(189)*.

4.3. Preferential Amplification

The most numerous microorganisms in a sample may not be the most strongly represented by PCR. Primers may have an unrecognized specificity for a subgroup of nucleic acid sequences and amplify these preferentially. In particular, this may relate to the %G–C, dilution, and the lengths of the primers and amplicons *(177)*. It has been reported that the amplification of the 16S rRNA gene from one genome could be completely inhibited by the presence of another genome *(190)*. This may be of particular relevance in the examination of samples containing mixed cultures using random or broad-range primers, where the precise sequence match is not known and varies with conditions *(39)*. Preferential amplification has also been demonstrated in systems such as 16S rDNA investigations *(191)*, and single-point mutations should not be excluded as a cause. Highly biased amplification has been reported where 16S rDNA from one species out of four was preferentially amplified. Touchdown PCR, denaturants, and cosolvents did not improve the situation. PCR of some sequences was thought to be inhibited by template flanking DNA segments, and the use of more than one primer set was recommended *(192)*.

Overamplification of certain templates in complex mixtures has been observed to relate to the G–C ratio of the template *(193)*. High G–C ratios encourage higher amplification efficiency. Other factors inherent to the organisms, genomes, genes, and primers involved may affect the selectivity of PCR (PCR selection). To some degree, these may be reproducible. Stochastic variations in the early cycles of PCR give rise to errors that should not be reproducible in replicate experiments (PCR drift). Such variations occur mainly from pipetting errors and instrumental variations. These considerations are of particular relevance in 16S studies using degenerate primers and to attempts at reliable clinical diagnosis and epidemiological typing of unculturable organisms *(165,193)*. Measures to minimize these errors have been discussed and include the selection of specific universal primers that avoid degeneracies, use of high template concentrations, combining several replicate reactions, and use of the minimum number of cycles necessary *(193)*. Currently, probing and *in situ* hybridization provide more informative quantitative data on microbial ecology, albeit with lower sensitivity than PCR.

The sources of error just discussed affect mixed cultures. Additional, more general, problems may affect methods used for either pure cultures or polymicrobial specimens. This spectrum of errors may arise from properties of the organisms and specimens examined, the methods used, human factors, and any combination of these. The points in the following subsections move sequentially through the factors that relate to properties of the organisms, the reactions, the analytical methodology, and interpretation. Many of these have been noted with application to multiple arbitrary amplicon profiling (MAAP) methods such as randomly amplified polymorphic DNA (RAPD) *(39)* and by other workers in association with a broad range of molecular methods.

4.4. Spontaneous Genetic Changes in Organisms

Spontaneous intramolecular genomic arrangements such as point mutations, insertions, and deletions may occur in organisms. This has been demonstrated to affect PFGE profiles of *C. coli* strains and may mislead when evaluating interstrain relationships. Five out of six isolates changed PFGE profiles when subcultured up to 50 times over a 6-month period irrespective of the restriction enzyme used *(185)*. Although this degree of repeated passage is unlikely in one laboratory, the rate of genetic change in the laboratory and in the environment may be different. Isolates recovered from samples that have passed through various animals over a period of time may exhibit differences from earlier isolates to which they were once identical. This variation has the potential to cause confusion in the investigation of long-term contamination in food factories, or infections in the population. The application of interpretative criteria is important to prevent such mutations giving misleading results *(170)*. Where restriction sites occur in hypervariable regions, a lack of reproducibility should be expected. One restriction site polymorphism should not be relied upon to differentiate subspecies types, and several restriction enzymes should be used to identify anomalies. Restriction may be inhibited at sites containing a high proportion of uracil *(79)*. Random mutations may change the electrophoretic pattern during an outbreak. The effects of mutations have been discussed, and at least 10 distinct fragments are necessary for reliable interpretation of PFGE gels *(170)*.

Small mutations at the 3' end of the primer binding site have been reported as a cause of typing error *(194)*. Genomic variability and the presence of several virus types simultaneously in an infected host have been observed in highly discriminatory molecular studies *(195)*.

Point mutations may have downstream effects on DNA replication and can induce compression of DNA. This could be misinterpreted as a true sequence variation, particularly in more sensitive mutation–detection typing techniques. Such errors should be suspected particularly where microdeletions and/or insertions occur in association with an upstream point mutation *(196)*.

Interpretation of molecular results therefore should not be restricted to the minute details of typing, but should also consider the larger picture using cultural and epidemiological methods.

4.5. Live or Dead?

The detection of nucleic acids from pathogens has been recognized as a source of interpretative error in molecular investigations of food and environmental samples. Pathogens detected may not be alive and may present no risk to humans. The viable but nonculturable state into which some organisms such as *Vibrio*, *Salmonella*, and *Campylobacter* can enter further complicates interpretation. rRNA is an attractive target for detection because of its universal and constituent expression and its high copy number. Its longevity is shorter than DNA and some investigators have used this as a measure of whether the organism detected is alive or dead, taking the detection of rRNA to indicate viability or very recent death. However, it has been shown that this assumption is only valid when cell death is caused by extreme conditions that also destroy RNases and disrupt ribosomes. Under less extreme conditions such as UV irradiation and most thermal food processes, rRNA is protected and degrades more slowly. Detection under these conditions is a poor indicator of viability *(197)*. The live/dead consideration also affects clinical investigations *(198)* and is of particular relevance to putative pathogens whose role in disease is not well understood. The importance of this factor is difficult to assess because of a lack of information. The DNA of many organisms has been found in diseased tissues and falsely implicated as the cause of a condition. Repeated positive samples and plausible epidemiological and serological evidence of causation are needed before the findings of PCR alone should be given credence *(146,168)*.

4.6. Preferential Cloning of PCR Products

It should not be assumed that all of a pool of randomly amplified sequences will be cloned with equal efficiency. Methods that rely on this approach to provide quantities of specific sequences may be misleading in certain circumstances that allow some sequences to be cloned less efficiently than others. PCR products from different sequences have been shown not be cloned with equal efficiency *(166)*. Where possible, it is necessary to assess the choice and pretreatment of cloning vectors for the sequences that are being cloned. Native restriction enzymes may selectively degrade certain DNA sequences. High copy numbers of certain sequences may be lethal to the host cell, and this can be a particular problem when working with virulence and pathogenesis determinants. The size limitations and sticky or blunt end requirements of the vector need to be considered. Large size variations and structural deformations such as hairpin loops should be considered as possible causes of preferential cloning and, in many cases, can be predicted. Point mutations should not be excluded as sources of this bias.

4.7. PCR-Mediated Chimeric Gene Amplification

Products of more than one base sequence can fuse and these chimeric products may come to dominate accurate sequences in later cycles. Detection of chimeras with high confidence is not straightforward, although methods exist that will detect many of the more obvious occurrences in small-subunit rRNA (SSU-rRNA) analysis of uncultivated organisms *(199)*. If chimerization occurs early in amplification, the dominant sequences will be entirely misleading. Incorrect sequence records may be entered into databases if this is not recognized. Detecting chimeras in closely related specimens is difficult.

4.8. Incorporation Errors Resulting from Enzyme Infidelity

Polymerases do not copy nucleic acid sequences perfectly. The error rates of different enzymes are known to vary, and high-fidelity options are available. Point mutations that occur early in amplification will produce erroneous sequence information and may affect restriction site analysis. Newly synthesized strands may not necessarily be perfect replicas of the original genetic material. Any errors that occur in the earliest cycles will be amplified so that erroneous sequences will outnumber accurate sequences. It is estimated that sequencing will not detect a mutation unless it represents >10% of PCR products in the reaction. This is only possible if less than five copies of template are present and misincorporation occurs in the first cycle *(200)*. Procedures such as cloning that select strands may magnify misincorporation errors.

4.9. Sequencing Errors

Errors may occur during sequencing *(169)*. Sequences may be chimeric or of mixed strands. Primers may be suboptimal because of inaccurate or inadequate information used in their design. Investigators may not be aware of errata correcting errors in published sequences.

Artifacts may arise in enriched DNA libraries and be overlooked because their homology is only partial. Probes based on these sequences will probably fail to produce products, or merely a smear. Hybridization-based subtraction of fragments of genomic DNA followed by cloning may be used to enrich clones with new low-abundance marker sequences that are not available from databases. This approach of enriching DNA libraries has been demonstrated to be a more reliable and cost-effective means of identifying marker sequences *(201)*.

4.10. Concentration-Dependent Effects

Variations in the concentration of reaction components may produce spurious bands that are not meaningful or reproducible, or absent bands. The effect of magnesium concentration and the duration of annealing and extension steps on cross-species PCR has been investigated *(202)*. The findings have implications for identification of unknown organisms using broad-range primers.

4.11. Comigration

Comigration of dissimilar sequences in gels has been identified as a source of errors *(187)*. These authors considered comigration to be particularly important in phylogenetic studies and species typing. Nucleic acid species of the same size move together through a gel, and specialized methods are needed to resolve sequence and conformational differences. This has implications for epidemiology because a single band consisting of several sequences where one of these is similar to that from another isolate could mislead the researcher into deducing a closer relationship than was the case. Denaturing polyacrylamide gels have the resolving power to avoid comigration problems in most cases, but agarose gels may not have sufficient resolution. Cloning and sequencing multiple clones may be necessary to assure the results in these cases.

Differential-display PCR (ddPCR) is a powerful but demanding technique that enables the identification and isolation of transcripts expressed at different times or sites in organisms. It is not generally used for typing purposes, but it has potential application in the molecular detection of unrecognized pathogens and the study of virulence and pathogenesis in research laboratories *(165)*. When comigrating species of the same size are assumed to have the same sequence, false positives may arise because of reamplified cDNA species that are not the originally selected transcript moving through the gel with the species of interest. It is necessary to resolve fragments of different sequences prior to cloning in order to avoid misleading results *(203)*. High-resolution polyacrylamide gels, cloning, and sequencing will detect erroneous results.

4.12. Contamination

As discussed earlier in the context of DNA amplification, nucleic acids originating outside the sample may also produce misleading results for typing methods. This has also been widely discussed in the literature and solutions described *(31)*. It has been estimated that 1–5% of published PCR work has been affected *(1)*.

In addition to false positives or reaction failure, enzyme contamination may be manifested as spurious background bands during amplification-based DNA fingerprinting using methods such as random amplified polymorphic DNA (RAPD). By nature of their low stringency, such typing techniques rely on both specific and nonspecific priming and combine the detection of artifactual variation with true polymorphism *(39)*.

4.13. Poor Quality Assurance

Being aware of potential sources of error is useful in deciding on appropriate procedures for quality assurance and control. It has been recommended that results are compared from different DNA extractions, amplifications, and cloning experiments to minimize the effects of the errors described. Probes against short-lived 16S rRNA sequences should be used to evaluate whether organisms detected by molecular methods are living and relevant to the investigation *(179)*.

The random distribution of microorganisms and the resulting variability of microbiology should not be forgotten *(204)*. This is especially important when analyzing the small quantities of sample that are used in molecular methods. The broad spread of enumeration results from different laboratories in national external quality assurance schemes demonstrates that precise and accurate results are not easy to obtain between laboratories. This is of particular importance when dealing with small sample volumes and low numbers of target cells where potential variation is greatest.

Contamination, concentration-dependent effects, spurious bands, inhibition, and other sources of error may complicate the interpretation of banding patterns and give rise to false results. Misleading or absent bands may arise from factors such as variations in enzyme purity or activity. Variations between brands of devices such as thermal cyclers and electrophoresis equipment and running conditions can produce results that are inconsistent. For these reasons, it is important that positive and negative controls are run with each test sample. Ten recommendations have been made for improving the consistency of arbitrary PCR *(39)*. These focus on purity, quantification of reactants, and standardization of conditions. Guidance on the interpretation of PFGE has been published *(170)*. A wide-ranging and comprehensive discussion of quality issues also exists *(169)*. It is beyond the scope of this chapter to review in depth the procedures for good quality assurance (QA) and quality control (QC) of molecular work. This has already been done, and it is essential that scientists put in place the quality practices detailed in these articles.

4.14. Reading and Interpretation Errors

Most molecular typing relies on reading patterns of bands on gels. Bands may shift and distort with variations in the running conditions. Some bands may be very faint and create uncertainty in reading. Photography is capable both of improving appearance and increasing error, so intelligent use must be made of camera and computer systems *(182)*. Inconsistency and error in reading gels can lead to false categorization of types. Frequent occurrence of faint bands may negate the usefulness of a typing method *(186)*.

Misinterpretation of typing may occur and it is important to focus on clinical problem-solving rather than modern methods alone. Typing is not mystical. Its goal is simply to provide evidence that epidemiologically related isolates are also related genetically, represent the same strain, and thus may have a common origin. In larger-scale studies, it may also suggest that genetically related isolates have a common epidemiological source. This may be harder to prove with certainty and gives potential for falsely positive associations that could have costly implications for the agricultural and food industries if food products are mistakenly implicated by overzealous investigators. An outbreak can occur with more than one epidemic strain, and an outbreak is not excluded because of strain heterogeneity *(205)*. Hepatitis C viral quasispecies, heterogenous mixtures of virus particles containing hypervariable regions, sequentially change during the natural course of infection and the resulting diversity of types will complicate epidemiological studies *(195)*. Limited genetic diversity within a species may result in indistinguishable genotypes between isolates that have no epidemiological relationship. This could lead to false associations being made. Epidemiological information is essential if typing data are to be correctly interpreted *(170)*. Some workers have been known to type rare serotypes of *Salmonella* or *E. coli* when the epidemiology and serotyping alone are clearly sufficient to implicate the isolates. Human DNA profiling also is not infallible and critical awareness should be maintained when interpreting typing results, particularly in epidemiological, clinical, and forensic situations *(180,206)*.

Epidemiological and taxonomic analysis can be compromised by misleading and invalid phylogenetic trees. Problems with data handling and analysis have been discussed *(207)*. As genetic information becomes increasingly more important than the biochemical and morphological observations that taxonomists have historically relied upon, it is important to ensure that errors do not arise in data recording, editing, processing, and interpretation. Just as powerful statistical software can mislead the uninformed user, so misuse of genetic software can allow errors to arise that might be difficult for other scientists and clinicians to detect. Inexperienced users may be unaware of errors caused by software algorithms that users with more understanding would recognize as aberrant.

5. Hybridization Microarrays

Hybridization microarray biochips or "gene chips" are microfabricated arrays of nucleic acids used for sequence identification by hybridization. They are becoming available commercially and offer great potential for low-cost rapid and specific mutation detection in the future, although the devices presently available are far from inexpensive *(208–210)*. Hybridization of target DNA to oligonucleotide probes is visualized in defined areas of fluorescence that can be read automatically *(211)*. Along with other microfabricated instruments such as PCR microchips *(212)* and microfluidics analyzers *(163)*, hybridization microarrays have obvious application in rapid and automated microbiological identification and typing. The potential for screening a food sample for a range of pathogens, their virulence factors, and epidemiological markers is enormous if the high cost can be reduced, as has happened with electronic microprocessors. However, six important validation issues must be addressed before these methods can be relied upon for diagnostic purposes *(213)*.

5.1. Standardization

Certified reference standards of nucleic acids must be available. Traceability to these standards will become an important issue as the requirement for laboratory accreditation becomes more widespread. Repeatability and interlaboratory comparisons will also be vital issues if results are to be able to withstand legal challenge.

5.2. Production

Arrays manufactured commercially or within laboratories must be capable of interlaboratory comparison of the specificity and sensitivity of both equipment and assay. Variations in thermal characteristics have been noted in PCR microchips, but these may be subject to control by calibration software *(212)*. Stability and shelf life must also be assured.

5.3. Hybridization

Considerable problems exist and often are not acknowledged regarding hybridization kinetics and steric hindrance. Variability in the melting temperatures of different sequences depending on their %G–C and secondary and tertiary structures cause much variation in interstrand binding. Appropriate ionic conditions and software may be important in generating and interpreting the data correctly *(214)*.

5.4. Detection

Internal quality standards must be used to ensure that fluorescence is measured correctly. DNA hybridization and PCR were once expected to be the ultimate identification methods. In practice, the sensitivity of these methods was lower than had been anticipated, and their effective application has been less universal than many had hoped. It is likely that the challenges of substrate interference and lack of sensitivity will remain when examining food samples with microarrays. Preliminary steps to ensure sufficient quantities of pure DNA may be required.

5.5. Quantitation

The relationship between signal strength and gene expression is nonlinear. Distorted results may appear when studying both large changes in expression in abundant species of mRNA and during low-level expression, which may be equally important. Accurate quantitation is important to the correct interpretation of results. As devices become increasingly small, probabilistic factors will become increasingly important.

5.6. Data Interpretation

Large amounts of data are generated by microarray assays and pose challenges for calibration and the organization, storage, distribution, and interpretation of data. Quality assurance will be a vital issue when large amounts of data are shared between workers who will rely heavily on information generated by others.

Although such systems are capable of being made in laboratories, it is likely that their full usefulness will only be realized by commercial production. It is incumbent on the companies developing these systems that they ensure the instruments can produce results that can be relied upon.

6. Conclusions

Although DNA amplification technologies continue to provide useful tools for the detection and investigation of microorganisms, their promise will not be completely fulfilled until improved and automated amplification and detection systems become available at affordable prices. This is most likely to happen through miniaturization, which unfortunately places limits on sensitivity and may require concentration steps.

Careful and well-organized laboratory work using chemical and physical measures can overcome false positives caused contamination. For DNA amplification technology to be widely applicable, methods must exist that allow the rapid and efficient removal of inhibitors and attenuators of amplification. Many reports have been made of different nucleic acid extraction methodologies. A range of techniques is being investigated, but it seems unlikely that a single method will emerge that is suitable for all sample types. The separation of cells from samples has been discussed using many techniques, including multiplexed methods *(20)*, general methods *(19,122)*, lectins *(81)*, phytohemagglutinin *(215)*, free-flow electrophoresis *(216)*, immunomagnetic separation (IMS) *(102)*, electroelution *(18)*, and electrofractionation *(217)*.

A large number of commercial rapid extraction methods are available. These mainly are based primarily on ion-exchange chromatography, size exclusion, and sorption. Such methods are advertised particularly for the purification of nucleic acids from relatively clean sources such as mammalian tissues, broths, and purified cultures. They do not often feature in studies where DNA is purified from more complex samples such as body fluids, food, soil, and surface water. Although rapid extraction methods may be suitable for many purposes, relatively few workers reporting inhibition have used such products as solutions to the inhibition. This probably reflects a lack of confidence in the suitability of such technology for these purposes, as most workers continue to prefer to use methods based on phenol/chloroform extraction.

Experimentation may also have shown that the more laborious extraction methods provide higher yields of purer DNA, particularly from complex samples such as sediments and foods. This has been my experience with several commercial systems. Immunomagnetic separation can be used for cells or nucleic acids in cell lysate and perhaps more complex samples. Although impaired by fats, it is perhaps the most promising general isolation method.

For some time, a confusing variety of methods will continue to be required for investigations in different sample types. This review shows that many studies have demonstrated a range of inhibitory compounds and that they achieve their effects at one or more of three sites. As better understanding of these points of inhibition is achieved, simplified and improved extraction methods may become available for various sample types. Improved understanding of the compounds and mechanisms of inhibition should enable the rationalization of approaches to purification. A smaller range of effective facilitating treatments may be developed that can be applied to nucleic acids from different source materials. This should improve both the sensitivity and applicability of these methods in every field.

The ease of PCR-based typing methods and the availability of sequence information via the Internet has opened molecular identification and typing technologies to less specialized scientists. Laboratory directors must carefully ensure that proper use is made of these techniques through stringent validation and quality assurance procedures so that false conclusions are not reached by inexperienced investigators. These could lead to expensive consequences in the food industry and cumulative errors in the literature of microbial ecology. In clinical laboratories, scientists and clinicians must guard against false conclusions that at best are time-wasting and that potentially might be life-threatening.

Misleading identification and typing results may arise from a number of sources. This is of particular importance if microorganisms are being identified from clinical specimens to guide therapeutic choice. Like biochemical microgallery tests, use of broad-range primers and sequencing gives an identification based on probabilities rather than certainty. The recent appearance of a commercial kit and instrumentation for 16S sequence analysis should improve standardization. It is important to assess the significance of findings in the light of various potential errors and whether even correctly identified nucleic acids indicate a true problem.

It is of particular importance that samples of clinical and epidemiological significance are examined in a careful manner with full cognisance of the pitfalls that may mislead the naïve investigator and lead to poor clinical decisions. It is also important that investigators are aware that ecological studies and database records may contain errors for the reasons discussed. Although the sources of error described are not all subject to direct control, the informed investigator can put in place appropriate confirmatory tests to monitor and verify the correctness of results. The data quality can therefore be better assured.

Molecular typing methods continue to be developed and to make a major contribution to epidemiology and clinical problem-solving. This is evidence that the problems described occur in a small proportion of cases where these techniques are used. Nevertheless, it is wise to consider the potential sources of errors so that anomalous results can be recognized. The eventual adoption of biochip devices will not solve all of the problems discussed, and intelligent consideration of quality issues and data interpretation will remain essential. These controls are vital to prevent potentially serious clinical and epidemiological misjudgements, as well as cumulative errors entering databases.

References

1. Cohen, P. (1998) Ghosts in the machine. *New Sci.* August 18–19.
2. Erlich, H. A., Gelfand, D., and Sninsky, J. J. (1991) Recent advances in the polymerase chain reaction. *Science* **252**, 1643–1651.
3. Mullis, K. P. and Faloona, F. F. (1987) Specific synthesis of DNA *in vitro* via a polymerase-catalysed chain reaction, in *Recombinant DNA, Methods in Enzymology,* vol. 155 (Wu, R., ed.), Academic Press, London, pp. 335–351.
4. Saiki, R. K., Gelfand, D. H., Stoeffel, S., Scharf, S. J., Higuchi, R., Horn, G. T., et al. (1988) Primer-directed enzymic amplification of DNA with a thermostable DNA polymerase. *Science* **239**, 487–491.
5. Lizardi, P. and Kramer, F. R. (1991) Exponential amplification of nucleic acids: new diagnostics using DNA polymerases and RNA replicases. *Trends Biotechnol.* **9**, 53–58.
6. Vaneechoutte, M. and Van Eldere, J. (1997) The possibilities and limitations of nucleic acid amplification technology in diagnostic microbiology. *J. Med. Microbiol.* **46**, 188–194.
7. Cahill, P., Foster, K., and Mahan, D. E. (1991). Polymerase chain reaction and Qβ replicase amplification. *Clin. Chem.* **37**, 1482–1485.
8. Birkenmeyer, L. and Armstrong, A. S. (1992) Preliminary evaluation of the ligase chain reaction for specific detection of *Neisseria gonorrhoeae*. *J. Clin. Microbiol.* **30**, 3089–3094.
9. Weiss, R. (1991) Hot prospect for new gene amplifier. *Science* **254**, 1292–1293.
10. Bush, C. E., Donovan, R. M., Peterson, W. R., Jennings, M. B., Bolton, V., Sherman, D. G., et al. (1992) Detection of human immunodeficiency virus type 1 RNA in plasma samples from high-risk pediatric patients by using the self-sustaining sequence replication reaction. *J. Clin. Microbiol.* **30**, 281–286.
11. Guatelli, J. C., Whitefield, K. M., Kwoh, D. Y., Barringer, K. J., Richman, D. D., and Gingeras, T. R. (1990) Isothermal, *in vitro* amplification of nucleic acids by a multienzyme reaction modeled after retroviral replication. *Proc. Natl. Sci. USA* **87**, 1874–1878.
12. Walker, G. T., Little, M. C., Nadeau, J. G., and Shank, D. D. (1992) Isothermal *in vitro* amplification of DNA by a restriction enzyme/DNA polymerase system. *Proc. Natl. Acad. Sci. USA* **89**, 392–396.
13. Walker, G. T., Nadeau, J. G., Linn, C. P., Devlin, R. F., and Dandliker, W. B. (1996) Strand displacement amplification (SDA) and transient-state fluorescence polarization detection of *Mycobacterium tuberculosis* DNA. *Clin. Chem.* **42**, 9–13.
14. Compton, J. (1991) Nucleic acid sequence-based amplification. *Nature* **350**, 91–94.
15. Uyttendaele, M., Schukkink, R., Gemen, B. V., and Debevere, J. (1996) Comparison of the nucleic acid amplification system NASBA and agar isolation for detection of pathogenic campylobacters in naturally contaminated poultry. *J. Food Protect.* **59**, 683–687.
16. Van Brunt, J. (1990) Amplifying genes: PCR and its alternatives. *Biotechnol.* **8**, 291–294.

17. Boom, R., Sol, C. J. A., Heijtink, R., Wertheim-van Dillen, P. M. E., and van der Noordaa, J. (1991) Rapid purification of hepatitis B virus DNA from serum. *J. Clin. Microbiol.* **29,** 1804–1811.
18. Rochelle, P. A. and Olson, B. H. (1991) A simple technique for electroelution of DNA from environmental samples. *BioTechniques* **11,** 724–728.
19. Roman, M. and Brown, P. R. (1992) Separation techniques for biotechnology in the 1990s. *J. Chromatogr.* **592,** 3–12.
20. Sharpe, A. N. (1991) On multiplexed separations in quick detection of microorganisms in foods. *Food Microbiol.* **8,** 167–170.
21. Wilson, I. G., Gilmour, A., and Cooper, J. E. (1993) Detection of enterotoxigenic microorganisms in foods by PCR, in *New Techniques in Food and Beverage Microbiology* (Kroll, R. G., Gilmour, A., and Sussman, M. eds.), Society for Applied Bacteriology/Blackwell, Oxford.
22. Rys, P. N. and Persing, D. H (1993) Preventing false positives: quantitative evaluation of three protocols for inactivation of polymerase chain reaction amplification products. *J. Clin. Microbiol.* **31,** 2356–2360.
23. Hughes, M. S., Beck, L.-A., and Skuce, R. A. (1994) Identification and elimination of DNA sequences in *Taq* DNA polymerase. *J. Clin. Microbiol.* **32,** 2007–2008.
24. Sarkar, G. and Sommer, S. S. (1990) Shedding light on PCR contamination. *Nature* **343,** 27.
25. Prince, A. M. and Andrus, L. (1992) PCR: how to kill unwanted DNA. *BioTechniques* **12,** 358–360.
26. DeFilippes, F. M. (1991) Decontaminating the polymerase chain reaction. *BioTechniques* **10,** 26–30.
27. Espy, M. J., Smith, T. F., and Persing, D. H. (1993) Dependence of polymerase chain reaction product inactivation protocols on amplicon length and sequence composition. *J. Clin. Chem.* **31,** 2361–2365.
28. Furrer, B., Candrian, U., Wieland, P., and Lüthy, J. (1990) Improving PCR efficiency. *Nature* **346,** 324.
29. Meier, A., Persing, D. H., Finken, M., and Bottger, E. C. (1993) Elimination of contaminating DNA within polymerase chain reaction reagents: implications for a general approach to detection of uncultured pathogens. *J. Clin. Microbiol.* **31,** 646–652.
30. Rossen, L., Norskov, P., Holmstrom, K., and Rasmussen, O. F. (1992) Inhibition of PCR by components of food samples, microbial diagnostic assays and DNA-extraction solutions. *Int. J. Food Microbiol.* **17,** 37–45.
31. Wilson, I. G. (1997) Inhibition and facilitation of nucleic acid amplification. *Appl. Environ. Microbiol.* **63,** 3741–3751.
32. Scheu, P. M., Berghof, K., and Stahl, U. (1998) Detection of pathogenic and spoilage micro-organisms in food with the polymerase chain reaction. *Food Microbiol.* **15,** 13–31.
33. Higuchi, R., von Beroldingen, C. H., Sensabaugh, G. F., and Erlich, H. A. (1988) DNA typing from single hairs. *Nature* **332,** 543–546.

34. Busch, M. P, Henrard, D. R., Hewlett, I. K., Mehaffey, W. F., Epstein, J. S., Allain, J-P., et al. and the Transfusion Safety Study Group (1992) Poor sensitivity, specificity and reproducibility of detection of HIV-1 DNA in serum by polymerase chain reaction. *J. Aquired Immune Defic. Syndrome* **5**, 872–877.
35. Nitschke, L., Kopf, M., and Lamers, M. C. (1993) Quick nested PCR screening of ES cell clones for gene targeting events. *BioTechniques* **16**, 914–916.
36. Parrish, K. D. and Greenberg E. P. (1995) A rapid method for extraction and purification of DNA from dental plaque. *Appl. Environ. Microbiol.* **61**, 4120–4123.
37. Wernars, K., Heuvelman, C. J., Chakraborty, T., and Notermans, S. H. W. (1991) Use of the polymerase chain reaction for direct detection of *Listeria monocytogenes* in soft cheese. *J. Appl. Bacteriol.* **70**, 121–126.
38. Wilson, I. G., Cooper, J. E., and Gilmour, A. (1991) Detection of enterotoxigenic *Staphylococcus aureus* in dried skimmed milk: use of the polymerase chain reaction for amplification and detection of staphylococcal enterotoxin genes *entB* and *entC1* the the thermonuclease gene *nuc. Appl. Environ. Microbiol.* **57**, 1793–1798.
39. Tyler, K. D., Wang, G., Tyler D., and Johnson, W. M. (1997) Factors affecting reliability and reproducibility of amplification-based DNA fingerprinting of representative bacterial pathogens. *J. Clin. Microbiol.* **35**, 339–346.
40. Kanungo, J. and Pandey, K. N. (1993) Kinasing PCR products for efficient blunt-end cloning and linker addition. *BioTechniques* **6**, 912–913.
41. Bickley, J., Short, J. K., McDowell, D. G., and Parkes, H. C. (1996) Polymerase chain reaction (PCR) detection of *Listeria monocytogenes* in diluted milk and reversal of PCR inhibition caused by calcium ions. *Lett. Appl. Microbiol.* **22**, 153–158.
42. Chaudhry, G. R., Toranzos, G. A., and Bhatt, A. R. (1989) Novel method of monitoring genetically engineered microorganisms in the environment. *Appl. Environ. Microbiol.* **55**, 1301–1304.
43. Chou, S. (1991). Optimizing polymerase chain reaction technology for clinical diagnosis. *Clin. Chem.* **37**, 1893–1894.
44. Wilson, I. G., Cooper, J. E., and Gilmour, A. (1994) Some factors inhibiting amplification of the *Staphylococcus aureus* enterotoxin C_1 (sec^+) by PCR. *Int. J. Food Microbiol.* **22**, 55–62.
45. De Lomas, J. G., Sunzeri, F. J., and Busch, M. P. (1992) False-negative results by polymerase chain reaction due to contamination by glove powder. *Transfusion* **32**, 83–85.
46. Starbuck, M. A. B., Hill, P. J., and Stewart, G. S. A. B. (1992) Ultra sensitive detection of *Listeria monocytogenes* in milk by the poymerase chain reaction. *Lett. Appl. Microbiol.* **15**, 248–252.
47. Herrick, J. B., Madsen, E. U., Batt, C. A., and Ghiorse, W. C. (1993) Polymerase chain reaction amplification of naphthalene-catabolic and 16S rRNA gene sequences from indigenous sediment bacteria. *Appl. Environ. Microbiol.* **59**, 687–694.
48. Bamford, K. B., Lutton, D. A., O'Loughlin, B., Coulter, W. A., and Collins, J. S. A. (1998) Nested primers improve sensitivity in the detection of *Helicobacter pylori* by the polymerase chain reaction. *J. Infect.* **36**, 105–110.

49. Drake, M., Small, C. L., Spence, K. D., and Swanson, B. G. (1996) Rapid detection and identification of *Lactobacillus* spp. in dairy products by using the polymerase chain reaction. *J. Food Protect.* **59,** 1031–1036.
50. Jaykus, L.-A., De Leon, R., and Sobsey, M. D. (1996) A virion concentration method for detection of human enteric viruses in oysters by PCR and oligoprobe hybridization. *Appl. Environ. Microbiol.* **62,** 2074–2080.
51. Alvarez, A. J., Buttner, M. P., Toranzos, G. A., Dvorsky, E. A., Toro, A., Heikes, T. B., et al. (1994) Use of solid-phase PCR for enhanced detection of airborne microorganisms. *Appl. Environ. Microbiol.* **60,** 374–376.
52. Tsai, Y. and Olson, B. H. (1992. Detection of low numbers of bacterial cells in soils and sediments by polymerase chain reaction. *Appl. Environ. Microbiol.* **58,** 754–757.
53. Tsai, Y. and Olson, B. H. (1992) Rapid method for separation of bacterial DNA from humic substances in sediments for polymerase chain reaction. *Appl. Environ. Microbiol.* **58,** 2292–2295.
54. Xia, J. Q., Yason, C. V., and Kibenge, F. S. B. (1995) Comparison of dot blot hybridization, polymerase chain reaction, and virus isolation for detection of bovine herpesvirus-1 (BHV-1) in artificially infected bovine semen. *Can. J. Vet. Res.* **59,** 102–109.
55. Long, G. W., Oprandy, J. J., Narayanan, R. B., Fortier A. H., Porter, K. R., and Nacy, C. A. (1993) Detection of *Francisella tularensis* in blood by polymerase chain reaction. *J. Clin. Microbiol.* **31,** 152–154.
56. Fehlmann, C., Krapf, R., and Solioz, M. (1993) Reverse transcriptase can block polymerase chain reaction. *Clin. Chem.* **39,** 368–369.
57. Young, C., Burghoff, R. L., Keim, L. G., Minak-Bernero, V., Lute, J. R., and Hinton, S. M. (1993) Polyvinylpyrrolidone–agarose gel electrophoresis purification of polymerase chain reaction-amplifiable DNA from soils. *Appl. Environ. Microbiol.* **59,** 1972–1974.
58. Steffan, R. J. and Atlas, R. M. (1988) DNA amplification to enhance detection of genetically engineered bacteria in environmental samples. *Appl. Environ. Microbiol.* **54,** 2185–2191.
59. Andersen, M. R. and Omiecinski, C. J. (1992) Direct extraction of bacterial plasmids from food for polymerase chain reaction amplification. *Appl. Environ. Microbiol.* **58,** 4080–4082.
60. Dickinson, J. H., Kroll, R. G., and Grant, K. A. (1995) The direct application of the polymerase chain reaction to DNA extracted from foods. *Lett. Appl. Microbiol.* **20,** 212–216.
61. Alvarez, A. J., Buttner, M. P., and Stetzenbach, L. D. (1995) PCR for bioaerosol monitoring: sensitivity and environmental interference. *Appl. Environ. Microbiol.* **61,** 3639–3644.
62. Lienert, K. and Fowler, J. C. (1992) Analysis of mixed human/microbial DNA samples: a validation study of two PCR AMP-FLP typing methods. *BioTechniques* **13,** 276–281.
63. Tebbe, C. C. and Vahjen W. (1993) Interference of humic acids and DNA extracted directly from soil in detection and transformation of recombinant DNA from bacteria and a yeast. *Appl. Environ. Microbiol.* **59,** 2657–2665.

64. St. Pierre, B. S., Neustock, P., Schramm, U., Wilhelm, D., Kirchner, H., and Bein, G. (1994) Seasonal breakdown of polymerase chain reaction. *Lancet* **343,** 673.
65. Weaver, J. W. and Rowe, M. T. (1997) Effect of non-target cells on the sensitivity of the PCR for *Escherichia coli* O157:H7. *Lett. Appl. Microbiol.* **25,** 109–112.
66. Vermunt, A. E. M., Franken, A. A. J. M., and Beumer, R. R. (1992) Isolation of salmonellas by immunomagnetic separation. *J. Appl. Bacteriol.* **72,** 112–118.
67. Jackson, C. J., Fox, A. J., and Jones, D. M. (1996) A novel polymerase chain reaction assay for the dectection and speciation of thermophilic *Camplylobacter* spp. *J. Appl. Bacteriol.* **81,** 467–473.
68. Rijpens, N. P., Jannes, G., Van Asbroeck, M., Rossau, R., and Herman, L. M. F. (1996) Direct detection of *Brucella* spp. in raw milk by PCR and reverse hybridisation with 16S-23S rRNA spacer probes. *Appl. Environ. Microbiol.* **62,** 1683–1688.
69. Mezei, L., Storts, D., Erickson D., Peyruchaud, S., and Smith, R. (1998) Web support with Amplification Assistant PCR troubleshooting program. *Promega Notes* **66,** 10–13.
70. Comey, C. T., Jung, J. M., and Budowle, B. (1991) Use of formamide to improve amplification of HLA DQa sequences. *BioTechniques* **10,** 60–61.
71. Sidhu, M. K., Liao,M., and Rashidbaigi, A. (1996) Dimethyl sulfoxide improves RNA amplification. *BioTechniques* **21,** 44–47.
72. Satsangi, J., Jewell, D. P., Welsh, K., Bunce, M., and Bell, J. I. (1994) Effect of heparin on polymerase chain reaction. *Lancet* **343,** 1509–1510.
73. Chen, Z., Swisshelm, K., and Sager, R. (1994) A cautionary note on reaction tubes for differential display and cDNA amplification in thermal cycling. *BioTechniques* **16,** 1002–1006.
74. Bej, A. K., Mahbubani, M. H., Dicesare, J. L., and Atlas, R. M. (1991) Polymerase chain reaction-gene probe detection of microorganisms by using filter-concentrated samples. *Appl. Environ. Microbiol.* **57,** 3529–3534.
75. Dohner D. E., Dehner, M. S., and Gelb, L. D. (1995) Inhibition of PCR by mineral oil exposed to UV irradiation for prolonged periods. *BioTechniques* **18,** 964–967.
76. Katzman, M. (1993) Use of oil overlays in "oil-free" PCR technology. *BioTechniques* **14,** 36–40.
77. Truscott, W. M. (1997) Glove associated cytotoxicity. *HES Supplement* August, 13–14.
78. Persing, D. H. (1991) Polymerase chain reaction: trenches to benches. *J. Clin. Microbiol.* **29,** 1281–1285.
79. Glenn, T. C., Waller, D. R., and Braun, M. J. (1994) Increasing proportions of uracil in DNA substrates increases inhibition of restriction enzyme digests. *BioTechniques* **17,** 1086–1090.
80. Wu, P., Daniel-Issakani, S., LaMarco, K., and Strulovici, B. (1997) An automated high throughput filtration assay: application to polymerase inhibitor identification. *Anal. Biochem.* **245,** 226–230.
81. Starnbach, M. N., Falkow, S., and Tomkins, L. S. (1989) Species-specific detection of *Legionella pneumophila* in water by DNA amplification and hybridization. *J. Clin. Microbiol.* **27,** 1257–1261.

82. Gannon, V. P. J., King, R. K., Kim, J. Y., and Goldsteyn Thomas, E. J. (1992) Rapid and sensitive method for detection of shiga-like toxin-producing *Escherichia coli* in ground beef using the polymerase chain reaction. *Appl. Environ. Microbiol.* **58**, 3809–3815.
83. Todd, D., Mawhinney, K. A., and McNulty, M. S. (1992) Detection and differentiation of chicken anemia virus isolates by using the polymerase chain reaction. *J. Clin. Microbiol.* **30**, 1661–1666.
84. Partis, L., Newton, K., Murby, J., and Wells, R. J. (1994) Inhibitory effects of enrichment media on the Accuprobe test for *Listeria monocytogenes*. *Appl. Environ. Microbiol.* **60**, 1693–1694.
85. Jinneman, K. C., Wetherington, J. H., Hill, W. E., Adams, A. M., Johnson, J. M., Tenge, B. J., et al. (1998) Template preparation for the detection and identification of *Cyclospora* sp. and *Eimeria* spp. oocysts directly from raspberries. *J. Food Protect.* **61**, 1497–1503.
86. Simon, M. C., Gray, D. I., and Cook, N. (1996) DNA extraction and PCR methods for the detection of *Listeria monocytogenes* in cold-smoked salmon. *Appl. Environ. Microbiol.* **62**, 822–824.
87. Jacobsen, C. S. and Rasmussen, O. F. (1992) Development and application of a new method to extract bacterial DNA from soil based on separation of bacteria from soil with cation-exchange resin. *Appl. Environ. Microbiol.* **58**, 2458–2462.
88. Lindahl, T. (1993) Instability and decay of the primary structure of DNA. *Nature* **362**, 709–715.
89. Gibson, J. R., Sutherland, K., and Owen, R. J. (1994) Inhibition of DNAse activity in PFGE analysis of DNA from *Campylobacter jejuni*. *Lett. Appl. Microbiol.* **19**, 357–358.
90. Dorward, D. W. and Garon., C. F. (1990) DNA is packaged within membrane-derived vesicles of Gram-negative but not Gram-positive bacteria. *Appl. Environ. Microbiol.* **56**, 1960–1962.
91. Alonso, R., Nicholson, P. S., and Pitt, T. L. (1993) Rapid extraction of high purity chromosomal DNA from *Serratia marcescens*. *Lett. Appl. Microbiol.* **16**, 77–79.
92. Raina, K. and Chandlee, J. M. (1996) Recovery of genomic DNA from a fungus (*Sclerotinia homoeocarpa*) with high polysaccharide content. *BioTechniques* **21**, 1030–1032.
93. Hochmeister, M. N., Budowle, B., Borer, U. V., and Dirnhofer, R. (1993) Effects of nonoxinol-9 on the ability to obtain DNA profiles from postcoital vaginal swabs. *J. Forensic Sci.* **38**, 442–447.
94. Ahokas, H. and Erkkila, M. J. (1993) Interference of PCR amplification by the polyamines, spermine and spermidine. *PCR Methods Applic.* **3**, 65–68.
95. Hall, L. M., Slee, E., and Jones, D. S. (1995) Overcoming polymerase chain reaction inhibition in old animal tissue samples using ethidium bromide. *Anal. Biochem.* **225**, 169–172.
96. Fendler, E. J., Dolan, M. J., and Williams, R. A. (1998) Handwashing and gloving for food protection. Part I: examination of the evidence. *Dairy Food Environ. Sanit.* **18**, 814–823.

97. Buffone, G. J., Demmler, G. J., Schimbor, C. M., and Greer, J. (1991) Improved amplification of cytomegalovirus DNA from urine after purification of DNA with glass beads. *Clin. Chem.* **37,** 1945–1949.
98. Noordhoek, G. T., Wolters, E. C., de Jonge, M. E. J., and Van Embden, J. D. A. (1991) Detection by polymerase chain reaction of *Treponema pallidum* DNA in cerebrospinal fluid from neurosyphilis patients before and after antibiotic treatment. *J. Clin. Microbiol.* **29,** 1976–1984.
99. Mayer, C. L. and Palmer, C. J. (1996). Evaluation of PCR, nested PCR, and fluorescent antibodies for detection of *Giardia* and *Cryptosporidium* species in wastewater. *Appl. Environ. Microbiol.* **62,** 2081–2085.
100. Goodyear, P. D., MacLaughlin-Black, S., and Mason, I. J. (1994) A reliable method for the removal of co-purifying PCR inhibitors from ancient DNA. *BioTechniques* **16,** 232–235.
101. Makino, S.-I., Okada, Y., and Maruyama, T. (1995) A new method for direct detection of *Listeria monocytogenes* from foods by PCR. *Appl. Environ. Microbiol.* **61,** 3745–3747.
102. Abrol, S. and Chaudhary, V. K. (1993) Excess PCR primers inhibit DNA cleavage by some restriction endonucleases. *BioTechniques* **15,** 630–632.
103. Arbuthnot, P. and Fitschen, W. (1992) Inhibition of primer extension by single mismatches during asymmetric re-amplification. *Biochem. J.* **288,** 1073–1074.
104. Akane, A., Matsubara, K., Nakamura, H., Takahashi, S., and Kimura, K. (1994) Identification of the heme compound copurified with deoxyribonucleic acid (DNA) from bloodstains, a major inhibitor of polymerase chain reaction (PCR) amplification. *J. Forensic Sci.* **39,** 362–372.
105. Cone, R. W., Hobson, A. C., and Huang, M. W. (1992) Coamplified positive control detects inhibition of polymerase chain reactions. *J. Clin. Microbiol.* **30,** 3185–3189.
106. Fiallo, P., Williams, D. L., Chan, G. P., and Gillis, T. P. (1992) Effects of fixation on polymerase chain reaction detection of *Mycobacterium leprae*. *J. Clin. Microbiol.* **30,** 3095–3098.
107. Fluit, A. C., Widjojoatmodjo, M. N., and Verhoef, J. (1995) Detection of *Salmonella* species in fecal samples by immunomagnetic separation and PCR. *J. Clin. Microbiol.* **33,** 1046–1047.
108. Grimprel, E., Sanchez, P. J., Wendel, G. D., Burstain, J. M., McCracken, G. H., Jr., Radolf, J. D., et al. (1991) Use of polymerase chain reaction and rabbit infectivity testing to detect *Treponema pallidum* in amniotic fluid, fetal and neonatal sera, and cerebrospinal fluid. *J. Clin. Microbiol.* **29,** 1711–1718.
109. Hay, P. E., Clarke, J. R., Strugnell, R. A., Taylor-Robinson, D., and Goldmeier, D. (1990) Use of the polymerase chain reaction to detect DNA sequences specific to pathogenic treponemes in cerebrospinal fluid. *FEMS Microbiol. Lett.* **68,** 233–238.
110. Kolk, A. H. J., Schuitema, A. R. J., Kuijper, S., van Leeuwen, J., Hermans, P. W. M., van Embden, J. D. A., et al. (1992) Detection of *Mycobacterium tuberculosis* in clinical samples by using polymerase chain reaction and a non-radioactive detection system. *J. Clin. Microbiol.* **30,** 2567–2575.

111. Widjojoatmodjo, M. N., Fluit, A. C., Torensma, R., Verdonk, G. P. H. T., and Verhoef, J. (1992) The magnetic immuno polymerase chain reaction assay for direct detection of salmonellae in fecal samples. *J. Clin. Microbiol.* **30,** 3195–3199.
112. Mercier, B., Gaucher, C., Feugeas, O., and Mazurier, C. (1990) Direct PCR from whole blood, without DNA extraction. *Nucleic Acids Res.* **18,** 5908.
113. Vettori, C., Paffetti, D., Pietramellara, G., Stotzky, G., and Gallori, E. (1996) Amplification of bacterial DNA bound on clay minerals by the random amplified polymorphic DNA (RAPD) technique. *FEMS Microbiol. Ecol.* **20,** 251–260.
114. Nilsson, H.-O., Aleljung, A., Nilsson, I., Tyszkiewicz, T., and Wadstrÿm, T. (1997) Immunomagnetic bead enrichment and PCR detection of *Helicobacter pylori* in human stools. *J. Microbiol. Methods* **27,** 73–79.
115. Saulnier, P. and Andremont, A. (1992) Detection of genes in feces by booster polymerase chain reaction. *J. Clin. Microbiol.* **30,** 2080–2083.
116. Powell, H. A., Gooding, C. M., Garrett, S. D., Lund, B. M., and McKee, R. A. (1994) Proteinase indibition of the detection of Listeria monocytogenes in milk using the polymerase chain reaction. *Lett. Appl. Microbiol.* **18,** 59–61.
117. Kreader, C. A. (1996) Relief of amplification inhibition in PCR with bovine serum albumin or T4 gene 32 protein. *Appl. Environ. Microbiol.* **62,** 1102–1106.
118. Wiedbrauk, D. L., Werner, J. C., and Drevon, A. M. (1995) Inhibition of PCR by aqueous and vitreous fluids. *J. Clin. Microbiol.* **33,** 2643–2646.
119. Lantz, P.-G., Matsson, M., Wadstrîm, T., and RÜdstrÿm, P. (1997) Removal of PCR inhibitors from human faecal samples through the use of an aqueous two-phase system for sample preparation prior to PCR. *J. Microbiol. Methods* **28,** 159–167.
120. Olive, D. M. (1989) Detection of enterotoxigenic *Escherichia coli* after polymerase chain reaction amplification with a thermostable DNA polymerase. *J. Clin. Microbiol.* **27,** 261–265.
121. Al-Soud, W. A. and Rüdström, P. (1998) Capacity of nine thermostable DNA polymerases to mediate DNA amplification in the presence of PCR-inhibiting samples. *Appl. Environ. Microbiol.* **64,** 3748–3753.
122. Clewley, J. P. (1995) Which amplification method? *PHLS Microbiol. Digest* **12,** 186.
123. Cotton, R. G. H. (1997) Slowly but surely towards better scanning for mutations. *Trends Genet.* **13,** 43–46.
124. Farber, J. M. (1996) An introduction to the hows and whys of molecular typing. *J. Food Protect.* **59,** 1091–1101.
125. Tang, Y.-W., Procop, G. W., and Persing, D. H. (1997) Molecular diagnostics of infectious disease. *Clin. Chem.* **43,** 2021–2038.
126. Threlfall, J. E., Ridley, A. M., and Hampton, M. D. (1996) Technical advances in the bacteriological laboratory: methods for DNA analysis. *PHLS Microbiol. Dig.* **13,** 138–141.
127. Wilson, I. G. Potential errors in molecular methods. Unpublished data.
128. Gendel, S. M. (1996) Computational analysis of the specificity of 16S rRNA-derived signature sequences for identifying food-related microbes. *Food Microbiol.* **13,** 1–15.

129. Wang, R.-F., Cao, W.-W., and Cerniglia, C. E. (1997) A universal protocol for PCR detection of 13 species of foodborne pathogens in foods. *J. Appl. Microbiol.* **83,** 727–736.
130. Burns, M. A., Johnson, B. N., Brahmasandra, S. N., Handique, K., Webster, J. R., Krishnan, M., et al. (1998) An integrated nanoliter DNA analysis device. *Science* **282,** 484–487.
131. Kricka, L. J. (1998) Miniaturization of analytical systems. *Clin. Chem.* **44,** 2008–2014.
132. Balter, M. (1998) Molecular methods fire up the hunt for emerging pathogens. *Science* **282,** 219–221.
133. Wilson, M. J., Weightman, A. J., and Wade, W. G. (1997) Applications of molecular ecology in the characterization of uncultured microorganisms associated with human disease. *Rev. Med. Microbiol.* **8,** 91–101.
134. Van Zwet, A. A., Thijs, J. C., Kooistra-Smid, A. M., Schirm, J., and Snijder, J. A. (1994) Use of PCR with feces for detection of *Helicobacter pylori* infections in patients. *J. Clin. Microbiol.* **32,** 1346–1348.
135. Kingsley, G. (1997) Microbial DNA in the synovium — a role in aetiology or a mere bystander? *Lancet* **349,** 1038–1039.
136. Neumaier, M., Braun, A., and Wagener C. (1998) Fundamentals of quality assessment of molecular amplification methods in clinical diagnostics. *Clin. Chem.* **44,** 12–26.
137. Tenover, F. C., Arbeit, R. D., Goering, R. V., Mickelsen, P. A., Murray, B. E., Persing, D. H., et al. (1995) Interpreting chromosomal DNA restriction patterns produced by pulsed-field gel electrophoresis: criteria for bacterial strain typing. *J. Clin. Microbiol.* **33,** 2233–2239.
138. Grant, I. R. (1998) Does *Mycobacterium paratuberculosis* survive pasteurization conditions? *Appl. Environ. Microbiol.* **64,** 2760–2761.
139. de Kok, J. B., Hendricks, J. C. M., van Solinge, W. W., Willems, H. L., Mensink, E. J., and Swinkels, D. W. (1998) Use of real-time quantitative PCR to compare isolation methods. *Clin. Chem.* **44,** 2201–2204.
140. Suzuki, M. T. and Giovannoni, S. J. (1996) Bias caused by template annealing in the amplfication of mixtures of 16S rRNA genes by PCR. *Appl. Environ. Microbiol.* **62,** 625–630.
141. Wang, G. C-Y., and Wang, Y. (1998) Frequency of formation of chimeric molecules as a consequence of PCR coamplification of 16S rRNA genes from mixed bacterial genomes. *Appl. Environ. Microbiol.* **63,** 4645–4650.
142. Zoetendal, E. G., Akkermanns, A. D. L., and de Vos, W. M. (1998) Temperature gradient gel electrophoresis analysis of 16S rRNA from human fecal samples reveals stable host-specific communities of active bacteria. *Appl. Environ. Microbiol.* **64,** 3854–3859.
143. Christian Hansen, M., Tolker-Nielsen, T., Givskov, M., and Molin, S. (1998) Biased 16S rDNA PCR amplification caused by interference from DNA flanking the template region. *FEMS Microbiol. Ecol.* **26,** 141–149.
144. Polz, M. F. and Cavanaugh, C. M. (1998) Bias in template-to-product ratios in multitemplate PCR. *Appl. Environ. Microbiol.* **64,** 3724–3730.

145. On, S. L. (1998) *In vitro* genotypic variation of *Campylobacter coli* documented by pulsed-field gel electrophoresis DNA profiling: implications for epidemiological studies. *FEMS Microbiol. Lett.* **165,** 341–346.
146. Adah, M. I., Rohwedder, A., Olaleyle, O. D., and Werchau, H. (1997) Nigerian rotavirus serotype G8 could not be typed by PCR due to nucleotide mutation at the 3' end of the primer binding site. *Arch. Virol.* **142,** 1881–1887.
147. Kerr, J. R. and Curran, M. D. (1996) Application of polymerase chain reaction-single stranded conformational polymorphism to microbiology. *Clin. Mol. Pathol.* **49,** 315–320.
148. Weinshenker, B. G., Hebrink, D. D., Gacy, A. M., and McMurray, C. T. (1998) DNA compression caused by an upstream point mutation. *BioTechniques* **25,** 68–72.
149. McKillop, J. L., Jaykus, L.-A., and Drake, M. (1998) rRNA stability in heat-killed and UV-irradiated enterotoxigenic *Staphylococcus aureus* and *Escherichia coli* O157:H7. *Appl. Environ. Microbiol.* **64,** 4264–4268.
150. Barer, M. R., Gribbon, L. T., Harwood, C. R., and Nwoguh, C. E. (1993) The viable but non-culturable hypothesis and medical bacteriology. *Rev. Med. Microbiol.* **4,** 183–191.
151. Kopczynski, E. C., Bateson, M., and Ward, D. M. (1994) Recognition of chimeric small-subunit ribosomal DNAs composed of genes from uncultivated micro-organisms. *Appl. Environ. Microbiol.* **60,** 746–748.
152. Anonymous (1998) QIA-hints. *QIAGEN News* **5,** 25.
153. Koblízíkova, A., Dolezel, J., and Macas, J. (1998) Subtraction with 3' modified oligonucleotides eliminates amplification artifacts in DNA libraries enriched for microsatellites. *BioTechniques* **25,** 32–38.
154. Ely, J. J., Reeves-Daniel, A., Campbell, M. L., Kohler, S., and Stone, W. H. (1998) Influence of magnesium ion concentration and PCR amplification conditions on cross-species PCR. *BioTechniques* **25,** 38–42.
155. Oakey, H. J., Gibson, L. F., and George, A. M. (1998) Co-migration of RAPD-PCR amplicons from *Aeromonas hydrophila*. *FEMS Microbiol. Lett.* **164,** 35–38.
156. Miele, G., MacRae, L., McBride, D., Manson, J., and Clinton, M. (1998) Elimination of false positives generated through PCR re-amplification of differential display cDNA. *BioTechniques* **25,** 138–144.
157. Wintzingerode, F., Göbel, U. B., and Stachebrandt, E. (1997) Determination of microbial diversity in environmental samples: pitfalls of PCR-based rRNA analysis. *FEMS Microbiol. Rev.* **21,** 213–229.
158. Tillet, H. E. and Lightfoot, N. E. (1995) Quality control in environmental microbiology compared with chemistry: what is homogenous and what is random? *Water Sci. Technol.* **31,** 471–477.
159. Burr, M. D. and Pepper, I. L. (1997) Variability in presence-absence scoring of AP PCR fingerprints affects computer matching of bacterial isolates. *J. Microbiol. Methods* **29,** 63–68.
160. Machado, J., Grimond, F., and Grimond, P. A. (1998) Computer identification of *Escherichia coli* rRNA gene restriction patterns. *Res. Microbiol.* **149,** 119–35.

161. Cimolai, N., Trombley, C., Wensley, D., and LeBlanc, J. (1997) Heterogenous *Serratia marcescens* genotypes from a nosocomial paediatric outbreak. *Chest* **111,** 194–197.
162. Singh, D. (1995) DNA profiling: "insurmountable proof" or exaggeration? *Med. Law* **14,** 445–451.
163. Welch, H. G. and Burke, W. (1998) Uncertainties in genetic testing for chronic disease. *JAMA* **280,** 1525–1527.
164. Clewley, J. P. (1998) A user's guide to producing and interpreting tree diagrams in taxonomy and phylogenetics. Part 4: practice. *Commun. Dis. Public Health* **1,** 285–287.
165. Angrist, M. (1997) Does Phaster mean better? *Clin. Chem.* **43,** 424–426.
166. Lipshutz, R. J., Morris, D., Chee, M., Hubbell, E., Kozal, M. J., Shah, N., et al. (1995) Using oligonucleotide probe arrays to assess genetic diversity. *BioTechniques* **19,** 442–447.
167. Service, R. F. (1998) Microchip arrays put DNA on the spot. *Science* **282,** 396–399.
168. Hacia, J. G., Brody, L. C., Chee, M. S., Fodor, S. P. A., and Collins, F. S. (1996) Detection of heterozygous mutations in BRCA1 using high density oligonucleotide arrays and two-colour fluorescence analysis. *Nat. Genet.* **14,** 441–447.
169. Belgrader, P., Bennet, W., Hadley, D., Long, G., Raymond, M., Milanovich, F., et al. (1998) Rapid pathogen detection using a microchip PCR array instrument. *Clin. Chem.* **44,** 2191–2194.
170. Parkes, H. (1998) DNA chips — a taste of the future. *VAM Bull.* No. 19 (Autumn), 14–18.
171. Guschin, D. Y., Mobarry, B. K., Prouodnikov, D., Stahl, D. A., Rittman, B. E., and Mirzabekov, A. D. (1997) Oligonucleotide microchips as genosensors for determinative and environmental studies in microbiology. *Appl. Environ. Microbiol.* **63,** 2397–2402.
172. Patchett, R. A., Kelly, A. F., and Kroll, R. G. (1991) The adsorption of bacteria to immobilized lectins. *J. Appl. Bacteriol.* **71,** 277–284.
173. Payne, M. J., Campbell, S., Patchett, R. A., and Kroll, R. G. (1992) The use of immobilized lectins in the separation of *Staphylococcus aureus, Escherichia coli, Listeria* and *Salmonella* spp. from pure cultures and foods. *J. Appl. Bacteriol.* **73,** 41–52.
174. Ehrlich-Kautzky, E., Shinomiya, N., and Marsh, D. G. (1991) Simplified method for isolation of white cells from whole blood suitable for direct polymerase chain reaction. *BioTechniques* **10,** 39–40.
175. Hannig, K. (1971) Free-flow electrophoresis (a technique for continuous preparative and analytical separation). *Methods Microbiol.* **58,** 513–548.
176. Shain, D. H., Yoo, J., Slaughter, R. G., Hayes, S. E., and Ji, T. H. (1992) Electrofractionation: a technique for detecting and recovering biomolecules. *Anal. Biochem.* **200,** 47–51.
177. Atmar, R. L., Metcalf, T. G., Neill, F. H., and Estes, M. K. (1993) Detection of enteric viruses in oysters by using the polymerase chain reaction. *Appl. Environ. Microbiol.* **59,** 631–635.

178. Grant, K. A., Dickinson, J. H., Payne, M. J., Campbell, S., Collins, M. D., and Kroll, R. G. (1993) Use of the polymerase chain reaction and 16S rRNA sequences for the rapid detection of *Brochothrix* spp. in foods. *J. Appl. Bacteriol.* **74,** 260–267.
179. Lantz, P., Tjerneld, F., Borch, E., Hahn-Hagerdal, B., and, Rådström, P. (1994) Enhanced sensitivity in PCR detection of *Listeria monocytogenes* in soft cheese through use of an aqueous two-phase system as a sample preparation method. *Appl. Environ. Microbiol.* **60,** 3416–3418.
180. McGregor, D. P., Forster, S., Steven, J., Adair, J., Leary, S. E. C., Leslie, D. L., et al. (1996) Simultaneous detection of microorganisms in soil suspension based on PCR amplification of bacterial 16S rRNA fragments. *Biotechniques* **21,** 463–471.
181. Jaykus, L-A., De Leon, R., and Sobsey, M. D. (1995) Development of a molecular method for the detection of enteric viruses in oysters. *J. Food Protect.* **58,** 1357–1362.
182. Herman, L. M. F., De Block, J. H. G. E., and Waes, G. M. A. V. J. (1995) A direct PCR detection method for *Clostridium tyrobutyricum* spores in up to 100 milliliters of raw milk. *Appl. Environ. Microbiol.* **61,** 4141–4146.
183. Lees, D. N., Henshilwood, K., Green, J., Gallimore, C. I., and Brown, D. W. G. (1995) Detection of small round structured viruses in shellfish by reverse transcription-PCR. *Appl. Environ. Microbiol.* **61,** 4418–4424.
184. Schwab, K. J., De Leon, R., and Sobsey, M. D. (1996) Immunoaffinity concentration of waterborne enteric viruses for detection by reverse transcriptase PCR. *Appl. Environ. Microbiol.* 62, 2086–2094.
185. Witham, P. K., Yamashiro, C. T., Livak, K. J., and Batt, C. A. (1996) A PCR-based assay for the detection of *Escherichia coli* shiga-like toxin genes in ground beef. *Appl. Environ. Microbiol.* **62,** 1347–1353.
186. Wilde, J., Eiden, J., and Yolken, R. (1990) Removal of inhibitory substances from human fecal specimens for detection of group A rotaviruses by reverse transcriptase and polymerase chain reactions. *J. Clin. Microbiol.* **28,** 1300–1307.
187. Pomp, D. and Medrano, J. F. (1991) Organic solvents as facilitators of polymerase chain reaction. *BioTechniques* **10,** 58–59.
188. De Leon, R., Matsui, S. M., Baric, R. S., Herrmann, J. E., Blacklow, N. R., Greenberg, H. B., and Sobsey, M. D. (1992) Detection of Norwalk virus in stool specimens by reverse transcriptase-polymerase chain reaction and nonradioactive oligoprobes. *J. Clin. Microbiol.* **30,** 3151–3157.
189. Lau, J. Y. N., Qian, K., Wu, P. C., and Davis, G. L. (1993) Ribonucleotide vanadyl complexes inhibit polymerase chain reaction. *Nucleic Acids Res.* **21,** 2777.
190. Schlapfer, G., Senn, H. P., Berger, R., and Just, M. (1993) Use of the polymerase chain reaction to detect *Bordetella pertussis* in patients with mild or atypical symptoms of infection. *Eur. J. Clin. Microbiol. Infect. Dis.* **12,** 459–463.
191. Varela, P., Pollevick, G. D., Rivas, M., Chinen, I., Binsztein, N., Frasch, A. C., et al. (1994) Direct detection of *Vibrio cholerae* in stool samples. *J. Clin. Microbiol.* **32,** 1246–1248.

192. Deneer, H. G. and Knight, I. (1994) Inhibition of the polymerase chain reaction by mucolytic agents. *Clin. Chem.* **40,** 171–172.
193. Kolk, A. H. J., Noordhoek, G. T., De Leeuw, O., Kuijper, S., and Van Embden, J. D. A. (1994) *Mycobacterium smegmatis* strain for detection of *Mycobacterium tuberculosis* by PCR used as internal control of inhibition of amplification and for quantification of bacteria. *J. Clin. Microbiol.* **32,** 1354–1356.
194. Amicosante, M., Richeldi, L., Trenti, G., Paone, G., Campa, M., Bisetti, A., and Saltini, C. (1995) Inactivation of polymerase inhibitors for *Mycobacterium tuberculosis* DNA amplification in sputum by using capture resin. *J. Clin. Microbiol.* **33,** 629–630.
195. Schirm, J., Oostendorp, L. A. B., and Mulder, J. G. (1995) Comparison of Amplicor, in-house PCR, and conventional culture for detection of *Mycobacterium tuberculosis* in clinical samples. *J. Clin. Microbiol.* **33,** 3221–3224.
196. An, Q., Liu, J, O'Brien, W., Radcliffe, G., Buxton D., Popoff, S., King, W., et al. (1995) Comparison of Qβ replicase-amplified assay with competitive PCR assay for *Chlamydia trachomatis*. *J. Clin. Microbiol.* **33,** 58–63.
197. Leng, X., Mosier, D. A., and Oberst, R. D. (1996) Simplified method for recovery and PCR detection of *Crytosporidium* DNA from bovine faeces. *Appl. Environ. Microbiol.* **62,** 643–647.
198. Toranzos, G. A. and Alvarez, A. J. (1992) Solid-phase polymerase chain reaction: applications for direct detection of enteric pathogens in waters. *Can. J. Microbiol.* **38,** 365–369.
199. Demeke, T. and Adams, R. P. (1992) The effects of plant polysaccharides and buffer additives on PCR. *BioTechniques* **12,** 332–334.
200. Smalla, K., Cresswell, N., Mendonca-Hagler, L. C., Wolters, A., and van Elsas, J. D. (1993) Rapid DNA extraction protocol from soil for polymerase chain reaction-mediated amplification. *J. Appl. Bacteriol.* **74,** 78–85.
201. Yoon, C. and Glawe, D. A. (1993) Pretreatment with RNase to improve PCR amplification of DNA using 10-mer primers. *BioTechniques* **6,** 908–910.
202. Moran, M. A., Torsvik, V. L., Torsvik, T., and Hodson, R. E. (1993) Direct extraction and purification of rRNA for ecological studies. *Appl. Environ. Microbiol.* **59,** 915–918.
203. Straub, T. M., Pepper, I. L., Abbaszadegan, M., and Gerba, C. P. (1994) A method to detect enteroviruses in sewage sludge-amended soil using the PCR. *Appl. Environ. Microbiol.* **60,** 1014–1017.
204. Straub, T. M., Pepper, I. L., and Gerba, C. P. (1995) Removal of PCR inhibiting substances in sewage sludge amended soil. *Water Sci. Technol.* **31,** 311–315.
205. Johnson, D. W., Pieniazek, N. J., Griffin, D. W., Misener, L., and Rose, J. B. (1995) Development of a PCR protocol for sensitive detection of *Cryptosporidium* oocysts in water samples. *Appl. Environ. Microbiol.* **61,** 3849–3855.
206. McHale, R. H., Stapleton, P. M., and Bergquist, P. L. (1991) Rapid preparation of blood and tissue samples for polymerase chain reaction. *BioTechniques* **10,** 20–22.

207. Jackson, C. R., Harper, J. P., Willoughby, D., Roden, E. E., and Churchill, P. (1997) A simple, efficient method for the separation of humic substances and DNA from environmental samples. *Appl. Environ. Microbiol.* **63,** 4993–4995.
208. Irving, W. L., Day, S., Eglin, R. P., Bennett, D. P., Jones, D. A., Nuttall, P., et al. (1992) HCV and PCR negativity. *Lancet* **339,** 1425.
209. Klein, D., Grody, W. W., Tabor, D. E., and Cederbaum, S. D. (1992) Pitfalls of restriction endonuclease digestion of point mutations. *Clin. Chem.* **38,** 1392–1394.
210. Das, S., Chan, S. L., Allen, B. W., Mitchison, D. A., and Lowrie, D. B. (1993) Application of DNA fingerprinting with IS986 to sequential mycobacterial isolates obtained from pulmonary tuberculosis patients in Hong Kong before, during and after short-course chemotherapy. *Tuberc. Lung Dis.* **74,** 47–51.
211. Coutinho, H. L. C., Handley, B. A., Kay, H. E., Stevenson, L., and Beringer, J. E. (1993) The effect of colony age on PCR fingerprinting. *Lett. Appl. Microbiol.* **17,** 282–284.
212. Greisen, K., Loeffelholz, M., Purohit, A., and Leong, D. (1994) PCR primers and probes for the 16S rRNA gene of most species of pathogenic bacteria, including bacteria found in cerebrospinal fluid. *J. Clin. Microbiol.* **32,** 335–351.
213. Schweder, M. E., Shatters, R. G., West S. H., and Smith, R. L. (1995) Effect of transition interval between melting and annealing temperatures on RAPD analyses. *BioTechniques* **19,** 38–42.
214. Dunbar, J., White, S., and Forney, L. (1997) Genetic diversity through the looking glass: effect of enrichment bias. *Appl. Environ. Microbiol.* **63,** 1326–1331.
215. Komatsoulis, G. A. and Waterman, M. S. (1997) A new computational method for detection of chimeric 16S rRNA artifacts generated by PCR amplification from mixed bacterial populations. *Appl. Environ. Microbiol.* **63,** 2338-2346.
216. Francis, K. P. and Stewart, G. S. A. B. (1997) Detection and speciation of bacteria through PCR using universal major cold-shock protein primer oligomers. *J. Ind. Microbiol. Technol.* **19,** 286-293.
217. Hennessy, L. K. Teare, J., and Ko., C. (1998) PCR condtions and DNA denaturants affect reproducibility of single-strand conformation polymorphism patterns for BRCA1 mutations. *Clin. Chem.* **44,** 879-882.

45

Problems with Genetically Modified Foods

Jose Manuel Bruno-Bárcena, M. Andrea Azcarate-Peril, and Faustino Siñeriz

During the last few years we have seen a public debate, involving every social statement, concerning food products and ingredients *(1)*. The appearance in the market of the first food products, or organisms, which have been improved by recombinant DNA technologies (rDNA), has been received by society in a negative manner. We are referring specifically to *genetically modified* (GM) food products. Several definitions can explain the concept of GM products or, in a general way, the imprecise word *transgenic*, a concept that comes from the seventies, even when the terms were not the same. Since then, scientists are capable of constructing recombinant DNA molecules and precisely moving them to another organism or to the original one, by direct genetic manipulation. People understand "transgenic" as food products or their ingredients, resulting from this kind of modification. In other words, "either add a gene (or a set of them) from a donor genome to a recipient one" and you will obtain it *(2)*.

However, the development and use of GM products in the pharmaceutical industry remains out of this debate. People do not question the application of DNA technologies in the medical field, but the public perception of what is a transgenic food is associated either with menace or disease, conditioning the final acceptance of these applications.

Certain products of animal or vegetable origin must be transformed by microbes to obtain the final food. In an empirical way, man has produced fermented food for thousands of years: production of yogurt, bread, and beer are examples of processes known since then. The microorganisms used for these products were extensively studied decades ago, and this knowledge allowed the controlled manufacture of fermented foods using "starters." The comprehension of the microorganisms acting in the production allowed the rationalization of the production avoiding the variability problems, and hence a controlled quality. The incorporation of useful characteristics in these microorganisms by means of recombinant technology will allow us to obtain foods with better flavor, taste, and quality.

The first nonpharmaceutical, genetically modified products from microorganisms to be commercialized were chymosin (used to replace rennet in cheese manufacture) and bovine somatotrophin (which increasees milk production) *(3)*. In 1994, the first genetically modified, whole foodstuff, was placed on the market: a tomato with a long shelf life. The introduction of these products caused a public reaction and their hazards were evaluated and finally accepted by the Food and Drug Administration, after their safety was confirmed. The agency decided that this GM food is no more toxic, allergenic or any less "substantially equivalent" than its standard counterpart.

Because of the foregoing, several issues must be considered (in the development of GM food products). The first step to obtain an acceptable product in the market is the application (and the evaluation) of an adequate scheme, which should include first the benefits for the consumer, which determine the public acceptance and, hence, the success of the developed product. To introduce a desirable trait to a useful strain, an adequate scheme should consider the benefits of the genetic modification of the product: the technical aspects of the application of recombinant technology, the regulation of the application, and the public acceptance.

The first aspect to be considered is the application of the improvement (i.e., the real benefits of the incorporation of new genetic material into the cell). The answer to this question comes from the actual problems in the industry. This is the case of the problems caused by the infection of starters by bacteriophages in the dairy industry. Bacteriophage infection of starter cultures is considered to be one of the most important factors resulting in slow acid production in large-scale dairy fermentations. Several strains of lactic acid bacteria contain resistance mechanisms against phages *(4–6)* and these systems have been isolated and characterized *(7,8)*. The incorporation of the genes encoding these defense mechanisms to starters along with rotation strategies would allow the solution of some of the problems caused by phages.

The application of rDNA methods requires the evaluation of the potential risk following a set of criteria such as those described by Verrips and van den Berg *(9)*, which must include the host (animal, plant, or microorganism that must be generally recognized as safe), the characteristics of the final product (i.e., will it contain rDNA? If it does, will this rDNA be present in living cells? If it does not, must this product be labeled as a transgenic or GM product?), and the position in the host of the new characteristic (interpreted as the molecular techniques used to integrate a gene in the chromosome, to maintain

Problems with Genetically Modified Foods 483

it in a plasmid, or to remove it from the chosen host). The transformation of the cell (by electroporation, conjugation, transduction, or microinjection) to introduce new genetic material that carries the gene encoding the protein of interest, or the DNA that will give the cell some advantage or to eliminate a catalytic feature, should evaluate a possible environmental risk *(10)*. The possibility of the occurrence of gene transfer in the environment to nonsecure organisms should influence the selection of the characteristic to be introduced in the host.

Environmental concerns are important from another point of view because transfer of DNA occurs among microorganisms present in the digestive tract of human beings and this phenomenon has been used as an argument against genetically modified organisms (GMO). The food products that we consume are of an animal or vegetable origin and, therefore, their cell components contain lipids, carbohydrates, proteins, and genetic material. It is most probable that the DNA included in food is digested and cannot be taken up by the intestinal cells. However, this assumption cannot be proved. If the DNA that we consume in food is not totally digested, genes encoding antibiotic resistance used as markers in GMO could be transferred to the normal (or to opportunistic) flora present in the human intestine, creating health problems. Consequently, it is necessary to eliminate the antibiotic resistance genes from the product after the research phase. There are two types of strategies to solve this problem, either the used vectors named "food-grade," which do not carry this kind of selection marker, or vectors that allow the elimination of these markers in one step after GMO commercialization.

The commercial use of genetically modified organisms involves social and ethical consequences, because rDNA in living cells is implicated. This is the third aspect to be considered in the development of genetically improved food products. Gaskell et al. *(11)* compared the public perceptions of applications of biotechnology and confirmed the existing differences between Europe and the United States. People on both continents encouraged the applications in the medicinal field, which means that they supported the use of genetically modified medicines, but there was a notable contrast in the case of food. Europeans consider "unnatural" and even dangerous the consumption of GM food. The responsibility for this conjecture may be the information that is given to the public, since there are a growing number of articles in the popular press that use exaggerated language (such as "Frankenfoods") or introduce unintentional errors. Sasson *(3)* described a set of criteria that seem to be necessary to achieve public acceptance of GM products. The most important item is the regulation of the genetically engineered product. Such regulation should be solidly scientifically based, promoting an understanding between the regulatory agency and the food industry and avoiding the creation of artificial barriers that could result in the presentation of equivocal information to the consumer. However, at the moment, that is not the case if we consider only the few fortunate results of the present regulations from the European Union that affect labeling *(12)*. In spite of this, we will still have hard work to persuade people and authorities about potential benefits of GM products.

Acknowledgments

The authors thank Dr. Daniel Ramón, Department of Biotechnology, Instituto de Agroquímica y Tecnología de Alimentos (IATA) Valencia, Spain, for providing information and for critically reading the manuscript. We are indebted to Dr. John F. T. Spencer, coeditor of this book, for his continuous help throughout the preparation of this manuscript. FS wishes to thank the Alexander von Humboldt Stiftung (Bonn) for its continuing support.

References

1. Frewer, L. (1999) Bioenhancement or playing God? Biotechnology and the future of food. *TIBTECH* **17,** 182–183.
2. Ramón, D. (1999) *Los Genes que Comemos. La Manipulación Genética de los Alimentos.* ALGAR, Valencia, Spain.
3. Sasson, A. (1998) Public acceptance in *Plant Biotechnology-Derived Products: Market-Value Estimates and Public Acceptance. IX International Congress on Plant Tissue and Cell Culture (International Association for Plant Tissue Culture, IAPTC, Jerusalem, Israel).* Kluwer Academic, Amsterdam.
4. Sanders, M. E. and Klaenhammer, T. R. (1980) Restriction and modification in group N Streptococci: effect of heat on development of modified lytic bacteriophage. *Appl. Environ. Microbiol.* **40,** 500–506.
5. Klaenhammer, T. R. and Sanozky, R. B. (1985) Conjugal transfer from *Streptococcus lactis* ME2 of plasmids encoding phage resistance and lactose-fermenting ability: evidence for a frequency conjugative plasmid responsible for abortive infection of virulent bacteriophage. *J. Gen. Microbiol.* **131,** 1531–1541.
6. Auad, L., Azcárate Peril, M. A., Ruiz Holgado, A. P., and Raya, R. R. (1998) Evidence of a restriction/modification system in *Lactobacillus delbrueckii* subsp. *lactis* CNRZ326. *Curr. Microbiol.* **36,** 271–273.
7. Hill, C., Pierce, K., and Klaenhammer, T. R. (1989) The conjugative plasmid pTR2030 encodes two bacteriophage defense mechanisms in lactococci, restriction modification and abortive infection. *Appl. Environ. Microbiol.* **55,** 2416–2419.
8. Ives, C., Sohail, A., and Brooks, J. E. (1995) The regulatory C proteins from different restriction-modification systems can cross-complement. *J. Bacteriol.* **177,** 6313–6315.
9. Verrips, C. T. and van den Berg, D. J. C. (1996) Barriers to application of genetically modified lactic acid bacteria. *Antonie van Leeuwenhoek* **70,** 299–316.
10. Schuler, T. H., Poppy, G. M., Kerry, B. R., and Denholm, I. (1999). Potential side effects of insect-resistant transgenic plants an arthropod natural enemies. *TIBTECH* **17,** 210–215.
11. Gaskell, G., Bauer, M. W., Durant, J., and Allum, N. C. (1999) Worlds apart? The reception of genetically modified foods in Europe and the U.S. *Science* **285,** 384–387.
12. Ramón, D., Calvo, M. D., and Peris, J. (1998) The new regulation for labeling genetically modified foods: a solution or a problem? *Nat. Biotechnol.* **16,** 889.

Index

A

APS (Ammonium persulfate), 192
Acetate, 191, 398, 401
Acetic acid, 198, 292
Acetic acid, tolerance, 209.223
Acetyl methyl carbinol, 21, 35
Achromobacter, 4
Acinetobacter, 4
Acriflavine HCl, 128,129, 131
Actomyosin, 261
Aesculin, 126, 128
Agar plate methods, 6,7
Agarose, 249, 250, 251, 254, 275, 278, 279,280
Agglutination tests, 89
Agitator, agitation, 397, 404, 406
Agrobacterium aurantiacum, 260
Alcohol, 288, 291, 292, 302, 303, 336,355–367
Alignment program, 39, 41
Alpha-aminobutyric acid,417
Amino acids, free (FAN), 283,287
Amino-N, 285–288
Ammonia N, 285, 416, 417
Ammonium citrate, 198
Ampicillin, 310
Amplicons, 76, 78, 82
Amplification of DNA, 40, 41, 75,76, 81, 102,109, 110, 245, 249
Amylases, amylolytic yeasts, 309
Aneuploidy, 274

Antibiotics, 27, 119, 129
  See Penicillin, streptomycin, chloramphenicol, chlortetracyclin, oxytetracyclin, vancomycin, trimethoprim, cyucloheximide, nalidixic acid
Antibody, 103
Antisera, 65,
  See Campylobacter
Apocarotenic acid (ethyl ester), 259,260,263–265, 267, 268
Arginine ,294
Arthroconidia, 212
Ascending velocity, 324, 327
Ascospore morphology, 252
Aspartic acid, 417
Aspergillus oryzae, 227
  A. niger, Aspergillus spp.
Assimilation,263
  Galactose, melibiose, trehalose
Astaxanthin, 259–261, 263, 264, 266–268
Aureobasisium pullulans, 310, 315

B

Bacillus thruingensis, 107
BGBL Broth; medium, 18, 19
Bacillus cereus, 11, 14,18, 23, 107,197, 227,228,395
Bacterial growth, 164, 384

Bacteriocins, 141–146, 143, 395
Bacteriophage, 203–207, 482
Baird-Parker test, agar, 18, 22
Barium chloride, 283, 284
Barium sulfate, 293
Basidiomycetous yeasts, 262, 264
Beer, 481
Bentonite, 290
Beta-1,3-laminaripentaohydrolase, 345
Beta-carotene, 260, 265, 266, 267, 269, 270
Beta-glucuronidase, 61, 64
Beta-mercaptoethanol, 198,199
Bile salts, 12, 13, 90
Bilirubin, 90
Biochemical tests, 246, 250, 253
Biomass, 263, 290, 357, 359, 361, 363, 364
Biotin-Strepatvidin, 107–109
Bixin, *Bixa orellana* 261, 262, 269
Bñakesleeana trispora, 259
Botrytis sp., 4
Botrytis cinerea, 285, 419
Bovine serum albumin (BSA), 231
Brain–heart infusion,17, 22
Bread, 481
Brilliant Green, 13, 62, 114
Buffers and solutions, 370, 372–381
Buoyant density centrifugation, 99
Buttermilk,210
Butylated hydroxytoluene, 270

C

CAMP test, 127, 129
CHEF, 254, 275, 280
CO_2, 222, 291, 416
Calculations, 7–9

Campylobacter jejuni, 95–106, 119, 120, 122
Campylobacter spp., 95–106, 99, 101, 103
Campylobacter; Campy-Cefex agar, 119–123, 122
Candida krusei, 252
Candida holmii, 43,
Candida milleri, 209
Candida parapsilosis, 211
Candida utilis
Candida vinaria, 398
Candida vini, 252
Candida cinnabarinus, 259
Canthaxanthin, 269, 260, 263, 264, 266–268, 270
Capsaxanthin, 262
Capsorubin, 262
Carotenoids, 259-271, 260
Carrots, 259
Cascade controller, 406-408
Casein, 228, 229, 231
Cassava (Manihot esculenta), 307, 308, 313,316
Catalase, 119, 129, 130, 375, 392
Cefoxime, 86
Cefotan, 128, 129
Ceftazidine, 128
Cell lysis, 100
Cheese, 90, 141, 183, 197
Chemostat, 321, 326, 327
Chicken rinse preparation, 100,101
Chloramphenicol, 310
Chloroform, 206, 261
Chromogenic *Salmonella* esterase agar, 14, 23
Chromogenic peptide (S-Ala), 197
Chromosomal DNA, 243, 245, 246, 249, 255

Index

Chromosomal size markers, 246
Chromosomes, chromosomal polymorphism, 274, 275, 277
Chymosin, 482
Clark–Lubs medium, 15, 16, 18
Clostridium sp., 4
Coagulase test, 21, 22
Coliform bacteria, 18–21, 29–36
Coliphage lambda, 203
Colistin SO_4, 127, 129
Colony-forming units, 27
Containment (known pathogens), 92
Coomassie Brilliant Blue, 150, 192
Creams, 11
Creatinine, 16
Crocus sativus, 262
Cryptococcus spp., 252–255
Crystal violet tetrazolium agar, 5, 9
Cultivation, 163–171, 193
Cycloheximide, 223
Cysteine, 199

D

DAPI (4,6-diamino-2-phenylindole), 336, 343
DMSO (dimethyl sulfoxide), 198, 263–269
DNA polymerase, 38, 64, 67, 98, 108, 110
DNA synthesizer, 38
DNA template, 109
DNS solution, 361, 362
DNase test, DNase test agar, 16, 21, 22
Data analysis, 387
Debaryomyces spp., 252, 255
Dekkera spp *(Brettanomyces* spp.), 255
Dendogram, 44, 390

Deoxynucleotide triphosphate (dNTP), 38
Diluents and dilutions, 3, 5, 6
Dilution (methods), 3–6
Dinotrosalicylic acid, 361
Diphenyl (for control of molds), 251
Dipodascus spp., 210
Dithiothreitol (DTT), 311
Double layer plaque assay, 204, 205
Drigalsky spatula, 8, 14, 16, 21
dTNP, 38, 40
Drosophila spp., 369
Dynabeads, 87, 88

E

ECP buffer, 275, 278, 279
EDTA, 245, 246, 250, 251, 278
ESP buffer, 275
Eggs, 11, 12
Electromorphs, 369, 387, 392
Electrophoresis, 137, 149, 155, 191, 193, 196, 246, 250, 251, 254, 280
Electrophoretic karyotyping, 246, 247, 251, 277, 278, 279
Embedded cells or spheroplasts, 278, 279
Enrichment, selective, 113, 114, 116
Enterobacter aerogenes, 22
Enterococci, 111, 112
Enterococci, media, 111
Enterohaemorrhagic strains, 12, 61
Enterotoxins, 12, 22, 23
Enzymes, 92
Eosin Y, 15
Ergosterol, 289
Erythritol, 233
Erythromycin, 310

Escherichia coli, esp. strain
  O157:H7, 61–66, 85, 94, 96
Ethanol, alcoholic fermentation,
  244, 245, 248, 285, 292,
Ethyl 4-dimethylamylobenzoate, 14
Ethyl acetate, 262
Ethyl carbamate, 420
Ethylamine hydrochloride, 222

F

FITC-casein assay, 199–201
FITC fluorescein isothiocyamanate,
  197, 335, 344
False negatives, 429
False positives, 428
Fat, interferrence by, 90
Fecal coliforms, 29, 32
Fermented foods, 135–140, 141–
  162, 163–172, 173–182
Ficoll, 311,314
Filamentous fungi, 311, 313, 314
Fish farming, 269
Flamingo, 269
Flavobacterium, 4
Flocculation, 300–302, 349
Flotation assay, yeast, 335, 337, 328
Flow cytometry, 335, 337,338
Fluorescein, 197, 201
Fluorescein isothiocyanate, 197,199,
  200, 344
Fluorescence labeling, 335–340,
  341–347, 337, 344
Foam, 349–352
Folin–Ciocalteau reagent, 229, 231
Formaldehyde, 283, 284,
Formate, 191, 398
Formol method, 283-296
Fosfomycin (phosphomycin), 128
Freeze-thawing, 389

Fungi, 308, 310, 316
Fungizone, 228
Fusion (Fungal nuclei with yeast
  protoplasts, 313-315

G

GYEP broth, 216
Galactomyces sp., 211, 213
Gas chromatography, 398
Gastroenteritis, 30
Gel mold, 371, 382, 383
Gelatinase reactioon medium, 15, 18
Genes (*E. coli*), 89–92
Geotrichum, 4
Giant colonies, 252
Glass beads, 389
Glucose, 222, 397, 415
Glucuronidase, 86, 89,
Gluten, 227
Glycerol, 233, 234, 237, 356
Glycerol-3-phosphate
  dehydrogenase, 356
Glycine, glycine-hydrazine buffer,
  322, 326
Grapes, grape must, 243, 273, 275,
  283–296, 297–306, 415–415–
  425
Growth media, 210
Growth temperature, 223

H

H_2S, 299–304
HPLC, 237–239, 262, 266–268
*Hae*III, *Cfo*I, *Sau*3A, 2
Haematococcus pluvialis, 260, 263,
  265, 267, 269
Hamburger, 11
Hanseniaspora spp (*Kloeckera*
  spp.), 255, 273, 283, 286

Index

Helicase, 275
Hemocytometer, 276
Hemolysin, 276
High cell density, 362
Homogenizer, homogenization (blender), 4, 25, 62
Homothallism, 274
Hybrids, hybridization, 109, 338, 345
Hydrophobicity, 350, 352

I

IMVIC test, 29, 30
ITS reagent, 40
Identification, 46, 63
Immobilized cells, 321, 322
Immunoassay, 67
Immunomagnetic separation, 85–94
Indole, 19, 20, 22
Induction, 203-206
Inhibitors, of amplification, 427–429
Inoculum, 95, 310, 362, 405
Intraspecific differences, 253
Invertase, 356
Iodine, 13, 262
Isobutanol, 196
Isolation, 244, 247, 251
Isolation and enumeration, yeasts, 243-257
Isopropanol, 245, 248
Issatchenkia orientalis, 209

K

KH_2PO_4 (NaH_2PO_4), 198
Kefir, 396
Killer factor (yeasts), 298–300, 303
Kluyveeromyces spp., 211, 215, 253, 255, 273

Kovac's reagent, 16, 31, 35

L

Lactic acid fermentation, 419, 420
Lactobacillus casei, 206
Lactobacillus delbrueckii ssp *bulgaricus*, 206
Lactobacillus helveticus, 207
Lactobacillus, lactobacilli/135, 141, 147, 167–171, 173, 183, 197, 203–207
Lactococcus lactis, 135, 303, 395, 396, 399, 402, 405, 407, 408
Lactose, 12–15, 31, 89, 113–115, 398
Lancefield Group D *Streptoccoccus* spp., 111
Lauryl sarcosine, 108, 246, 251
Lauryl sulfate broth (LST), 29, 30
Least squares method, 398, 400, 401
Lecithin, 18, 20
Levine agar, 15, 31
Lipids, 289
Listeria monocytogenes, 53–57, 67–83, 125–131, 126, 127, 395
Lithium Cl, 16, 126, 129
Low-temperature incubator, 5
Lutein, 266
Lycopene, 264, 267, 269
Lysogeny, 203
Lysozyme, 369
Lytic enzymes, 275, 279
Lyticase, 254

M

M-FC broth, 29
MEE (multilocus enzyme electrophoresis), 369–393

MPN (most probable number), 20, 24
MRS broth (For growing lactobacilli), 198–201
Magnetic capture, 86–88, 107–110
Maintenance (lactic acid bacteria), 163–171
Maize, 209
Malachite Green, 15
Maltose, 396–398
Mannitol, 14, 23
Margarine, 396
Material balance equations, 401
Mating, 336
Mayonnaise, 11
McClary's agar (sporulation of yeasts), 245
MeOsuc-Arg-Pro-Tyr-p-nitroanilide, 201
Meat extract, 198
Meats (cold or raw), 53, 125, 329
Media; selective and nonselective, 13, 14, 16, 17, 27, 119, 204, 336, 342, 360
Media, differential, 114, 115
Meiosis, 277
Membrane filter, 361
Mercaptoethanol, 237, 343
Mesophilic microorganisms, 25, 26
Methanol, 198
Methyl red, 15, 20, 31, 35
Methylene blue, 15, 301, 361, 362
Methyumbelliferyl-beta-glucuronide, 64
Metschnikowia pulcherrima, 252, 255
$MgSO_4 \cdot 7H_2O$, 198
Micromanipulator, 276
Micropipettor, 39, 41

Milk, flavored dairy products, 11, 85, 95, 396
Mitochondrial DNA, 243
Mitomycin C, 203–206
Mixed culture systems, 195–411, 402, 405
$MnSO_4 \cdot 4H_2O$, 198
Molasses, 360
Mold, 27, 28, 285, 292
Morphological characteristics, yeasts, 212, 244
Mossel agar, 14, 18, 23
Most probable number (MPN), 20, 23, 34
Motility, 127, 129
Mucor spp., 227
Muller valve, 301
Multilocus Enzyme electrophoresis, 369–393
Multiplex PCR, 62, 63, 86, 89, 91
Must (grape), 287, 288
Mutation, 355
Mycobacterium paratuberculosis, 107
Mycotoxins, 285

N

N-succinyl-Ala-Ala-Pro-Phe-p-nitroanilide 198
NSBA (nucleic acid-based amplification), 67–83
NaCl, 12–17, 396
Nalidixic acid, 120–126
Ninhydrin, 421
Nisin, 395, 396–411
Nitrate, 311, 416
Nitrogen, 284, 285–288
Nonsoluble solids, 290
Novobiocin, 14, 114

Index

Novozym, 245, 278
Nuclear magnetic resonance (NMR), 261, 270
Nuclei, 313, 314, 316
Nucleic acid preparation, 313, 314
Nutrient addition— timing, 288

O

OPA method, 151, 157, 198, 199
Oligonucleotides, 79, 80
Oligonucleotide synthesizer, 62–64, 72, 79
Olives, 224
o-phthaldialdehyde (OPA), 197, 198, 199
Osmotic buffer, 237, 342
Osmotic stabilizer, 237, 342, 345
Osmotolerant yeasts, 233, 241
Oxford agar, 129
Oxgall-Brilliant Green, 30
Oxygen/SO$_2$, 238, 240, 289
Oxytetracyclin, 27, 251

P

PAGE, SDS-PAGE, 191, 193, 312
PALCAM agar (for *Listeria*), 128, 129
PALCAM media, 125, 128
PCR (polymerase chain reaction), 37, 61, 427 *et seq*
   Amplification, rDNA, 102, 103, 245, 246, 249, 250
   Buffers, 245, 246
   RFLP, 73-75, 89, 101-104, 427-480
   Sampling errors, differential enrixhment, recovery duriing culture, 24, 450, 456
   Comigration, 460
   Concentration effect, 460
   Conclusions, 464-466
   Contamination, 461
   Data interpretation, 464
   Detection, 464
   Endogenous and exogenous contamination, 433
   Factors affecting detection, 429
   Failure of lysis, 443
   Hybridization, 463
   Incorporation errors resulting from enzyme infidelity, 469
   Inhibitors and facilitators in food and clinical samples, 434–440
   Inhibitors and facilitators in environmental samples, 441, 442
   Live or dead?, 458
   Misidentificaction, bias and error in molecular methods 448–450
   Nucleic acid degradation and capture, 444–447
   Mediated chimeric gene amplification, 459
   Poor quality assuraance, 461
   Preferential amplification, 456
   Production, 463
   Quantitation, 464
   Reading and interpretation error, 462
   Reported errors in molecular methods, 451–456
   Selective lysis of bacteria, 456
   Sequencing errors, 460
   Standardization, 463

Hybridization microarrays, 463
Mechanisms of inhibition, 434-448
Reaction conditions, 431
p-anisaldehyde, 266, 269
Packed-bed system, 327
p-nitroanilide (pNA), 200, 201
Paprika, 282
Paradimethyl-aminobenzaldehyde, 15
Parameters, 323
Penicillium sp.,4
Peptone, 3, 5, 6, 12–14, 16–18, 25, 26
Pesticides and fungicides, 293
PH290
Phaffia rhodozyma (Xanthophyllomyces dendrothous) 259,260, 263–269
Phages, 203–207,482
Phenol red, 23
Phenylmethylsulfonyl fluoride (PMSF), 229, 231, 311
Phosphate buffer, 198
Phosphoramidate, 38
Phycomyces blakesleeanus, 261, 287, 269
Phyogenetic tree, 43, 44
Phylogeny inference, 39, 45
Physiological tests, 246, 250
Pichia (Hansenula) anomala, 252, 254, 256
Pichia membranaefaciens, 252
Plate count agar, 5, 10
Plating, selective, 114, 116
Polyethylene glycol (PEG), 236, 237, 311
Polyethyleneimine, 327

Polyhydroxy alcohols, 233–241
Polymyxin B, 15, 128
Polypeptone, 198, 396
Pomace, 289
Porosity, 323
Potassium acetate, 245, 248
Potassium nitrate, 222
Poultry, 4, 11, 98, 99, 101, 119, 125
Presumptive test, coliforms, 111
Pretreatment solution, 311, 314
Primers, 39, 64, 72, 73, 79, 80, 82, 88, 97, 98, 245
   L. monnocytogenes, Campylobacter spp. , 87, 95, 119
Proline, 199, 283, 294, 416
Propionic acid, 360, 363
Proteases (Acid, neutral, alkaline), 227–229, 231
Protein, 191, 193, 195, 196
Proteinase K, 246, 251, 279
Proteolysis, 197–202
Proteus spp., 197
Protoplast, protoplasting, 309, 311, 313, 314, 336, 342–344
Pseudomonas spp., 4, 197
Psychrotrophic organisms, 3, 4, 7, 8
Pulsed field gel electrophoresis (PFGE), 250, 251, 275, 277, 278, 280
Pyruvate, 119

R

RNA polymerase, 68
Rnase A, 245
Reactor configuration, 321–328, 323, 324, 326
Reducing sugar, 362, 364
Regeneration, 345

Rennet, 482
Restriction analysis, 249
Restriction enzymes, *Hae*1, *Cfo*1, *Msp*1, *Rsa*1, 252
Reverse transcriptase, 68, 96
Rhodamine B, 6G, 336, 343
Rhodococcus equi, 125
Rhodosporidium, *252*
Rhodotorula sp., 211, 252, 255, 261,
Ribonuclease H, 68
Ribosomal DNA, 244, 245, 249
Rose Bengal, 228, 360

S

SO_2, 298, 299, 301, 303
SS agar, *Shigella*, 13, 17, 23
Saccharomyces bayanus, 252
Saccharomyces cerevisiae, 44–46, 239, 243, 251, 255, 286, 292, 297, 309, 355, 361
Saccharomyces diastaticus, 356
Saccharomyces exiguus, 43, 209, 212
Saccharomyces pastorianus, 45
Saccharomycodes ludwigii, 256
Safranin, 15, 23
Salmonella sp., 12, 13, 17, 18, 22, 23, 66,
Sampling, 228
Schizosaccharomyces spp., 46, 254, 256
Seafood, 125
Selenite broth, 13, 17
Septicaemia, 125
Sequence analysis, 37, 67
Serine, 294
Sheep, cattle, chickens, 125
Shiga-like toxinsSLT1, SLT II, 61, 65

Silage, 209, 236
Simmons citrate agar, 16
Skim milk, 198
Sodium acetate, 136, 142, 164, 174, 192, 198, 204
Sodium citrate test, 20, 21
Sodium dodecyl sulfate (SDS), 198
Sodium hydroxide (NaOH), 283, 284
Sodium tetratborate (borax), 198
Soluble starch, 219–221
Somatotrophin, 482
Sonicator, 369, 384
Solvents, 263, 264
Sorbitol, 64, 235
Soybeans, 228
Spheroplasts, 245, 275, 278, 279
Spiking, 101
Spoilage yeasts, 37–51
Spores, sporulation, 274, 275, 276
Sporidiobolus spp., 252, 254, 256
Sporobolomyces spp., 252
Sporulation media, 245, 275
Spreaders (colonies), 8
Stability, 275, 277, 279
Stacking gel, 194, 195
Staining tray, 371
Stains, staining, 371, 381
Standard plate count, 4, 5, 7, 8
Staphylococci, 16, 107, 395
Staphylococcus aureus, 16, 21, 22, 125, 129, 130, 395, 430
Starch, starch gel, 307–317, 384, 385, 389, 391
Sterols, 289
Stomacher, stomacher bags, 3, 62, 145
Streptavidin, 108, 109
Streptococcus sp., 4, 111, 112

Stgreptococcus viridosporus, 327
Streptomycin, 397
Stuck fermentations, 286, 289,292, 415
Suc-Ala-Glu-Pro-Phe-p-nitroanilide, 201
Succinate, 191
Sucrose, 356
Sudan Black (stain), 5, 23
Sugar cane syrup, 358, 360, 362–364
Sulfur dioxide SO_2, 283, 289
Suuport materials–384, 389
   Supports, 322, 327Surface spread plate, 7, 8
   Symba process, 308

T

TAE buffer, 245, 249
TBE buffer, 245, 276
TE buffer, 245, 249
TEMED (Tetramethylethylenediamine, 192, 195
TRITC (tetramethyl rhodamine isothiocyanate), 335
Taq polymerase, 245–249
Temperature, temperature shock, 290, 356–367, 363, 364
Terminator cycle sequencing kit, 38
Tetrathioate broth, 12, 17
Thermal cycler, 38, 62, 64, 68, 72
Thermoduric organisms, 11
Thermonuclease test, 22
Thermophilic *Campylobacter* sp., 95–106, 103
Thermotolerance, in yeast, 356–367
Thin-layer chromatography, 236, 262, 264, 268, 269

Threonine, 294, 417
Toluidine Blue, 22
Tomato, 482
Torulaspora sp.,44, 235, 239
Toxicity, 291, 292
Toxins, 61, 62, 64
Toxins (killer factors), 292, 298, 300, 303
Transduction, 203
Transilluminator, 231
Trichloroacetic acid (TCA), 198, 199
Trichosporon spp.,212, 213, 221
Triolein, 269
Tris-hydroxymethyl-aminomethane, 198, 199, 201
Trypticase soy broth, 86, 87,
Tryptone, tryptone water, 14, 15, 16, 19, 87, 223
Tryptophan, 199
Tween-80, 198
Tyrosine, 232

U

Urease, urea, 222, 223

V

Verotoxin, VTEC, VT genes, 85, 86, 92
Viability, 276
Vibrio sp., 66
Vineyard, 284
Vitamins, 210, 222, 288, 289
Vitis vinifera, 418
Voges–Proskauer test, 20, 21
Volume, of reactor, 323

W

Williopsis saturnus, 46
Wilson–Blair medium, 13, 17, 23

Index

Wine, 243–257, 297–306, 307–317, 335–350, 341–347, 416, 418, 420

X

Xanthophylls, 259, 261, 262
Xylitol, 234, 236, 239

Y

YEPD medium, 301
YM agar, broth, 310
Yeast nitrogen base, 210
Yeast extract, 164, 174, 192, 198, 210, 204, 311
Yeast hulls, 289
Yeast species, 27, 28, 37–51, 211, 216, 235, 243-257, 309
Yeasts and fermented foods, 211, 212
Yeasts (other)
  Debaryomyces hansenii, 252
  D. hansenii, 252
  D. udenii, 252
  D. vanrijii, 252
  Kloeckera apiculata, 273, 286, 288 292
  Kluyveromyces lactis, 253
  Pichia canadensis, 254
  Saccharomyces bayanus, 252
  S. pastorianus, 252
  Saccharomyces sensu stricty, 254, 256
  Saccharomyces spp., 286, 288
  Schizosaccharomyces pombe, 251, 253, 254
  Torulaspora delbrueckii, 252, 253
  T. pretoriensis, 253
  Zygosaccharomyces microellipsoides, 253
  Z. bailii, 252, 253
  Z. bisporus, 253
  Z. rouxii, 252, 253,
Yogurt, 209

Z

Zymomonas mobilis, 314, 315